CORE MEDIA COLLECTION FOR SECONDARY SCHOOLS

Second Edition

CORE MEDIA COLLECTION FOR SECONDARY SCHOOLS

Second Edition

Lucy Gregor Brown

Assisted by

Betty McDavid,
Technical Librarian,
Mt. Diablo Unified School District

R. R. Bowker Company
New York & London, 1979

Published by R. R. Bowker Company
1180 Avenue of the Americas, New York, N.Y. 10036

Library of Congress Cataloging in Publication Data

Brown, Lucy Gregor.
   Core media collection for secondary schools.

   Includes indexes.
   1. Audio-visual materials—Catalogs. I. Title.
LB 1043.Z9B76   1979   016.3731′33   79-6969
ISBN 0-8352-1162-2

CONTENTS

CONTENTS

PREFACE

The purpose of *Core Media Collection for Secondary Schools* is to provide a qualitative selection guide to nonprint media titles. The majority of the titles in this book are for sound or captioned filmstrips, kits, recordings, and 16mm films; however, 8mm loops, slides, and some video cassettes are also listed. They cover a wide variety of subject and ability levels.

In many instances, only the series or set title has been listed as a main entry, but parts of the set or series are available separately and should be considered by checking the "Contents" paragraph of the main entry. Including these separate items, over 3,000 titles are listed. Although not listed separately, these titles are readily accessible through the expanded use of 3,200 subject headings. All titles—as sets, series, or individually—should be considered for purchase based on needs and budget requirements.

This collection was developed with the school curriculum needs of students in grades 7–12 in mind, but it is expected that librarians in public libraries will also find the recommendations and ordering information helpful in developing their collections. Very few titles in this collection appear in the first edition of this book. This second edition is meant to supplement the first *Core Media Collection for Secondary Schools*, but is also a complete collection in itself. School librarians and teachers will find media noted as appropriate for grade levels as well as suitable for various subjects.

Materials included have been favorably reviewed, are award winners, or have been evaluated for their authenticity, technical quality, student level, interest and motivation, accuracy in content, and validity in treatment. It is recognized that not all the listed titles are in the award-winning class, just as all entries in print collections are not "classics." However, all entries have generally met the criteria stated and have proven to be of interest to students as well as being directly pertinent to the secondary curriculum.

This list is not intended to replace the selection process by an individual librarian or media specialist. The variables considered in selecting books for a particular library collection apply equally to selecting nonprint material and eliminate the possibility of developing a collection applicable to all schools. Individuals will want to apply their own selection criteria, responsive to local conditions of curriculum design, budget, student interest, nonprint materials currently owned, facilities, and equipment. This guide, featuring selections based on authoritative sources, author experience, and widely accepted criteria, will provide a substantial reduction in time investment in the evaluation process for the acquisition of a viable and accountable collection.

Only a few titles produced before 1973 are listed, since many titles are by nature subject to rapid obsolescence through advances in technological fields and because of frequent changes in topical areas. Such materials were closely examined before inclusion. Some older materials in such areas as literature, art, and music are included since their date of production is less relevant. Other older but excellent materials are not included because they have been withdrawn by producers or have become impossible to locate with the changes in acquisitions of companies over the past few years. The new "owners" of material are mistaken when they make it impossible to locate known good materials. Some new titles have been included even though no reviews were available at the time of publication, if the titles were known and evaluated by the author and met the other prescribed criteria or were known to be 1979 spring award winners.

Inclusion of all other materials in the list was based on favorable reviews in professional review journals and authoritative content area publications such as *English Journal*, *Booklist*, *Science News*, and/or award winners in major festivals. The 38 librarians and interested and knowledgeable teachers who have been involved in reviewing nonprint materials for a number of major producers over the past eight years and have had these reviews published by the Mt. Diablo Unified School District have been included as a source. Finally, some materials have been included based on the author's experience derived from many years working with nonprint materials as a school librarian, as a program administrator of library services in both small and large public school districts, and as an instructor of courses on the evaluation and selection of nonprint materials for the University of California, Berkeley, and San Jose State University.

American Library Association standards, informal discussions at professional meetings with librarians who manage effective media libraries, plus reports from producer/distributor representatives were instrumental in determining the quantities of each of the various medium formats included. Since this list is intended to be a *practical* guide and resource for secondary school librarians and media specialists, it intentionally deviates from the quantities recommended in standards.

It may be noted that few 8mm single-concept loop films are included. Lack of review recommendations, as well as the discontinuance of these loops by a number of producer/distributors, is a major reason for the limited number. However, the interest shown this medium requires their inclusion. The loop film perhaps has not been fully recognized as a valuable teaching tool. If the old adage that "a picture is worth a thousand words" has merit, then a four-minute *moving* picture is well worth the current price, if the need for "motion" is applicable to the concept being presented.

The number of musical recordings included is not indicative of the percentage of recordings that most media libraries should and do purchase. Selections in this medium are so vast and so influenced by the school curriculum, ever-changing student interests, and local budgets that only some basic music recordings are noted. Reviews for recorded music are sketchy and often incomplete.

Professional background and experience provide me with an understanding of the problems a librarian faces in setting up a new media library or in updating an older print collection. It is the intent of this book to provide reliable titles to make the selection task easier and to provide cataloging information to make the nonprint media compatible with the print collection. It is suggested that the reader consult "How to Use This Book" immediately following.

The 1970s witnessed a revolution in the learning resources area. The book remains a basic tool, but has been joined by a myriad of other media. The ranks are ever growing. At the same time, the traditional three Rs are being reemphasized, with more attention to a variety of instructional designs that require a great many materials. The school library in its multimedia form is needed more today than ever before.

The fourth R looms great on the horizon. It cannot be stressed enough. Its name is Retrieval of Information, and it is a skill that must be emphasized early in a student's training. Cataloging of the materials for school media libraries needs to partake more freely of new information science techniques. There must be more points of entry to the cataloging system, even at the risk of redundancy, which has been formerly eschewed by catalogers for economy and a certain intellectual discipline.

Newer technology makes it possible to duplicate catalog cards, store data electronically, and share cataloging information efficiently. There is no longer the need to be parsimonious with catalog information because every character on a card was made by the expensive key stroke of a manual typewriter. Our responsibility to today's student is to increase access to all forms of information. The professional cataloger must be flexible, sensitive, and responsible to the needs of users in order to meet the challenge of today's multimedia world.

A special thanks for the cooperation of the many people who supported the compilation of this book. Librarians were generous with their time, assistance, and sharing of evaluations. Producers and distributors were responsive in supplying requested information promptly and completely. The editorial staff at Bowker was helpful and understanding. Betty McDavid, who did a very professional job in the descriptive cataloging, and Chris P. Williams, who programmed the computer (and almost became a librarian in the process), both are deserving of recognition. Most of all, my thanks to my husband, Tom, and my family, who allowed me the time and gave their support to complete this project. Grateful acknowledgment is also due to Lorraine Hodge, my typist, who kept the project moving with all her library clerical experience.

Lucy Gregor Brown
Program Administrator
Library Services
Mt. Diablo Unified School District

HOW TO USE THIS BOOK

It is extremely important for the reader to examine the Preface of this book to understand completely the purpose and scope of the titles included in this collection. The organization of entries under a subject index and a title index will be explained here, as well as some suggestions on how most effectively to use this core media collection of recommended titles. Rather than specific titles, librarians and media specialists are usually looking for titles on a specific subject to meet their nonprint needs. Therefore, information for selecting, ordering, and cataloging will be found in the main part of the book—"Media Indexed by Subject." However, the person looking for a particular title can locate it readily in the "Media Indexed by Title" section; subject headings are indicated for each title listed, directing the reader to the "Media Indexed by Subject" section for complete information.

ORGANIZATION OF ENTRIES

Titles are annotated under subjects, with headings based on the *Sears List of Subject Headings* (11th edition). The main entry from each title is listed under its appropriate subject heading, except in cases where the whole of a title is attributable to a single author, composer, artist, or performer. The main entry contains adequate information to be used as both a selection guide and ordering source. The inclusion of the Dewey Decimal number in the main subject entry will assist in the cataloging process for those so involved. Additional subject headings may be necessary for specific school situations, however.

Many main entry titles represent a set or series that includes several individual titles. Usually these individual titles have not been listed separately, but full bibliographic information is given under the set or series title. Most of these individual titles are available from the producer/distributor separately, and if only one or two titles are desired from a set or series, they should be purchased this way. These individual titles will show up in the entry as "Contents." Each main entry is arranged as follows:

TITLE (Medium)
Producer/Distributor, Release Date, Series Title, Collation, Order Number (Price), Grade Level, Recommending Sources, Dewey Classification, Contents, and Annotation.

The title is followed by its format, spelled out in full (for example, Filmstrip—Sound). Terms used to designate media are those advocated by the *Anglo-Ameri-*

can Cataloging Rules. The producer/distributor is in abbreviated form; full names and addresses are given in the Producer/Distributor Directory. (The addresses of some producers were not available; see Producer/Distributor Directory for explanation of listings.) The order number is given if the producer uses one or it was available, and the prices are the most current prior to the printing of this book or that could be located. The prices are "list" and do not reflect discounts for quantity orders or prompt payment, providing the producers/distributors give such discounts. Price variables due to different formats (phonodiscs or cassettes) are indicated. Few black and white 16mm films have been included and only where they are of historical value and available only in black and white. Many of the color 16mm films are available in black and white from the producer, but are not recommended here. Note: Probable general price increases are expected after January 1980, so these prices should be only a guide for planning budgets or purchases.

Production and/or release dates on nonprint materials are often not available or are unreliable. Every effort has been made to make these dates as accurate as possible, or they have not been given. The collation includes sufficient physical description to indicate what the type of hardware required may be, in addition to the shelving and circulation requirements.

The recommended range of grade levels is directed toward grades 7–12, the lowest and highest levels of comprehension or interest for each title. These should not be taken as absolutes, for some materials have no real grade limitations and others will be identified for a wider audience.

Code letters refer to the recommending sources for each title as well as its known awards. A maximum of three codes has been listed even though some titles have received more. A list of the codes along with the names and addresses of the recommending sources follows these explanatory notes.

Following the codes in each entry, a Dewey Decimal classification is given without specifying the media designation. We recommend spelling out the name of the medium rather than using codes, which act as a barrier for the user in locating the needed information, not to mention the lack of consistency in the use of codes available. Additional information about cataloging will be found later in this section.

Supplementary notes, containing such information as the names of performers and the individual titles within a series or set, may be included. A brief annotation is included for explanation of the content or treatment of the material, but is sometimes omitted if it is not absolutely valuable.

DESCRIPTIVE CATALOGING AND CLASSIFICATION

Cataloging for items included in this volume is based on the theory and application of rules as found in *Anglo-American Cataloging Rules* (1967), *Dewey Decimal Classification and Relative Index* (18th edition), *Sears List of Subject Headings* (11th edition), *Nonbook Materials: The Organization of Integrated Collections* (1973), and *ALA Rules for Filing Catalog Cards* (2nd edition). However, the author sometimes departs from these sources for very practical reasons, the most important being service to the ultimate user of library materials—the student, the teacher, the parent volunteer, and others who may not be familiar with library procedure. Of course, librarians, audiovisual specialists, and library technicians are understood to be potential users of the book as well. But for generations library staff members have communicated with one another through esoteric cataloging rules, tending to leave the patron outside the conversation. It is our conviction that simplified, clear cataloging will greatly aid *all* users of library materials.

All library materials (print and nonprint) must be made available to the users through that library index—The Catalog—whether in card form, book style, microfiche, or other format. Not all materials on the Civil War are found on the shelf in the 973.7 section! Reference to the catalog must be encouraged. Straightforward, practical cataloging will enhance catalogs and thus the learning environment. Cataloging that has as one of its basic rationales the curriculum of the school will be most useful. Better accountability will be realized and cost effectiveness improved if all items in a multimedia collection can be quickly and accurately located by every user.

Classification and cataloging rationale for this book is discussed below in several parts: Main entry, Classification, Subject cataloging, Cross-references.

Main Entry. All entries are by title. The main entry containing all information describing each item will be found under its principal subject. Other references to each title will refer back to the principal (or main) subject. Relevant entries are included for author, artist, composer, and so on. In some cases, where individual titles are unique, title entries will aid the user in locating them. That is, these titles will also appear as separate subject headings.

Classification. The Dewey Decimal system was chosen because it is so frequently used by school and public libraries in the United States and abroad. Accession numbers are still used by some nonprint libraries as a classifying device. They lack the flexibility inherent in the expandable Dewey system, however. Shelving materials by accession numbers is very rigid, and when materials are not clustered in relative subject areas (as in Dewey), the advantages of user browsing are lost. Serendipity, that great boon to library work, cannot operate as well in the accession number system as it does under Dewey. Interfiling of print and nonprint materials can be accomplished only when the same classifying scheme is used. Where interfiling can be achieved, the best use of all materials will take place.

Through the maintenance of an authority file, the author has been able to be consistent in the use of the Dewey numbers with subject heading choices.

Some special classifications have been adopted for this book. All fiction is designated by the three letters "Fic." This classification includes materials of a more elementary level, as it is believed that students working with this type of material for remedial work or for sheer enjoyment will find it more acceptable. The Dewey classification 920 has been adopted for all collected biographies, while 921 is used for all individual biographies.

Subject Cataloging. Sears subject headings were chosen partly for their universality, but also because of their compatibility with curriculum areas. Many nonprint libraries still use lists of curriculum headings as subjects for cataloging. These are judged by the author to be too easily outdated and too limiting in their scope, since they are, by necessity, very general.

Subject headings have been given to all items listed in this book, including fiction, poetry, drama, and so on. This practice increases the usefulness of all materials in the curriculum. Librarians should carry through with this idea to the local card catalog. Sometimes both specific and general headings are given for an item, contrary to the instructions in *Sears.* It is felt that there is a need to show as complete a picture of the library holdings as possible to the user, who may not be ready to research all possible specifics nor have the patience to do so. Therefore, an item about eagles may be cataloged under that heading, and also under BIRDS, or perhaps under BIRDS OF PREY. *Sears* is very flexible and allows for the free addition of many specific topics not covered in its chosen list. The author has added such cataloger-prepared headings. The authority file keeps usage consistent. Other additions have been made to give access to items that do not fit under the standard headings. It is hoped that the chosen terminology will allow persons to discover these items easily. Representative of such headings are COMPARISONS, METAMORPHOSIS, and STRING ART.

The form headings in *Sears* are advocated for application to items about the literary form, such as books and software describing fiction, essays, and poetry. Many teachers are interested in these materials on form, but they are equally interested in examples of the form to be used in classroom presentations and for individual study by young people. FICTION, of course, is no problem to locate. But AUTOBIOGRAPHIES is a different matter! And so the author has bent the rules a little to allow retrieval of materials through the use of form headings to meet a definite curriculum need.

Because this book has been prepared by computer filing and printing, certain adaptations of the basic library filing rules have had to be made. Machines cannot make the same kind of intellectual decisions that the human mind can perform during manual filing. It has been useful to refer to Hines and Harris's *Computer Filing of Index, Bibliographic, and Catalog Entries* (Bro-Dart Foundation, 1966). The most noticeable application will be found by the reader in the filing of headings indicating a chronology. Here the dates precede the descriptive words, e.g., U.S.—HISTORY—1775-1783—REVOLUTION. The computer knows that the Revolution came before the Civil War and should be filed accordingly. This practice also can aid manual filing and searching, and can be recommended for all catalog preparation.

Cross-References. Three types of cross-references will be found in this book. There are two "See" references. One occurs when the reader has looked under a term that has not been chosen as a subject heading. In that case, the reader will be referred to the proper term; for instance, GANGS. *See* Juvenile delinquency.

The other "See" reference occurs when a title, but not full information, is given under a subject heading. The reader will be referred to the subject heading where the full (main) entry is listed; e.g., JOHNNY APPLESEED AND PAUL BUNYAN (Phonodisc or Cassette). *See* Folklore—U.S.

There is a "See also" reference. It occurs when the reader has looked under a preferred term and finds other related subjects are also suggested; such as ANIMALS. *See also* names of individual types, e.g., Bears, Lions, etc. This is a great aid in providing in-depth coverage for a particular field.

Betty McDavid
Technical Librarian
Mt. Diablo Unified School District

RECOMMENDING SOURCES

AAS   American Association for the Advancement of Science. 1515 Massachusetts Ave. N.W., Washington, DC 20005.

AAW   Academy Award Winner. Academy of Motion Picture Arts and Sciences, 9038 Melrose Ave., Los Angeles, CA 90069.

ABT   *American Biology Teacher*. National Association of Biology Teachers, 11250 Roger Bacon Dr., Reston, VA 22090.

ACI   American Council on Consumer Interests. 238 Stanley Hall, University of Missouri, Columbia, MO 65201.

AFF   American Film Festival Award. Educational Film Library Association, 17 W. 60 St., New York, NY 10023.

AIF   Atlanta International Film Festival. Drawer 13258K, Atlanta, GA 30324.

ALA   Featured at ALA Conference. 50 E. Huron St., Chicago, IL 60611.

ALR   *Annotated List of Recordings in Language Arts*. Schreiber, Morris (*see* NCT), 1111 Kenyon Rd., Urbana, IL 61801.

ARA   *Arts and Activities*. Publishers Development Corporation, 8150 Central Pk. Ave., Skokie, IL 60076.

ATE   *Arithmetic Teacher*. National Council of Teachers of Mathematics, 1906 Association Dr., Reston, VA 22091.

AUR   *Author's Recommendation*. Lucille Gregor Brown, Box 615, Alamo, CA 94507.

AVI   *AV Instruction*. Association for Educational Communication & Technology, 1201 16th St. N.W., Washington, DC 20036.

BKL   *The Booklist*. American Library Association, 50 E. Huron St., Chicago, IL 60611.

CCI   Council on Consumer Interests. *See* ACI

CFF   Columbus Film Festival. Chamber of Commerce, Kresge Bldg., Rm. 408, 83 S. High St., Columbus, OH 43215.

CGE   *CINE Golden Eagle*. Council of International Nontheatrical Events, 1201 16th St. N.W., Washington, DC 20036.

CHH   *Children's House*. Box 111, Caldwell, NJ 07006.

CHT   *Children Today*. U.S. Government Printing Office, c/o Superintendent of Documents, Washington, DC 20402.

CIF   Chicago International Film Festival. Chicago Film Board, 415 N. Dearborn St., Chicago, IL 60610.

CPR   *Curriculum Product Review*. McGraw-Hill, Inc., 230 W. Monroe St., Suite 1100, Chicago, IL 60606.

CRA   Chris Award. Film Council of Greater Columbus, 8 E. Broad St., Suite 706, Columbus, OH 43215.

CRC   *Children's Record Critique*. Free Public Library, Louisville, KY 40203.

CRH   *Craft Horizon*. American Crafts Council, 22 W. 55 St., New York, NY 10019.

EFL   *Educational Film Library Association Evaluations.* Educational Film Library Association, 17 W. 60 St., New York, NY 10023.

EGT   *English Journal.* National Council of Teachers of English, 1111 Kenyon Rd., Urbana, IL 61801.

ELE   Elementary English. *See* LAM

ESL   *The Elementary School Library Collection.* Bro-Dart Foundation, 1609 Memorial Ave., Williamsport, PA 17701.

EYR   *Early Years.* 11 Hale Lane, Box 1223, Darien, CT 06820.

FHE   *Forecast for Home Economics.* Scholastic Magazine, 50 W. 44 St., New York, NY 10036.

FLI   *Film Information Communication Commission.* National Council of Churches, 475 Riverside Dr., Rm. 853, New York, NY 10027.

FLN   *Film News.* Film News Company, 250 W. 57 St., New York, NY 10019.

FLQ   *Film Library Quarterly.* University of California Press, Berkeley, CA 94720.

GCA   Gold Camera Award. U.S. Industrial Film Festival, 161 E. Grand Ave., Suite 216, Chicago, IL 60611.

GGF   Golden Gate Film Festival. San Francisco International Film Festival, 425 California St., San Francisco, CA 94104.

GRT   *Grade Teacher.* Macmillan Professional Magazines, Inc., One Fawcett Pl., Greenwich, CT 06830.

HIS   Hispania Holy Cross College. Worcester, MA 01610.

HOB   *The Horn Book.* The Horn Book, Inc., 585 Boylston St., Boston, MA 02116.

HST   *History Teacher.* History Teacher Association, Memorial Library, Notre Dame, IN 46556.

IAF   International Animation Film Festival. 45 W. 47 St., New York, NY 10036.

IDF   U.S. Industrial Film Festival. 161 E. Grand Ave., Suite 216, Chicago, IL 60611.

IFF   International Film Festival. *See* IFT

IFP   Information Film Producers of America Film Festival. Box 1470, Hollywood, CA 90028.

IFT   International Film and TV Award. International FTF Corporation, 251 W. 57 St., New York, NY 10019.

INS   *The Instructor.* Instructor Publications, Inc., 7 Bank St., Dansville, NY 14437.

JEH   *Journal of Environmental Health.* 1600 Pennsylvania Ave., Denver, CO 80203.

JLD   *Journal of Learning Disabilities.* 101 E. Ontario St., Chicago, IL 60611.

JNE   *Journal of Nutrition Education.* 2140 Shattuck Ave., Suite 1110, Berkeley, CA 94704.

JOR   *Journal of Reading.* International Reading Association, 800 Barksdale Rd., Newark, DE 19711.

LAM   *Language Arts Magazine.* National Council of Teachers of English, 1111 Kenyon Rd., Urbana, IL 61801.

LBJ   *Library Journal.* R. R. Bowker Company, 1180 Ave. of the Americas, New York, NY 10036.

LFR   *Landers Film Reviews.* Lander's Associates Awards of Merit, Box 6970, Los Angeles, CA 90069.

LGB   *Learning Magazine*, "Best of the Year." Education Today Company, Inc., 530 University Ave., Palo Alto, CA 94301.

LNG   *Learning Magazine*. Education Today Company, Inc., 530 University Ave., Palo Alto, CA 94301.

LTP   *Listening Post*. Bro-Dart Inc., 1236 S. Hatcher St., La Puente, CA 91748.

M&M   *Media and Methods*. North American Publishing Company, 401 N. Broad St., Philadelphia, PA 19108.

MDI   *Media Index*. 343 Manville Rd., Pleasantville, NY 10570.

MDM   *Media Mix*. Claretian Publications, 221 W. Madison St., Chicago, IL 60606.

MDU   *Validated Nonprint Review List*. Library Media Services Division, Mt. Diablo Unified S.D., 1936 Carlotta Dr., Concord, CA 94519.

MED   *Media Digest*. National Education Film Center, 4321 Sykesville Rd., Finksburg, MD 21048.

MEJ   *Music Educator's Journal*. Music Educator's National Conference, 1902 Association Dr., Reston, VA 22091.

MER   *Media Review*. University of Chicago Laboratory Schools, 1362 E. 59 St., Chicago, IL 60637.

MMB   *Media & Methods*, "Best of the Year." 401 N. Broad St., Philadelphia, PA 19108.

MMT   *K-Eight*. North American Publishing Company, 401 N. Broad St., Philadelphia, PA 19108.

MNL   *Mass Media Newsletter*. Mass Media Newsletter Association, 2116 N. Charles St., Baltimore, MD 21218.

MTE   *Mathematics Teacher*. National Council of Teachers of Mathematics, 1906 Association Dr., Reston, VA 22091.

NCF   National Council on Family Relations. 1219 University Ave. S.E., Minneapolis, MN 55414.

NCT   National Council of Teachers of English. 1111 Kenyon Rd., Urbana, IL 61801.

NEA   National Educational Film Festival Award. 5555 Ascot Dr., Oakland, CA 94611.

NEF   National Educational Film Festival. 5555 Ascot Dr., Oakland, CA 94611.

NST   National Science Teachers Association. 1742 Connecticut Ave. N.W., Washington, DC 20009.

NVG   National Vocational Guidance Association. 1607 New Hamsphire Ave. NW, Washington, DC 20009.

NYF   New York Film and TV Festival. 1865 Broadway, New York, NY 10023.

NYR   *Recordings for Children*. New York Library Association, Box 521, Woodside, NY 11377.

NYT   *New York Times*. 229 W. 43 St., New York, NY 10036.

PRV   *Previews*. R. R. Bowker Company, 1180 Ave. of the Americas, New York, NY 10036.

PVB   *Previews*, "Best of the Year." R. R. Bowker Company, 1180 Ave. of the Americas, New York, NY 10036.

RTE   *Reading Teacher*. International Reading Association, 800 Barksdale Ave., Newark, DE 19711.

SAC   *Science and Children*. 1742 Connecticut Ave. N.W., Washington, DC 20009.

SAM   *Scientific American*. 415 Madison Ave., New York, NY 10017.

SCN   *Science News*. 1719 N St. NW, Washington, DC 20036.

SCT   *Science Teacher*. 1742 Connecticut Ave. N.W., Washington, DC 20009.

SLJ   *School Library Journal*. R. R. Bowker Company, 1180 Ave. of the Americas, New York, NY 10036.

SOC   *Social Education*. National Council for Social Education, 2030 M St. N.W., Suite 400, Washington, DC 20036.

SPN   *School Product News*. Pentom/IPC Inc., 1111 Chester Ave., Cleveland, OH 44114.

STE   *Scholastic Teacher*. Scholastic Magazines, Inc., 50 W. 44 St., New York, NY 10036.

TCT   *Today's Catholic Teacher*. 2451 E. River Rd., Suite 200, Dayton, OH 05439.

TEA   *Teacher*. Paul Beauvais, publisher, One Fawcett Pl., Greenwich, CT 06830.

TED   *Today's Education*. National Education Association, 1201 16th St. N.W., Washington, DC 20036.

WLB   *Wilson Library Journal*. 950 University Ave., Bronx, NY 10452.

CORE MEDIA COLLECTION FOR SECONDARY SCHOOLS

Second Edition

MEDIA INDEXED BY SUBJECT

ABORTION

THE RIGHT TO LIFE? (Filmstrip—Sound). Current Affairs Films, 1977. 1 color, sound filmstrip, 74 fr.; 1 cassette, 16 min.; Teacher's Guide. #569 ($24.00). Gr. 7–A. PRV (11/77). 179

Even though the Supreme Court ruled in 1973 that abortions during the first three months of pregnancy were legal, anti-abortion groups continue their protests, in many places with increasing support, and they are urging a Constitutional amendment to counter the Supreme Court ruling. This filmstrip examines the meaning of ''right to life'' from different points of view and discusses the possible outcome of the controversy.

ACADEMY AWARDS

SENTINELS OF SILENCE (Motion Picture—16mm—Sound). *See* Mexico—Antiquities

ACTING

ACTING FOR FILM (LONG CHRISTMAS DINNER) (Motion Picture—16mm—Sound). EBEC, 1976 (Short Play Showcase). 16mm color, sound film, 14 min. #47819 ($240.00). Gr. 10–C. EGT (1976), LFR (1977), PRV (1976). 791.4

The actors in the dramatization discuss the differences between acting for a stage presentation and acting for a film production, explaining the reasons for their preferences.

AMERICAN THEATRE (Filmstrip—Sound). *See* Theater—U.S.

CHAPLIN—A CHARACTER IS BORN (Motion Picture—16mm—Sound). *See* Chaplin, Charles Spencer

ACTORS AND ACTRESSES

BEAH RICHARDS: A BLACK WOMAN SPEAKS (Videocassette). *See* Richards, Beah

ACUPUNCTURE

ACUPUNCTURE: AN EXPLORATION (Motion Picture—16mm—Sound). FilmFair Communications, 1973. 16mm color, sound film, 16½ min. ($210). Gr. 7–A. BKL (12/15/73), CFF (1974), CGE (1974). 615

The purpose of this film is to introduce acupuncture, which is part of a 5,000-year-old Chinese science and art of internal medicine, describe experiments being held, to learn why it works, and to show its relation to ancient and modern medical thought. The film describes current American experiments which confirm the accuracy of the ancient Chinese acupuncture charts and show that the effective points group themselves into a pattern.

ADAMS, JOHN

THE FOUNDING FATHERS IN PERSON (Cassette). *See* U.S.—History—Biography

ADAMSON, JOY

ELSA BORN FREE (Filmstrip—Sound). *See* Lions

PIPPA THE CHEETAH (Filmstrip—Sound). *See* Cheetahs

ADAPTATION (BIOLOGY)

ANIMAL LIFE SERIES, SET TWO (Film Loop—8mm—Silent). McGraw-Hill Films, 1969. 7 silent, 8mm technicolor loops, 3–4 min. ea. #668558-6 ($154). Gr. 5–12. INS (2/72). 574.5

Contents: 1. Moving in Water. 2. Moving in Air. 3. Moving on Land. 4. Escaping Enemies. 5. Special Defenses. 6. Plant Eaters. 7. Animal Eaters.

This set shows how different animals in the same environment have similar adaptations for movement.

THE CACTUS: ADAPTATIONS FOR SURVIVAL (Motion Picture—16mm—Sound). *See* Cactus

1

ADOLESCENCE

ADOLESCENCE, LOVE AND DATING
(Kit). Butterick Publishing, 1976 (Family
Life). 6 color, sound filmstrips, 65–75 fr.; 6
cassettes or discs, 10–12 min.; 11 Spirit Masters; Teacher's Guides. #C–603–9E Cassettes ($125). #R–602–0E Discs ($125). Gr.
7–A. PRV (9/77). 301.42

Contents: 1. & 2. Adolescence. 3. & 4.
Love. 5. & 6. Dating.

The adolescent is neither child nor adult. In
this set, students are helped to better understand and deal with their growing freedoms
and options.

ADOLESCENCE TO ADULTHOOD: RITES
OF PASSAGE (Filmstrip—Sound). Sunburst Communications, 1974. 2 color sound
filmstrips, 76–84 fr.; 2 cassettes or discs, $9^{1}/_{2}$–
11 min.; Teacher's Guide. #213 ($59). Gr.
9–12. BKL (6/75), LGB (12/75), PVB (4/76).
301.43

Contents: 1: What Is an Adult? 2: Modern
Rites of Passage.

Examines the ways in which society, particularly our own society, confers adult status
on its young people. Explores the concepts
of maturity and adulthood. Asks students to
determine what they feel constitutes adult
behavior and to evaluate how their views differ from the opinions of their parents.

ADOLESCENT CONFLICT: PARENTS VS.
TEENS (Filmstrip—Sound). Sunburst Com-
munications, 1977. 2 color, sound filmstrips,
70–80 fr.; 2 cassettes or discs, 12–14 min.;
Teacher's Guide. #244–98 ($55). Gr. 7–12.
BKL (5/1/78), PRV (12/78). 301.42

Contents: 1. Understanding the conflict. 2.
Dealing with conflict.

Presents an objective look at the parent and
the adolescent points of view. Also suggests
that the parent, child, and adult within each
individual requires compromise on the part
of both parent and adolescent.

ADOLESCENT RESPONSIBILITIES:
CRAIG AND MARK (Motion Picture—
16mm—Sound). EBEC, 1973. 16mm color,
sound film, 28 min. #3204 ($360). Gr. 7–12.
EFL (1975), LFR (1974), PRV (1974). 301.43

In this film, the family is undertaking a difficult decision: should they leave their life in
the big city for a (hopefully) more idyllic ex-
istence in Colorado? In their efforts to arrive
at an answer, the family raises questions
about work and money, responsibilities and
privileges.

BECOMING AN ADULT: THE PSYCHO-
LOGICAL TASKS OF ADOLESCENCE
(Filmstrip—Sound). Human Relations Media, 1977. 3 color, sound filmstrips, 69–78 fr.;
3 cassettes or discs, $10^{1}/_{2}$–14 min.; Teacher's Guide. #603 ($90). Gr. 9–12. BKL (5/1/
78), FHE (5/78), NCF (1978). 155.5

Contents: 1. Attaining Sexual Maturity. 2.
Breaking Away from Parents. 3. Personal
Values and Life Roles.

Identifies critical life tasks of adolescence:
attaining sexual maturity, breaking away
from parents, discovering personal values,
and choosing a role in society. Students are
challenged to evaluate the status of these important tasks and formulate plans for achieving them.

COPING WITH PARENTS (Motion Picture—
16mm—Sound). See Conflict of generations

THE INDIVIDUAL (Kit). See Self

LOOKING TOWARDS ADULTHOOD
(Filmstrip—Sound). Parents' Magazine
Films, 1976 (Parents and Teenagers, Portraits and Self-Portraits). 5 color, sound filmstrips 38–62 fr. ea.; 1 disc or 3 cassettes, 6–
10 min. ea.; 5 Audio Scripts; Guide. ($65).
Gr. 7–12. PRV (3/78). 301.43

Parents and teenagers share their concerns
and air the conflicts which develop as the
adolescent moves towards adulthood.
School and career are considered from both
perspectives.

MATURITY: OPTIONS AND CON-
SEQUENCES (Filmstrip—Sound). Knowledge Aid/United Learning, 1977. 6 color,
sound filmstrips, 46–68 fr.; 6 cassettes, 7:17–
11 min.; 1 Teacher's Guide. ($92). Gr. 7–12.
PRV (10/19/77). 301.43

Contents: 1. Honesty. 2. Getting Married. 3.
Sexual Ethics. 4. Teenage Mother. 5. Teenage Parents. 6. Teenage Alcoholic.

Six episodes present realistic and accurate
situations involving young people faced with
critical choices—choices that often mean the
difference between a full, happy life, and a
life of helplessness and despair. The sound
filmstrip presentations are designed to point
out alternatives so the endings are open.

PORTRAIT OF PARENTS (Filmstrip—
Sound). Parents' Magazine Films, 1976 (Parents and Teenagers, Portraits and Self Portraits). 5 color, sound filmstrips, 38–62 fr.
ea.; 1 disc or 3 cassettes, 6–10 min. ea.;
Audio Scripts; Teacher's Guide. ($65). Gr.
7–12. PRV (3/78). 301.42

This set presents parents as they examine
their own needs and the issues which cause
conflicts within themselves and their families. Single parents, working mothers, and a
mother returning to school are some of the
situations families find themselves in that
can create conflicts.

PORTRAIT OF TEENAGERS (Filmstrip—
Sound). Parents' Magazine Films, 1976 (Parents and Teenagers, Portraits and Self Portraits). 5 color, sound filmstrips, 38–62 fr.
ea.; 1 disc or 3 cassettes, 6–10 min. ea; audio
scripts; Teacher's Guide. ($65). Gr. 7–12.
PRV (3/78). 301.42

Parents and teenagers talk about what it means to be a teenager—the problems and pressures, and the difficulties in resolving them.

THE STRUGGLE FOR INDEPENDENCE (Filmstrip—Sound). Parents' Magazine Films, 1976 (Parents and Teenagers, Portraits and Self-Portraits). 5 color, sound filmstrips, 38–62 fr. ea.; 3 cassettes or 1 disc, 6–10 min. ea.; Audio Scripts; Teacher's Guide. ($65). Gr. 7–12. PRV (3/78). 301.43

Explored in this set are the conflicts that arise within teenagers, and between teenagers and their parents, as adolescents try to take control of their lives and abandon the attitudes and influence of their parents.

TEENAGE RELATIONSHIPS: VENESSA AND HER FRIENDS (Motion Picture—16mm—Sound). EBEC, 1973. 16mm color, sound film, 19 min. #3202 ($255). Gr. 7–12. EFL (2/75), LFR (2/74), PRV (1974). 155.5

This portrait of tenth-grade student Venessa explores her relationships with Richard, a nonromantic boyfriend, Linda, her closest girlfriend, and Casey, her "sweetheart." A consultation with an older friend, Jessica, helps Venessa arrive at her own decisions. The film raises questions about group dating . . . the importance of the crowd . . . responsibilities of friendship.

ADOLESCENCE—FICTION

THE OUTSIDERS (Phonodisc or Cassette). *See* Juvenile delinquency—Fiction

SADDLE UP! (Kit). *See* Horses—Fiction

ADVENTURE AND ADVENTURERS—FICTION

THE ADVENTURES OF HUCKLEBERRY FINN BY MARK TWAIN (Phonodisc or Cassette). Listening Library, 1974. 3 cassettes or discs, approx. 173 min. Cassettes ($22.95), Discs ($20.95). Gr. 7–12. LTP (8/75), PRV (1/76). Fic

Huck runs away to escape his drunken father and decides the best means of escape is on a raft down the Mississippi. He joins up with Jim, a runaway slave, and their adventures are both exciting and moving since Huck develops a new relationship and understanding. They later meet Tom Sawyer and in a final escapade all turns out well.

HUCKLEBERRY FINN (Cassette). *See* U.S.—Social life and customs—Fiction

HUCKLEBERRY FINN (Phonodisc or Cassette). *See* U.S.—Social life and customs—Fiction

STORIES OF ADVENTURE AND HEROISM (Filmstrip—Sound). *See* Reading materials

SWISS FAMILY ROBINSON (Phonodisc or Cassette). Caedmon Records, 1975. 1 cassette or disc. #CDL51485 Cassette ($7.95), #TC1485 Disc ($6.98). Gr. 4–8. PRV (4/76). Fic

An abridgement of Johann Wyss's famous novel is presented by Anthony Quayle. This is the story of the family washed up on an island and forced to utilize their imagination and knowledge for survival.

ADVERTISING

THE BUY LINE (Motion Picture—16mm—Sound). FilmFair Communications, 1972. 16mm color, sound film, 14 min. ($180). Gr. 7–A. BKL (5/1/73), FHE (9/73), LFR (3/73). 659.1

This film provides examples of current manipulative, deceptive, misleading, and informative advertising with stress placed on how to read the psychological appeal of advertisements.

SIXTY SECOND SPOT (Motion Picture—16mm—Sound). *See* Television advertising

SOOPERGOOP (Motion Picture—16mm—Sound). Churchill Films, 1975. 16mm, color, sound, animated film, 13 min.; Study Guide. ($205). Gr. 1–A. PRV (10/76). 659.1

A fast animated story in which two irreverent characters concoct a TV commercial for a sweet cereal. Reveals selling techniques and commercialism behind the fun.

THE THIRTY SECOND DREAM (Motion Picture—16mm—Sound). Lexington Recording Company/Mass Media Ministries, 1977. 16mm color, sound film, 15 min. ($250). Gr. 6–12. BKL (9/1/78). 659.14

This film is a montage of visual and sound images from television commercials designed to illustrate the creativity, subtlety, and emotional force of television advertising. It shows how advertisers play on fears of inadequacy and fantasies of wish-fulfillment in four essential areas: family, intimacy, vitality, and success.

AERONAUTICS

AIRCRAFT: THEIR POWER AND CONTROL (Filmstrip—Sound). Prentice-Hall Media, 1973. 6 color, sound filmstrips; 6 cassettes or discs; 12 Activity Cards; Teacher's Guide. #HAC5960 Cassettes ($108), #HAR5060 Discs ($108). Gr. 4–8. STE (9/73). 629.13

Contents: 1. Lift and Thrust. 2. Controlling an Airplane. 3. How Helicopters Fly. 4. Jet Power. 5. Jet Flight 923. 6. Rocket Power.

Introduces basic principles of flight and explains how various types of aircraft are controlled and propelled.

AERONAUTICS (cont.)

QUEST FOR FLIGHT (Motion Picture—16mm—Sound). *See* Flight

AERONAUTICS—BIOGRAPHY

THE WRIGHT BROTHERS (Cassette). Ivan Berg Associates/Jeffrey Norton Publishers, 1977 (History Makers). 1 cassette, approx. 78 min. #41022 ($11.95). Gr. 7–A. BKL (4/15/78), PRV (5/19/78). 920

This program vividly traces the lives of these famous brothers—from their early days when they ran a bicycle shop; their days as inventors when they were often scoffed at and disbelieved; their successful development of flying machines; and their eventual acknowledgment by royalty and learned scientists throughout the world.

AFRICA

AFRICA: LEARNING ABOUT THE CONTINENT (Filmstrip—Sound). Society for Visual Education, 1978. 4 color, sound filmstrips, 64 fr.; 4 cassettes, 10 min,; Guide. #LG283–SBTC Cassettes ($74), #LG283–SBR Discs ($74). Gr. 7–12. BKL (7/1/78). 916

Contents: 1. Looking at the Land. 2. Studying the Population. 3. Using the Resources. 4. Working in Africa.

The vastness, complexity, and diversity of the African continent and its people are illustrated. The geographic features are explored, as well as the effect on the lives of the African people. The problems of divergent ethnic groups and their struggle for survival are discussed. Job opportunities and occupational changes are examined.

AFRICA: PORTRAIT OF A CONTINENT (Kit). Educational Enrichment Materials, 1976. 6 color, sound filmstrips, av. 56–75 fr.; 6 cassettes or discs 13–18 min.; 5 Wall Charts; 6 Spirit Duplicating Masters; 1 paperback book. Teacher's Guide. #51003 ($127). Gr. 4–9. BKL (1977), LGB (1977). 916

Contents: 1. The Land and Its Resources. 2. The History of a Continent. 3. Religion and Culture. 4. The Tribal Way of Life. 5. The New Society. 6. Problems and Prospects.

An excellent overview of the continent of Africa. Well-written scripts are objective and show a respect for the people and their culture. The fine visual photos of objects are unusually clear. This very informative kit will hold the attention of the individual viewer or classroom audience.

AFRICAN CLIFF DWELLERS: THE DOGON PEOPLE OF MALI, PART ONE (Kit). EMC, 1970. 1 color, sound filmstrip, 74 fr.; 1 cassette, 14 min.; 3 Charts; Teacher's Guide. #SS-20400 ($50). Gr. 4–9. BKL (10/72), INS (2/72), M&M (10/70). 916

Contents: 1. Home, Masks and Ancestors: Village Life. 2. Crafts, Culture, and the Environment: The Dry Season.

This program takes a close look at people who live in total harmony with their environment.

AFRICAN CLIFF DWELLERS: THE DOGON PEOPLE OF MALI, PART TWO (Kit). EMC, 1970. 1 color, sound filmstrip, 67 fr.; 1 cassette, 12 min.; 3 Charts; Teacher's Guide. #SS-20400 ($50). Gr. 4–9. BKL (10/72), INS (2/72), M&M (10/70). 916

Contents: 1. Crafts, Culture, and the Environment: The Dry Season.

This program takes a close look at people who live in total harmony with their environment.

ELSA BORN FREE (Filmstrip—Sound). *See* Lions

NIGERIA AND THE IVORY COAST: ENTERING THE 21ST CENTURY (Filmstrip—Sound). Multi-Media Productions, 1977. 2 color, sound filmstrips, 43–48 fr.; 2 cassettes, 9–11 min,; Teacher's Guide. #7207C ($19.95). Gr. 9–12. PRV (2/79). 916

Nigeria, a former British colony, and the Ivory Coast, with its French heritage, are used to illustrate some of the problems and challenges of today's developing countries.

PIPPA THE CHEETAH (Filmstrip—Sound). *See* Cheetahs

AFRICA—CIVILIZATION

THE CREATIVE HERITAGE OF AFRICA: AN INTRODUCTION TO AFRICAN ART (Kit). *See* Art, African

AFRICA, WEST

FAMILIES OF WEST AFRICA: FARMERS AND FISHERMEN (Filmstrip—Sound). EBEC, 1974. 4 color sound filmstrips; av. 79 fr.; 4 cassettes av. 14 min. ea.; Teacher's Guide. #6498 Discs ($57.95), #6498K Cassettes ($57.95). Gr. 3–9. ESL (1977), PRV (5/76). 916.6

Contents: 1. Cocoa Farmer of Ghana. 2. Farmer of Mali. 3. Fisherman of Liberia. 4. Rural Medic of the Ivory Coast.

Rural families in four West African countries meet the challenges of survival, using methods dictated by their individual climates and geography.

WEST AFRICAN ARTISTS AND THEIR ART (Filmstrip—Sound). *See* Art, African

AGING

YOUTH, MATURITY, OLD AGE, AND DEATH (Filmstrip—Sound). *See* Pantomimes

AGRICULTURE—EXPERIMENTS

THE NEW ALCHEMISTS (Motion Picture—
16mm—Sound). National Film Board of
Canada/Benchmark Films, 1975. 16mm col-
or, sound film, 29 min. ($395). Gr. 7–A. AAS
(12/77), AFF (1976), JNE (10/77). 630.2
A small group of young scientists and families successfully work an experimental plant
and fish farm using organic fertilizers in
an efficient self-contained ecosystem with
solar heat and windmill for energy.

AGRICULTURE—INDIA

AGRICULTURAL REFORM IN INDIA: A
CASE STUDY (Filmstrip—Sound). Zenger
Productions/Sunburst Communications,
1975. 1 filmstrip; 1 cassette or disc; Teacher's Guide. #226 ($29). Gr. 9–C. BKL (7/75),
CPR (9/75). 630.954
The problems and consequences of introducing a new high-yield variety of wheat
into India's agriculture system are docu-
mented.

AGRICULTURE—U.S.

THE FARMER IN A CHANGING AMERI-
CA (Motion Picture—16mm—Sound).
EBEC, 1973. 16mm color, sound film, 27
min. #3196 ($360). Gr. 7–C. BKL (1974),
LFR (1973), PRV (1974). 630.973
Such things as air-conditioned combines or
seeds coated with their own fertilizer and
pesticides would make today's mechanized
farms seem strange to the farmers who set-
tled America. How farming has come to be a
huge business is the story in this film.
Glimpses of projected developments, such
as Skylab reports on crop conditions and
domes in which farmers will dial their own
environments, suggests that change will continue to be a dominant factor on the farms of
industrial America.

HARVEST (Motion Picture—16mm—Sound).
Centron Educational Films, 1977. 16mm col-
or, sound film, 8 min. ($145). Gr. 7–A. CFF
(1977), CGE (1977), IFT (1977). 630.973
This theatrical featurette chronicles the outdoor life of today's farmer, documents his
struggle with the elements, and underscores
the responsibility of this grower-of-food-for-
a-hungry-world. The film is visually oriented
and sparsely narrated.

AIR

ATMOSPHERE IN MOTION (Motion Picture—16mm—Sound). *See* Atmosphere

AIRPLANES—MODELS

MODEL AIRPLANES (Kit). Children's
Press, 1976 (Ready, Get-Set, Go). 1 cassette;
1 Hardback Book; Teacher's Guide.

#07558–6 ($11.95). Gr. 1–8. PRV (9/15/77).
629.122
Students enjoy the photographs and understand the instructions that explain how to
choose, build, and fly a model airplane in
this high-interest, low reading ability set.

ALASKA

ALASKA: THE BIG LAND AND ITS
PEOPLE (Filmstrip—Sound). EBEC, 1975.
5 color, sound filmstrips, av. 100 fr. ea.; 5
discs or cassettes, 12 min. ea.; Teacher's
Guide. #6906K Cassettes ($72.50), #6906
Discs ($72.50). Gr. 4–8. BKL (10/75), ESL
(1977), PRV (9/75, 76), 917.98
Contents: 1. Alaska: The Big Land. 2.
Alaska's Economy: Development or Exploitation? 3. The Core Area: Anchorage
and Fairbanks. 4. The Life of the Eskimo:
Hooper Bay, Alaska. 5. The Life of the Indian: Arctic Village, Alaska.
This series explores the cultures of the Eskimo and the Alaskan Indian—two groups
who live their lives poised between the old
ways and the new and find that neither life-
style solves all their problems. It shows
Alaska's problems as well as its promise as it
traces the history of the 49th state down to
its place in the current American scene.

SEEING ALASKA (Filmstrip—Sound). Coronet Instructional Media, 1973. 4 color, sound
filmstrips, 50–53 fr.; 2 discs or 4 cassettes, 8–
13 min.; Teacher's Guide. #S246 ($65),
#M246 ($65). Gr. 4–8. PRV (5/74). 917.98
Contents: 1. Land and Resources. 2. Fishing, Hunting, and Farming. 3. Industry and
Commerce. 4. History and People.
In a land with abundant natural resources
but rapid population growth, there is a controversy between conservationists and the
business interests concerning transportation
and oil development. This set explores the
issues involved.

ALCOHOL

ALCOHOL AND ALCOHOLISM (Filmstrip—Sound). *See* Alcoholism

ALCOHOL AND ALCOHOLISM: THE
DRUG AND THE DISEASE (Filmstrip—
Sound). *See* Alcoholism

ALCOHOL: FACTS, MYTHS, AND DECI-
SIONS (Filmstrip—Sound). Sunburst Communications, 1976. 3 color, sound filmstrips,
57–80 fr.; 3 cassettes or discs, av. 11 min.
ea.; Teacher's Guide. #299 ($85). Gr. 9–12.
BKL (6/15/77), MMB (4/77), NCF (1977).
613.8
Contents: 1. The Facts. 2. Choices. 3.
Where to Go for Help.
This research-based program provides honest answers to questions teenagers ask about
alcohol use and abuse. Is alcohol a stimulant

ALCOHOL (cont.)

or a depressant? How does it affect the brain? How long does it take for the body to burn up alcohol? Answers to these and other questions provide students with an understanding of alcohol in terms of their own experience.

WHAT ARE YOU GOING TO DO ABOUT ALCOHOL? (Filmstrip—Sound). Guidance Associates, 1975. 2 color, sound filmstrips, av. 68 fr. ea.; 2 discs or cassettes, av. 8 min. ea.; Teacher's Guide. Discs #9A–301–182 ($52.50), Cassettes #9A–301–190 ($52.50). Gr. 5–8. PRV (4/76). 301.47

In Part I the basic facts of the physical and psychological effects of alcohol are presented, exploring factors motivating alcohol use. Part II includes dramatized vignettes about the peer pressure, advertising, parental models, and drunk driving as they relate to alcohol use.

ALCOHOL—PHYSIOLOGICAL EFFECTS

ALCOHOL PROBLEM: WHAT DO YOU THINK? (Motion Picture—16mm—Sound). EBEC, 1975. 16mm color, sound film, 18 min. #3194 ($240). Gr. 7–C. AVI (1973), CPR (1974), EFL (1974). 616.861

Social aspects of drinking and the history of alcohol usage are presented, and the economic importance of the alcohol industry is discussed. The chemistry of alcohol and the physiological effects of drinking are analyzed. How alcohol affects the coordination of a baseball player at bat and how a breathalyzer is used are illustrated. Experiments with mice show the physical dependency and withdrawal symptoms caused by alcoholism. Youthful alcoholics are used to provoke discussion.

THE D. W. I. (DRIVING WHILE IN-TOXICATED) DECISION (Motion Picture—16mm—Sound). *See* Safety education

PHYSIOLOGY OF SMOKING AND DRINKING (Filmstrip—Sound). Sunburst Communications, 1973. 2 color, sound filmstrips, 73–76 fr.; 2 cassettes or discs, 15 min. ea.; Teacher's Guide. #125 ($59). Gr. 8–12. LGB (1974), PRV (3/74). 616.86

Contents: 1. Smoking: Physiological Effects. 2. Drinking: Physiological Effects.

Outlines the newest facts about smoking and drinking, without making value judgments, by presenting their effects on the human body.

ALCOHOLISM

ALCOHOL AND ALCOHOLISM (Filmstrip—Sound). Imperial Educational Resources, 1977. 4 color, sound filmstrips, av. 70–76 fr.; 4 discs or cassettes, av. 10–13 min. #3KG–68300 Cassettes ($62), #RG–68300 Discs ($56). Gr. 5–12. BKL (12/15/77). 616.861

Contents: 1. What Is Alcohol? 2. Why People Drink. 3. Teenagers and Drinking. 4. The Counterattack on Despair.

Diagrams, ancient artwork, and photos depict the tradition of the pleasures and problems of drinking alcohol.

ALCOHOL AND ALCOHOLISM: THE DRUG AND THE DISEASE (Filmstrip—Sound). University Films/McGraw-Hill Films, 1975. 4 color, sound filmstrips, av. 66 fr.; 4 cassettes or discs, 12 min.; Teacher's Guide. ($72). Gr. 5–12. ABT (5/77), NST (9/76), PRV (3/77). 616.861

Contents: 1. Alcohol and the Human Body. 2. Alcohol Abuse and Society. 3. Alcoholism and Youth. 4. Alcoholism: Danger Signals.

A comprehensive look at America's most widely used and abused drug. Each filmstrip probes a particular aspect of the problem of student drinking, to create a thought-provoking forum for discussion and to give young people the information they need to formulate healthy personal attitudes toward the use of alcohol.

ALCOHOL, DRUGS OR ALTERNATIVES (Motion Picture—16mm—Sound). Sandler Institutional Films, 1975. 16mm color, sound film, approx. 21 min. ($330). Gr. 7–A. AUR (1978). 301.47

Explores alternatives to dependence upon drugs and alcohol with observations from halfway house residents. Discusses causes of feelings of inadequacy and considers how this influences actions. Presents techniques used as substitutes for dependence upon artificial stimulants and depressants. Hosted by Christopher George and Tommy Smothers.

ALCOHOL: PINK ELEPHANT (Motion Picture—16mm—Sound). EBEC, 1976. 16mm color, sound film, 15 min. #3389 ($185). Gr. 7–12. LFR (1977). 616.8

Solid factual information about the causes and cure of alcoholism is presented against a background of fast-paced action. Light touches of humor and occasional slapstick keep the elephant's comments from turning into a lecture.

CHILDREN OF ALCOHOLIC PARENTS (Filmstrip—Sound). Multi-Media Productions, 1977. 2 color, sound filmstrips, 42–51 fr.; 2 cassettes, 8–9 min. #7213 ($19.95). Gr. 9–12. PRV (12/19/78). 616.8

This set presents the case of the child living with an alcoholic parent (or parents). The basics of the problem are outlined, and the film is intended as an introduction to the difficulties such children face.

ALCOTT, LOUISA MAY

CHILDHOOD OF FAMOUS WOMEN, VOLUME THREE (Cassette). *See* Women—Biography

LITTLE WOMEN (Phonodisc or Cassette). *See* Family life—Fiction

ALEXANDER, LLOYD

LLOYD ALEXANDER (Filmstrip—Sound). *See* Authors

ALGAE

THE PROTISTS (Filmstrip—Sound). *See* Microorganisms

ALGERIA

ALGIERS: A STEP INTO THE FUTURE, A STEP INTO THE PAST (Filmstrip—Sound). National Film Board of Canada/Donars Productions, 1976. 1 color, sound filmstrip, 108 fr.; 1 cassette, 13 min.; 1 Guide. ($25). Gr. 5–8. BKL (7/1/77). 916.5

Helps promote a better understanding of Algeria: homes, food, leisure activities, jobs, religion, education, culture, and history are all discussed. This an interesting look at a little-known nation, clinging to old ways and discovering the new.

ALIENATION (SOCIAL PSYCHOLOGY)—FICTION

THE PIGMAN (Phonodisc or Cassette). *See* Friendship—Fiction

ALLEN, WOODY

WOODY ALLEN (Cassette). Tapes for Readers, 1978. 1 cassette, 20 min. #ENT-001 ($10.95). Gr. 10–A. BKL (1/15/79). 921

This tape is a serious interview with this favorite comedian as he discusses his favorite literary and film figures. Obsessions with sex and death are mentioned, as is the idea of humanity.

ALLIGATORS

ALLIGATOR—BIRTH AND SURVIVAL (Film Loop—8mm—Silent). Walt Disney Educational Media, 1966. 8mm color, silent film loop, approx. 4 min. #62–5053L ($30). Gr. K–12. MDU (1978). 589.14

Baby alligators are viewed as they hatch and head for water. They are shown running into danger from an adult alligator and a hungry raccoon.

ALLUSIONS

SPLENDOR FROM OLYMPUS (Kit). *See* Mythology, classical

AMERICA—ANTIQUITIES

DIGGING UP AMERICA'S PAST (Filmstrip—Sound). National Geographic, 1977. 5 color, sound filmstrips, 42–50 fr.; 5 cassettes or discs, 11–14 min.; Teacher's Guide. #03240 Cassettes ($74.50), #03239 Disc ($74.50). Gr. 5–12. PRV (3/78), PVB (4/78). 970.01

Contents: 1. North America before Columbus. 2. Middle America before Cortes. 3. South America before Pizarro. 4. The First Europeans in the Americas. 5. Colonization and After.

Covers archeological techniques and discoveries in the Americas.

AMERICA—EXPLORATION

DIGGING UP AMERICA'S PAST (Filmstrip—Sound). *See* America—Antiquities

DISCOVERY AND EXPLORATION (Filmstrip—Sound). McGraw-Hill Films, 1972 (American Heritage). 5 color, sound filmstrips, av. 50 fr. ea.; 5 cassettes or discs, 7 min. #102448-4 Cassettes ($95), #103682-2 Discs ($95). Gr. 7–A. PRV (11/73). 970.01

Contents: 1. Terra Incognita. 2. Mundus Novus. 3. Noche Triste. 4. El Dorado. 5. De Revoluntionibus.

Through paintings and authentic period maps, this set traces the exploits of Christopher Columbus, Vasco da Gama, Amerigo Vespucci, and others, as well as surveys the Conquistadors.

AMERICAN DRAMA

THE LONG CHRISTMAS DINNER (Motion Picture—16mm—Sound). EBEC, 1976. 16mm color, sound film, 37 min. #47818 ($510). Gr. 9–A. BKL (1976), EGT (1976), PRV (1976). 812

An example of nonrepresentational, symbolic theater, this play represents ninety Christmas dinners in the Bayard household. All action takes place around the dining room table. The portals to right and left, through which the characters pass, represent birth and death. By stylizing dialogue, dispensing with "real" scenery, ignoring the illusion of time, and emphasizing universal experiences, the author expresses his view that life is cyclical.

TENNESSEE WILLIAMS: THEATER IN PROCESS (Motion Picture—16mm—Sound). *See* Theater—Production and direction

AMERICAN FICTION

BARTLEBY (Motion Picture—16mm—Sound). EBEC, 1969 (Short Story Showcase). 16mm color, sound film, 28 min. #47753 ($359). Gr. 7–C. M&M (1974). Fic

Herman Melville's enigmatic, haunting story of the man who "preferred not to" has been faithfully but imaginatively translated onto film, with authentic locales, period sets, and unusually sensitive performances.

CANNERY ROW, LIFE AND DEATH OF AN INDUSTRY (Filmstrip—Sound). *See* Monterey, California—History

A DISCUSSION OF BARTLEBY (Motion Picture—16mm—Sound). *See* Motion pictures—History and criticism

A DISCUSSION OF DR. HEIDEGGER'S EXPERIMENT (Motion Picture—16mm—Sound). *See* Motion pictures—History and criticism

A DISCUSSION OF MY OLD MAN (Motion Picture—16mm—Sound). *See* Motion pictures—History and criticism

A DISCUSSION OF THE HUNT (Motion Picture—16mm—Sound). *See* Motion pictures—History and criticism

A DISCUSSION OF THE LADY OR THE TIGER? (Motion Picture—16mm—Sound). *See* Motion pictures—History and criticism

A DISCUSSION OF THE LOTTERY (Motion Picture—16mm—Sound). *See* Motion pictures—History and criticism

A DISCUSSION OF THE SECRET SHARER (Motion Picture—16mm—Sound). *See* Motion pictures—History and criticism

DOCTOR HEIDEGGER'S EXPERIMENT (Motion Picture—16mm—Sound). EBEC, 1969 (Short Story Showcase). 16mm color, sound film, 22 min; Guide #47751 ($296). Gr. 7–C. LFR (1970). Fic

This film dovetails with the study of Nathaniel Hawthorne, since it deals with two of the author's favorite themes: the consequences of tampering with nature and of rejecting conventional morality. The story's "science-fiction" format makes it particularly appealing to today's youngsters.

THE HUNT (Motion Picture—16mm—Sound). EBEC, 1975 (Short Story Showcase). 16mm color, sound film, 30 min. #47798 ($430). Gr. 7–C. AFF (1975), LFR (1975). Fic

This film is adapted from Richard Connell's short story by the same name. The suspense of the hunter becoming the hunted provides a cliff-hanger situation when the hunter meets an old man jaded so with killing that he now only stalks human prey.

THE LADY OR THE TIGER? (Motion Picture—16mm—Sound). EBEC, 1969 (Short Story Showcase). 16mm color, sound film, 16 min. #47755 ($232.50). Gr. 7–C. CIF (1971), M&M (11/74), LFR (1972). Fic

This 1882 classic short story by Frank Stockton has been reset in the space age preserving all the whimsy and "open-ended" suspense of the original.

THE LOTTERY BY SHIRLEY JACKSON (Motion Picture—16mm—Sound). EBEC, 1969 (Short Story Showcase). 16mm color, sound film, 18 min. #47757 ($265). Gr. 7–C. M&M (11/74), STE (1974). Fic

This film of Shirley Jackson's short story is as powerful as the story on which it is based. It is a dramatization on the theme of society and the individual.

MY OLD MAN (Motion Picture—16mm—Sound). EBEC, 1971 (Short Story Showcase). 16mm color, sound film, 27 min. #47759 ($359). Gr. 7–A. BKL (1972), M&M (1971). Fic

Almost every youngster identifies with Joe Butler and his conflict between accepting reality and preserving his illusions. Ernest Hemingway captured this strong and poignant appeal in this remarkable short story. The film was produced in Paris, using race track scenes and backgrounds.

THE PORTABLE PHONOGRAPH (Motion Picture—16mm—Sound). EBEC, 1977 (Short Story Showcase). 16mm color, sound film, 24 min. #47827 ($390). Gr. 7–A. BKL (1978), EGT (1978), PRV (1979). Fic

The scene is a desolate landscape and then faintly, a human voice reads from Shakespeare's *The Tempest*. Four war survivors have gathered in a dugout to hear the portable phonograph one of them has saved. An example of doomsday science fiction, but Walter Van Tilburg Clark also shows what the Humanities are all about in this poignant filmed dramatization.

THE RAZOR'S EDGE (Cassette). *See* Self-realization—Fiction

SCOURBY READS HEMINGWAY (Phonodisc or Cassette). Listening Library, 1978. 2 cassettes or discs, approx 60 min. with notes #CX3124-2 Cassettes ($15.90), #AA33124-2 Discs ($15.90). Gr. 7–A. BKL (1/15/79). Fic

Contents: 1. The Short Happy Life of Francis Macomber. 2. The Snows of Kilimanjaro.

Alexander Scourby's reading of Hemingway captures the zest and sensitivity of Hemingway as he brings us two great short stories by this literary giant.

THE SECRET SHARER (Motion Picture—16mm—Sound). EBEC, 1973. 16mm color, sound film, 30 min. #47785 ($430). Gr. 7–C. BKL (1973), M&M (1974), STE (1973). Fic

This short story rediscovers the haunting tale of a young sea captain whose inner conflicts pit his conscience against the safety of the ship he commands and the men he leads—a struggle precipitated by the "secret sharer" of his cabin. An authentic, turn-of-the-century sailing ship, with shadowy nooks and corners, sets the mood for this probing drama.

STORY INTO FILM: CLARK'S THE POR-TABLE PHONOGRAPH (Motion Picture—16mm—Sound). *See* Motion pictures—History and criticism

AMERICAN LITERATURE—HISTORY AND CRITICISM

FIVE MODERN NOVELISTS (Filmstrip—Sound). EBEC, 1975 (The American Experience in Literature). 5 color, sound filmstrips, av. 100 fr. ea.; 5 discs or cassettes, 17 min. ea.; Guide. #6911K Cassettes ($82.95), #6911 Discs ($82.95). Gr. 9C. BKL (10/1/75), M&M (6/77), MDU (1/76). 810.9
Contents: 1. Sinclair Lewis. 2. F. Scott Fitzgerald. 3. Ernest Hemingway. 4. John Steinbeck. 5. William Faulkner.
From the 1920s to the 1950s the novel was America's most popular literary form, and these five writers were the reason for that popularity. Color and historical photos dramatically portray their environment and period about which they wrote.

NINETEENTH CENTURY POETS (Filmstrip—Sound). EBEC, 1979 (The American Experience in Literature). 4 color, sound filmstrips, av. 80 fr. ea.; 4 cassettes, 17 min. ea.; Guide. #17042K ($66.50). Gr. 9–C. AUR (1979). 810.9
Contents: 1. Henry Wadsworth Longfellow. 2. Whittier, Holmes, and Lowell. 3. Walt Whitman. 4. Emily Dickinson.
From the traditional works of Henry Wadsworth Longfellow, to the quiet and introspective genius of Emily Dickinson, this set presents major American poets of the 19th century. The social and critical poems of the New England group of John Greenleaf Whittier, Oliver Wendell Holmes, and James Russell Lowell are presented, as are the innovative works of Walt Whitman, who, more than any other, provided the foundations of a modern American poetry. These critical biographies present the poets in their milieus.

POETS OF THE TWENTIETH CENTURY (Filmstrip—Sound). EBEC, 1976 (The American Experience in Literature). 5 color, sound filmstrips, av. 90 fr. ea.; 5 cassettes or discs, 17 min. ea.; Guide. ($82.95). Gr. 9–C. BKL (4/1/77), EGT (11/77), PRV (10/77). 810.9
Contents: 1. Robert Frost. 2. Carl Sandburg. 3. Marianne Moore. 4. E. E. Cummings. 5. Langston Hughes.
This set is an analysis of the works, lives, and times of five of America's greatest literary masters. These five poets were instrumental in shaping the directions and dimensions of American poetry in this century: Langston Hughes, the voice of black America; Robert Frost, four-time winner of the Pulitzer Prize for Poetry; Carl Sandburg, the voice of the common man, the working class; Marianne Moore, the most honored woman poet of her generation; E. E. Cummings, famed for his innovative use of the language and for his typographic uniqueness.

THE ROMANTIC AGE (Filmstrip—Sound). EBEC, 1977 (The American Experience in Literature). 5 color, sound filmstrips, av. 90 fr. ea.; 5 cassettes or discs, 20 min. ea.; Guide. #6962K Cassettes ($82.95), #6962 Discs ($82.95). Gr. 9–C. BKL (2/78), EGT (1/77), PRV (10/78). 810.9
Contents: 1. Edgar Allan Poe. 2. Nathaniel Hawthorne. 3. Herman Melville. 4. Henry David Thoreau. 5. Ralph Waldo Emerson.
Within a span of a few years in the mid-19th century, these five writers penned some of the greatest stories, novels, and poems the world has ever known. Using portraits and photographs of these authors, the filmstrips reconstruct their lives and time. Contemporary photography shows such settings as Salem, Massachusetts, Concord, Massachusetts, Walden Pond, the New England seaports, and the Berkshire Mountains. Historical visuals range from South Pacific whaling scenes to views of the English countryside.

AMERICAN LITERATURE—STUDY AND TEACHING

AMERICANS ON AMERICA: OUR IDENTITY AND SELF IMAGE (Slides/Cassettes). Center for Humanities, 1976. 157 slides in 2 carousel cartridges; 2 cassettes; 2 discs; Teacher's Guide. #1006 ($139.50). Gr. 7–12. BKL (7/1/76), PRV (11/77). 810.7
Contents: 1. Freedom and Equality. 2. Conflicting Images.
This program traces the two concepts of freedom and equality as they appear time and again in American history and literature. It moves from a humorous introductory sketch by Bill Cosby depicting basic differences between American colonists and British soldiers to a descriptive account of how freedom and opportunity become integral segments of American self-image.

COPING WITH CONFLICT: AS EXPRESSED IN LITERATURE (Filmstrip—Sound). Sunburst Communications, 1973. 2 color, sound filmstrips, 77 fr. ea.; 2 cassettes or discs, 15 min. ea.; Teacher's Guide. #115 ($59). Gr. 7–12. BKL (10/11/73). 810.7
Contents: 1. Physical Struggles. 2. Emotional Conflict.
Recreates personal conflict from the works of Richard Wright, Marjorie K. Rawlings, Emily Dickinson, Mother Jones and Richard Olivas, and Chief Joseph of the Nez Perce

AMERICAN LITERATURE—STUDY AND TEACHING (cont.)

Indian Tribe. This presentation shows how different individuals have reacted to conflict and how they have succeeded or failed to resolve it.

THE DILEMMA OF PROGRESS (Filmstrip—Sound). McGraw-Hill Films, 1974 (Themes and Values in America). 2 color, sound filmstrips, av. 50 fr. ea.; 2 cassettes or discs, 9 min. ea.; Guide. #102571–5 Cassettes ($38), #103950–3 Discs ($28). Gr. 9–C. BKL (1975), PRV (10/75). 810.7

Contents: 1. The American City—Grand or Grotesque? 2. Modern Technology—Miracle or Monster?

Henry Adams probes our coming technological dilemma in this set as it develops contasting aspects of the values of "Technological Progress" vs. "Humanistic Values." He is joined by other literary notables in this project.

THE FALL OF THE HOUSE OF USHER (Motion Picture—16mm—Sound). Avatar Learning/EBEC 1975. 16mm color, sound film, 12 min. #47823 ($205). Gr. 7–C. BKL (1977), EFL (1976), LFR (1976). 810.7

Science fiction writer Ray Bradbury comments on the story, compares this screenplay to the written work, and discusses the gothic tradition and Poe's influence on contemporary science fiction.

QUESTIONING THE WORK ETHIC (Filmstrip—Sound). McGraw-Hill Films, 1974 (Themes and Values in America). 2 color, sound filmstrips, av. 50 fr. ea.; 2 cassettes or discs, 9 min. ea.; Guide. #102568–5 Cassettes ($38), #103945–7 Discs ($38). Gr. 9–C. BKL (1975), PRV (10/75). 810.7

Contents: 1. The Measure of Success. 2. Doing What You Love to Do.

Kurt Vonnegut looks at our work habits in this set as it develops contrasting values in the area of "The Work Ethic" vs. "Personal Fulfillment." Other literary contributions are used to help with this development.

SEEKING THE GOOD LIFE (Filmstrip—Sound). McGraw-Hill Films, 1974 (Themes and Values in America). 2 color, sound filmstrips, av. 50 fr. ea.; 2 cassettes or discs, 9 min. ea.; Guide. #102562–6 Cassettes ($38), #103935–X Discs ($38). Gr. 9–C. BKL (1975), PRV (10/75). 810.7

Contents: 1. Happiness Is a Solid Gold Cadillac. 2. Something to Believe In.

This first set in the series uses contributions from a variety of literary genres and eras as it develops contrasting aspects of a value area—"Materialism" vs. "Idealism." Sinclair Lewis explodes the material myopia of Middle America.

AMERICAN NATIONAL CHARACTERISTICS. *See* National characteristics, American

AMERICAN POETRY

AMERICAN POETRY TO 1900 (Phonodiscs). Lexington Recording, 2 discs. ($16.95). Gr. 4–12. ALR. 808.81

Poems by twenty American authors from Ann Bradstreet to Walt Whitman are read by Nancy Marchand and others.

JAMES DICKEY: POET (LORD LET ME DIE, BUT NOT DIE OUT) (Motion Picture—16mm—Sound). *See* Dickey, James

KARL SHAPIRO'S AMERICA (Motion Picture—16mm—sound). Pyramid Films, 1976. 16mm color, sound film, 13 min.; ³/₄ in. videocassette, also available. ($200). Gr. 8–A. EFL (1977), CGE (1977), M&M (1977). 811

Pultizer Prize winning poet, Karl Shapiro, comments on his philosophy and his work. Several poems are heard as collage animation, photographs, American art, and live action interpret their meaning visually.

PAUL REVERE'S RIDE AND HIS OWN STORY (Cassette). Children's Classics on Tape, 1974. 1 cassette; Teacher's Guide. #130 ($9.50). Gr. 4–8. PRV (5/74). 811

Longfellow's poem, "Paul Revere's Ride," is read, followed by Revere's own story of the famous night, as told 30 years later in a letter to a friend.

THE RAVEN (Motion Picture—16mm—Sound). Texture Films, 1976. 16mm b/w, sound film, 11 min. ($160). Gr. 8–A. BKL (1976). 811

A daring, deeply emotional interpretation of the classic poem by Edgar A. Poe, using the engravings of Gustave Dore. Poe wrote "The Raven" when he was haunted by fear that his wife was dying. The film captures Poe's obsession and his tormented imagination.

SPOON RIVER ANTHOLOGY (Filmstrip—Sound). Coronet Instructional Media, 1972. 2 color, sound filmstrips, av. 43 fr.; 1 disc or 2 cassettes, av. 17¹/₂ min. #M193 Cassettes ($35), #S193 Discs ($35). Gr. 7–12. CRA (1974), M&M (1974), PRV (1973). 811

An unscrupled thief robbed Spoon River of its darkest secrets. Citizens of a small town speak from their graves in this collection of 28 poems by Edgar Lee Masters. Brooding tombstones provide the background for acid but perceptive commentaries on customs and behavior in 19th-century America.

AMERICAN WIT AND HUMOR

WILL ROGERS' NINETEEN TWENTIES (Motion Picture—16mm—Sound). *See* U.S.—History—1919–1933

AMISH

THE AMISH: A PEOPLE OF PRESERVATION (Motion Picture—16mm—Sound).

Heritage Productions/EBEC, 1976. 16mm color, sound film, 28 min. #3399 ($380). Gr. 4–A. CGE (1976), EGT (1977), NEF (1978). 289.7.

This documentary captures the sensitivity and humility of the Amish people, while examining their religious beliefs, closeness to nature, and strong sense of community. Viewers get an intimate look at a people who differ from most of society in fundamental ways, while proudly preserving their own values.

AMPHIBIANS

AMPHIBIANS (Filmstrip—Sound). Educational Development, 1973. 4 color, sound filmstrips, av. 42 fr.; 4 cassettes or discs, 11–13 min. #L01–C Cassettes ($51.80), #401–R Discs ($48). Gr. 5–8. BKL (7/1/74). 597.6

Contents: 1. Amphibians of North America. 2. Frogs and Toads. 3. Salamanders. 4. Catching and Caring for Amphibians.

This set gives a simple definition of vertebrates, explains how amphibians differ from vertebrates, points out the habits and characteristics of frogs and toads, examines the life cycle of a typical amphibian, and describes putting an aquarium and a terrarium together. Photographs and drawings are used as illustrations.

ENDANGERED SPECIES: REPTILES AND AMPHIBIANS (Study Print). *See* Rare reptiles

ANATOMY

HUMAN BODY AND HOW IT WORKS (Filmstrip—Sound). Troll Associates, 1974. 6 color, sound filmstrips av. 45 fr. ea.; 3 cassettes av. 16 min. ea.; Teacher's Guide. ($78). Gr. 6–9. PRV (4/75). 611

Contents: 1. Respiratory System. 2. Circulatory System. 3. Muscle and Skeletal System. 4. Nervous System. 5. Digestive System. 6. Reproductive System.

Introduces the basic functions of the human body. Diagrams with photographs are used. The material is in a precise, logical, easy to understand format. (Note: The strip on reproduction is tastefully and scientifically done, but careful previewing may be needed before use in some communities.)

THE HUMAN BODY, SET ONE (Filmstrip—Sound). University Films/McGraw-Hill Films, 1975. 4 color, sound filmstrips, av. 59 fr. ea.; 4 cassettes or discs, 11 min. ea.; Teacher's Guide. Cassettes #102579-0 ($72), Discs #103964-3 ($72). Gr. 4–8. BKL (11/1/76), SAC (1/77). 611

Contents: 1. The Teeth. 2. The Skin, Hair, and Nails. 3. The Eyes. 4. The Ears.

An introduction to human physiology from the outside in. The filmstrips view anatomical structure and function as they relate to the countless activities we do every day. The physiological answers are concise and are visualized in ways students can easily relate to and remember. Specially designed transparency illustrations build anatomical structures layer by layer to clarify relationships.

THE HUMAN BODY, SET TWO (Filmstrip—Sound). University Films/McGraw-Hill Films, 1975. 5 color, sound filmstrips, av. 59 fr. ea.; 5 cassettes or discs, 11 min. ea.; Teacher's Guide. Cassettes 102584-7 ($90), Discs #103973-2 ($90). Gr. 4–8. ABT (5/77). 611

Contents: 1. The Respiratory System. 2. The Nervous System. 3. The Digestive System. 4. The Circulatory System. 5. The Muscular and Skeletal System.

The filmstrips view anatomical structure and function as they relate to the countless activities we do every day. The physiological answers are concise and are visualized in ways students can easily relate to and remember. Specially designed transparency illustrations build anatomical structures layer by layer to clarify relationships.

THE HUMAN BODY (Filmstrip—Sound). Instant Miracles/Time-Life Multimedia, 1976. 6 color, sound filmstrips, 69–85 fr. ea.; 6 cassettes or discs, 8–12 min. ea.; Guide. #ES-15–C Cassettes ($150.), #ES-15–T Discs ($150). Gr. 10–A. BKL (10/15/77). 611

Contents: 1. Respiration and Circulation. 2. Bone, Muscle, and Skin. 3. The Skin. 4. Reproduction. 5. Digestion. 6. The Nervous System.

Microphotography permits the viewer to observe in the human body the specialized processes that makes people function. Simple charts and photographs of a seminude young man and woman, which may offend some viewers.

THE HUMAN MACHINE (Filmstrip—Sound). Coronet Instructional Media, 1974. 8 color, sound filmstrips, av. 53 fr.; 4 discs or 8 cassettes, av. 13½ min. #M247 Cassettes ($120), #S247 Discs ($120). Gr. 7–12. CPR (1975), PRV (1975), SCT (1975). 611

Contents: 1. Its Complex Systems. 2. Bones, Muscles, and Skin. 3. The Nervous System. 4. The Heart and Circulatory System. 5. The Respiratory System. 6. The Digestive System. 7. The Excretory System. 8. The Reproduction System.

Photography, art, music, and electronic sounds convey a sense of wonder about the bodily process.

THE MENTAL/SOCIAL ME (Kit). Nystrom, 1975 (Human Environment). 6 color, sound filmstrips, 86–110 fr. ea.; 6 cassettes, 6–9 min. ea.; 24 Duplicating Masters; 16 Idea Cards; Teacher's Guide. #HE-300 ($175). Gr. 7–12. BKL (12/15/76). 611

Contents: 1. Awareness. 2. Between Two Worlds. 3. Am I Part of a Problem? 4. Riches

ANATOMY (cont.)

Beyond Measure. 5. Reflections. 6. Understanding.

The relationship of one person's behavior to another's and their reactions to each other is dramatized to promote an understanding of the dynamics of interpersonal relations.

ANATOMY, COMPARATIVE

BIOLOGICAL DISSECTION (Filmstrip—Sound). *See* Dissection

DISSECTION OF A FROG (Filmstrip—Sound). *See* Dissection

AN INSIDE LOOK AT ANIMALS (Study Print). Kenneth E. Clouse, 1974. 14 heavy stock cardboard, unmounted prints, 11 in. × 14 in. ($16). Gr. 5–9. PRV (11/75). 591.4
These are x-ray photographs with color added to enhance interest and highlight details of the skeletal structure. Included are such things as the top and side view of a frog, the bat (which shows the fingers on its wings), the remnants of a pelvic girdle on a snake, and eggs inside a turtle's body. On the reverse side of each print is a black and white photo of the animal, a labeled diagram pointing out all the details revealed by the x ray, a list of key vocabulary words, and a 400 word text which explains the x ray, the animal's group, and other facts about the animal.

ANDERSEN, HANS CHRISTIAN

THE LITTLE MATCH GIRL AND OTHER TALES (Phonodisc or Cassette). *See* Fairy tales

THE LITTLE MERMAID (Phonodisc or Cassette). *See* Fairy tales

ANGLO-SAXON LITERATURE

BEOWULF AND THE MONSTERS (Cassette). *See* Beowulf

ANIMALS

ANIMAL KINGDOM, SET ONE (Filmstrip—Sound). Random House, 1977. 2 color, sound filmstrips, 52–60 fr.; 2 discs or cassettes, 10–11 min. Teacher's Guide. #05071– 1 Cassettes ($39), #05070–3 Discs ($39). Gr 4–8. LGB (1977), PRV (1/78). 591.5
Contents: 1. Sharks. 2. Lions.
This set was produced in collaboration with the New York Zoological Society and the New York Aquarium. The sets are brief and interesting. Photographs and drawings are used to present factual information in cooperation with clearly narrated script.

BEAR COUNTRY AND BEAVER VALLEY (Filmstrip—Sound). *See* Natural history

ENDANGERED ANIMALS: WILL THEY SURVIVE? (Motion Picture—16mm—Sound). *See* Rare animals

ENDANGERED SPECIES: MAMMALS (Study Print). *See* Rare animals

AN INSIDE LOOK AT ANIMALS (Study Print). *See* Anatomy, Comparative

JAGUAR: MOTHER AND CUBS (Film Loop—8mm—Silent). *See* Jaguars

POUCHED ANIMALS AND THEIR YOUNG (Film Loop—8mm—Silent). Walt Disney Educational Media, 1966. 8mm color, silent film loop, approx. 4 min. #62–5253L ($30). Gr. K–12. MDU (1978). 599.2
Shows the childbearing habits of these animals. Closeups of a baby wombat struggling to enter its mother's pouch. A phalanger climbs in a tree, the young clinging to its back.

See also Names of orders and classes of animals, i.e., Mammals; and also names of individual animals, e.g., Bears, Lions, etc.

ANIMALS—ANTARCTIC

CREATURES OF THE ANTARCTICA (Filmstrip—Sound). *See* Antarctic regions

ANIMALS—FICTION

THE INCREDIBLE JOURNEY BY SHEILA BURNFORD (Filmstrip—Sound). Current Affairs Films, 1978. 1 color, sound filmstrip; 1 cassette; Teacher's Edition of the Book; Discussion Guide and Testing Materials. #654 ($30). Gr. 7–12. BKL (12/15/78). Fic
A rather unusual assortment of three house pets—consisting of a young Labrador retriever, an old English bull terrier and a Siamese cat—are the heroes of this novel, as they make an "incredible journey" across the Canadian wilderness to rejoin their human family.

RUDYARD KIPLING STORIES (Filmstrip—Sound). Xerox Educational Publications, 1976. 6 color, sound filmstrips; 6 cassettes; 3 Teaching Guides. #SC02700 ($110). Gr. K–8. CGE (1976), LGB (12/77). Fic
Contents: 1. The White Seal. 2. Rikki-Tikki-Tavi. 3. Mowgli's Brothers.
Adapted from the film version of the award winning Kipling film.

TWO FOR ADVENTURE (Kit). EMC, 1976. 2 color, sound filmstrips, 66–68 fr.; 2 cassettes, 2 Paperback Books; Activities and Duplicating Masters; Teacher's Guide. #EL-235000 ($63). Gr. 5–12. PRV (11/76). Fic
Contents: 1. The Incredible Journey. 2. Gifts of an Eagle.
A sound filmstrip accompanies each book but does not tell the story. Rather, the back-

ground of the story is presented as motivation to read the books. A dramatic reading cassette accompanies each book in which parts of each book are read by a narrator.

THE WORLD OF JUNGLE BOOKS, SET ONE (Kit). Spoken Arts, 1976. 4 cassettes, 22–28 min.; 40 Books (4 Titles); 1 Guide. #SAC-6505 ($59.95). Gr. 5–A. BKL (9/15/77). Fic

Contents: 1. Mowgli's Brothers. 2. Rikki-Tikki-Tavi. 3. Tiger! Tiger! 4. Toomai of the Elephants.

This cassette-paperback set effectively presents Rudyard Kipling's characters in these high adventure stories.

THE WORLD OF JUST SO STORIES, SET ONE (Kit). Spoken Arts, 1976. 4 cassettes, av. 7–9 min.; 40 Books (4 Titles); 1 Guide. #SAC-6503 ($59.95). Gr K–12. BKL (10/15/77). Fic

Contents: 1. How the Whale Got His Throat. 2. The Sing Song of Old Man Kangaroo. 3. How the Camel Got His Hump. 4. How the Rhinoceros Got His Skin.

This production recounts how several jungle animals gained their special characteristics as told in Kipling's tales recording the oral folk traditions of India.

ANIMALS—HABITATIONS

INSECT COMMUNITIES (Film Loop—8mm—Silent). *See* Insects

ANIMALS—HABITS AND BEHAVIOR

ANIMALS, ANIMALS (Filmstrip—Sound). Society for Visual Education, 1976. 6 color, sound filmstrips. av. 66 fr. ea.; 6 cassettes or phonodiscs, av. 9¼ min. ea.; Teacher's Guide. A442-SATC ($95), A442-SAR ($95). Gr. 5–12. BKL (2/15/76). 591.5

Contents: 1. Animal Societies. 2. How Animals Build. 3. How Animals Protect Themselves. 4. How Animals Migrate. 5. How Animals Communicate. 6. How Animals Feed Themselves.

Animals, Animals is designed to help children develop the ability to observe the characteristics of animals at close range in order to recognize similarities among certain groupings. Students can then conclude that adaptations enable some animals to survive and thrive in well-defined habitats. While developing an appreciation for the wild animals of the world, children will also be developing skills in identifying and classifying animals more readily.

BEAVER (Film Loop—8mm—Silent). *See* Beavers

BEAVER DAM AND LODGE (Film Loop—8mm—Silent). *See* Beavers

FUR, FINS, TEETH, AND TAILS (Filmstrip—Sound). Adrian Vance Productions, 1977.

6 color, sound filmstrips, av. 52–75 fr.; 6 cassettes, av. 7–10 min. ($90). Gr. 5–9. BKL (11/1/77). 591.5

Contents: 1. Animal Clothes. 2. Animal Motion. 3. Animal Teeth. 4. Animal Tails. 5. Animals in Balance. 6. Classification.

Natural sounds add to this introduction to biology. Consideration is given to the adaptation and variation of several physical characteristics: skin, teeth, tails, and locomotion—throughout the animal kingdom.

KOALA BEAR (Film Loop—8mm—Silent). *See* Koala bears

LIFE CYCLE (Filmstrip—Sound). National Geographic, 1974. 5 color, sound filmstrips, 60–72 fr.; 5 discs or cassettes, 12–14 min.; Teacher's Guide. #03760 Discs ($74.50), #03761 Cassettes ($74.50). Gr. 4–12. PRV (1975), PVB (5/75), 591.5

Contents: 1. Mammals. 2. Fishes. 3. Birds. 4. Amphibians and Reptiles. 5. Insects.

Dramatic biographies of various animal species . . . from birth to death. Describes the dominant and distinguishing characteristics of each.

LIFE CYCLE OF COMMON ANIMALS, GROUP 2 (Filmstrip—Sound). *See* Sheep

LION (Motion Picture—16mm—Sound). *See* Lions

LION: MOTHER AND CUBS (Film Loop—8mm—Silent). *See* Lions

PLAYING IT SAFE WITH ANIMALS (Filmstrip—Sound). Marshfilm, 1976, (Safety). 1 color, sound filmstrip, 50 fr.; 1 cassette or disc, 15 min.; Teacher's Guide. ($21). Gr. K–A. INS (1/78), NYF (1976). 591.5

This filmstrip deals with what may happen when wild animals such as raccoons, coyotes, or possums are kept as pets; describes reactions and behavior of sick or injured animals, and presents first aid treatment for animal, snake, and spider bites and wasp and bee stings.

POLAR BEAR: MOTHER AND CUBS (Film Loop—8mm—Silent). *See* Bears

PRIDE OF LIONS (Film Loop—8mm—Silent). *See* Lions

ANIMALS—HIBERNATION

INVESTIGATING HIBERNATION: THE GOLDEN-MANTLED SQUIRREL (Motion Picture—16mm—Sound). EBEC, 1972. 16mm color, sound film, 15 min. #3150 ($185). Gr. 7–12. BKL (1973), LFR (1973), SCT (1973). 591.5

Offspring of captured Golden-Mantled Squirrels are raised under controlled laboratory conditions. As autumn approaches the young squirrels go into hibernation though never exposed to their natural environment or colder temperatures. Their heartbeats

ANIMALS—HIBERNATION (cont.)

slow from 400 to 2 beats per minute and body temperatures lower from 37 to 2 degrees centigrade.

ANIMALS—IDENTIFICATION

FUR, FINS, TEETH, AND TAILS (Filmstrip—Sound). *See* Animals—Habits and behavior

ANIMALS—INFANCY

MAMMALS (Film Loop—8mm—Captioned). BFA Educational Media, 1973 (Animal Behavior). 8mm color, captioned film loop, approx. 4 min. #41821 ($30). Gr 4-9. BKL (5/1/76). 591.3

Demonstrates experiments, behavior, and characteristics of mammals.

ANIMALS—MIGRATION

THE YEAR OF THE WILDEBEEST (Motion Picture—16mm—Sound). *See* Gnu

ANIMATED MOTION PICTURES. *See* Motion pictures, Animated

ANIMATION. *See* Motion pictures, Animated

ANTARCTIC REGIONS

ANTARCTICA: THE WHITE CONTINENT (Filmstrip—Sound). Lyceum/Mook & Blanchard, 1974. 2 color, sound filmstrips, 44-48 fr.; 2 cassettes or discs, 7-10 min. Teacher's Guide. #LY35673C Cassettes ($46), #LY35673R Discs ($37). Gr. 5-A. FLN (4/5/74), PRV (10/74), PVB (7/75). 919.8

Contents: 1. Challenge of the Antarctic. 2. Antarctica Today.

Helps to understand the Antarctic environment and the reasons for exploration and research.

CREATURES OF THE ANTARCTICA (Filmstrip—Sound). Lyceum/Mook & Blanchard, 1974. 1 color, sound filmstrip, 46 fr.; 1 cassette or disc, 8½ min.; Teacher's Guide. #LY35174C Cassette ($25), #LY35174R Disc ($19). Gr. 5-A. FLN (9/74), INS (2/75), PRV (2/74). 919.8

The filmstrip helps students understand and appreciate antarctic animal and marine life. Photographs and little-known facts describe the inhabitants of the area. Penguins, seals, and other marine life stimulate scientific research.

ANTHROPOLOGY

ANTHROPOLOGIST AT WORK (Filmstrip—Sound). Coronet Instructional Media, 1976. 4 color, sound filmstrips, av. 80 fr. ea.; 4 cassettes or discs, av. 11 min. ea. #M711 Cassettes ($69), #S711 Discs ($69). Gr. 7-10. BKL (12/1/76). 573.023

Contents: 1. The Paleontologist. 2. The Archeologist. 3. The Historian. 4. The Ethnologist.

First-person accounts of how anthropologists inquire into the nature of humankind, past and present. Location photography includes East Africa, the Middle East, Latin America, and California. Culture, site, artifacts, hypothesis, grid and other key concepts are explored.

THE TASADAY: STONE AGE PEOPLE IN A SPACE AGE WORLD (Filmstrip—Sound). *See* Man, Nonliterate

WHAT DOES IT MEAN TO BE HUMAN? (Slides/Cassettes). Center for Humanities, 1976. 240 slides in 3 Carousel cartridges; 3 cassettes; 3 discs; Teacher's Guide. #0284 ($179.50). Gr. 9-C. BKL (3/1/78), IDF (1977), M&M (1977). 573

Contents: 1. Coordination of the Brain and Hand. 2. Human Beings Are Social. 3. Civilization as a Human Invention.

This set focuses on the coordination of the brain and hand, which gives us dexterity, and that of the brain and vocal organs, which makes speech possible. It also explains that human beings are by nature social, and explores certain basic characteristics that human societies share. Civilization as a distinctly human invention and civilized societies different from folk, or tribal, societies are shown.

APPALACHIAN MOUNTAINS

THE MOUNTAIN PEOPLE (Motion Picture—16mm—Sound). Granada International Television/Wombat Productions, 1975. 16mm color, sound film, 24 min. ($360). Gr. 7-A. AFF (1975), CRA (1975), FLN (11/02/76). 917.4

The Appalachian poor are given new dimension. Here are people made noticeable not only by their situation but also by their pride and inner strength.

ARAB COUNTRIES

THE ARAB WORLD (Kit). EMC, 1974. 4 color, sound filmstrips; 4 discs or cassettes; 3 paperback books; political map; Teacher's Guide. #SS-212000 ($92). Gr. 9-12. FLN (9/74), INS (5/75), SLJ (9/75). 915.3

Contents: 1. The Land and the Heritage. 2. Oil and Water: Keys to the Future. 3. Nomads, Villagers, City Dwellers. 4. A Time of Change.

This set highlights the comparison between diverse but unified nations. The environment, history, and culture are explored. A traditional and changing life-style is presented and change is a key concept.

EYES ON THE ARAB WORLD (Kit). EMC, 1975. 4 color, sound filmstrips, 86–103 fr. ea.; 4 cassettes, 10–16 min. ea.; 3 paperback books; political map; Teacher's Guide. #SS-215000 ($92). Gr. 10–C. BKL (10/1/75), INS (5/75), PVB (05/75). 915.3

Contents: 1. The Mark of History. 2. Oil: Key to World Power. 3. From Nomad to City Dweller. 4. Contrasting Society.

A structured study of a key region: its historical background, people, contrasting lifestyles, and the impact of modernization of Arab society. Mini-case studies provide contrasting views of traditional and modern patterns of living.

ARAB COUNTRIES—CIVILIZATION

THE ARAB CIVILIZATION (Filmstrip—Sound). Eye Gate Media, 1978. 4 color, sound filmstrips, 72–101 fr.; 2 cassettes, 13–15 min.; 1 Guide. #E816 ($55). Gr 9–C. BKL (10/1/78). 953

Contents: 1. The Rise of Islam. 2. From Mecca to Baghdad. 3. The Great Islamic Empire. 4. From World Empires to Modern Nations.

This set examines the rise and flourishing of Arab/Islamic culture. This culture is traced from its beginnings through its golden age and expansion into one of the world's greatest empires. The emergence of the Arab world as a major force in the modern world is also viewed.

ARCHEOLOGY

ARCHEOLOGICAL DATING: RETRACING TIME (Motion Picture—16mm—Sound). EBEC, 1975. 16mm color, sound film, 18 min. #3480 ($240). Gr. 9–A. AAS (1977), BKL (1978), EFL (1977). 913

Ancient ruins are messages from the past. But they mean little unless tied to dates. This film explores relative and absolute dating techniques archeologists use to date their finds.

ART OF PERSEPOLIS (Slides). *See* Iran—Antiquities

THE BIG DIG (Motion Picture—16mm—Sound). *See* Israel—Antiquities

DIGGING UP AMERICA'S PAST (Filmstrip—Sound). *See* America—Antiquities

KING TUTANKHAMUN: HIS TOMB AND HIS TREASURE (Kit). *See* Egypt—Antiquities

ONE IMIX, EIGHT POP (Cassette). *See* Mexico—History

SENTINELS OF SILENCE (Motion Picture—16mm—Sound). *See* Mexico—Antiquities

ARITHMETIC

DECIMAL NUMERATION SYSTEM (Filmstrip—Sound). Viewlex Educational Media, 1973. 3 color, sound filmstrips, 25–57 fr.; 3 discs or cassettes, 13–18 min.; Teacher's Guide; Worksheet Pads. #5507-9 ($48). Gr 6–9. PRV (10/74), PVB (5/75). 513.2

Contents: 1. Development of Our Number System. 2. Decimal System. 3. Decimal Fractions.

Each filmstrip presents a lesson on a single topic related to the Decimal System of Numeration. Concepts are introduced in sequences. Line drawings and diagrams in colors stand out against a black background.

UNDERSTANDING AND USING DECIMALS (Filmstrip—Sound). Pathescope Educational Media, 1976. 10 color, sound filmstrips; 10 cassettes; 68 Spirit Masters; Teacher's Manual. #441 ($225). Gr. 4–8. MTE (1/77). 513.2

Contents: 1. Reading and Writing Decimals. 2. Comparing Decimals. 3. Adding Decimals. 4. Subtracting Decimals. 5. Multiplying Decimals—Part 1. 6. Multiplying Decimals—Part 2. 7. Rounding Decimals. 8. Dividing Decimals by a Whole Number. 9. Dividing Decimals by a Decimal. 10. Changing Fractions to Decimals.

Is aimed at giving students a thorough understanding of decimals and mastery of their use in real life situations.

UNDERSTANDING AND USING PERCENT (Filmstrip—Sound). Pathescope Educational Media, 1976. 10 color, sound filmstrips; 10 cassettes; 68 Spirit Masters; Teacher's Guide. #442 ($225). Gr. 4–8. MTE (5/77). 513.2

Contents: 1. The Meaning of Percent. 2. Changing Percent to Fractions. 3. Changing Percent to Decimals. 4. Changing Decimals to Percent. 5. Changing Fractions to Percent. 6. Finding Percentage: Commission. 7. Finding Percentage: Discount. 8. Finding Percentage: Interest. 9. Finding the Rate of Percent. 10. Finding the Base.

Based on the concept that "percent means hundredths," this set is aimed at giving students a thorough understanding and mastery of percentage.

ARITHMETIC—STUDY AND TEACHING

CONSUMER MATH (Kit). Newsweek Educational Division, 1976. 2 color, sound filmstrips, 90–95 fr.; 2 cassettes or discs, 15–16 min.; 13 Duplicating Masters; 10 Study Prints; 1 Teacher's Guide. #602C Cassettes ($65), #602 Discs ($60). Gr. 7–12. BKL (1/15/77). 513.2

This program deals with basic, mathematical principles such as units of measure, percentage, credit, and interest. It is aimed at the average consumer who wants to get the best value for his or her money. Helps the viewer find the true cost of a purchase and how to

ARITHMETIC—STUDY AND TEACHING (cont.)

tell the difference between reality and appearances and how to learn the true interest.

STUMBLING BLOCKS IN ARITHMETIC (Filmstrip—Sound). Pathescope Educational Media, 1973. 3 color, sound filmstrips; 3 cassettes; Teacher's Manual. #304 ($55). Gr. 4–8. MTE (1/74). 372.7

Contents: 1. Regrouping in Subtraction. 2. The Two-Place Multiplier. 3. The Two-Place Divisor.

Covers three major areas of difficulty and makes mathematical abstractions real and understandable. Good questions are raised and answered. Each concept is presented at the concrete level first, then at the representational level and finally at the abstract level. Live photography and cartoons are used with a story approach and touches of humor.

STUMBLING BLOCKS IN FRACTIONS (Filmstrip—Sound). Pathescope Educational Media, 1973. 4 color, sound filmstrips; 4 cassettes; Teacher's Manual. #305 ($70). Gr. 4–8. MTE (1/74). 372.7

Contents: 1. The Language of Fractions. 2. Equivalent Fractions. 3. Addition and Subtraction of Fractions. 4. Multiplication and Division of Fractions.

Four major problems in dealing with fractions are covered in this set. The students learn to understand the problems and then solve them with the aid of live photography and cartoon characters.

UNDERSTANDING AND USING WHOLE NUMBERS (Kit). Pathescope Educational Media, 1977 (Computational Skills in Mathematics). 10 filmstrips; 10 cassettes, $15^1/_2$–$20^1/_2$ min. ea.; 72 Spirit Masters; Teacher's Manual. #443 ($225). Gr. 7–12. MTE (12/77). 513.2

Contents: 1. Understanding Our Numeration System. 2. Adding Whole Numbers. 3. Subtracting Whole Numbers—Part 1. 4. Subtracting Whole Numbers—Part 2. 5. Multiplying Whole Numbers—Part 1. 6. Multiplying Whole Numbers—Part 2. 7. Multiplying Whole Numbers—Part 3. 8. Dividing Whole Numbers—Part 1. 9. Dividing Whole Numbers—Part 2. 10. Dividing Whole Numbers—Part 3.

This multimedia kit has been designed to help students who do not have immediate recall of basic facts. The program is correlated with any math course. Can be used in remedial situations as well as for whole-class instruction.

ARMSTRONG, WILLIAM

WILLIAM H. ARMSTRONG (Filmstrip—Sound). Miller–Brody Productions, 1977 (Meet the Newbery Author). 1 color, sound filmstrip, 87 fr.; 1 cassette or disc, 9 min. #MNA1011C Cassette ($32), #MNA1011 Disc ($32). Gr. 4–8. BKL (1/1/78). 921

The quietly busy life of author William H. Armstrong is depicted in this filmstrip that catches the man in his many roles as teacher, writer, shepherd, gardener, cabinetmaker, and real estate agent. Photos, family snapshots, and book illustrations add another dimension to the few remarks on Armstrong's childhood, education, and Connecticut home built by his own hands.

ARNOLD, BENEDICT

BENEDICT ARNOLD: TRAITOR OR PATRIOT? (Filmstrip—Sound). Current Affairs Films, 1977. 2 color, sound filmstrips, 2 cassettes—"Pro-and-Con" and Discussion Guide. #585 ($30). Gr 7–A. PRV (4/78). 921

Benedict Arnold, one of the greatest early American military leaders and patriots, was court martialed before the War for Independence was over. His court martial is explored in this filmstrip program, revealing not only the true nature of the man and his crimes, but also the distinct differences between civil and military law.

ART, AFRICAN

THE CREATIVE HERITAGE OF AFRICA: AN INTRODUCTION TO AFRICAN ART (Kit). EBEC, 1972 (Creative Heritage of Africa). 58 color slides; discs or cassettes; 8 Display Cards; Guide. #6630 ($72.50). Gr. 7–12. BKL (10/15/73), LFR (4/74), PRV (10/73). 709.6

This combination of slides, recorded lecture, and color reproductions presents African art and its role in African culture. Traditional African art was a functional and necessary part of everyday life. Since no written language existed, art—especially sculpture—served to record laws, beliefs and history. Authentic West African music supplements the lecture.

WEST AFRICAN ARTISTS AND THEIR ART (Filmstrip—Sound). EBEC, 1973. 7 color, sound filmstrips, av. 75 fr. ea.; 7 discs or cassettes, 8 min. ea.; Teacher's Guide. #6464K Cassettes ($101.50), #6464 Discs ($101.50). Gr 4–8. BKL (7/15/73), PRV (11/73). 709.6

Contents: 1. Kumasi Brass Caster. 2. Dschang Woodcarver. 3. Ghana Dancer. 4. Mali Mask Carver. 5. Cameroon Blacksmith. 6. Gambian Weaver. 7. Fumban Sculptors.

Vivid documentaries of artists and craftsmen at work transport students to many parts of West Africa.

ART, AMERICAN

AMERICA IN ART: THE AMERICAN REVOLUTION (Filmstrip—Sound). *See* U.S.—History—1775-1783—Revolution

AMERICAN CIVILIZATION: 1783–1840 (Filmstrip—Sound). Sunburst Communications, 1973. 3 color, sound filmstrips, 71–80 fr.; 3 cassettes or discs, 13–15 min.; Teacher's Guide. #201 ($85). Gr. 8–C. BKL (12/1/73), STE (9/73). 709.73

Contents: 1. Portrait of a Young Nation. 2. Young America Admires the Ancients. 3. The Arts and the Common Man.

Examines the history and ideals of our nation's early years through the subjects and styles of its artists. Correlates history, painting, architecture, sculpture, and music, as it shows early America's sense of purpose and its fundamental respect for the common man. The works of West, Copley, Trumbull, Peale, Mills, Phillips, Prior, Catlin, and Neagle are used as examples.

AMERICAN CIVILIZATION: 1840–1876 (Filmstrip—Sound). Sunburst Communications, 1973. 4 color, sound filmstrips, 74–80 fr.; 4 cassettes or discs, 15 min. ea.; Teacher's Guide. #202 ($99). Gr. 8–C. BKL (2/1/74), MMB (11/75), STE (9/73). 709.73

Contents: 1. America the Beautiful: The Land as Inspiration. 2. The Arts Reflect Daily Life. 3. Architecture as a Language. 4. Bridges to the 20th Century.

The program studies the mood of American society preceding, during and after the Civil War. It integrates history, painting, architecture, sculpture and music, as it outlines the rapid changes characteristic of these years.

ART CAREERS (Motion Picture—16mm—Sound). Handel Film, 1977 (Art in America Part V). 16mm color, sound film, 30 min. ($400). Gr 9–C. CFF (1978), CGE (1978), NEF (1978). 704.023

Careers available and avenues open to people who are interested in the field of art are examined. Advice is given by successful people in fine arts, illustrations, graphic design, architecture, interior design, industrial design, photography, textile design, crafts, fashion design, teaching, and museum work.

ART, DECORATIVE

EDIBLE ART (Filmstrip—Sound). *See* Cookery

ART—EXHIBITIONS

ART BY TALENTED TEENAGERS (Filmstrip). Scholastic Book Services, 1977. 3 color filmstrips, 56–58 fr. #SF 505 ($18). Gr. 9–12. PRV (12/78). 708

Contents: 1. Painting. 2. Drawing and Prints. 3. Three Dimensional Art.

Presents selections from Scholastic Magazine's 50th National High School Art Awards Program. There is no guide nor narration, but the work speaks for itself. Works in oils, acrylics, watercolor, pencil, charcoal, ink, pastel, pottery, jewelry, and sculpture are included.

ART—EXHIBITIONS—FICTION

CLOSED MONDAYS (Motion Picture—16mm—Sound). Pyramid Films, 1974. 16mm color, sound film, 8 min. ³/₄ in. videocassette, also available. ($150). Gr. 7–A. AAW (1975), FLN (1975), PRV (1975). Fic

As an eccentric old man wanders into an art museum after hours, the paintings and sculpture on exhibit come to life. Three-dimensional clay animation brings a new perspective to art as part of life.

ART—HISTORY

THE IMPRESSIONIST EPOCH (Filmstrip—Sound). *See* Impressionism (art)

SURREALISM (AND DADA) (Motion Picture—16mm—Sound). *See* Surrealism

TREASURES OF KING TUT (Videocassette). *See* Tutankhamun

ART, JAPANESE

THE ARTS OF JAPAN—slide set (slides). Educational Dimensions Group, 1973. 20 cardboard mounted, color slides in plastic storage page; Teacher's Guide. #926 ($25). Gr. 6–A. BKL (5/75), PRV (1/75), PVB (5/75). 709.52

Designed to provide a chronological view of Japanese art, this set of slides includes sculpture, architecture, painting, and the crafts of each major historical period of Japanese art.

ART, MEDIEVAL

THE ART OF THE MIDDLE AGES (Filmstrip—Sound). Educational Dimensions Group, 1976 (How to Look At). 1 color, sound filmstrip; 1 cassette #669 ($30). Gr. 7–12. BKL (12/15/77). 709

This set explains how and why Gothic artists developed unique ways of expressing their ideas and faith. Gothic art attained an elegance and grace that is expressed in the architecture of France, England, and Italy and in the arts and crafts of religion of that period. The historical background helps the viewer understand and appreciate the period.

ART, MODERN

BUT IS IT ART? (Filmstrip—Sound). Encore Visual Ed., 1977. 2 color, sound filmstrips, 39–47 fr.; 2 cassettes, 6–9 min.; Teacher's Guide. #89 ($35). Gr. 6–A. PRV (1/79). 709.04

Contents: 1. A Look at Today's Art. 2. Art—Another Point of View.

ART, MODERN (cont.)

Contemporary art expresses new ideas based upon our own artistic heritage and trends. This set stresses the point that art does not have to represent the obvious.

ART, RENAISSANCE

THE ART OF THE RENAISSANCE (Filmstrip—Sound). Educational Dimensions Group, 1977 (How to Look At). 1 color, sound filmstrip; 1 cassette #670 ($30). Gr. 7–12. BKL (7/77), IFT (1977). 709

During the 15th century, a "rebirth" of classical and ancient tradition captured the imaginations of artists, architects, writers, and politicians. This set presents the achievements of the great artists as well as the unknown craftspeople whose contributions to world art are still felt today.

ART—STUDY AND TEACHING

BAUHAUS (Filmstrip—Sound). International Film Bureau, 2 color, silent filmstrips, 28–31 fr.; 2 Guides. ($15). Gr. 6–A. MDU (1974). 701

Contents: 1. Learning through Doing. 2. The World as a Total Work of Art.

These silent filmstrips present the overall view of the Bauhaus School of Art. Probably more useful for teachers than students.

SOFT SCULPTURE (Filmstrip—Sound). See Sculpture

ART—TECHNIQUE

WATCHING ARTISTS AT WORK (Filmstrip—Sound). Imperial Educational Resources, 1972. 4 color, sound filmstrips, 49–53 fr.; 4 discs or cassettes. #3RG32500 Discs ($56), #3KG32500 Cassettes ($62). Gr. 4–12. BKL (2/1/73). 702.8

Contents: 1. Creating an Abstract Watercolor. 2. Creating a Realistic Oil. 3. Creating a Ceramic Sculpture. 4. Creating a Contemporary Print.

These filmstrips show the creation of four original works of art done in different media. The artists explain methods and procedures and discuss their thoughts and feelings about their work.

ART AND NATURE

FOG (Motion Picture—16mm—Sound). EBEC, 1972. 16mm color, sound film, 9 min. #3052 ($115). Gr. K–A. AFF (1972), BKL (1972), LNG (1973). 704.94

This film captures the poetic and artistic mood of fog as the scientific phenomena of its textures, patterns, and motion are explored. Carl Sandburg's poem "Fog," was the inspiration for this nonnarrated film. Views of fog rolling into valleys, clouding sea and ships, and obscuring landmarks such as the Golden Gate Bridge encourage observation of environmental changes. The film also stimulates creative expression such as poetry, music, artwork.

ART AND SOCIETY

THE SPIRIT OF ROMANTICISM (Motion Picture—16mm—Sound). See Romanticism

ART APPRECIATION

THE ART OF SEEING (Filmstrip—Sound). Warren Schloat Productions/Prentice-Hall Media, 1968. 6 color, sound filmstrips; 6 cassettes or discs.; Teacher's Guide. #KAC 250 Cassettes ($126), #KAR 250 Discs ($126). Gr. 4–12. AFF, EFL, SLJ. 701.8

Contents: 1. How to Use Your Eyes, Part I. 2. How to Use Your Eyes, Part II. 3. Lines. 4. Shapes. 5. Colors. 6. Space.

This program explores the artist's visual vocabulary.

BUT IS IT ART? (Filmstrip—Sound). See Art, modern

OUR VISUAL WORLD (Filmstrip—Sound). ACI Media/Paramount Communications, 1974. 4 color, sound filmstrips, av. 70 fr.; 4 cassettes, 8–10 min.; Teacher's Guide. #9119 ($78). Gr. 4–12. PRV (5/75), PVB (4/76). 701.8

Contents: 1. Line, Surface, and Volume. 2. Shape. 3. Light and Color. 4. Pattern and Texture.

This set stimulates visual awareness to the world around us. It helps to see things as if for the first time. Students are led to consideration of the artistic process and the ways in which works of art can alter the way we see the world.

RIGHT ON/BE FREE (Motion Picture—16mm—Sound). See Black artists

ARTISANS

IN PRAISE OF HANDS (Motion Picture—16mm—Sound). See Handicraft

ARTISTS

B. KLIBAN (Cassette). See Kliban, B.

GRAVITY IS MY ENEMY (Motion Picture—16mm—Sound). See Hicks, Mark

ARTS AND CRAFTS. See Handicraft; and also specific kinds, e.g., Batik

ASIA

FAMILIES OF ASIA (Filmstrip—Sound). EBEC, 1975. 6 color, sound filmstrips, av. 75 fr. ea.; 6 discs or cassettes, 8 min. ea.;

Teacher's Guide. Discs #6910 ($86.95), Cassettes #6910K ($86.95). Gr. 1–8. BKL (11/75), ESL (1977), PRV (9/75, 76). 915

Contents: 1. The Families of Hong Kong. 2. Family of Bangladesh. 3. Family of India. 4. Family of Japan. 5. Family of Java. 6. Family of Thailand.

Diverse ways of life in Asia are often too large, too foreign for the western students to grasp. But when approached through a single family with children, the life-styles of Bangladesh, Hong Kong and many other Asian nations become more comprehensible. Colorful, on-location photographs and natural sounds depict everyday life in six different countries.

SOUTH ASIA: REGION IN TRANSITION (Filmstrip—Sound). EBEC, 1976. 5 color, sound filmstrips, av. 76 fr. ea.; 5 cassettes or discs, 10 min. ea.; Teacher's Guide. #6924K Cassettes ($72.50), #6924 Discs ($72.50). Gr. 5–9. BKL (10/1/77), MTU (4/77). 915

Contents: 1. The Winning of Independence. 2. Religion and Change. 3. An Indian Village: Model for Change. 4. The Urban Workers. 5. South Asia: Key Decisions.

This is an assessment of how India, Pakistan and the other countries of the Indian subcontinent are meeting economic change. The filmstrips examine the changing urban scene, inspect a model farming community and portray the struggle against a three-pronged threat; poverty, over-population and lack of technology.

ASIA, SOUTHEAST

SOUTHEAST ASIA (Filmstrip—Sound). Educational Design, 1976, (Global Culture Series). 4 color, sound filmstrips, av. 75 fr.; 2 cassettes, av. 13 min.; 1 Guide. #EDI-460 ($79). Gr. 5–9. BKL (10/15/77). 915.9

Contents: 1. Many Ways of Life. 2. Beliefs and Bread. 3. Fun and the Future. 4. The Factors in Common.

Presents the basic geographical, historical, economic, and cultural information of these nations. Unusual and uncommon insights into the people's lives are revealed.

ASIMOV, ISAAC

ENCOUNTERS WITH TOMORROW: SCIENCE FICTION AND HUMAN VALUES (Filmstrip—Sound). *See* Science fiction—History and criticism

ASTRONOMY

BLACK HOLES OF GRAVITY (Motion Picture—16mm—Sound). British Broadcasting/Time-Life Multimedia, 1973. 16mm color, sound film, 56 min. #V1876 ($600). Gr. 9–A. BKL (1/15/79). 521.5

This film explores past and present thinking about gravity's effect on the structure of the universe. The chain of thought that led Newton to state his Inverse-Square Law of force is traced and Einstein's General Theory of Relativity which gave credence to the idea of black holes . . . is presented and defined.

WORLDS IN COLLISION (Motion Picture—16mm—Sound). British Broadcasting/Time-Life Multimedia, 1978. 16mm color, sound film, or videocassette, 47 min. #1887 Video ($275), #F2127 Film ($550). Gr. 9–A. BKL (1/15/79). 521.5

This film focuses on Immanuel Velikovsky's book of the same title and his theory that occurrences detailed in ancient myths and features of Earth's makeup can be traced to a near collision with the planet Venus around 1500 B.C. Animated sequences of this event are intercut with ancient manuscripts and artwork that, to Velikovsky, document his ideas.

ATHLETES

BLACK AMERICAN ATHLETES (Kit). EMC, 4 Paperback Books; 4 Read-along Cassettes; Student Activities; Teacher's Guide. #ELC–234000 ($55). Gr. 4–12. BKL (7/76), SLJ (9/76). 920

Contents: 1. Lee Elder: The Daring Dream. 2. Julius Erving: Dr. J. and Julius W. 3. Madeline Manning Jackson. 4. Arthur Ashe: Alone in the Crowd.

Four black athletes who have reached superstar status . . . sometimes against great odds. They stress the importance of effort and discipline to reach the top.

HOCKEY HEROES (Kit). EMC, 1974, 4 Paperback Books; 4 Read-along Cassettes; Teacher's Guide; Student Activities. #ELC–217000 ($55). Gr. 4–12. SLJ (12/74). 920

Contents: 1. Stan Mikita: Tough Kid Who Grew Up. 2. Bobby Hull: Superstar. 3. Gil Perreault: Makes It Happen. 4. Frank Mahovlich: The Big M.

Behind the scene stories of four noted hockey favorites. These high-interest books demonstrate the patience, courage, and perseverance needed to succeed in this competitive sport.

SPORTS CLOSE-UPS 1 (Kit). EMC, 1973. 5 Paperback books; 5 Read-along Cassettes, approx. 35 min.; Student Activities; Teacher's Guide. #ELC–212000 ($71.50). Gr. 3–8. BKL (7/73). 920

Contents: 1. Willie Mays: Most Valuable Player. 2. Johnny Unitas and the Long Pass. 3. Mickey Mantle Slugs It Out. 4. Hank Aaron: Home Run Superstar. 5. Jim Brown Runs with the Ball.

High-interest reading materials for young sports buffs.

SPORTS CLOSE-UPS 3 (Kit). EMC, 1973. 5 Paperback Books; 5 Read-along Cassettes;

ATHLETES (cont.)

Teacher's Guide; Student Activities.
#ELC–212000 ($71.50). Gr. 4–12. BKL (6/75). 920

Contents: 1. O. J. Simpson: Juice on the Gridiron. 2. Bobby Hull: Hockey's Golden Jet. 3. Lee Trevino: The Golf Explosion. 4. Billie Jean King: Tennis Champion. 5. Roy Campanella: Brave Man of Baseball.

A best-selling series of photo illustrated biographies, with read-along cassettes, examines the lives and careers of these outstanding athletes.

SPORTS CLOSE–UPS 4 (Kit). EMC, 1974. 7 Paperback Books; 7 Read-along Cassettes; Teacher's Guide; Student Activities. #ELC–228000 ($94). Gr. 4–12. BKL (5/76). 920

Contents: 1. Evonne Goolagong: Smasher from Australia. 2. Frank Robinson: Slugging Toward Glory. 3. Evel Knievel: Daredevil Stuntman. 4. Phil Esposito: The Big Bruin. 5. Joe Namath: High-Flying Quarterback. 6. Muhammad Ali: Boxing Superstar. 7. Vince Lombardi: The Immortal Coach.

The lives and careers of seven outstanding athletes show the talent and competitive drive that brought them to stardom.

WOMEN WHO WIN, SET 1 (Kit). EMC, 1974. 4 Paperback Books; 4 Read-along Cassettes, 29–31 min.; Student Activities; Teacher's Guide. #ELC–221000 ($55). Gr. 4–12. BKL (9/74), INS (11/75), SLJ (11/74). 920

Contents: 1. Olga Korbut: Tears and Triumphs. 2. Shane Gould: Olympic Swimmer. 3. Janet Lynn: Sunshine on Ice. 4. Chris Evert: Tennis Pro.

There are four biographies in this set, introducing outstanding young women who have set their own goals, performed, competed, and reached the top of their chosen field.

WOMEN WHO WIN, SET 2 (Kit). EMC, 1974. 4 Paperback Books; 4 Read-along Cassettes; Student Activities; Teacher's Guide. #ELC–125000 ($55). Gr. 4–12. BKL (5/01/75). 920

Contents: 1. Evonne Goolagong: Smiles and Smashes. 2. Laura Baugh: Golf's Golden Girl. 3. Wilma Rudolph: Run for Glory. 4. Cathy Rigby: On the Beam.

Four women superstars of today are discussed—biographies of top headliners who generate drama and excitement in today's sports world.

WOMEN WHO WIN, SET 3 (Kit). EMC, 1975. 4 Paperback Books; 4 Read-along Books; Teacher's Guide; Student Activities. #ELC–126000 ($55). Gr. 4–12. PRV (2/76). 920

Contents: 1. Mary Dickie: Speed Records and Spaghetti. 2. Annemarie Proell: Queen of the Mountain. 3. Joan Moore Rice: The Olympic Dream. 4. Rosemary Casals: The Rebel Rosebud.

Young readers will thrill to the stories of these superb athletes from a wide world of sports.

WOMEN WHO WIN, SET 4 (Kit). EMC, 1975. 4 Paperback Books; 4 Read-along Cassettes; Students Activities; Teacher's Guide. #ELC–232000 ($55). Gr. 4–12. BKL (5/75), INS (11/75). 920

Contents: 1. Martina Navratilova: Tennis Fury. 2. Robyn Smith: In Silks. 3. Cindy Nelson: North Country Skier. 4. Robin Campbell: Joy in the Morning.

Stories of four superb athletes involved in a wide variety of sports today. Young women who have reached the top in their chosen field through discipline, practice, and commitment.

ATMOSPHERE

ATMOSPHERE IN MOTION (Motion Picture—16mm—Sound). EBEC, 1973. 16mm color, sound film, 20 min. #3330 ($275). Gr. 5–10. BKL (1974), LFR (1974), PRV (11/74). 551.5

The invisible forces that keep the atmosphere in continuous circulation can be visualized in this film. Convection cells, the Coriolis Effect, the composition and structure of the atmosphere are revealed through the use of demonstrations, laboratory models, and special effects animation. The film emphasizes human relationship to the atmosphere and stresses our responsibility for preserving its quality.

ATOM SMASHERS

FERMILAB (Filmstrip—Sound). Hawkhill Associates, 1977. 1 color, sound filmstrip, 111 fr.; 1 cassette, 23 min; Teacher's Guide. ($26.50). Gr. 7–C. BKL (7/1/77). 539.7

The Fermi National Accelerator Laboratory in Batavia, Illinois, is the focus of this production. Photographs, diagrams, and narration clarify the complex experiments performed in the subterranean accelerator ring—the ''largest man made machine in the world.'' Fermilab is used to probe the secrets of atoms and this filmstrip provides an understanding of what basic research in physics is all about.

ATOMIC BOMB

THE ATOM BOMB (Filmstrip—Sound). The Great American Film Factory, 1977. 1 color, sound filmstrip, 110 fr.; 1 cassette, 20 min. #FS–21 ($35). Gr. 6–10. BKL (2/77). 623.4

The story of how the nuclear age began, including the science fiction stories of H. G. Wells and the first 1945 test blast in New

Mexico. Includes many rare photographs of early scientific work. Focus is on how the power of the atom was realized.

ATOMS

CHEMISTRY: DISSECTING THE ATOM (Filmstrip—Sound). *See* Chemistry

MATTER (Filmstrip—Sound). *See* Nuclear physics

STANDING WAVES AND THE PRINCIPLES OF SUPERPOSITION (Motion Picture—16mm—Sound). *See* Waves

ATTUCKS, CRISPUS

FAMOUS PATRIOTS OF THE AMERICAN REVOLUTION (Filmstrip—Sound). *See* U.S.—History—Biography

ATWOOD, ANN

THE GODS WERE TALL AND GREEN (Filmstrip—Sound). *See* Forests and forestry

HAIKU: THE HIDDEN GLIMMERING (Filmstrip—Sound). *See* Haiku

HAIKU: THE MOOD OF EARTH (Filmstrip—Sound). *See* Haiku

MY FORTY YEARS WITH BEAVERS (Filmstrip—Sound). *See* Beavers

SEA, SAND THE SHORE (Filmstrip—Sound). *See* Ocean

THE WILD YOUNG DESERT SERIES (Filmstrip—Sound). *See* Deserts

AUDIOVISUAL MATERIALS

MEDIA: RESOURCES FOR DISCOVERY (Filmstrip—Sound). EBEC, 1974. 8 color, sound filmstrips, av. 95 fr. ea.; 8 discs or cassettes, 11 min. ea.; Teacher's Guide. English/Spanish edition available. #6902 Discs ($132.95), #6902K Cassettes ($132.95). Gr. 4–8. BKL (6/1/75), CPR (11/75), LFR (12/75). 28.7

Contents: 1. The World of Media. 2. Media Organization: Fiction. 3. Media Organization: Nonfiction. 4. Indexes to Media. 5. The Encyclopedia in the World of Media. 6. Resources for Reference. 7. Choosing the Medium. 8. One Search, One Report.

Designed to make students self-sufficient and at ease in the modern library media center, this series demonstrates what a media center is and how to use it.

AUDUBON, JOHN JAMES

AUDUBON'S SHORE BIRDS (Motion Picture—16mm—Sound). *See* Water birds

AUSTEN, JANE

EPISODES FROM CLASSIC NOVELS (Filmstrip—Sound). *See* Literature—Collections

AUSTRALIA

AUSTRALIA AND NEW ZEALAND (Filmstrip—Sound). EBEC, 1972. 6 color, sound filmstrips, av. 81 fr. ea.; 6 discs or cassettes, 13 min. ea.; Teacher's Guide. #6456K Cassettes ($86.95), #6456 Discs ($86.95). Gr. 4–8. BKL (10/15/73), LFR (10/73), PRV (4/73). 919.4

Contents: 1. The Agricultural Achievement. 2. Toward Industrialization. 3. Australia: The Island Continent. 4. New Zealand: Land of the Long White Cloud. 5. The Australians. 6. The New Zealanders.

Life "down under" has a flavor all its own. From bustling coastal cities to lonely sheep stations in the sun-scorched Outback, Australians are seen to mingle their British heritage with the pioneer's independence and zest for life. In New Zealand, the Polynesian Maoris and their European-descended neighbors share in the country's development—an outstanding example of racial harmony and equality.

AUTHORS

ELEANOR ESTES (Filmstrip—Sound). Miller-Brody Productions, 1974, (Meet the Newbery Authors). 1 color, sound filmstrip, approx. 96 fr.; 1 cassette or disc, approx. 17 min. #8–MNA1001C Discs ($32), #8–MNA1001 Cassettes ($32). Gr. 4–9. PRV (2/75), PVB (5/75), 921

An on-screen interview captures moments in her childhood, influences on her writings.

ELIZABETH YATES (Filmstrips—Sound). *See* Yates, Elizabeth

JEAN CRAIGHEAD GEORGE (Filmstrip—Sound). Miller-Brody Productions, 1974 (Meet the Newbery Authors). 1 color, sound filmstrip, approx. 100 fr.; 1 cassette or disc, approx. 19 min. #8–MNA1003 Disc ($32), #8–MNA1003C Cassette ($32). Gr. 4–9. PRV (2/75), PVB (5/75). 921

Her passion for nature—from spiders to the beauty of a single leaf—is made clear in this filmstrip.

LAURA: LITTLE HOUSE, BIG PRAIRIE (Filmstrip—Sound). *See* Wilder, Laura Ingalls

LLOYD ALEXANDER (Filmstrip—Sound). Miller-Brody Productions, 1974 (Meet the Newbery Authors). 1 color, sound filmstrip, 98 fr.; 1 cassette or disc, approx. 18 min. #8–MNA1002 Disc ($32), #8–MNA1002C Cassette ($32). Gr. 4–9. PRV (2/75). PVB (5/75). 921

AUTHORS (cont.)

Besides his agonies over writing, the author's love of Mozart, cats, the epic hero, and more are discussed.

NOVELISTS AND THEIR TIMES (Kit). Educational Enrichment Materials, 1976. 4 color, sound filmstrips, 75–100 fr. ea.; 4 cassettes or discs, 15–20 min.; 4 Wall Charts; Teacher's Guide. #51001 C or R ($84). Gr. 7–12. PRV (4/77). 920
Contents: 1. Dickens: Victorian England. 2. Balzac: Napoleonic France. 3. Twain: Democratic America. 4. Dostoevski: Czarist Russia.
Designed to help students see how society, historical events, and personal qualities inspired four major writers, and in turn, how they influence all writers. Filled with excerpts from their most memorable works, this program examines the similar concerns and devotion to realism shared by these four legendary authors.

VIRGINIA HAMILTON (Filmstrip—Sound). *See* Hamilton, Virginia

WILLIAM H. ARMSTRONG (Filmstrip—Sound). *See* Armstrong, William

AUTHORS, AMERICAN

AYN RAND—INTERVIEW (Videocassette). *See* Rand, Ayn

CHILDREN OF THE NORTHLIGHTS (Motion Picture—16mm—Sound). Weston Woods Studios, 1977 (Signature Collection). 16mm color, sound film (also available as video-cassette), 20 min. #426 ($245). Gr. 3–A. CGE (1977). 920
Ingri and Edgar Parin d'Aulaire, winners of the 1940 Caldecott Award for *Abraham Lincoln*, are visited in their studio-home, where they discuss the philosophy behind their work and their unique process of creating illustrations by using lithographic stones. Animated sequences of their illustrations and delightful excerpts from their own home movies are incorporated into this portrait, which also follows their most recent book, *The Terrible Troll Bird*, from drawing board to printing press.

EZRA JACK KEATS (Motion Picture—16mm—Sound). *See* Keats, Ezra Jack

FIVE MODERN NOVELISTS (Filmstrip—Sound). *See* American literature—History and criticism

MARK TWAIN (Filmstrip—Sound). *See* Twain, Mark

MARK TWAIN: MISSISSIPPI RENAISSANCE MAN (Filmstrip—Sound). *See* Twain, Mark

ROOTS WITH ALEX HALEY (Cassette). *See* Haley, Alex

WOMEN WRITERS: VOICES OF DISSENT (Filmstrip—Sound). *See* Women authors

AUTHORS, ENGLISH

CHARLES DICKENS (Cassette). *See* Dickens, Charles

GREAT WRITERS OF THE BRITISH ISLES, SET I (Filmstrip—Sound). Pathescope Educational Media, 1969, (Great Writers). 6 color, sound filmstrips, 56–57 fr. ea.; 6 cassettes, 17–22 min ea.; Teacher's Guide. #526 ($100). Gr. 7–12. BKL (1/15/70). 920
Contents: 1. Geoffrey Chaucer 2. Oliver Goldsmith. 3. William Shakespeare Part I—His Life and Times. 4. William Shakespeare Part II—Theater and Plays. 5. Sir Walter Scott. 6. The Brontes.
A colorful, pictorial study of the lives and works of five major authors. Each personality is considered from the historical, geographical, and other significant standpoints. The narrative is coded to the language superimposed on each visual. It reviews the lives and time of the authors. Each author's writings are examined, ideas and philosophies explained; the audience for whom they wrote identified; their style and the nature of their stories analyzed; and the influence of their works on others traced.

GREAT WRITERS OF THE BRITISH ISLES, SET II (Filmstrip—Sound). Pathescope Educational Media, 1969 (Great Writers). 6 color, sound filmstrips, 53–56 fr. ea.; 6 cassettes, 13-1/2–26 min. ea.; Teacher's Guide. #527 ($100). Gr. 7–12. BKL (2/1/70). 920
Contents: 1. Charles Dickens Part I—His Life. 2. Charles Dickens Part II—His Work. 3. George Eliot. 4. Thomas Hardy. 5. Robert Louis Stevenson. 6. Winston S. Churchill.
This set tells about the lives and works of five outstanding British authors. Each author's writings are examined; ideas and philosophies explained; the audience for whom they wrote identified; their style and the nature of their stories analyzed; and the influence of their works on others traced.

THE LETTERS OF VIRGINIA WOOLF (Cassette). *See* Woolf, Virginia

MR. SHEPARD AND MR. MILNE (Motion Picture—16mm—Sound). Weston Woods Studios, 1973. 16mm color, sound film, 29 min. #415 ($365). Gr. 3–A. CFF (1974), FLN (2/74), FLQ (1973). 920
Between 1923 and 1928, two men collaborated on a series of four books—*When We Were Very Young, Winnie-the-Pooh, The House at Pooh Corner,* and *Now We Are Six. Mr. Shepard and Mr. Milne* tells the story of the men and how the books come into being. In this film, C. R. Milne reads his father's words as well as selections from the books, and Ernest Shepard reminisces about

his experiences with Milne as the camera visits some of the places which were put into the stories and poems.

SHAKESPEARE (Filmstrip—Sound). *See* Shakespeare, William

WILLIAM SHAKESPEARE (Cassette). *See* Shakespeare, William

AUTHORSHIP

WRITING SKILLS—THE FINAL TOUCH: EDITING, REWRITING & POLISHING (Slides/Cassettes). *See* English language—Composition and exercises

AUTOBIOGRAPHIES

THE DIARY OF ANNE FRANK (Filmstrip—Sound). Films, 3 color, sound filmstrips; 3 cassettes, 20 min. ea.; Plot Synopsis; 2 Theme Strips. ($69). Gr. 4–8. M&M (4/76). 921

Adapted from the film directed by George Stevens, the story is familiar. This condensation of the original soundtrack preserves the integrity of the story.

A MISERABLE MERRY CHRISTMAS (Motion Picture—16mm—Sound). *See* Christmas

AUTOMOBILE RACING

AUTO RACING: SOMETHING FOR EVERYONE (Kit). Walt Disney Educational Media, 1976. 6 color, sound filmstrips, av. 72–91 fr.; 6 discs or cassettes, av. 8–10 min.; 6 Reading Books; Teacher's Guide. #63–8040 ($129), Individual Sound Filmstrip ($25). Gr. 6–12. BKL (12/15/77). 796.7

Contents: 1. It All Started with a Road Race. 2. The "Indy" Story. 3. Made in U.S.A.: Stock Car and Drag Racing. 4. Safety in Racing. 5. Off Road Racing. 6. Careers in Racing.

The filmstrips and books provide information on the various kinds of racing. In addition, they detail the development of standard accessories originally intended for racing and alert the viewer to occupations related to car racing. The illustrated text offers expanded information, including additional information on careers.

DRAGSTRIP CHALLENGE (Kit). Insight Media Programs, 1974. 1 color, sound filmstrip, 110 fr.; 1 disc or cassette, 13 min.; 1 book (also available with 6 paperback books). Cassette ($25.45), Disc ($23.45). Gr. 4–9. PRV (4/75), PVB (5/75). 796.7

Two real dragstrip racers introduce readers to the sport. A word-for-word reading of the first two chapters of the book is on the reverse of each disc or cassette for those with reading difficulties. There is also a glossary of racing terminology.

THE MIGHTY MIDGETS (Kit). Bowmar Publishing, 1967. 1 color, sound filmstrip, 37 fr.; 1 cassette or disc; 7 Student Readers and Reading Programs; Teacher's Guide. #4536 Disc ($38.95), #4537 Cassette ($38.95). Gr. 3–12. MDU, SLJ. 629.22

A high-interest level, easy readability unit on midget racers.

RACING NUMBERS (Kit). Children's Press, 1976 (Ready, Get-Set, Go). 1 cassette; 1 hardback book; Teacher's Guide. #07557–8 ($11.95). Gr. 1–8. PRV (9/15/77). 796.7

Color photographs of soap box racers, antique cars, motorcycles, and other racing machines combine with a simple text to help youngsters learn number recognition in this high-interest, low-reading ability kit.

AUTOMOBILES

BATE'S CAR (Motion Picture—16mm—Sound). *See* Fuel

TOMMY'S FIRST CAR (BUYING A USED CAR) (Motion Picture—16mm—Sound). *See* Consumer education

AUTOMOBILES—LAW AND LEGISLATION

SPEEDING? (Motion Picture—16mm—Sound). Direct Cinema, 1978. 16 mm color, sound film, 21 min. #002 ($325). Gr. 9–A. BKL (2/1/79). 345.024

Actors are cast as traffic offenders voicing their opinions of speeding violations. The attitudes frequently displayed by violators are dramatized as well as the responses from the policeman ticketing such drivers. The various safety issues are presented through the dramatizations.

AUTOMOBILES—MAINTENANCE AND REPAIR

AIR CONDITIONING SERVICEMAN, SET SEVEN (Film Loop—8mm—Silent). McGraw-Hill Films, 1973 (Automotive Damage Correction). 6 silent, 8mm technicolor, $3\frac{1}{2}$–4 min. ea.; Guide. #103902–3 ($132). Gr. 9–A. WLB (6/74). 629.287

Contents: 1. Identifying the Basic Components of an Air Conditioning System. 2. Inspecting the System Visually. 3. Understanding the Test Gauges. 4. Connecting Gauges and Discharging the System. 5. Evacuating the System. 6. Connecting the Refrigerant Can and Tap.

These loops demonstrate basic skills needed for actual jobs in the air conditioning work areas of the automotive trade.

AIR CONDITIONING SERVICEMAN, SET EIGHT (Film Loop—8mm—Silent). McGraw-Hill Films, 1973 (Automotive Damage Correction). 5 silent, 8mm technicolor, $3\frac{1}{2}$–4 min. ea.; Guide. #103909–0 ($110). Gr. 9–A. WLB (6/74). 629.287

AUTOMOBILES—MAINTENANCE AND REPAIR (cont.)

Contents: 1. Installing Gauge Set and Purging Hoses. 2. Charging the System. 3. Testing the System for Leaks. 4. Performance Testing the System. 5. Trouble Shooting the System.

These loops demonstrate basic skills needed for actual jobs in the air conditioning work areas of the automotive trade.

AUTO-BODY SHEET METAL MAN, SET ONE (Film Loop—8mm—Silent). McGraw-Hill Films, 1973 (Automotive Damage Correction). 6 silent, 8mm technicolor, $3^{1}/_{2}$–4 min. ea.; Guide. #103726–8 ($132). Gr. 9–A. WLB (6/74). 629.287

Contents: 1. Roughing-Out a Dent. 2. Picking and Filing a Damaged Area. 3. Grinding and Sanding Metal. 4. Using a Dent Puller. 5. Applying Plastic Filler. 6. Finishing Plastic Filler.

These loops demonstrate basic skills needed for actual jobs in the body work areas of the automotive trade.

AUTO-BODY SHEET METAL MAN, SET TWO (Film Loop—8mm—Silent). McGraw-Hill Films, 1973 (Automotive Damage Correction). 8 silent, 8mm technicolor, $3^{1}/_{2}$–4 min. ea.; Guide. #103733–0 ($176). Gr. 9–A. WLB. (6/74). 629.287

Contents: 1. Making a Template. 2. Cutting Out a Metal Patch. 3. Fastening a Patch. 4. Filling a Patched Area. 5. Straightening a Torn Section. 6. Welding a Tear In Metal. 7. Metal Finishing a Weld. 8. Torch Soldering.

These loops demonstrate basic skills needed for actual jobs in the body work areas of the automotive trade.

AUTO PAINTER HELPER, SET THREE (Film Loop—8mm—Silent). McGraw-Hill Films, 1973) (Automotive Damage Correction). 7 silent, 8mm technicolor, $3^{1}/_{2}$–4 min. ea.; Guide. #103742–X ($154). Gr. 9–A. WLB (6/74). 629.287

Contents: 1. Degreasing Surfaces. 2. Masking. 3. Sandpaper Types and Uses. 4. Featheredging. 5. Priming and Glazing. 6. Block Sanding. 7. Preparing the Primed Spot Color.

These loops demonstrate basic skills needed for actual jobs in the paint work areas of the automotive trade.

AUTO PAINTER, SET FOUR (Film Loop—8mm—Silent). McGraw-Hill Films, 1973 (Automotive Damage Correction). 7 silent, 8mm technicolor, $3^{1}/_{2}$–4 min. ea.; Guide. #103750–0 ($154). Gr. 9–A. WLB (6/74). 629.287

Contents: 1. Spray Gun Operation. 2. Finding the Color Number. 3. Mixing and Thinning Color. 4. Applying Color (Lacquer and Acrylic). 5. Color Sanding. 6. Compounding a Freshly Painted Panel. 7. Buffing and Polishing a Painted Panel.

These loops demonstrate basic skills needed for actual jobs in the paint work areas of the automotive trade.

AUTOMOBILE GLASS MAN, SET NINE (Film Loop—8mm—Silent). McGraw-Hill Films, 1973 (Automotive Damage Correction). 7 silent, 8mm technicolor, $3^{1}/_{2}$–4 min. ea.; Guide. #106364–1 ($154). Gr. 9–A. WLB (6/74). 629.287

Contents: 1. Identifying and Selecting Sealers. 2. Removing Window Moldings. 3. Replacing a Water Shield. 4. Sealing a Window or Backlight. 5. Sealing a Rear Quarter Window. 6. Sealing a Tailgate and Glass.

These loops demonstrate basic skills needed for actual jobs in the glass work areas of the automotive trade.

AUTOMOBILE GLASS MAN, SET TEN (Film Loop—8mm—Silent). McGraw-Hill Films, 1973 (Automotive Damage Correction). 7 silent, 8mm technicolor, $3^{1}/_{2}$–4 min. ea.; Guide. #106372–2 ($154). Gr. 9–A. WLB (6/74). 629-287

Contents: 1. Removing a Molded Rubber Installation. 2. Removing a Caulked Installation Using a Hot Knife. 3. Removing a Caulked Installation Using a Wire. 4. Preparing Window Openings and Fitting the Glass. 5. Installing a Molded Rubber and Window. 6. Installing Adhesive Caulking and Window. 7. Installing Ribbon Sealer and Window.

These loops demonstrate basic skills needed for actual jobs in the glass areas of the automotive trade.

AUTOMOBILE UPHOLSTERY REPAIRMAN, SET ELEVEN (Film Loop—8mm—Silent). McGraw-Hill Films, 1973 (Automotive Damage Correction). 7 silent, 8mm technicolor, $3^{1}/_{2}$–4 min. ea.; Guide. #106380–3 ($154). Gr. 9–A. WLB (6/74). 629.287

Contents: 1. Renewing Rubber and Vinyl. 2. Washing and Dying Fabric. 3. Repairing a Vinyl Top Installation. 4. Repairing a Torn Seat. 5. Repairing a Torn Headliner. 6. Covering an Arm Rest. 7. Replacing Interior Trim Panels.

These loops demonstrate basic skills needed for actual jobs in the upholstery work areas of the automotive trade.

FIBERGLASS REPAIRMAN, SET FIVE (Film Loop—8mm—Silent). McGraw-Hill Films, 1973 (Automotive Damage Correction). 7 silent, 8mm technicolor, $3^{1}/_{2}$–4 min. ea.; Guide. #103885–X ($154). Gr. 9–A. WLB (6/74). 629.287

Contents: 1. Preparing a Damaged Panel. 2. Preparing Resin Mix. 3. Repairing a Crack or Tear. 4. Filling and Finishing a Repaired Section. 5. Patching a Hole. 6. Preparing and Applying Cloth to Metal. 7. Mixing and Applying "Chopped" Fiberglass Filler.

These loops demonstrate basic skills needed for actual jobs in the fiberglass work areas of the automotive trade.

FIBERGLASS REPAIRMAN, SET SIX (Film Loop—8mm—Silent). McGraw-Hill Films, 1973 (Automotive Damage Correction). 8 silent 8mm, technicolor, 3$\frac{1}{2}$–4 min. ea.; Guide. #103893–0 ($176). Gr. 9–A. WLB (6/74). 629.287

Contents: 1. Cutting Out a Section. 2. Joining Repaired Sections. 3. Reinforcing and Filling Joined Panels. 4. Reinforcing Hinge and Torque Boxes. 5. Reconstructing a Mashed Crown. 6. Reinforcing a Mashed Crown. 7. Making a Core with Fiberglass Material. 8. Installing and Building-Up a Fiberglass Core.

These loops demonstrate basic skills needed for actual jobs in the fiberglass work areas of the automotive trade.

JOBS IN AUTOMOTIVE SERVICES (Filmstrip—Sound). SA Films/Coronet Instructional Media, 1974. 8 color, sound filmstrips, av. 50 fr.; 4 discs or 8 cassettes, av. 8-$\frac{1}{2}$ min. Guide. #M273 Cassettes ($120), #S273 Discs ($120). Gr. 7–A. BKL (3/1/75). 629.287

Contents: 1. Tune-Up Trainee. 2. General Automobile Technician Apprentice. 3. Truck Service Technician Apprentices. 4. Front End Trainee. 5. Transmission Trainee. 6. Body and Fender Apprentice. 7. Air Conditioning Trainee. 8. Brake Trainee.

On-the-job interviews take students for a look under the entry-level automotive repair jobs. The challenges of the jobs are presented along with information on work hazards, union membership, opportunities for training and advancement.

AUTOMOBILES—REPAIRING. *See* Automobiles—Maintenance and repair

AUTOMOBILES—TRAILERS

EFFECTIVE TRAILERING (Filmstrip—Sound). Professional Arts, 1977. 1 color, sound filmstrip, 140 fr.; 1 cassette, 20 min. ($45.). Gr. 9–12. PRV (1/79). 629.22

A "system concept" of common sense safety tips. Five elements of this system are presented in depth and then reviewed. It is nontechnical and provides basic information for beginning drivers.

BABOONS

BABOONS AND THEIR YOUNG (Film Loop—8mm—Silent). Walt Disney Educational Media, 1966. 8mm color, silent film loop, approx. 4 min. #62–5219L ($30). Gr. K–12. MDU (1974). 599.8

The maternal behavior of the baboon is illustrated in a variety of sequences, both on the ground and in trees. Both parents are shown caring for their young.

BACKPACKING

BACKPACKING BASICS (Slides/Cassettes). Backpacker Magazine/Center for Humani-

ties, 1977. 65 slides in 1 carousel cartridge; 1 cassette; 1 disc; Teacher's Guide. #1034–1200 ($69.50). Gr. 9–A. BKL (3/1/78), IDF (1977), PRV (10/77). 796.5

In this program, Bill Kensley, publisher of *Backpacker* magazine, outlines the many pleasures and some of the potential dangers for new hikers along the trail. He points out their responsibility in preserving the delicate ecology of our remaining wilderness. Information is provided on the selection of backpacks, tents, sleeping bags, and other gear.

BACKPACKING: REVISED (Filmstrip—Sound). The Great American Film Factory, 1978 (Outdoor Education). 1 color, sound filmstrip; 1 cassette, 18 min. #FS–1 ($35). Gr. 5–A. PVB (4/19/76). 796.5

Completely new audio program and many new visuals updates this backpacking trip. Things any beginner needs to know about backpacking are presented and the important rules of safety, ecology, and trail manners. Set in a "how-to" format.

BAEZ, JOAN

FOLK SONGS IN AMERICAN HISTORY, SET ONE: 1700–1864 (Filmstrip—Sound). *See* Folk songs—U.S.

FOLK SONGS IN AMERICAN HISTORY, SET TWO: 1865–1967 (Filmstrip—Sound). *See* Folk songs—U.S.

BALLADS, AMERICAN

AMERICAN HISTORY: IN BALLAD AND SONG, VOLUME ONE (Phonodiscs). Folkways Records, 1954. 1 disc, 12 in., 33-$\frac{1}{3}$ rpm; Guide. #5801 ($6.98). Gr. 3–12. MDU (1978). 784.75

Pete Seeger, Woody Guthrie, and others sing about American history in song.

AMERICAN HISTORY: IN BALLAD AND SONG, VOLUME TWO (Phonodiscs). Folkways Records, 1963. 1 disc, 12 in., 33-$\frac{1}{3}$ rpm; Guide. #5802 ($6.98). Gr. 3–12. MDU (1978). 784.75

Peter Seeger, Woody Guthrie, and others sing about American history in song.

BALTIMORE, LEE

LEE BALTIMORE: NINETY NINE YEARS (Motion Picture—16mm—Sound). Odyssey Prod./EBEC, 1976. 16mm color, sound film, 17 min. #3491 ($240). Gr. 7–A. BKL (11/15/76), EFL (1977), EGT (1977). 921

Lee Baltimore—farmer, philosopher, and son of slaves—is 99 years old. He tells of a life simple, poor, yet rich in memories and love of the land that sustains him. His wife is dead and he is often lonely, but never despondent. He has kept his faith in the goodness of life. And he has his piano, which he

BALTIMORE, LEE (cont.)

taught himself to play two years ago at the age of 97.

BALZAC, HONORE

LA GRANDE BRETECHE (Motion Picture—16mm—Sound). *See* Horror—Fiction

NOVELISTS AND THEIR TIMES (Kit). *See* Authors

BANKS AND BANKING

CAREERS IN BANKING AND INSURANCE (Filmstrip—Sound). Pathescope Educational Media, 1975. 2 color, sound filmstrips, 81–84 fr ea.; 2 cassettes, 11–12 min. ea.; Teacher's Guide. #727 ($50). Gr. 9–C. BKL (6/1/76). 332.1023
Positions found in the financial field require honesty, mathematical ability, an eye for detail and an understanding of the needs of people. The variety of careers in the two businesses are discussed and the required skills and education are explored.

BARLING, BOB

PETROGLYPHS: ANCIENT ART OF THE MOJAVE (Filmstrip—Sound). *See* Indians of North America—Art

BARLING, TILLY

PETROGLYPHS: ANCIENT ART OF THE MOJAVE (Filmstrip—Sound). *See* Indians of North America—Art

BARTLEBY

A DISCUSSION OF BARTLEBY (Motion Picture—16mm—Sound). *See* Motion pictures—History and criticism

BARTOK, BELA

CONCERTO FOR ORCHESTRA (Phonodiscs). *See* Concertos

BASHFULNESS

SHYNESS: REASONS AND REMEDIES (Filmstrip—Sound). Human Relations Media, 1975. 2 color, sound filmstrips, 72–77 fr.; 2 cassettes or discs, 12–13 min.; Teacher's Guide. #617 ($60). Gr. 7–12. BKL (4/76), MNL (4/76), PRV (9/76). 152.4
Contents: 1. Understanding Shyness. 2. Living with Shyness.
An incisive study into the causes, effects, and management of shyness.

BASKETBALL

BASIC DRIBBLE/CONTROL DRIBBLE/SPEED (Film Loop—8mm—Silent). Athletic Institute (Basketball—Women's). 8mm, color, silent film loop, approx. 4 min.; Guide. #TRWB-1 ($22.95). Gr. 3–12. AUR (1978). 796.323
A single concept loop demonstrating the basic technique.

CHEST PASS/BOUNCE PASS (Film Loop—8mm—Silent). Athletic Institute (Basketball—Women's). 8mm color, silent film loop, approx. 4 min.; Guide. #WB-4 ($22.95). Gr. 3–12. AUR (1978). 796.323
A single concept loop demonstrating the basic technique.

CHEST PASS/OVERHEAD PASS (Film Loop—8mm—Silent). Athletic Institute (Basketball). 8mm color, silent film loop, approx. 4 min.; Guide. #M-4 ($22.95). Gr. 3–12. AUR (1978). 796.323
A single concept loop demonstrating the basic technique.

CROSSOVER CHANGE/REVERSE PIVOT CHANGE (Film Loop—8mm—Silent). Athletic Institute (Basketball). 8mm color, silent film loop, approx. 4 min.; Guide. #M-2 ($22.95). Gr. 3–12. AUR (1978). 796.323
A single concept loop demonstrating the basic technique.

CROSSOVER DRIBBLE/REVERSE DRIBBLE (Film Loop—8mm—Silent). Athletic Institute (Basketball—Women's). 8mm color, silent film loop, approx. 4 min.; Guide. #WB-2 ($22.95). Gr. 3–12. AUR (1978). 796.323
A single concept loop demonstrates the basic technique.

DRIVE/CROSSOVER DRIVE (Film Loop—8mm—Silent). Athletic Institute (Basketball). 8mm color, silent film loop, approx. 4 min.; Guide. #M-3 ($22.95). Gr. 3–12. AUR (1978). 796.323
A single concept loop demonstrating the basic technique.

INSIDE POWER SHOT (Film Loop—8mm—Silent). Athletic Institute (Basketball). 8mm color, silent film loop, approx. 4 min.; Guide. #M-6 ($22.95). Gr. 3–12. AUR (1978). 796.323
A single concept loop demonstrating the basic technique.

JUMP SHOT (Film Loop—8mm—Silent). Athletic Institute (Basketball). 8mm color, silent film loop, approx. 4 min.; Guide. #M-7 ($22.95). Gr. 3–12. AUR (1978). 796.323
A single concept loop demonstrates the basic technique.

JUMP SHOT/ONE HAND SET/TURN-
AROUND JUMP SHOT (Film Loop—
8mm—Silent). Athletic Institute (Basketball—Women's). 8mm color, silent film
loop, approx. 4 min.; Guide. #WB-7
($22.95). Gr. 3–12. AUR (1978). 796.323
A single concept loop demonstrates the basic techniques of jump shots.

LAY UP SHOT (Film Loop—8mm—Silent).
Athletic Institute, 1970 (Basketball—Wom-
en's). 8mm color, silent film loop, approx. 4
min.; Guide. #WB-6 ($22.95). Gr. 3–12.
AUR (1978). 796.323
A single concept loop demonstrates the basic technique.

ONE-ON-ONE DRIVE (Film Loop—8mm—
Silent). Athletic Institute (Basketball—
Women's). 8mm color, silent film loop, approx. 4 min.; Guide. #WB-3 ($22.95). Gr. 3–
12. AUR (1978). 796.323
A single concept loop demonstrates the basic technique.

OVERARM PASS/OVERHEAD PASS/
UNDERHAND PASS (Film Loop—8mm—
Silent). Athletic Institute, 1970 (Basketball—Women's). 8mm color, silent film
loop, approx. 4 min.; Guide. #WB-5
($22.95). Gr. 3–12. AUR (1978). 796.323
A single concept loop demonstrates the basic technique.

REBOUNDING (Film Loop—8mm—Silent).
Athletic Institute (Basketball). 8mm, color,
silent film loop, approx. 4 min.; Guide. #M-
9 ($22.95). Gr. 3–12. AUR (1978). 796.323
A single concept loop demonstrates the basic technique.

REBOUNDING/BLOCKING OUT (Film
Loop—8mm—Silent). Athletic Institute
(Basketball—Women's). 8mm, color, silent
film loop, approx. 4 min.; Guide. #WB-8
($22.95). Gr. 3–12. AUR (1978). 796.323
A single concept loop demonstrates the bas-
ic technique.

SPEED DRIBBLE/CONTROL DRIBBLE
(Film Loop—8mm—Silent). Athletic Institute (Basketball). 8mm, color, silent film
loop. approx. 4 min.; Guide. #TRM-1
($22.95). Gr. 3–12. AUR (1978). 796.323
A single concept loop demonstrating the basic technique.

TURN AROUND JUMP SHOT (Film Loop—
8mm—Silent). Athletic Institute (Basketball). 8mm, color, silent film loop, approx. 4
min.; Guide. #M-8 ($22.95). Gr. 3–12. AUR
(1978). 796.323
A single concept loop demonstrates the basic technique.

BATIK

CRAYON BATIK MAGIC (Filmstrip—
Sound). Warner Educational Productions,
1973. 1 filmstrip, 60 fr.; 1 cassette, 10-1/2
min.; Teaching Guide. #890 ($30.50). Gr. 6–
A. BKL (5/1/74), PRV (9/75). 746.6
A batik method which uses melted wax cray-
ons as a replacement for liquid dyes is demonstrated. A practical presentation on
planning designs and color schemes as well
as creative projects.

CREATIVE BATIK (Filmstrip—Sound).
Warner Educational Productions, 1973. 2
color, sound filmstrips, 57–67 fr.; 1 cassette,
9-11 1/2 min.; Teaching Guide. #670
($47.50). Gr. 7–A. BKL (5/1/74), PRV (9/75).
746.6
Introduces the skill of batik competently.
The origin of the craft is noted in this task
oriented demonstration. It teaches waxing,
dyeing, crackling, and use of the tjanting, as
well as planning design and color schemes.
The history and modern application of batik
and the materials and tools used are described.

CREATIVE TIE/DYE (Filmstrip—Sound).
Warner Educational Productions, 1973. 2
color, sound filmstrips, 58–72 fr.; 1 cassette,
9–12 min.; Teaching Guide. #450 ($47.50).
Gr. 7–A. BKL (5/1/74), PRV (9/75). 746.6
A practical presentation in the art of tying
and dyeing. Shows popular knots, ties and
folds. Techniques for dyeing are described
and innovative uses in clothing and home
decorations are included.

BATS

BATS (Film Loop—8mm—Silent). Walt Disney Educational Media, 1966. 8mm color, silent film loop, approx. 4 min. #62-51076
($30). Gr. K-12. AUR (1974). 599.4
Millions of bats are seen awakening in a
cave. Shows a close-up view of a bat hanging
up-side down. A hawk is shown trying to
catch one of the bats as they fly off into the
evening sky.

BATTERED CHILD SYNDROME. See Child
abuse

BAUM, WILLI

BIRDS OF A FEATHER (Motion Picture—
16mm—Sound). See Birds—Fiction

BEARS

POLAR BEAR: MOTHER AND CUBS (Film
Loop—8mm—Silent). Walt Disney Educational Media, 1966. 8mm color, silent film
loop, approx. 4 min. #62-5355L ($30). Gr.
K-12. MDU (1974). 599.7
Arctic life of mother polar bear and her cubs
is viewed from within a snow cave, including
close-ups of cubs nursing. Mother follows
young bears outside and observes as they
play on a snowslide.

BEARS—FICTION

THE GRIZZLY (Kit). Insight Media Programs, 1976. 1 color, sound filmstrip, 80 fr.; 1 disc or cassette, 15 min.; 8 Paperback Books; 1 Teacher's Guide. ($41.95). Gr. 5–9. BKL (11/15/76). Fic

Student copies of Annabel and Edgar Johnson's novel are included with the filmstrip adaptation of a boy reunited with his father after a five-year separation. Their trout fishing experience in a remote area ends with an encounter with a bear that injures the father. The son finds his valor and develops self-reliance, thereby earning his father's respect. Father-son interdependency and communication are stressed.

BEAVERS

BEAVER (Film Loop—8mm—Silent). Walt Disney Educational Media, 1966. 8mm color, silent film loop, approx. 4 min. #62–5009L ($30). Gr. K–12. MDU (1974). 599.3

The beaver is seen swimming and shown inside its lodge with muskrats, which share its home. A beaver choosing its life mate is shown.

BEAVER DAM AND LODGE (Film Loop—8mm—Silent). Walt Disney Educational Media, 1966. 8mm color, silent film loop, approx. 4 min. #62–5010L ($30). Gr. K–12. MDU (1974). 599.3

A beaver is studied as it constructs its dam and lodge. Pictures the beaver gnawing down a tree, the inside of the completed lodge, and its cleaning and grooming habits.

MY FORTY YEARS WITH BEAVERS (Filmstrip—Sound). Lyceum/Mook & Blanchard, 1975. 2 color, sound filmstrips, 60–65 fr.; 2 cassettes or discs, 14 min. #LY35374SC Cassettes ($46), #LY35374SR Discs ($37). Gr. 3–A. BKL (12/15/75), LNG (11/75). 599.3

Contents: 1. A Kinship with Wild Beavers. 2. A Houseful of Beavers.

Introduces students to the patient observation and study of nature as they follow the unique experience of Dorothy Richards as she studies and becomes friends with a pair of beavers. Sensitive photographs are included by Ann Atwood.

BEES

BEEKEEPING (Filmstrip—Sound). Hawkhill Associates, 1977. 1 color, sound filmstrip, 107 fr.; 1 cassette, 15 min.; Teacher's Guide. ($26.50). Gr. 4–C. BKL (12/1/77), LNG (5/19/78), SCT (1978). 638

An introduction to bees and to beekeeping. The life cycle of the beehive is shown, as well as the various steps involved in professional beekeeping. Also suggestions for starting your own hive are given.

THE HONEY BEE (Film Loop—8mm—Captioned). BFA Educational Media, 1973 (Animal Behavior). 8mm color, captioned film loop, approx. 4 min. #481414 ($30). Gr. 4–9. BKL (5/1/76). 595.79

Demonstrates experiments, behavior, and characteristics of the honey bees.

LIFE OF A WORKER BEE (Film Loop—8mm—Silent). BFA Educational Media, 1973 (Animal Behavior). 8mm color, silent film loop, approx. 4 min. #481415 ($30). Gr. 4–9. BKL (5/1/76). 595.79

Demonstrates experiments, behavior, and characteristics of the worker bees.

BEETHOVEN, LUDWIG VAN

SYMPHONIC MOVEMENTS NO. 2 (Phonodiscs). See Symphonies

BEHAVIOR. See Animals—Habits and behavior; Human behavior

BEHAVIOR MODIFICATION

CHANGING HUMAN BEHAVIOR (Filmstrip—Sound). Human Relations Media, 1978. 3 color, sound filmstrips, 66–78 fr.; 3 cassettes, 12½–15 min.; Teacher's Guide. #637–A ($90). Gr. 10–C. BKL (2/1/79). 159.1943

Contents: 1. Principles of Learning. 2. Dealing with Problems. 3. Blueprint for a New World.

Studies theories of human behavior to understand how behavior is learned and how it can be changed or modified.

BELLAK, LEOPOLD

GROWING OLD WITH GRACE (Cassette). See Old age

BEOWULF

BEOWULF AND THE MONSTERS (Cassette). Children's Classics on Tape, 1973. 1 cassette with detailed study guide; 45 minutes. ($10.95). Gr. 4–9. BKL (7/15/74), ESL (10/77), PRV (1/75). 829.3

Contents: 1. Story of Beowulf. 2. History of the Anglo-Saxons and original Beowulf.

On Side 1 Beowulf goes to the great hall of King Hrothgar to kill the monster Grendel, who has terrorized the kingdom. Beowulf battles the Water Witch and a fire breathing dragon. On Side 2 gives history of the Anglo-Saxons and Beowulf story. The original Beowulf manuscript—and how it survived. Story and historical narrative by Ruth Lind.

BETTMAN, OTTO

THE GOOD OLD DAYS—THEY WERE TERRIBLE (Cassette). See History—Philosophy

BEUERMAN, LEO

LEO BEUERMAN (Motion Picture—16mm—
Sound). Centron Films, 1969. 16mm color,
sound film, 13 min.; Leader's Guide. ($205).
Gr. 4–A. AFF, BKL (12/1/70), M&M (2/74),
921

Documents the life of Leo Beuerman, an unusual man physically handicapped since
birth, describing his outlook on life and his
attitudes towards life and his fellow man.

BIBLE—OLD TESTAMENT

HISTORY, POETRY AND DRAMA IN THE
OLD TESTAMENT (Motion Picture—
16mm—Sound). EBEC, 1973 (Bible as Literature). 16mm color, sound film, 24 min.
#47795 ($330). Gr. 7–A. BKL (1974), EGT
(1975), EFL (1976). 221

This film examines Joshua, Samuel, and
Kings as historical documents, the Book of
Proverbs as lyric poetry, and the Prophetical
Books as protest literature. The film concludes with the story of Job, treating the
timeless moral question: What is man's solution to the problem of evil? Students will see
the universality of the Bible as a major
source of themes for literature and the fine
arts.

SAGA AND STORY IN THE OLD TESTA-
MENT (Motion Picture—16mm—Sound).
EBEC, 1973 (Bible as Literature). 16mm col-
or, sound film, 27 min. #47794 ($360). Gr. 7–
A. EFL (1976), EGT (1975), LFR (1974). 221

Treating the Bible as a collection of literary
masterpieces, this film skillfully weaves together paintings, sculpture, music, and drama to reenact the stories of the Creation, the
Expulsion, Noah's Ark, David's Killing of
Goliath. The film discusses the influence of
the King James version of the Bible in the
development of the English language and literature.

BIBLE AS LITERATURE

HISTORY, POETRY AND DRAMA IN THE
OLD TESTAMENT (Motion Picture—
16mm—Sound). See Bible—Old Testament

SAGA AND STORY IN THE OLD TESTA-
MENT (Motion Picture—16mm—Sound).
See Bible—Old Testament

BICULTURALISM

FROM HOME TO SCHOOL (Filmstrip—
Sound). See Family life

THE MANY AMERICANS, UNIT ONE
(Filmstrip—Sound). Learning Corporation
of America, 1976. 4 color, sound filmstrips,
av. 150 fr.; 4 cassettes, 10–13 min.; Teacher's Guide. #76–730197 ($94). Gr. 4–8.
M&M (4/77), PRV (12/77), PVB (4/78).
301.45

Contents: 1. Geronimo Jones. 2. Felipa:
North of the Border. 3. Todd: Growing Up
in Appalachia. 4. Matthew Aliuk: Eskimo in
Two Worlds

These sets reveal the customs, attitudes, and
social problems of bicultural groups in
America. Each presents the moving story of
a conflict in the life of an ethnic minority
child who seeks to assimilate two worlds.
(Based on films.)

THE MANY AMERICANS, UNIT TWO
(Filmstrip—Sound). Learning Corporation
of America, 1976. 4 color, sound filmstrips,
av. 150 fr.; 4 cassettes, 10–13 min.; Teacher's Guide. #76–730198 ($94). Gr. 4–8.
M&M (4/77), PRV (12/77), PVB (4/78),
301.45

Contents: 1. Lee Suzuki: Home in Hawaii.
2. Miguel: Up from Puerto Rico. 3. Siu Mei
Wong: Who Shall I Be? 4. William: From
Georgia to Harlem.

Each filmstrip presents the story of conflict
of an ethnic minority child who seeks to assimilate two worlds. (Based on films.)

OUR LANGUAGE, OUR CULTURE, OUR-
SELVES (Filmstrip—Sound). See Family
life

PARENT-SCHOOL RELATIONSHIPS
(Filmstrip—Sound). See Family life

BICYCLES AND BICYCLING

BICYCLE SAFELY (Motion Picture—
16mm—Sound). FilmFair Communications,
1974. 16mm color, sound film, 12 min.
#D257 ($150). Gr. 3–12. BKL (6/74), LFR
(9/74), PRV (6/74). 614.8

This film shows the major causes of bicycle
accidents and establishes the safety practices that can avoid them.

BICYCLING ON THE SAFE SIDE (Motion
Picture—16mm—Sound). Ramsgate Films,
1974. 16mm color, sound film, 16 min.;
Study Guide. ($225). Gr. 4–A. BKL (10/15/
74), LGB (1974-75), PRV (2/75). 796.6

This film utilizes the popular ten speed bike
to illustrate principles of bicycle safety, as
well as rider comfort and efficiency.

BILINGUALISM. See Education, Bilingual

BILL OF RIGHTS. See U.S. Constitution—
Amendments

BIOGRAPHY

BIOGRAPHY: BACKGROUND FOR IN-
SPIRATION (Filmstrip—Sound). Library
Filmstrip Center, 1977. 1 color, sound film-
strip, 55 fr.; 1 disc or cassette, 15-1/2 min.
($26). Gr. 9–A. PRV (2/79). 028.7

A brief survey of the many books and sets of
books in which students can locate biographical information.

BIOGRAPHY—COLLECTIONS

BLACK AMERICAN ATHLETES (Kit). *See* Athletes

HISPANIC HEROES OF THE U.S.A. (Kit). EMC, 1976. 4 Paperback Books; 4 Read-along Cassettes; Teacher's Guide; Student Activities. #ELC-236000 ($55). Gr. 4–12. BKL (7/76). 920

Contents: Book I—1. Raul H. Castro: Adversity Is My Angel. 2. Tommy Nunez: NBA Ref. 3. Presenting: Vikki Carr. Book II—1. Henry B. Gonzalez: Greater Justice for All. 2. Trini Lopez: The Latin Sound. 3. Edward Roybal: Awaken the Sleeping Giant. Book III—1. Carmen Rose Maymi: To Serve American Women. 2. Robert Clemente: Death of a Proud Man. 3. Jose Feliciano: One Voice, One Guitar. Book IV—1. Tony Perez: The Silent Superstar. 2. Lee Trevino: Supermex. 3. Jim Plunkett: He Didn't Drop Out.

Inspiring biographies of twelve famous Americans of Hispanic heritage, grouped three to a book, representing the broad contribution of Spanish-speaking Americans to the culture of this country.

HOCKEY HEROES (Kit). *See* Athletes

LEADERS, DREAMERS AND HEROES (Cassette). Troll Associates, 1972. 7 cassettes, approx. 10 min. ea. ($38.50). Gr. 4–8. BKL (7/15/74). 920

Contents: 1. Daniel Boone. 2. Davy Crockett. 3. Francis Marion. 4. Lewis and Clark. 5. Kit Carson. 6. Buffalo Bill. 7. John Paul Jones.

A series of brief biographies of American heroes who emerged in the late Colonial period and during the Westward expansion.

SPORTS CLOSE-UPS 1 (Kit). *See* Athletes

SPORTS CLOSE-UPS 3 (Kit). *See* Athletes

SPORTS CLOSE-UPS 4 (Kit). *See* Athletes

BIOLOGY

THE MIRACLE OF ALL LIFE: VALUES IN BIOLOGY (Slides/Cassettes). Science and Mankind, 1975. 160 slides; 2 cassettes or 2 discs ($68.50). Gr. 9–12. SCT (11/76). 574.5

This set describes the interrelatedness of all life on earth. Its primary objective is to develop appreciation for all life forms and their ability to adapt to different environments.

See also Anatomy, Comparative; Cells; Embryology; Fresh-water biology; Marine biology; Natural history; Reproduction

BIOLOGY—PERIODICITY

BIOLOGICAL RHYTHMS: STUDIES IN CHRONOBIOLOGY (Motion Picture—16mm—Sound). EBEC, 1977. 16mm color, sound film, 22 min. #3568 ($330). Gr. 7–12.

BKL (1977), LFR (1978), PRV (1979). 574.1

This film documents recent experiments designed to discover the source of internal rhythms, how they affect behavior and efficiency in plants, animals, and humans, and how they can be changed. Experimental subjects include one-celled Gonyaulax, bean plants, fiddler crabs, homing pigeons, and human beings.

BIOLOGY CLASSIFICATION. *See* Classification—Biological

BIORHYTHMS. *See* Biology—Periodicity

BIRDS

ANTARCTIC PENGUINS (Film Loop—8mm—Silent). *See* Penguins

BIRDS FEEDING THEIR YOUNG (Film Loop—8mm—Silent). Walt Disney Educational Media, 1966. 8mm color, silent film loop, approx. 4 min. #62–5918L ($30). Gr. K–12. MDU (1974). 598.2

The grosbeak, woodpecker, hummingbird, and waxwing are closely studied as they gather food and feed their young. Close-up shots show parents feeding worms and caterpillars to their young.

CHUPAROSAS: THE HUMMINGBIRD TWINS (Filmstrip—Sound). *See* Hummingbirds

ENDANGERED SPECIES: BIRDS (Study Print). *See* Rare birds

INVESTIGATING BIRDS (Filmstrip—Sound). Coronet Instructional Media, 1975. 6 color, sound filmstrips, av. 59 fr.; 3 discs or 6 cassettes, av. 11 min. #S255 Discs ($95), #M255 Cassettes ($95). Gr. 4–9. BKL (12/75), INS (1975), PVB (1975–76). 598.2

Contents: 1. Birds of the Forest. 2. Birds of the Meadow. 3. Birds of the Sea. 4. Birds of the Southern Swamps and Marshes. 5. Birds of Ponds and Lakes. 6. Birds Near Your Home.

The adaptations, characteristics, and behavior of birds are presented. Authentic bird calls accompany the natural photographs of the birds.

PELICANS (Film Loop—8mm—Silent). *See* Pelicans

See also Names of birds, e.g., Eagles, Hummingbirds, etc., and classes of birds, e.g., Birds of prey, Water birds, etc.

BIRDS—FICTION

BIRDS OF A FEATHER (Motion Picture—16mm—Sound). Bosustow Productions, 1972. 16mm color, sound, animated film, 6-1/2 mins.; Reader's Guide. ($125). Gr. 2–A. CGE (1973), LBJ (1973), NEF (1973). Fic

A plain little bird is attracted by plumes in the hat of a passer-by. He plucks them and proudly attaches them, greatly enhancing his natural plumage. A bird catcher swoops him up and places him in a cage with similarly adorned exotic birds. They know he is an imposter and remove his glorious feathers. Seeing he is just an ordinary little fellow, the bird catcher lets him go. But once he has been reunited with his flock, he immediately begins the next scheme of mischievous fun. Based on the book by Willi Baum.

THE HOARDER (Motion Picture—16mm—Sound). *See* Fables

BIRDS—MIGRATION

FLIGHT FOR SURVIVAL (THE MIGRATION OF BIRDS) (Motion Picture—16mm—Sound). EBEC, 1975. 16mm color, sound film, 14 min. #3509 ($205). Gr. 5–9. LFR (1977), NEF (1978), PRV (1978). 598.2
This film explores the navigational and homing instincts of various migratory bird species. Viewers learn about different techniques used to map bird migration routes, and they see recent experiments aimed at discovering the means by which birds navigate over very long distances.

BIRDS OF PREY

BIRD OF FREEDOM (Motion Picture—16mm—Sound). *See* Eagles

BIRDS OF PREY (Filmstrip—Sound). Coronet Instructional Media, 1976. 4 color, sound filmstrips, 47–55 fr.; 4 cassettes or discs, 10–12 min.; 1 Guide. #M257 Cassettes ($65), #S257 Discs ($65). Gr. 4–10. BKL (5/15/77). 598.9
Contents: 1. Their Characteristics. 2. Hawks, Eagles, and Vultures. 3. Owls. 4. Their Survival.
First-hand observation of the physical characteristics and behavior of the great birds and the study of the destructive effects of human behavior on bird species.

BISON

BISON AND THEIR YOUNG (Film Loop—8mm—Silent). Walt Disney Educational Media, 1966. 8mm color, silent film loop, approx. 4 min. #62–5152L ($30). Gr. K–12. MDU (1974). 599.7
Bison and their young are seen feeding on a plain. Adults are shown as well as newborn calves trying to stand and nurse.

BUFFALO: AN ECOLOGICAL SUCCESS STORY (Motion Picture—16mm—Sound). EBEC, 1972. 16mm color, sound film, 13 min. #3148 ($185). Gr. 7–12. AAS (1975), BKL (1973), EFL (12/19/74). 599.7
This film documents the story of the buffalo and the efforts that have been made to preserve and extend this almost extinct species. It recaptures a vision of the vast herds that once roamed North America; of how Indian tribes depended upon the animal for food, clothing, shelter, weapons, and fuel; of how settlers, railroads, and professional hunters nearly decimated this important animal resource.

BLACK ARTISTS

RIGHT ON/BE FREE (Motion Picture—16mm—Sound). FilmFair Communications, 1971. 16mm color, sound film, 15 min. ($200). Gr. 5–A. LFR (9/19/71), PRV (11/19/73). 709.2
This film shows the energy, vitality, and strong sense of identity of the Black American artist. Rather than concentrating on one art form, it expresses the black mood and temperament in music, poetry, painting, and dance.

BLACKS

SOUL FOOD (Kit). *See* Cookery, Black

BLACKS—BIOGRAPHY

BEAH RICHARDS: A BLACK WOMAN SPEAKS (Videocassette). *See* Richards, Beah

LEE BALTIMORE: NINETY NINE YEARS (Motion Picture—16mm—Sound). *See* Baltimore, Lee

LORRAINE HANSBERRY (Motion Picture—16mm—Sound). *See* Hansberry, Lorraine

MARTIN LUTHER KING: THE MAN AND HIS MEANING (Filmstrip—Sound). *See* King, Martin Luther

ROOTS WITH ALEX HALEY (Cassette). *See* Haley, Alex

BLACKS—FICTION

PHILIP HALL LIKES ME: I RECKON MAYBE (Phonodisc or Cassette). *See* Friendship—Fiction

BLACKS—HISTORY

I SHALL MOULDER BEFORE I SHALL BE TAKEN (Motion Picture—16mm—Sound). *See* Surinam—History

BLIND

ELIZABETH (Motion Picture—16mm—Sound). *See* Handicapped children

BLOCK, HERBERT

SATIRICAL JOURNALISM (Cassette). *See* Cartoons and caricatures

BLOOD—CIRCULATION

THE HEART AND THE CIRCULATORY
SYSTEM (Motion Picture—16mm—Sound).
Media Four Productions/EBEC, 1974. 16mm
color, sound film, 16 min. #3349 ($220). Gr.
7-9. LFR (1975), PRV (1975), SCT (1976).
612.1

Traces the scientific discoveries that have
led to our current understanding of the circulatory system. The structures and processes of this vital system are shown through
the use of animation, cinemicrography, cine-
angiography, and recently developed techniques for studying the diffusion of
substances from the bloodstream to the
cells. The film presents some of the causes of
heart disease, and shows how young people
can reduce the risk of developing this disease in later life.

BLUE RIDGE MOUNTAINS

THE BLUE RIDGE: AMERICA'S FIRST
FRONTIER (Filmstrip—Sound). Lyceum/
Mook & Blanchard, 1973 (Lyceum). 1 color,
sound filmstrip, 64 f.; 1 cassette or disc, 10³⁄₄
min. #LT35273C Cassette ($25),
#LY35273R Disc ($19.50). Gr. 4-A. BKL (7/
1/73), INS (12/73), PRV (12/75). 917.55

Dramatic photographs and almost poetic
commentary describes the terrains, plants,
and wildlife of the Blue Ridge area. It traces
the historical exploration and development
of the region and shows the use of the land
and development of crafts using the natural
environment.

BLUE RIDGE MOUNTAINS—BIOGRAPHY

AUNT ARIE (Motion Picture—16mm—
Sound). *See* Carpenter, Aunt Arie

BLUES (SONGS, ETC.)

GOT SOMETHING TO TELL YOU (Slides/
Cassettes). Center for Southern Folklore,
1977. 130 color, slides in Carousel tray, 1
cassette, av. 15 min.; 1 transcript. ($120).
Gr. 10-A. BKL (9/1/78). 784.7

B. B. King provides background music and
narrates this overview of the development of
the blues.

BOATBUILDING

MALCOLM BREWER: BOAT BUILDER
(Motion Picture—16mm—Sound). *See*
Brewer, Malcolm

BODY TEMPERATURE

REGULATING BODY TEMPERATURE
(SECOND EDITION) (Motion Picture—
16mm—Sound). EBEC, 1972. 16 mm color,
sound film, 22 min. #3155 ($290). Gr. 9-C.
FLN (1973), PRV (1973), SCT (1973). 574.19

Vertebrates maintain a stable body-temper-
ature despite considerable surrounding temperature changes. The film investigates this
phenomenon in experiments with both human and animal subjects. An unusual experiment demonstrates that the hypothalamus is
the seat of the body's temperature control
center.

BONES

AN INSIDE LOOK AT ANIMALS (Study
Print). *See* Anatomy, Comparative

BOOKS

HOW A PICTURE BOOK IS MADE (Filmstrip—Sound). Weston Woods Studios,
1976. 1 color, sound filmstrip, 66 fr.; 1 cassette, 9-1/2 min.; 1 Guide. #SF451C ($25).
Gr. K-A. BKL (1/15/77). 686

Author-illustrator Steven Kellogg describes
the steps involved in creating "The Island of
the Skog," a picture book relating the adventures of a group of mice on a far away
island. He begins with the idea and follows
the book's development through writing and
rewriting to reworking after consulting with
his publisher's editor. He explains the creation of the illustrations, the characters and
the dummy for the book. He follows the
steps to the final printing and binding.

BOOKS—REVIEWS

THE HOBBIT (Kit). *See* Fantasy—Fiction

BOOKS AND READING—BEST BOOKS

I COULDN'T PUT IT DOWN: HOOKED ON
READING, SET TWO (Kit). Center for Humanities,.1978. 138 color slides in 2 Carousel
trays; 2 discs or 4 cassettes, 27 min.; 4 Paperback Books; 4 Activity Books; Guide.
#320 ($139.50). Gr. 5-12. BKL (9/1/78).
028.52

Contents: 1. The Red Badge of Courage. 2.
Treasure Island. 3. The Call of the Wild. 4.
The Adventures of Tom Sawyer.

Four classic stories of adventure are presented in comic-style format, permitting the
use of selected vocabulary to advance plot-
lines. Each story is stopped at a crucial moment, motivating students to read the work
itself.

I COULDN'T PUT IT DOWN: HOOKED ON
READING, SET THREE (Kit). Center for
Humanities, 1978. 154 color slides in 2
Carousel trays; 4 cassettes or 2 discs, 30
min.; 4 activity books; 4 paperback books;
Guide. #335 ($139.50). Gr. 5-12. BKL
(139.50). 028.52

Contents: 1. Frankenstein. 2. The Strange
Case of Dr. Jekyll and Mr. Hyde. 3. The Invisible Man. 4. Dracula.

Another collection of classic stories of suspense presented in comic style format, permitting the use of selected vocabulary to advance the plotline. Each story is stopped at a crucial moment, motivating students to read the work itself.

BOTANY

THE PLANTS, SERIES TWO (Filmstrip—Sound). EBEC, 1976. 5 color, sound filmstrips, 79 fr. ea.; 5 cassettes or discs, 15 min. ea.; #6933K Cassettes ($72.50), #6933 Discs ($72.50). Gr. 9–12. BKL (7/78), LFR (4/78). 581

Contents: 1. Mosses and Liverworts. 2. Ferns and Their Allies. 3. Gymnosperms. 4. Angiosperms: Diversity and Reproduction. 5. Angiosperms: Roots, Stems, and Leaves.

The wide variety of plant forms comprise more than 250,000 species. This series highlights the diversity of plant reproductive methods from simple spores, to "naked seeds," to sophisticated fertilization strategies. Vascular systems are examined as they evolved from simple "canals" into highly specialized tissues and structures.

BOTANY—EXPERIMENTS

ISOLATION AND FUNCTION OF AUXIN (Film Loop—8mm—Silent). *See* Phototropism

MECHANICS OF PHOTOTROPISM (Film Loop—8mm—Captioned). *See* Phototropism

THE NATURE OF PHOTOTROPISM (Film Loop—8mm—Captioned). *See* Phototropism

BOXING

BOXING (Kit). Educational Activities, 1973. 1 color, sound filmstrip, 32 fr.; 1 disc or cassette, 12 min.; 10 student books; Teacher's Guide. #450 Disc ($35), #450 Cassette ($36). Gr. 4–9. PRV (11/74), PVB (5/75). 796.83
A professional tells and illustrates the whys and wherefores of boxing. A mix of "vocabulary" extension, picture-reading practice, and improvement of comphrehension as students are motivated by this high-interest, low vocabulary kit. Action photography and interesting, informative narration add to its value.

GOODNIGHT MISS ANN (Motion Picture—16mm—sound). Pyramid Films, 1978. 16mm color, sound film, 28 min. #1407 ($375). Gr. 9–A. BKL (1/15/79), CGE (1978). 796.83
Boxing has always been a way up for the disadvantaged. A stream of hopeful boxers come to the Main Stream Gym in downtown Los Angeles to try to win. Old fighters turned trainers, cynical officials, dedicated managers, and beginners all provide a penetrating commentary on boxing.

BOXING—FICTION

THE CONTENDER BY ROBERT LIPSYTE (Filmstrip—Sound). Current Affairs Films, 1978. 1 color, sound filmstrip; 1 cassette; Teacher's edition of the book and testing materials; Discussion Guide. #LP–622 ($30). Gr. 6–12. BKL (11/1/78). Fic
The hero of this contemporary juvenile novel is a black youth from an urban ghetto who tries to be "somebody special" by becoming a boxer. He succeeds, though not exactly in the way he expected.

BOYS

THERE'S A NEW YOU COMIN' FOR BOYS (Filmstrip—Sound). Marshfilm, 1972 (The Human Growth Series). 1 color, sound filmstrip, 50 fr.; 1 disc or cassette, 15 min.; Teacher's Guide. #1113 ($21). Gr. 5–8. AIF (1973), BKL (10/15/73), PRV (12/73). 613.6
This filmstrip offers a look at what constitutes both physical and emotional maturity. Health and hygiene are stressed, including a basic study of male anatomy.

BRADBURY, RAY

A DISCUSSION OF THE FALL OF THE HOUSE OF USHER (Motion Picture—16mm—Sound). *See* Motion pictures—History and criticism

FAHRENHEIT 451 (Kit). *See* Science fiction—History and criticism

THE FALL OF THE HOUSE OF USHER (Motion Picture—16mm—Sound). *See* American literature—Study and teaching

THE FOGHORN BY RAY BRADBURY (Filmstrip—Sound). *See* Science fiction

THE ILLUSTRATED MAN (Filmstrip—Sound). *See* Science fiction

THE MARTIAN CHRONICLES (Phonodiscs). *See* Science fiction

THE VELDT (Filmstrip—Sound). *See* Science fiction

BRADY, MATHEW B.

MATHEW BRADY: PHOTOGRAPHER OF AN ERA (Motion Picture—16mm—Sound). *See* Photography—History

BRAIN

EXPLORING THE BRAIN: THE NEWEST FRONTIER (Filmstrip—Sound). Human Relations Media, 1978. 5 color, sound filmstrips, 45–94 fr.; 5 cassettes or discs, 8–18 min.; Teacher's Guide. #625 ($140). Gr.

BRAIN (cont.)

8–C. BKL (7/1/78), PRV (1/79). 612.8
Contents: 1. Geography of the Brain. 2.
Memory. 3. Split-Brain Research. 4. Electrical Stimulation of the Brain. 5. Biofeedback.
Probes the latest scientific discoveries on the
nature and function of the human brain. Explains how the brain controls thinking, learning, remembering, forgetting, creativity,
language, imagination, emotions, and personality. Includes landmark experiments
and examines ongoing research into some of
the brain's continuing mysteries.

BRASILIA

TOMORROW'S CITIES TODAY (Filmstrip—
Sound). *See* Cities and towns

BRAZIL

BRAZIL (Filmstrip—Sound). EBEC, 1973. 6
color, sound filmstrips, av. 92 fr. ea.; 6 discs
or cassettes, 12 min. ea.; Teacher's Guide.
Discs #6466 ($86.95), Cassettes #6466K
($86.95). Gr. 4–8. CPR (4/74), INS (2/75),
PRV (11/74–75). 918.1
Contents: 1. The People of Brazil. 2. The
Land of Brazil. 3. The Amazon Basin. 4.
Flight to the Cities. 5. Coffee Fazenda in
Brazil. 6. Two Brazilian Cities (Salvador and
Sao Paulo).
Viewers are transported to semi-desert
backlands and prosperous coffee fazendas,
to Amazon fishing villages, and highly industrialized cities. They meet people of every
race and discover that in Brazil racial democracy is a reality. There are intriguing
glimpses of festivals and soccer games and
searching looks at the problems that this advancing nation has yet to solve.

BREWER, MALCOLM

MALCOLM BREWER: BOAT BUILDER
(Motion Picture—16mm—Sound). Oyddsey
Productions/EBEC, 1976. 16mm color,
sound film, 18 min. #3492 ($240). Gr. 7–A.
BKL (1976), EFL (1977), EGT (1977). 921
Malcolm Brewer, 77 years old, still practices
the craft of building boats by hand in his
shop on the coast of Maine. As he works, he
tells of a life built on self-reliance and joy in
hard work. As he talks, Mr. Brewer, son of a
lobster fisherman, recreates the generations
of stolid New Englanders.

BRONTE (FAMILY)

GREAT WRITERS OF THE BRITISH
ISLES, SET I (Filmstrip—Sound). *See* Authors, English

NO COWARD SOUL: A PORTRAIT OF
THE BRONTES (Cassette). *See* English
drama

BRONZES

THE BRONZE ZOO (Motion Picture—
16mm—Sound). Texture Films, 1974. 16mm
color, sound film, 16 min.; Guide. ($225). Gr.
7–A. AFF (1974), ALA (1975), BKL (11/15/
74). 731
The film shows Shay Rieger sketching a yak,
constructing the armature of chicken wire
and plumbing pipes, molding with plaster of
paris and finally watching foundry workers
cast it in bronze by the ''lost wax'' process.
The narration describing what is taking place
is done by Shay Rieger.

BROUN, HEYWOOD HALE

THE AGE OF SENSATION (Cassette). *See*
Youth—Attitudes

AN ALMANAC OF WORDS AT PLAY (Cassette). *See* English language—Usage

CONVERSATION WITH LOREN EISELEY
(Cassette). *See* Eiseley, Loren

THE FOUNDING FATHERS IN PERSON
(Cassette). *See* U.S.—History—Biography

THE GOOD OLD DAYS—THEY WERE
TERRIBLE (Cassette). *See* History—Philosophy

THE LONG THIRST: PROHIBITION IN
AMERICA (Cassette). *See* Prohibition

PRESIDENTIAL STYLES: SOME GIANTS
AND A PYGMY (Cassette). *See* Presidents—U.S.

ROOTS WITH ALEX HALEY (Cassette).
See Haley, Alex

SATIRICAL JOURNALISM (Cassette). *See*
Cartoons and caricatures

BROWER, DAVID

TALKING WITH THOREAU (Motion Picture—16mm—Sound). *See* Philosophy,
American

BROWN, MARCIA

THE CRYSTAL CAVERN (Filmstrip—
Sound). *See* Caves

BRUBECK, DAVE

ART OF DAVE BRUBECK: THE FAN-
TASY YEARS (Phonodiscs). *See* Jazz music

BUFFALO. *See* Bison

BUNYAN, PAUL

JOHNNY APPLESEED AND PAUL BUN-
YAN (Phonodisc or Cassette). *See* Folklore—U.S.

BURNFORD, SHEILA

THE INCREDIBLE JOURNEY BY SHEILA BURNFORD (Filmstrip—Sound). *See* Animals—Fiction

TWO FOR ADVENTURE (Kit). *See* Animals—Fiction

BUSINESS

CAREERS IN BUSINESS ADMINISTRATION (Filmstrip—Sound). Associated Press/Pathescope Educational Media, 1973 (Careers In). 2 color, sound filmstrips, 78–80 fr.; 2 cassettes, 11–13 min.; Teacher's Manual. #705 ($50). Gr. 9–12. PRV (1/19/75). 658.0023
Contents: Part 1. Part 2.
Describes the types of jobs available by using interviews to determine necessary training, job satisfactions, and opportunities.

HAS BIG BUSINESS GOTTEN TOO BIG? (Filmstrip—Sound). *See* Economics

JERRY'S RESTAURANT (Motion Picture—16mm—Sound). *See* Individuality

PARTNERS, UNIT 3 (Motion Picture—16mm—Sound). *See* Occupations

BUSINESS—LAW AND LEGISLATION

THERE IS A LAW AGAINST IT (Motion Picture—16mm—Sound). *See* Consumer education

THIS IS FRAUD! (Motion Picture—16mm—Sound). *See* Consumer education

BUTLER, ROBERT

GROWING OLD WITH GRACE (Cassette). *See* Old age

BUTTERFLIES

BUTTERFLY: THE MONARCH'S LIFE CYCLE (Motion Picture—16mm—Sound). International Film Bureau, 1975. 16mm color, sound film, 10½ min.; Teacher's Guide. #1IFB 652 ($145). Gr. 4–8. PRV (11/76). 595.78
Good photographic closeups of the egg, the young caterpillar emerging from the egg and passing through the repeated stages of eating, growing and molting; building the button with which to attach itself for the pupa stage: shedding the last skin and changing within the chrysalis; and finally the adult emerging to dry its wings before flight.

DON'T (Motion Picture—16mm—Sound). Phoenix Films, 1974. 16mm color, sound film, 19 min. #2072 ($275). Gr. 7–A. AAW (1975), FLI (2/76). 595.78
The lyric passage of a Monarch butterfly, beginning with its birth, through its delicate

metamorphosis from caterpillar to butterfly and on its journey from country to city.

LIFE CYCLE OF COMMON ANIMALS, GROUP 3 (Filmstrip—Sound). Imperial Educational Resources, 1973. 2 color, sound filmstrips, 36–39 fr.; 2 cassettes or discs, 10–12 min. #3RG 40300 Discs ($30), #3KG 40300 Cassettes ($33). Gr. 4–9. PRV (2/74), PVB (5/74), STE (4/5/74). 595.78
Contents: 1. The Monarch Butterfly, Part One. 2. The Monarch Butterfly, Part Two.
Portrays the significant events in the life history of the Monarch butterfly.

METAMORPHOSIS (Motion Picture—16mm—Sound). Texture Films, 1977. 16mm color, sound film, 10 min. ($175). Gr. 2–8. BKL (1/15/78), CGE (1977). 595.78
A young people's science film about a young girl watching the wonderful transformation of a caterpillar into a butterfly.

MONARCH: STORY OF A BUTTERFLY (Film Loop—8mm—Silent). Troll Associates. 3 8mm color, silent film loops, approx. 4 min. ($74.85). Gr. 1–8. AUR (1978). 595.78
Contents: 1. Monarch: Eggs to Catepillar. 2. Monarch: Caterpillar to Pupa. 3. Chrysalis to Butterfly.
These loops document in close-up detail the life cycle of the Monarch butterfly.

CACTUS

THE CACTUS: ADAPTATIONS FOR SURVIVAL (Motion Picture—16mm—Sound). EBEC, 1973. 16mm color, sound film, 14 min. #3195 ($185). Gr. 7–A. BKL (1974), LFR (1974), NST (1975). 583.47
How has the cactus been able to adapt so successfully to harsh environments? The answer is revealed through this close-up study of the physical characteristics and life cycles of this unique family of plants. Laboratory experiments demonstrate the plant's water-conserving characteristic . . . while ultra close-ups of spines, areoles, and flowers emphasize the startling beauty of the cactus. A dramatization of the spread of the prickly pear in Australia points out the ecological dangers when man introduces plants into areas foreign to that of their origin.

CALCULATING MACHINES

COMPUTERS: FROM PEBBLES TO PROGRAMS (Filmstrip—Sound). *See* Computers

INTRODUCING THE ELECTRONIC CALCULATOR (Filmstrip—Sound). BFA Educational Media, 1975. 2 color, sound filmstrips; 2 cassettes or discs; Teacher's Guide. #VGB000 Cassettes ($36), #VGA000 Discs ($36). Gr. 4–8. MTE (5/77). 681.14

CALCULATING MACHINES (cont.)

Contents: 1. A Boxful of Magic. 2. Electronic Wizard.

An overview to help students understand the capabilities of the calculator and to teach them to use it to solve mathematical problems.

CALCULUS

INTRODUCTION TO CALCULUS: SEQUENCES AND CONVERGENCE (Filmstrip—Sound). EBEC, 1971. 6 color, sound filmstrips, av. 57 fr. ea.; 6 cassettes or discs, 17 min. ea.; Guide. #6444K Cassettes ($86.95), #6444 Discs ($86.95). Gr. 10–C. BKL (2/73), MTE (2/73). 515

Contents: 1. Imagining Sequences. 2. Getting Down to Terms. 3. The Tail End of a Sequence. 4. Convergence to Zero. 5. Limits Different from Zero. 6. Bounded Increasing Sequences.

The idea of infinite sequence is defined and clarified in this set. A visual point of view is emphasized throughout, supported by graphical representation of sequences. Students learn to discern and describe the ultimate behavior of terms in sequences, leading naturally to the idea of convergence. Criteria for convergence are developed and applied to specific sequences.

CALDECOTT MEDAL BOOKS

ARROW TO THE SUN (Motion Picture—16mm—Sound). *See* Indians of North America—Legends

GAIL E. HALEY: WOOD AND LINOLEUM ILLUSTRATION (Filmstrip—Sound). *See* Haley, Gail E.

CALIFORNIA—GOLD DISCOVERIES

GOLD TO BUILD A NATION: THE UNITED STATES BEFORE THE CIVIL WAR (Filmstrip—Sound). Multi-Media Prods., 1977. 3 color, sound filmstrips, 76–84 fr.; 3 cassettes, 14–15 min.; Teacher's Manual and Script. #7218C ($36). Gr. 9–12. BKL (4/1/78). 929.4

Traces the history of mining development in the West, life in the California mines during the Gold Rush, and the great impact California's gold had upon the nation's economic, social, and political life, particularly in the decade before the Civil War.

CAMPING

BACKPACKING BASICS (Slides/Cassettes). *See* Backpacking

BACKPACKING: REVISED (Filmstrip—Sound). *See* Backpacking

CAMPING (Filmstrip—Sound). Great American Film Factory, 1975. 1 color, sound filmstrip, 72 fr.; 1 cassette, 19$^1/_2$ min. #FS–3 ($35). Gr. 5–A. BKL (2/1/76), PRV (1/77), PVB (4/77). 796.54

Supplies information about camping emphasizing the idea of taking nothing from the wilderness in order to provide an unspoiled site for those who follow. Comparative data about site choices, cooking, and tents, the set provides rules and helpful hints.

FOUR WHEELS (Filmstrip—Sound). Great American Film Factory, 1977 (Outdoor Education). 1 color, sound filmstrip, 95 fr.; 1 cassette, 17 min. #FS–8 ($35). Gr. 5–12. PRV (2/79). 796.54

A brief history of the desert, with a tribute to its beauty and mystery, is shown. Two campers are getting their trucks in order for a "four wheeling" trip. They drive across the desert and set up camp at night.

WEATHER IN THE WILDERNESS (Filmstrip—Sound). *See* Survival

CANADA

CANADA (Filmstrip—Sound). National Geographic, 1976. 5 color, sound filmstrips; 5 cassettes or discs, 12–14 min. (also available in French-English). #03782 Discs ($74.50), #03783 Cassettes ($74.50). Gr. 5–12. M&M (4/77). 917.1

Contents: 1. The Atlantic Provinces. 2. Quebec. 3. Ontario. 4. The Prairie Provinces and Northwest Territories. 5. British Columbia and the Yukon Territory.

Canada's national life and breathtaking geography are presented in this set. Its abundance of resources and industries are shown.

CANADA: LAND OF NEW WEALTH (Filmstrip—Sound). EBEC, 1976. 5 color, sound filmstrips, av. 77 fr. ea.; 5 discs or cassettes, 10 min. ea.; Teacher's Guide. Discs #6946 ($72.50), Cassettes #6946K ($72.50). Gr. 4–12. ESL (1977), MTD (4/77). 917.1

Contents: 1. The Physical Base. 2. Natural Resources. 3. Food Production. 4. Industry. 5. Cities.

Following World War II, Canada's history has been one of sudden and dramatic change brought about by rapid industrialization. This series examines that transition and the factors responsible.

THE CANADIANS (Filmstrip—Sound). EBEC, 1976. 5 color, sound filmstrips, av. 77 fr. ea.; 5 discs or cassettes, 11 min. ea.; Teacher's Guide. Discs #6954 ($72.50), Cassettes #6954K ($72.50). Gr. 4–9. BKL (6/15/77), MTD (4/77). 917.1

Contents: 1. The Two Founding Nations. 2. Canadian, Canadien: A Portrait of Two Families. 3. The Many Canadians. 4. Canada and the World. 5. Unity in Diversity.

Although Canada has long been a land of many cultures, its recent industrial progress has added to the problems which accompany such ethnic diversity. This series traces the early settlement patterns by the French and English, portrays some current manifestations of the dual heritage, and examines the continuing influence of the original natives— the Indians and Eskimos.

DOCUMENTARY ON CANADA (Filmstrip—Sound). Society For Visual Education, 1973. 6 color, sound filmstrips, av. 89 fr.; 3 discs or cassettes, av. 19 min.; 1 Guide. #274-SAR Discs ($83), #274-SATC Cassette ($83). Gr. K–A. BKL (4/73). 917.1

Contents: 1. The Fishermen of Nova Scotia. 2. Quebec City and the French Canadians. 3. Pinawa, Manitoba, Suburb for Atomic Energy. 4. Yellowknife: Capitol. 5. Canada's Arctic Settlements. 6. The Port of Vancouver: Canada's Pacific Gateway.

This set examines a cross section of the country of Canada in the light of modern problems. Each filmstrip describes the unique background of a particular region and focuses on people and their lives.

A PORTRAIT OF CANADA (Filmstrip—Sound). Educational Enrichment Materials, 1977. 6 color, sound filmstrips; 6 cassettes; 2 Wall Posters; Teacher's Guide. #51038 ($125). Gr. 6–12. BKL (3/15/78). 917.1

Contents: 1. The Land and Its Resources. 2. The History of a Nation. 3. The Queen's Choice. 4. Multi-Cultures—Multi-Heritages. 5. Canada's Dilemma: Economics Nationalism. 6. Canadian Industry: An Economics Success Story.

Designed to help students understand and appreciate Canada's uniqueness, as well as its importance to the United States both politically and economically. Explores the special relationship that the United States and Canada continue to share as neighbors, allies, and friends. Examines the history, physical resources, industry, culture, and present-day challenges that face our neighbor to the north.

CANADA—HISTORY

GREAT GRANDMOTHER (Motion Picture—16mm—Sound). Filmwest Assoc., National Film Board of Canada/New Day Films, 1976. 16mm color, sound film, 28 min. ($375). Gr. 9–A. BKL (1/15/79). 971

A collection of archival stills, dramatized re-enactments, interview script reading of primary sources, all help document the toil of the women of this period and their contribution to the founding of a nation. The bleak homesteads on the Canadian prairie remind us of those early women settlers and their courage, loneliness, and endless drudgery as well as their strength.

PAUL KANE GOES WEST (Motion Picture—16mm—Sound). *See* Indians of North America—Canada

CANADIAN FICTION

GERTRUDE THE GOVERNESS AND OTHER WORKS (Phonodisc or Cassette). Caedmon Records, 1977. 1 cassette or disc, av. 60 min., with notes. #CDL51559 Cassettes ($7.98), #TC1559 Discs ($7.98). Gr. 10–A. BKL (1/15/79). Fic.

Canadian humorist Stephen Leacock is known for his satire, irony, and just plain fun. Actor Christopher Plummer reads his works effectively.

CANCER

CANCER (Filmstrip—Sound). Educational Dimensions Group, 1978. 2 color, sound filmstrips, 60 fr. ea.; 2 cassettes, 14 min. ea.; 1 Guide. #1216 ($45). Gr. 9–A. BKL (10/1/78). 616.9

Contents: 1. What Is Cancer? 2. Diagnosis and treatment.

This set examines the group of diseases known as cancer through photomicrography of normal and cancerous cells, simple diagrams, and photos of research and medical processes. Technical terms, such as metastasis, are defined. Emphasis is put on early diagnosis and alerting the student to the seven warning signs and the tests and examinations which make this early diagnosis possible.

CANDIDE

VOLTAIRE PRESENTS CANDIDE: AN INTRODUCTION TO THE AGE OF ENLIGHTENMENT (Motion Picture—16mm—Sound). *See* French fiction

CAPITAL PUNISHMENT

CHANGING VIEWS ON CAPITAL PUNISHMENT (Filmstrip—Sound). New York Times Filmstrips/Educational Enrichment Materials, 1977. 1 color, sound filmstrip, 80 fr.; 1 cassette, 15 min.; Teacher's Guide. #32099 ($25). Gr. 7–12. BKL (7/15/78), PRV (12/19/78). 179

Capital punishment is presented in a straightforward manner. Archival paintings and photographs of prison facilities provide background while the narration describes capital punishment's historical basis throughout the world.

CAPITALS (CITIES)

THE IDEOLOGICAL CAPITALS OF THE WORLD (Filmstrip—Sound). *See* Cities and towns

CAPITALS (CITIES) (cont.)

THE ISLAND NATION CAPITALS (Filmstrip—Sound). *See* Cities and towns

TOMORROW'S CITIES TODAY (Filmstrip—Sound). *See* Cities and towns

CAREERS. *See* Occupations; Vocational guidance; Work

CARPENTER, AUNT ARIE

AUNT ARIE (Motion Picture—16mm—Sound). EBEC, 1975 (The American Character). 16mm color, sound film, 18 min. #3418 ($240). Gr. 7–A. CFF (1975), EGT (1976), M&M (1977). 921

This film pays tribute to an American cultural region and documents the life of a passing generation. Spry and independent Aunt Arie Carpenter, who is nearly 87, lives in the Blue Ridge Mountains of North Carolina. She tells of her childhood, her marriage, and how she helped her husband build their house and farm the red clay soil. Aunt Arie is a living history as she tells her stories—bequeathing to the present generation an oral heritage rich in folkways, folktales, and hard-won wisdom.

CARRADINE, JOHN

THE LEGEND OF SLEEPY HOLLOW (Motion Picture—16mm—Sound). *See* Legends—U.S.

CARROLL, LEWIS

ALICE IN WONDERLAND (Kit). *See* Fantasy—Fiction

EPISODES FROM FAMOUS STORIES (Filmstrip—Sound). *See* Literature—Collections

THROUGH THE LOOKING GLASS (Phonodiscs). *See* Fantasy—Fiction

CARTOONS AND CARICATURES

B. KLIBAN (Cassette). *See* Kliban, B.

BARTLEBY (Motion Picture—16mm—Sound). *See* American fiction

SATIRICAL JOURNALISM (Cassette). Cinema Sound/Jeffrey Norton Publishers, 1975 (Avid Reader). 1 cassette, approx. 55 min. #40090 ($11.95). Gr. 10–A. BKL (7/15/77). 070.412

Moderator Heywood Hale Broun interviews political cartoonists Herbert Block and David Levine in this recording of a radio broadcast. From the discussion, one can trace the history of cartooning, from Daumier through the work of Thomas Nast to the daily cartoons of Herb Block, and sense the transition in cartooning style, from illustrated editorials to the present.

CATHER, WILLA

EPISODES FROM CLASSIC NOVELS (Filmstrip—Sound). *See* Literature—Collections

WOMEN WRITERS: VOICES OF DISSENT (Filmstrip—Sound). *See* Women authors

CATHERINE DE'MEDICI

CATHERINE DE'MEDICI (Cassette). Ivan Berg Associates/Jeffrey Norton Publishers, 1977 (History Makers). 1 cassette, 59 min. #41005 ($11.95). Gr. 10–A. BKL (1/15/78). 921

This biography of Catherine de'Medici presents the intrigues of history through the interpretations of letters and documents. Interspersed with the narration are brief dramatizations of important events in which she played a major part in the 16th-century European scene.

CATLIN, GEORGE

AMERICAN CIVILIZATION: 1783–1840 (Filmstrip—Sound). *See* Art, American

CATS

WHAT IS A CAT? (Motion Picture—16mm—Sound). An Amitai Film/FilmFair Communications, 1972. 16mm color, sound film, 13½ min. ($185). Gr. K–A. BKL (7/73), CGE (1972), PRV (5/74). 636.8

Shows how history has seen the cat as an object of worship, fear and admiration, and touches on good cat care and shows the habits that make the cat a delightful companion and an eternally free spirit.

CAVES

CAVES: THE DARK WILDERNESS (Motion Picture—16mm—Sound). Avatar Learning, Inc. EBEC, 1976 (Wide World of Adventure). 16mm color, sound film, 24 min. #3398 ($320). Gr. 5–A. AVI (1977), BKL (1977), PRV (9/19/77). 551.4

Examines the history of caves, the life forms that inhabit them and the mystery surrounding these natural underground chambers. The film focuses on early cavemen, bat colonies, and blind creatures in cave depths that never see the light of day.

THE CRYSTAL CAVERN (Filmstrip—Sound). Lyceum/Mook & Blanchard, 1974. 1 color, sound filmstrips, 61 fr.; 1 cassette or disc, 11 min.; Teacher's Guide. #LY35274C Cassette ($25), #LY35274R Disc ($19.50). Gr. 6–A. BKL (10/15/74), CPR (12/75), PRV (3/75). 796.525

Marcia Brown, twice winner of the Caldecott award, with beautiful photography and imaginative prose takes us through the Crystal Cavern of Xanadu. The classic poem

"Kubla Khan" by Samuel Coleridge is used as a point of departure. The objective of this filmstrip is to stimulate students' imagination through nature.

UNDERGROUND WILDERNESS (Filmstrip—Sound). Great American Film Factory, 1977 (Outdoor Education). 1 color, sound filmstrip, 110 fr.; 1 cassette, 20 min. #FS-9 ($35). Gr. 5–A. BKL (3/19/78). 796.525

Caves and caving have become a sport, but there is a danger to caves if people do not respect their delicate nature. Caves have always captured our imagination, from myths and legends to Huckleberry Finn. Impressions are explored as a group of cave explorers discover an uncharted cave. Carefully composed photos of the underground world are presented.

CELLS

DIFFUSION AND OSMOSIS (2ND EDITION) (Motion Picture—16mm—Sound). *See* Osmosis

THE LIVING CELL: AN INTRODUCTION (Motion Picture—16mm—Sound). EBEC, 1974. 16 mm color, sound film, 20 min. #3342 ($255). Gr. 9–12. AAS (1975), BKL (1975), SCT (1976). 574.87

Using dramatic photomicroscopy, the film examines the structures and biochemical processes that occur in all living cells. Included is the historical background leading to the discovery of the cell, and an examination of topics including cellular differentiation, cell structure, and the relationship between cells and viruses. Attention is given to the nucleus and its role in the synthesis of proteins.

THE LIVING CELL: DNA (Motion Picture—16mm—Sound). EBEC, 1976. 16mm color, sound film, 20 min. #3477 ($275). Gr. 9–12. ABT (1978), BKL (1977), EFL (1976). 574.8

Introduces DNA and the genetic material of all living things and investigates its molecular structure. Shows how DNA can be studied in a laboratory. The film reveals how the information in a DNA molecule is duplicated and passed on to new cells. Indicates the chemical changes that occur in the DNA molecule to the observable features of evolution.

CENTRAL AMERICA

CENTRAL AMERICA: FINDING NEW WAYS (Motion Picture—16mm—Sound). EBEC, 1974. 16mm color, sound film, 17 min. #3306 ($220). Gr. 4–9. LFR (5/19/75), PRV (4/75). 917.28

Linking North and South America are six republics with two merging cultures—Spanish and Mayan. This film shows how the poor live and work, and the many programs instituted by the government to redistribute land, modernize farming, diversify crops and develop industries.

CENTRAL EUROPE

EAST CENTRAL EUROPE, GROUP ONE (Filmstrip—Sound). Society for Visual Education. 1972. 6 color, sound filmstrips, av. 68 fr.; 3 discs or cassettes, av. 19³/₄ min. #292 SAR Discs ($83), #292 SATC Cassettes ($83). Gr. 4–9. BKL (2/73), PRV (2/73). 914.3

Contents: 1. The Culture of Czechoslovakia. 2. Czech Agriculture. 3. Hungary's Collective Farms. 4. The Danube: Hungary's "Seacoast." 5. Poland's Paradox, Religion and Communism. 6. Poland's Educational System.

Contemporary photos, historical maps, and charts blend with an authoritative narration to create a fascinating overview of six East Central European countries. Viewers study geography, culture, industry, agriculture, religion, education, and politics of these important nations.

CHAGRIN, JULIAN

THE CONCERT (Motion Picture—16mm—Sound). *See* Pantomimes

CHAPLIN, CHARLES SPENCER

CHAPLIN—A CHARACTER IS BORN (Motion Picture—16mm—Sound). S–L Film Productions, 1977. 16mm b/w, sound film, 40 min. ($270). Gr. 7–A. AFF (1977), FLN (1/2/77). 791.43

The film traces the evolution of the character created by Charlie Chaplin from his earliest film through the high points of his early achievements. Included are selections from many of his classic films.

CHAPMAN, JOHN—FICTION

JOHNNY APPLESEED AND PAUL BUNYAN (Phonodisc or Cassette). *See* Folklore—U.S.

CHAUCER, GEOFFREY

GREAT WRITERS OF THE BRITISH ISLES, SET I (Filmstrip—Sound). *See* Authors, English

CHEETAHS

CHEETAH (Motion Picture—16mm—Sound). EBEC, 1971, (Silent Safari). 16mm color, sound, nonnarrated film, 11 min.; Teacher's Guide. #3125 ($150). Gr. 2–9. AIF (1977), BKL (10/1/76), CGE (1977). 599.7

Shows the fleet cheetah, the fastest mammal, in action as it hunts its favorite prey, the Thompson's gazelle. Unsuccessful, it playfully attacks a topi and then dodges a

CHEETAHS (cont.)

scavenging lion. Hunt scenes show how the cheetah uses its tail for balance.

PIPPA THE CHEETAH (Filmstrip—Sound). Benchmark Films, 1978 (Joy Adamson's Africa). 3 color, sound filmstrips, 82–104 fr.; 3 cassettes, 13–16 min. ($80). Gr. 4–A. BKL (7/1/78), PRV (9/19/78). 599.7

The story of a tame cheetah cub retrained to live free in the wild.

CHEMICAL APPARATUS

HANDLING, TRANSFERRING, AND FILTERING OF CHEMICALS (Motion Picture—16mm—Sound). Prentice-Hall Media, 1970 (Basic Lab Techniques in Chemistry). 16mm color, sound film, 11 min. #KC182 ($120). Gr. 9–12. SCT (1971), STE (1972). 542

Illustrates the fundamentals of chemistry lab procedures and emphasizes the need for safety and accuracy.

MEASURING VOLUMES OF LIQUIDS (Motion Picture—16mm—Sound). Prentice-Hall Media, 1970 (Basic Lab Techniques in Chemistry). 16mm color, sound film, 11 min. #KC182 ($120). Gr. 9–12. SCT (1971), STE (1972). 542

Illustrates the fundamentals of chemistry lab procedures and emphasizes the need for safety and accuracy.

TECHNIQUES OF TITRATION (Motion Picture—16mm—Sound). Prentice-Hall Media, 1970 (Basic Lab Techniques in Chemistry). 16 mm color, sound film, 11 min. #KC182 ($120). Gr. 9–12. SCT (1971), STE (1972). 542

Illustrates the fundamentals of chemistry lab procedures and emphasizes the need for safety and accuracy.

USE OF TRIPLE BEAM BALANCE (Motion Picture—16mm—Sound). Prentice-Hall Media, 1970 (Basic Lab Techniques in Chemistry). 16mm color, sound film, 11 min. #KC182 ($120). Gr. 9–C. SCT (1971), STE (1972). 542

Illustrates the fundamentals of chemistry lab procedures and emphasizes the need for safety and accuracy.

USING THE BUNSEN BURNER AND WORKING WITH GLASS (Motion Picture—16mm—Sound). Prentice-Hall Media, 1970 (Basic Lab Techniques in Chemistry). 16mm color, sound film, 11 min. #KC182 ($120). Gr. 9–C. SCT (1971), STE (1972). 542

Illustrates the fundamentals of chemistry lab procedures and emphasizes the need for safety and accuracy.

CHEMISTRY

CHEMISTRY: DISSECTING THE ATOM (Filmstrip—Sound). Bergwall Productions, 1975. 7 color, sound filmstrips; 7 cassettes. ($145). Gr 9–12. SCT (1/77). 541

Atomic structure and electron configurations are introduced. A second sequence option is featured which allows students to move directly to valence electrons and dot structures for isolated atoms, without detailed exposure to subshell and orbital notation.

CHEMISTRY IN NATURE (Motion Picture—16mm—Sound). Centron Educational Films, 1974. 16mm color, sound film, 16 min. ($250). Gr. 7–C. AAS (1975), SCT (4/75). 540

Shows students how investigations are carried out to discover possible answers to problems of chemistry in nature. One example used water from a lake that had formed in the crater of a dormant volcano in Japan.

CRYSTALS AND THEIR STRUCTURES (Motion Picture—16mm—Sound). See Crystallography

CHEMISTRY—STUDY AND TEACHING

HANDLING, TRANSFERRING, AND FILTERING OF CHEMICALS (Motion Picture—16mm—Sound). See Chemical apparatus

MEASURING VOLUMES OF LIQUIDS (Motion Picture—16mm—Sound). See Chemical apparatus

TECHNIQUES OF TITRATION (Motion Picture—16mm—Sound). See Chemical apparatus

USE OF TRIPLE BEAM BALANCE (Motion Picture—16mm—Sound). See Chemical apparatus

USING THE BUNSEN BURNER AND WORKING WITH GLASS (Motion Picture—16mm—Sound). See Chemical apparatus

CHEROKEE

THE NATIVE LAND (Motion Picture—16mm—Sound). Atlantis Productions, 1977. 16mm color, sound film, 17 min. ($250). Gr. 11–A. BKL (9/1/78). 970.3

Life as it is presently and the future as it is pessimistically perceived by the native American Indian narrator is presented in this film.

CHICAGO

SWEET HOME, CHICAGO (Filmstrip—Sound). Hawkhill Associates, 1977. 1 color, sound filmstrip, 108 fr.; 1 cassette, 12 min.; Teacher's Guide. ($26.50). Gr. 4–C. BKL (1/15/78). 917.7311

An enthusiastic statement of the positive aspects of Chicago. The sound-filmstrip alternates a lyrical narration with the verses of a blues arrangement of the old song, "Sweet Home, Chicago." An excellent introduction to the urban scene in contemporary United States.

CHICAGO—SOCIAL CONDITIONS

THE WOODLAWN ORGANIZATION (Filmstrip—Sound). Hawkhill Associates, 1977. 1 color, sound filmstrip, 103 fr.; 1 cassette, 17½ min.; Teacher's Guide. ($26.50). Gr. 8–C. ALA (1978), BKL (1978). 309.1773

Traces the history and present day workings of a community self-help organization on the South Side of Chicago. Founded by activist Saul Alinsky, the Woodlawn Organization is an outstanding example of how an exemplary community group works to rebuild a depressed urban neighborhood.

CHICKENS

BIRTH AND CARE OF BABY CHICKS (Film Loop—8mm—Captioned). BFA Educational Media, 1973. 8mm color, captioned film loop, approx. 4 min. #481419 ($30). Gr. 4–9. BKL (5/1/76). 636.5

Demonstrates experiments, behavior, and characteristics of baby chicks.

EMBRYOLOGY OF THE CHICK (Film Loop—8mm—Captioned). See Embryology

CHILD ABUSE

CHILD ABUSE: AMERICA'S HIDDEN EPIDEMIC (Filmstrip—Sound). Multi-Media Productions, 1977. 2 color, sound filmstrips, 49–57 fr.; 1 cassette, 13 min. #7212C ($19.95). Gr. 6–A. PRV (12/78). 362.7

A presentation of a sensitive subject without sensationalism. The horrors of child abuse are reflected without scenes of violence. A careful distinction is drawn between abuse and discipline. Statistics are given as well as sociological and psychological contributing causes.

CHILD DEVELOPMENT

THE CHILD AND THE FAMILY (Kit). Parents' Magazine Films, 1978 (The Forgotten Years: Understanding Children 6–12). 5 color, sound filmstrips, 3 cassettes, or 1 disc; 5 Audio Script Booklets; Guide. ($65). Gr. 10–A. BKL (12/78). 155.4

Children enter a period during middle childhood where they seem to be pushing themselves away from the family and developing a strong sense of peer group loyalty. This set examines this changing relationship. It also examines various family structures and lifestyles.

CHILD CARE AND DEVELOPMENT, SET ONE (Filmstrip—Sound). See Children—Care and hygiene

CHILD CARE AND DEVELOPMENT, SET TWO (Filmstrip—Sound). See Children—Care and hygiene

THE DEVELOPMENT OF FEELINGS IN CHILDREN (Motion Picture—16mm—Sound). Parents' Magazine Films, 1974. 16mm color, sound film, 45 min. ($295). Gr. 7–A. CGE (1975), PRV (5/19/76). 155.4

How emotions such as fear, anger, sadness, jealousy, love, and hate develop in and are expressed by children is examined in detail. Beginning with an infant's demands for basic needs' satisfaction, the film goes on to explore a child's feelings about family, friends, separation, death, illness, socialization, etc. Older children and adults will be better equipped to understand a child's true feelings, as well as help him or her define and effectively express those feelings after viewing the film.

FOOD AND NUTRITION (Filmstrip—Sound). Parent's Magazine Films, 1975, (Child Development and Child Health). 5 color, sound filmstrips, 46–65 fr.; 1 disc or 3 cassettes, 6–8 min.; 5 Scripts; Teacher's Guide. ($65). Gr. 9–A. PRV (1/19/76). 612.4

Contents: 1. The Foundation of Health. 2. Good Nutrition Before Birth. 3. The Balanced Diet. 4. Nutrition: Good and Bad. 5. Judging for Yourself.

An explanation of why proper food and nutrition are essential to the healthy development of children. Discusses how a proper diet during pregnancy affects almost every aspect of the unborn's development and later growth. Certain dietary inadequacies in young children can lead to a variety of ailments. Advice is offered on shopping effectively and economically.

THE GROWTH OF INTELLIGENCE (Kit). Parents' Magazine Films, 1978 (The Forgotten Years: Understanding Children 6–12). 5 color, sound filmstrips; 3 cassettes or 1 disc; 5 audio script booklets; Guide. ($65). Gr. 10–A. BKL (12/78). 155.4

The sequence of intellectual development is discussed in this set, showing how the parents lay the foundation for their child's learning. The abilities and interests of children between the ages of six and twelve are explained.

INFANT CARE AND DEVELOPMENT, SET ONE (Filmstrip—Sound). See Children—Care and hygiene

CHILD DEVELOPMENT (cont.)

INFANT CARE AND DEVELOPMENT,
SET TWO (Filmstrip—Sound). *See* Children—Care and hygiene

PHYSICAL DEVELOPMENT (Kit). Parents'
Magazine Films, 1978 (The Forgotten Years:
Understanding Children 6–12). 5 color,
sound filmstrips; 3 cassettes or 1 disc; 5
Audio Script Booklets; Guide. ($65). Gr 10–
A. BKL (12/78): 155.4

This set explores the impact of physical development during middle childhood on the
emotional, social, and intellectual characteristics children will possess as adults.

THE PRENATAL PERIOD AND INFANCY
(Kit). Butterick Publishing, 1976 (The Child
Development Series). 4 color, sound filmstrips, 60–70 fr. ea.; 4 cassettes or discs, 7–
10 min. ea.; 8 Spirit Masters. #C–611–XE
Cassettes ($85), #R–610–1E Discs ($85). Gr.
10–A. FHE (6/78), JNE (4/6/78), PRV (12/
78). 649

Students follow a couple through their decision to have a child, conception, the development of the fetus, and the first months of
infancy. They observe the growth pattern
from helplessness to the two-year-old beginning to explore.

THE PRESCHOOLER (Kit). Butterick Publishing, 1976 (The Child Development Series). 4 color, sound filmstrips, 60–79 fr. ea.;
4 cassettes or discs, 7–10 min. ea.; 8 Spirit
Masters; Teacher's Guide. #C–613–2E Cassettes ($85), #614–4E Discs ($85). Gr. 10–A.
FHE (6/78), JNE (4/5/78), PRV (12/78). 649

This set covers children from ages three to
five. It shows how they begin to move away
from the home and relate to others in a social
environment. Students watch children express themselves through fantasy play, interact with peers, learn rules, become leaders,
followers, "loners" even. Emotional problems are observed and the preschoolers' understanding of time, space, direction, size
and shape.

THE SCHOOL AGE CHILD (Kit). Butterick
Publishing, 1976 (The Child Development
Series). 4 color, sound filmstrips, 60–79 fr.
ea.; 4 cassettes or discs, 7–10 min. ea.; 8
Spirit Masters; Teacher's Guide. #C–617–
9E Cassettes ($85), #R–616–0E Discs ($85).
Gr. 10–A. FHE (6/78), JNE (4/6/78), PRV
(12/78). 649

The period from five to ten is a time of leaving the nest. In this set, students discover
how five-to-ten-year-olds stop viewing the
world solely from an egocentric point of
view and how they begin to understand relationships and extend their experiences.

THE SCHOOL EXPERIENCE (Kit). Parents'
Magazine Films, 1978 (The Forgotten Years:
Understanding Children 6–12). 5 color,
sound filmstrips; 3 cassettes or 1 disc; 5
Audio Script Booklets; Guide. ($65). Gr. 10–
A. BKL (12/78). 155.4

Parents and teachers can provide a complete
educational experience for children by allowing for creative, social, physical, and intellectual growth. Also examined are the
factors which influence a child's ability to
learn, ways to stimulate a child's interest
and curiosity, and how to provide the best
learning environment.

A SENSE OF SELF (Kit). Parents' Magazine
Films, 1978 (The Forgotten Years: Understanding Children 6–12). 5 color, sound filmstrips; 3 cassettes or 1 disc; 5 Audio Script
Booklets; Guide. ($65). Gr. 10–A. BKL (12/
78). 155.4

Factors which influence a child's self-image
are discussed and how parents can help
make that image a positive one is presented.
Certain levels of physical, social, emotional,
and intellectual maturity must be reached for
children to function well in society. This set
examines those developmental accomplishments expected during the middle years.

THE TODDLER (Kit). Butterick Publishing,
1976 (The Child Development Series). 4 color, sound filmstrips, 60–79 fr. ea.; 4 cassettes
or discs, 7–10 min. ea; 8 Spirit Masters;
Teacher's Guide. #C–613–6E Cassettes
($85), #R–612–8E ($85). Gr. 10–A. FHE (6/
78), JNE (4/6/78), PRV (12/78). 649

Students observe how the toddler works towards responsibility and self-reliance, establishing a unique identity. The importance of
child's play is emphasized. The development
of language is also presented.

CHILDBIRTH

PREGNANCY (Filmstrip—Sound). *See* Pregnancy

PREPARING FOR PARENTHOOD (Filmstrip—Sound). *See* Pregnancy

CHILDREN—CARE AND HYGIENE

CHILD CARE AND DEVELOPMENT, SET
ONE (Filmstrip—Sound). Project 7 Films/
McGraw-Hill Films, 1971. 4 color, sound
filmstrips, av. 70 fr. ea.; 4 cassettes or discs,
av. 6 min. ea.; Guide. #102231–7 Cassettes
($68), #102028–4 Discs ($68). Gr. 7–A. PRV
(2/73). 649

Contents: 1. Food Needs of Children. 2.
Clothing Needs of Children. 3. Children's
Play. 4. Caring for Children: An Important
Job.

Through typical family situations, viewers
become involved in the drama of family life:
caring for the child's emotional and physical
needs, all are examined as the responsibilities of parents.

CHILD CARE AND DEVELOPMENT, SET
TWO (Filmstrip—Sound). Project 7 Films/

McGraw-Hill Films, 1971. 4 color, sound filmstrips, av. 70 fr. ea.; 4 cassettes or discs, av. 6 min. ea.; Guide #102236-8 Cassettes ($68), #102028-4 Discs ($68). Gr. 7-A. PRV (2/73). 649

Contents: 1. Discipline and Punishment. 2. Influences on Children. 3. Anxieties of Children. 4. Intellectual Development of Children.

Through typical family situations, viewers become involved in the drama of family life: directing intellectual and social growth, and disciplining and punishing as needed, all are examined as the responsibilities of parents.

COOPERATION AMONG FAMILY, STAFF, AND COMMUNITY (Filmstrip—Sound). Parents' Magazine Films, 1978 (Working with Children). 5 color, sound filmstrips av. 60 fr. ea.; 3 cassettes or 1 disc; Booklet; Teacher's Guide. ($65). Gr. 9-A. BKL (9/15/78), PRV (1/79). 649

Interaction among the child care staff, the family and the community is important in providing quality child care. Cooperation with the family will establish a sense of continuity for the child and will enable caregivers to detect any special problems.

DEALING WITH DAILY SITUATIONS (Filmstrip—Sound). Parents' Magazine Films, 1978 (Working with Children). 5 color, sound filmstrips, 60-100 fr. ea.; 3 cassettes or 1 disc; Booklet; Teacher's Guide. ($65). Gr. 9-A. BKL (9/15/78), PRV (1/79). 649

This set outlines five key classroom techniques to use in daily situations. Basics of keeping children safe and healthy, establishing learning goals, guidance and discipline, observing children and fostering their healthy self-concepts are discussed in detail.

ENCOURAGING HEALTHY DEVELOPMENT IN CHILDREN (Filmstrip—Sound). Parents' Magazine Films, 1978 (Working with Children). 5 color, sound filmstrips, 60-100 fr. ea.; 3 cassettes or 1 disc; Booklet; Teacher's Guide. ($65). Gr. 9-A. BKL (9/15/78), PRV (1/79). 649

Specific developmental accomplishments that may be expected from children from birth to age six are explored. Various phases and milestones of child development enable the caregiver to determine responses and behavior at different ages. Caregivers should encourage verbal and physical growth, trust and affection and social and environmental awareness.

FOOD AND NUTRITION (Filmstrip—Sound). See Child development

INFANT CARE AND DEVELOPMENT, SET ONE (Filmstrip—Sound). Project 7 Films/McGraw-Hill Films, 1971. 4 color, sound filmstrips, av. 65 fr. ea.; 4 cassettes or discs, av. 12 min. ea.; Guide. #102221-X

Cassettes ($68), #102001-2 Discs ($68). Gr. 5-12. JNE (3/73), PRV (3/73). 649

Contents: 1. Prenatal Care and Planning. 2. The Family and the New Baby. 3. Parental Responsibility. 4. A Baby's Day.

Realistic family situations are presented in which young parents, their friends, and doctors discuss common questions and concerns.

INFANT CARE AND DEVELOPMENT, SET TWO (Filmstrip—Sound). Project 7 Films/McGraw-Hill Films, 1971. 4 color, sound filmstrips, av. 65 fr. ea.; 4 cassettes or discs, av. 12 min. ea.; Guide. #102226-0 Cassettes ($68), #102010-1 Discs ($68). Gr. 5-12. JNE (3/73), PRV (3/73). 649

Contents: 1. Infants and Learning. 2. Growth in the First Year. 3. Infants' Food Needs. 4. Breast Feeding and Bottle Feeding.

Realistic family situations are presented in which young parents, their friends, and doctors discuss common questions and concerns.

THE PRENATAL PERIOD AND INFANCY (Kit). See Child development

THE PRESCHOOLER (Kit). See Child development

THE SCHOOL AGE CHILD (Kit). See Child development

THE TODDLER (Kit). See Child development

UNDERSTANDING THE RESPONSIBILITIES OF CHILD CARE (Kit). Parents' Magazine Films, 1978 (Working with Children). 5 color, sound filmstrips, av. 60-100 fr. ea.; 3 cassettes or 1 disc; Booklet; Teacher's Guide. ($65). Gr. 9-A. BKL (9/15/78), PRV (1/78). 649

This set examines the broad range of opportunities available in child care programs and the reasons behind the increasing demand for adults who are able to provide for the needs of the growing child. The qualifications for applicants on what an employer looks for in a potential caregiver are explored.

CHILDREN AS ARTISTS

ART BY TALENTED TEENAGERS (Filmstrip). See Art—Exhibitions

LIFE TIMES NINE (Motion Picture—16mm—Sound). See Life

CHINA

CHINA MULTIMEDIA PROGRAM (Kit). Nystrom, 1973. 5 color, sound filmstrips; 2 cassettes; Set of Activity Sheets; 2 Transparencies; 10 Student Readers; Teacher's Guide. ($165). Gr 7-12 SOC (1975), TEA (1/75). 915.1

CHINA (cont.)

Contents: 1. History. 2. Agriculture. 3. Industry. 4. Daily life. 5. Summary and Analysis.

The China program utilizes a variety of instructional materials, to show the complex transformations that have taken place in Chinese society under Communist leadership.

CHINA—CIVILIZATION

CHINA: A NETWORK OF COMMUNES (Motion Picture—16mm—Sound). EBEC, 1977. 16mm color, sound film, 15 min. #3518 ($220). Gr 7–12. LFR (1978), PRV (1978). 915.103

Today in China, over 50,000 rural and urban communes organize workers into cooperative work/living environments. This film by Jens Bjerre visits farms, a Manchurian steel mill, and a Peking textile factory to examine communal organization, the workers and what they produce, the handling of child care, birth control and medical care (including acupuncture), and the duties of Peking "street committees." The film explores the success of the system and China's goals for the future.

CHINA: EDUCATION FOR A NEW SOCIETY (Motion Picture—16mm—Sound). EBEC, 1976. 16mm color, sound film, 15 min. #3383 ($205). Gr 7–12. LFR (3/4/77), TED (1978). 915.103

Unusual footage captures the broad scope of education in the People's Republic of China. Students from kindergarten through university levels are photographed in a wide variety of activities, each invariably a reflection of Chairman Mao-Tse-Tung's political and economic ideology.

CHINA—FOREIGN RELATIONS

CHINA (Cassette). *See* U.S.—Foreign relations

CHINA—HISTORY

ASIAN MAN: CHINA (Kit). EBEC, 1977 (Asian Man). 6 color, sound filmstrips, 82–114 fr.; 6 cassettes, 13–18 min.; 10 student booklets; 1 cassette, 23 min.; 1 Teacher's Guide. #6999K ($175). Gr. 9–A. BKL (4/15/78), FLN (1/78), LAM (1/79). 951

Contents: 1. China: The Middle Kingdom. 2. Confucius and the Peaceful Empire. 3. Tao: The Harmony of the Universe. 4. Buddhism: The Way of Compassion. 5. Ch'i: The Arts of China. 6. Wei Min: For the People.

Chinese civilization has flourished for over 4,000 years. This kit views China's development through six thematic units. Each unit is introduced by a sound filmstrip, then examined through student handbook readings and lessons in the Teacher's Guide. Listening se-

lections are provided in an audio cassette. *Asian Man: China* shows Chinese traditions and cultures in the light of the people they serve.

CHINESE IN THE U.S.

CHINESE FOOD (Kit). *See* Cookery, Chinese

CHORUSES AND PART SONGS

LET'S SING A ROUND (Phonodiscs). Bowmar Publishing, 1 disc. #214 ($6.95). Gr 3–8. AUR (1978). 784.1

Songs are introduced in unison, then as rounds. Such songs as "Kookaburra," "Are You Sleeping?" "White Coral Bells," and others are included.

CHRISTMAS

A MISERABLE MERRY CHRISTMAS (Motion Picture—16mm—Sound). WNET/13 (Glenn Jordan) Educational Broadcasting/ EBEC, 1973. 16mm color, sound film, 15 min. #3300 ($185). Gr. 1–12. BKL (12/74), LFR (12/74), PRV (2/11/75). 394.26

Based on a chapter from the *Autobiography of Lincoln Steffens,* this film tells a true story about a boy who refused to compromise. It also raises questions about the spirit and values of Christmas.

THE STORY OF CHRISTMAS (Motion Picture—16mm—Sound). National Film Board of Canada/Films, 1976. 16mm color, sound, narrationless film, 8 min. #101–0074 ($145). Gr 6–A. BKL (6/76), FLN. 394.2

Using a medieval setting and music of the same period, this is a narrationless animated film. Paper cut-out animation and Elizabethan music recreate the Christmas story.

CHRISTMAS—FICTION

A CHRISTMAS CAROL (Phonodisc or Cassette). Listening Library, 1973. 3 discs or cassettes, 150 min. #AA–3386/88 Discs ($20.95), #CX–386/88 Cassettes ($22.95). Gr 4–A. BKL (3/15/74), NYR, PRV. Fic

Patrick Horgan reads this classic story by Charles Dickens. Tiny Tim and Scrooge come alive in this Christmas favorite.

A CHRISTMAS CAROL (Kit). Jabberwocky Cassette Classics, 1972. 1 cassette, 60 min.; 6 Read-along scripts. #1121 ($15). Gr. 5–A. BKL (1975), JOR (1974), PRV (11/73). Fic

A female narrator adds counterpoint to the cast of characters Dickens has brought together on Christmas Eve. The dramatization is lively and faithful to the original. Sound effects are used to enhance the mood of the tale.

THE LITTLEST ANGEL (Phonodisc or Cassette). Caedmon Records, 1973. 1 cassette or disc, approx. 50 min. #CDL51384 Cassette ($7.95), #TC1384 Disc ($6.98). Gr. K–8. INS (1973). 394.26

Contents; 1. The Littlest Angel. 2. The Bells of Christmas.

Here are two of the best loved Christmas stories of all times, read by Dame Judith Anderson. Music composed and performed by Dick Hyman.

CHURCHILL, WINSTON

GREAT WRITERS OF THE BRITISH ISLES, SET II (Filmstrip—Sound). *See* Authors, English

CICADAS

INSECT LIFE CYCLE (PERIODICAL CICADA) (Motion Picture—16mm—Sound). EBEC, 1975. 16mm color, sound film, 12 min. #3428 ($150). Gr. 7–9. AAS (1977), NST (1977), SAC (1977). 597.52

The life of this insect with its strange cycle of death and rebirth is shown in detail. The camera traces it from the egg laying, through the 17 years underground and the emergence from the nymph stage, to the mating of the mature adult. A dramatic, time-lapse sequence lets viewers watch a nymph emerge from its skin, harden, darken and develop the wings of a mature adult.

CITIES AND TOWNS

BOOMSVILLE (Motion Picture—16mm—Sound). National Film Board of Canada/Learning Corporation of America, 1970. 16mm, color, animated, sound film, 11 min. ($170). Gr. 3–A. AFF (1970), BKL (10/1/70), STE (10/71). 301.3

An animated overview of the growth of cities, showing what people have done to the environment. Recreates human interaction with our surroundings, tracing the process by which we made it a frantic, congested "boomsville."

CITIES OF AMERICA, PART ONE (Filmstrip—Sound). *See* U.S.—Description and travel

CITIES OF AMERICA, PART TWO (Filmstrip—Sound). *See* U.S.—Description and travel

CITY AND TOWN (Filmstrip—Sound). Educational Direction/Learning Tree Filmstrips, 1974. 4 color, sound filmstrips, av. 60 fr. ea.; 2 cassettes, av. 6 min. ea.; Teacher's Guide. #LT428 ($52). Gr. 2–8. ESL (1976), PRV (10/74). 301.3

Contents: 1. The Megalopolis. 2. The City. 3. The Town. 4. The Small Town.

The revealing full color photographs in this set take the student to, and demonstrate the wide variety of, the places we all live in— from the megalopolis to the most remote areas.

ECONOMICS OF THE CITY (Filmstrip—Sound). BFA Educational Media, 1972 (Man and the Cities). 6 color, sound filmstrips; 6 cassettes or discs. #VN-4000 Cassettes ($104), #VN-3000 Discs ($104). Gr. 7–9. BKL (10/15/72), PRV (1972). 301.3

Contents: 1. The Life Cycle of the City. 2. Economics of Change: Westwood Village. 3. A City Needs Goods. 4. A City Needs Services. 5. Specialization and Mass Production. 6. Cities are Run by People.

This set examines the city as a living thing, illustrating that to the extent a city lives, its people's spirit lives.

FROM CAVE TO CITY (Motion Picture—16mm—Sound). *See* Civilization

THE IDEOLOGICAL CAPITALS OF THE WORLD (Filmstrip—Sound). Pathescope Educational Media, 1974 (Great Cities in Transition). 2 color, sound filmstrips, 82 fr.; 2 cassettes, 12–14 min. #775 ($50). Gr. 7–12. BKL (9/74). 910.03

Contents: 1. Moscow: The Communist Monolith. 2. Washington, D.C.: Focus on Democracy.

An examination of Moscow and Washington, D.C., cities of ultimate national and strong international power. This role affects the development, function, and progress of each city. It also affects the social, cultural, and political lives of their citizens.

IMAGE OF THE CITY (Motion Picture—16mm—Sound). *See* City planning

THE ISLAND NATION CAPITALS (Filmstrip—Sound). Pathescope Educational Media, 1973 (Great Cities in Transition). 2 color, sound filmstrips, 94 fr.; 2 cassettes, 15–23 min.; Teacher's Guide. #751 ($50). Gr. 7–12. BKL (9/74), IFT (1973), STE (4/74). 910.03

Contents: 1. London: A City with a Plan. 2. Tokyo: A Problem of People and Space.

A comparative examination of how two cities, London and Tokyo, are coping with the problems of housing, pollution and transportation.

NIGHT PEOPLE'S DAY (Motion Picture—16mm—Sound). FilmFair Communications, 1971. 16mm color, sound film, 10½ min. ($150). Gr. K–8. BKL (9/15/71), CGE (1973), LFR (5/71). 301.3

This film explores the city at night with emphasis on the people who work nights and their occupations. Night people tell why they prefer to work at night.

OUT OF THE HOLOCAUST (Filmstrip—Sound). Pathescope Educational Media, 1974 (Great Cities in Transition). 2 color, sound filmstrips, 75–79 fr.; 2 cassettes, 12–14

CITIES AND TOWNS (cont.)

min.; Teacher's Guide. #752 ($50). Gr. 8–12. BKL (9/74). 910.03

Contents: 1. Hiroshima: Up from the Ashes. 2. Dresden: Reborn into Tradition.

Old black and white photographs of the destruction of Hiroshima and Dresden during World War II are intermixed with modern color stills that show each city as it has been rebuilt.

PLANT A SEED (Motion Picture—16mm—Sound). Phoenix Films, 1977. 16mm color, sound film, 3 min. ($75). Gr. 3–A. M&M (9/77). 301.3

This film shows what any individual can do to help bring floral beauty into the urban setting.

TOMORROW'S CITIES TODAY (Filmstrip—Sound). Pathescope Educational Media, 1974 (Great Cities in Transition). 2 color, sound filmstrips, 81–84 fr.; 2 cassettes, 11–13 min.; Teacher's Guide. #753 ($50). Gr. 7–12. BKL (9/74). 910.03

Contents: 1. Brasilia: Building the New. 2. Montreal: Reconstructing the Old.

Old Montreal, using radical new concepts, has undergone tremendous physical changes in recent years. Brasilia, however, has no long history. It is an entirely new city created to provide an inland capital for Brazil and to lead the whole nation into a new era of development. Both Brasilia and Montreal are planned for the future, but what is it like to live in them now?

WHY CITIES? (Filmstrip—Sound). Walt Disney Educational Media, 1976. 5 color, sound filmstrips, 68–80 fr. ea.; 5 cassettes or discs, 11–16½ min. ea.; Teacher's Guide. #63–8039 ($90). Gr. 7–12. BKL (9/1/77), PRV (5/78). 301.3

Contents: 1. Perspectives. 2. Progression. 3. Pleasures. 4. Problems. 5. Prognosis.

This set traces the history of the city from 3000 B.C. to the present and includes possible alternatives for the future. It is the center of government, culture, commerce, crime, crowds, and the first to benefit from new technology. Because most of today's population lives in or near cities, their problems have become societal problems.

CITIES AND TOWNS—RUINED, EXTINCT, ETC.

GHOST TOWNS OF THE WESTWARD MARCH (Motion Picture—16mm—Sound). *See* The West (U.S.)—History

CITIES AND TOWNS—U.S.

THE AMERICAN CITY (Kit). Films/EBEC, 1977 (Ongoing Issues in American Society). 3 color, sound filmstrips, 79–95 fr.; 3 cassettes, 12–14 min.; 16 Spirit Masters; 16 mini-prints; 30 Resource Readers; Teacher's Guide. #6976K ($133). Gr. 7–12. PRV (10/78). 301.3

Contents: 1. Boston, 1765: The Stamp Act Riot. 2. Chicago, 1890's: Machine Politics. 3. Los Angeles, 1965: The Watts Riot.

The changing character of American society over the past 250 years is reflected in this study of three cities. Boston typifies New England colonial life in the agricultural 18th century. Chicago demonstrates change in the industrialized 19th century. Los Angeles epitomizes our dependence on the automobile in the urbanized 20th century. This unit considers the quality of city life—past, present, and future—with an eye to further changes needed.

THE CITY AT THE END OF THE CENTURY (Motion Picture—16mm—Sound). Arthur Mokin Productions, 1977. 16mm color, sound film, 19 min. ($315). Gr. 7–A. AAS (9/78), BKL (12/1/77), LFR (5/6/77). 301.3

This film addresses our nation's urban crisis by asking the question: "Can our cities be saved—or should our cities be saved?" The film traces the historic and traditional functions of the city in the growth of civilization. It probes the origins, characteristics, challenges, and value of cities. New York is used as a prototype of all cities. (It is not about cities of the future.)

RISE OF THE AMERICAN CITY (Motion Picture—16mm—Sound). EBEC, 1971. 16mm color, sound film, 32 min. #2850 ($430). Gr. 7–12. BKL (1971), EFL (1972), STE (1974). 301.3

Contents: The surging growth of our great cities forces them to confront a host of problems: pollution, poverty, hunger, violence, and social change. Century-old institutions often seem inadequate to deal with human problems in an age of technology. John Lindsay, former Mayor of New York, and Godfrey Cambridge, well-known black comedian and social observer, help us see what the problems really are—and some of the steps we can take to solve them.

CITIZENSHIP

FREEDOM RIVER (Motion Picture—16mm—Sound). *See* Freedom

CITY LIFE

OUR CHANGING CITIES: CAN THEY BE SAVED? (Motion Picture—16mm—Sound). EBEC, 1972. 16mm color, sound film, 17 min. #3168 ($220). Gr. 8–A. BKL (1973), PRV (1974), TEA (1974). 301.3

This film focuses on the crises of the city; its unemployment, social conflicts, poverty, and crime. Using drawings and photography, the film finds the roots of these problems in the technological revolution that has made it difficult for the unskilled to find work. As the camera scans high-rises, welfare applicants, and slums, it reveals how the failure to deal with technological changes is causing a critical breakdown in our cities.

SWEET HOME, CHICAGO (Filmstrip—Sound). *See* Chicago

THE WOODLAWN ORGANIZATION (Filmstrip—Sound). *See* Chicago—Social conditions

CITY PLANNING

IMAGE OF THE CITY (Motion Picture—16mm—Sound). EBEC, 1973. 16mm color, sound film, 16 min. #3186 ($220). Gr. 7–C. BKL (1974), LFR (1975), PRV (1975). 309.2
City planners and developers are faced with an intricately complex subject: the modern city. The film proves that such problems as ethnic housing patterns, crowd estimations, and traffic loading can be measured through photography-related techniques. A segment portrays the use of multiband cameras, radar images, computer-generated graphics, satellite observations, and thermograms. The film also displays a unique history of photographs that have influenced urban political action and scientific studies of the city.

CIVICS. *See* Citizenship; Political Science; U.S.—Politics and government

CIVIL RIGHTS

FREEDOM RIVER (Motion Picture—16mm—Sound). *See* Freedom

HUMAN RIGHTS: WHO SPEAKS FOR MAN? (Filmstrip—Sound). *See* Human relations

KING: A FILMED RECORD, MONTGOMERY TO MEMPHIS (Videocassette). *See* King, Martin Luther

MARTIN LUTHER KING JR.—THE ASSASSIN YEARS (Motion Picture—16mm—Sound). *See* King, Martin Luther

MARTIN LUTHER KING: THE MAN AND HIS MEANING (Filmstrip—Sound). *See* King, Martin Luther

YOUR FREEDOM AND THE FIRST AMENDMENT (Filmstrip—Sound). *See* U.S. Constitution—Amendments

CIVIL WAR—UNITED STATES. *See* U.S.—History—1861–1865—Civil war

CIVILIZATION

COMPARATIVE CULTURES AND GEOGRAPHY, SET ONE (Filmstrip—Sound). Learning Corporation of America, 1973. 4 color, sound filmstrips, av. 150 fr.; 4 cassettes. ($94). Gr. 5–10. EGT (2/75), M&M (4/77), PRV (9/74). 901.9

Contents: 1. Two Cities: London and New York. 2. Two Factories: Japanese and American. 3. Two Towns: Gubbio, Italy and Chillicothe, Ohio. 4. Two Families: African and American.
This set of filmstrips reveals how the interplay of people and nature has evolved in our present day world. Moving back and forth between continents, each filmstrip provides comparison of contrasting cultures within similar geographical regions.

COMPARATIVE CULTURES AND GEOGRAPHY, SET TWO (Filmstrip—Sound). Learning Corporation of America, 1973. 4 color, sound filmstrips, av. 150 fr.; 4 cassettes. ($94). Gr. 5–10. EGT (2/75), M&M (4/77), PRV (9/74). 901.9

Contents: 1. Two Deserts: Sahara and Sonora. 2. Two Farms: Hungary and Wisconsin. 3. Two Grasslands: Texas and Iran. 4. Two Mountainlands: Alps and Andes.
This set presents similarities and differences among societies in various cultural regions. Central focus is on human beings' adaptation to natural environment with culture.

FROM CAVE TO CITY (Motion Picture—16mm—Sound). FilmFair Communications, 1973. 16mm color, sound film, 10 min. #C193 ($135). Gr. 4–A. CRA (1973), LFR (10/73), PRV (4/74). 901.9
This unique film traces human evolution from a cave dweller—hunter and gatherer—to our present status as a member of a highly complex urban society. The emphasis is on human needs and goals, and the implicit questions raised deal with our future.

PATTERNS OF CIVILIZATION: LESSONS FROM THE PAST (Filmstrip—Sound). Pathescope Educational Media, 1973. 5 color, sound filmstrips 64–79 fr.; 5 cassettes, 15–19 min.; Teacher's Guide. #517 ($90). Gr. 7–12. PRV (12/74). 901.9

Contents: 1. The Meaning of Ancient Greece—Today. 2. The Meaning of Ancient Rome—Today. 3. The Meaning of the Islamic Europe—Today. 4. The Meaning of the Mongol Empire—Today. 5. The Meaning of the Chinese Empire—Today.
The set views five great civilizations which showed disturbing parallels to present day problems and challenges. The framework is Toynbee's thesis that civilizations prosper or perish according to the quality of their response to great challenges. Open-ended questions are used to evaluate past failures.

WHAT DOES IT MEAN TO BE HUMAN? (Slides/Cassettes). *See* Anthropology

WHY CITIES? (Filmstrip—Sound). *See* Cities and towns

CIVILIZATION, MODERN

THE GOOD OLD DAYS—THEY WERE TERRIBLE (Cassette). *See* History—Philosophy

CIVILIZATION, MODERN (cont.)

VOLTAIRE PRESENTS CANDIDE: AN INTRODUCTION TO THE AGE OF ENLIGHTENMENT (Motion Picture—16mm—Sound). *See* French fiction

CLARK, KENNETH

IN THE BEGINNING (Motion Picture—16mm—Sound). *See* Egypt—Antiquities

CLARK, WALTER VAN TILBURG

THE PORTABLE PHONOGRAPH (Motion Picture—16mm—Sound). *See* American fiction

STORY INTO FILM: CLARK'S THE PORTABLE PHONOGRAPH (Motion Picture—16mm—Sound). *See* Motion pictures—History and criticism

CLASSICAL MYTHOLOGY. *See* Mythology, Classical

CLASSIFICATION, BIOLOGICAL

THE PROTISTS (Filmstrip—Sound). *See* Microorganisms

CLASSIFICATION, DEWEY DECIMAL

DEWEY DECIMAL CLASSIFICATION (Filmstrip—Sound). Library Filmstrip Center, 1976. 1 color, sound filmstrip, 50 fr.; 1 cassette or disc, 11 min. #76-730017 Disc ($24), #76-730017 Cassette ($26). Gr. 6-A. AFF (1977), PRV (4/78). 028.4

This filmstrip takes a step by step approach to the use of the card catalog. The cards and books illustrated are classified by the Dewey Decimal Classification system. The types of cards used in the card catalog are explained with special emphasis on the author, title, and subject cards. The lettering and numbering of trays, guide headings, call numbers, cross references, and how to locate a book on the shelf are explained.

CLEMENS, SAMUEL LANGHORNE. *See* Twain, Mark

CLIFF DWELLERS AND CLIFF DWELLINGS

AFRICAN CLIFF DWELLERS: THE DOGON PEOPLE OF MALI, PART ONE (Kit). *See* Africa

AFRICAN CLIFF DWELLERS: THE DOGON PEOPLE OF MALI, PART TWO (Kit). *See* Africa

CANYON DE CHELLY CLIFF DWELLINGS (Film Loop—8mm—Silent). Thorne Films/Prentice-Hall Media, 1972 (Southwest Indians). 8mm color, silent film loop, ap-

prox. 4 min. #HAT 270 ($26). Gr. 4-A. PRV (4/73). 970.1

Shows the cliff dwellings of the Indians of the Southwest.

INDIAN CLIFF DWELLINGS AT MESA VERDE—SLIDE SET (Slides). Donars Productions, 1973. 48 cardboard mounted, color slides in booklet of plastic storage pages; Notes. ($30). Gr. 4-8. BKL (3/1/74). 970.1

Shows in detail the structures of the cliff dwellings at Mesa Verde, Colorado, emphasizing prominent features of these structures. Also presents photographs of artifacts and a mummy found at the site and depicts possible methods of construction and areas of present archeological interest.

MESA VERDE (Study Print). EBEC, 1975. 8 13 in. x 18 in. color study prints; Study Guide. #6840 ($19.50). Gr. 5-A. MDU (5/76), PRV (11/77), 970.1

Contents: 1. Cliff Palace. 2. House of Many Windows. 3. Spruce Tree House. 4. Cliff Palace in Winter. 5. Long House. 6. Plaza of the Kivas—Spruce Tree House. 7. The Kiva Story. 8. Square Tower House.

Eight colorful study prints detail the land and the lives of the Anasazi—a mysterious Indian people of the American Southwest whose highly developed civilization ended abruptly around A.D. 1300.

PUYE DE CHELLY CLIFF DWELLINGS (Film Loop—8mm—Silent). Thorne Films/Prentice-Hall Media, 1972 (Southwest Indians). 8mm color, silent film loop, approx. 4 min. #HAT 271 ($28). Gr. 4-A. AUR (1978). 970.4

This film loop shows the cliff dwellings of the Puye De Chelly Indians of the Southwest.

CLIMBING PLANTS

KUDZU (Motion Picture—16mm—Sound). *See* Weeds

CLOTHING AND DRESS

AMERICAN MAN: TWO HUNDRED YEARS OF AUTHENTIC FASHION (Kit). Butterick Publishing, 1977. 2 color, sound filmstrips, 74-76 fr.; 2 cassettes, 12-14 min.; 1 Wall Chart; Teacher's Guide. #C422-4 ($49). Gr. 7-A. BKL (11/1/77), PRV (12/78). 391

Contents: 1. 1776-1876. 2. 1876-Present.

The development of male attire in relation to daily life and historical events. The influences on and of fads, trends, seasons, grooming, and other countries are followed through American history. Photographs, drawings, and silhouettes are used.

THE CLOTHES WE WEAR (Filmstrip—Sound). Burt Munk/Society for Visual Education, 1976. 6 color, sound filmstrips, 37-53 fr.; 6 cassettes or discs, 6-11 min.; 1 Guide.

#A213–SAR Discs ($90), #A213–SATC Cassettes ($90). Gr. 4–8. BKL (1/15/77). 646
Contents: 1. Natural Fibers. 2. Synthetic Fibers. 3. Fiber to Fabric. 4. Cloth to Clothes. 5. Shoes and How They Are Made. 6. Clothing Care.
A class travels to various sources of fibers to observe their development into clothing.

CLOTHING (Kit). Butterick Publishing, 1975 (Independent Living). 4 color, sound filmstrips, approx. 70 fr.; 4 cassettes or discs, 8–9 min.; 6 Spirit Masters; Teacher's Guide. #C–508–3E Cassette ($85), #R–507–5E Disc ($58). Gr. 8–12. FHE (9/19/76). 646
Contents: 1. The Clothes We Wear. 2. Your Clothing Personality. 3. Your Clothing Selection. 4. Spending Your Clothing Dollar.
Designed to help in the selection of personal wardrobes, appropriate to lifestyle, personality, activities, and budget of the individual.

FINISHING TOUCHES (Kit). *See* Sewing

THE GREAT COVER UP (Motion Picture—16mm—Sound). HSM Films/Texture Films, 1977. 16 mm color, sound film, 12 min. ($185). Gr. 8–A. ALA (1978), BKL (1978), CGE (1978). 391
The Great Cover Up focuses on clothing as an expression of our most crucial concerns: sexuality, politics, aesthetics. In live-action and animation, the film makes a number of observations about revealing and concealing clothing and about the responses clothing has made to war and peace, intellectual ferment, social upheaval, the industrial age, contraception, and the generation gap.

PLANNING TO SEW (Kit). *See* Sewing

SETTING UP TO SEW (Kit). *See* Sewing

STARTING TO SEW (Kit). *See* Sewing

TAKING SHAPE (Kit). *See* Sewing

A TRIP TO THE FABRIC STORE (Kit). *See* Sewing

CLOTHING TRADE

CAREERS IN THE FASHION INDUSTRY (Kit). Joshua Tree Productions/Butterick Publishing, 1973. 8 color, sound filmstrips, av. 75 fr. ea.; 8 cassettes or discs, av. 10 min. ea.; 22 Reproducible Student Guides; Teacher's Guide. #C–201–7E Cassettes ($135), #R–200–9E Discs ($135). Gr. 7–12. PRV (9/74), PVB (5/75). 687
This industry offers opportunities for people of almost every ability and interest. Viewers learn about the designing, producing, promoting, and selling of clothes. They will hear about the career rewards and drawbacks in each of the major industry areas.

CLUBB, EDMUND O.

CHINA (Cassette). *See* U.S.—Foreign relations

COLERIDGE, SAMUEL

COLERIDGE: THE FOUNTAIN AND THE CAVE (Motion Picture—16mm—Sound). *See* English poetry

THE CRYSTAL CAVERN (Filmstrip—Sound). *See* Caves

COLLAGE

HORSE FLICKERS (Motion Picture—16mm—Sound). Texture Films, 1977. 16mm color, sound film, 10 min. ($165). Gr. 7–A. BKL (1978), LFR (1977). 745.59
A series of images of horses, lovely, funny, and inventive are conjured up by the filmmaker on a large, empty barn floor. Made of feathers and spangles, of newspapers and sleeping bags, of human beings and broomsticks, they are a blend of imagination, humor, whimsy, and art.

HOW TO DO: SLIDE COLLAGE (Filmstrip—Sound). Educational Dimensions Group, 1977. 2 color, sound filmstrips, av. 60 fr.; 2 cassettes, av. 16–18 min. #667 ($49). Gr. 6–A. BKL (7/1/77). 745.59
A new versatile art of slide collage where photographic and drawing skills are not needed. After the film is prepared, a variety of effects can be achieved with simple, inexpensive resources around you, i.e., felt-tip pens, tissue paper, feathers, spices, thread, etc.

COLLECTIVE BARGAINING

LABOR UNIONS: WHAT YOU SHOULD KNOW (Filmstrip—Sound). *See* Labor unions

COLLECTIVE SETTLEMENTS—CHINA

CHINA: A NETWORK OF COMMUNES (Motion Picture—16mm—Sound). *See* China—Civilization

COLLECTIVE SETTLEMENTS—ISRAEL

ISRAELI BOY: LIFE ON A KIBBUTZ (Motion Picture—16mm—Sound). *See* Israel

COLLIER, JAMES LINCOLN

MY BROTHER SAM IS DEAD (Phonodisc or Cassette). *See* U.S.—History—Revolution, 1775–1783—Fiction

COLONIAL HISTORY (UNITED STATES).
See U.S.—History—1600–1775—Colonial period

COLOR

COLOR AND ITS PERCEPTION (Filmstrip—Sound). Coronet Instructional Media, 1971. 3 color, sound filmstrips, 25–28 fr.; 3

COLOR (cont.)

cassettes or 2 discs, 8-^{1}/$_2$ min. #S217 ($50). Gr. 6–12. BKL (1972). PRV (1972). 535.6
Contents: 1. Color and Its Perception, Part I. 2. Color and Its Perception, Part II. 3. Color and Its Perception, Part III.
Photographs, drawings, charts and special camera techniques demonstrate the visual and mental effects of color.

A LIGHT BEAM NAMED RAY (Motion Picture—16mm—Sound). Handel Film Corporation, 1972, (Special #5). 16mm color, sound film, 20 min.; Film Guide. ($250). Gr. 4–12. AAS (12/75), FLN (10/73), LFR (1973). 535.6
An animated light beam, "Ray," explains the fundamentals of light, color and optics. The film also tells about the invisible neighbors of light and introduces a superlight—the laser.

COLOR OF PEOPLE

WHY SKIN HAS MANY COLORS (Filmstrip—Sound). Sunburst Communications, 1973. 2 color, sound filmstrips, 63–67 fr.; 2 cassettes or discs, 13 min. ea.; Teacher's Guide. #100 ($59). Gr. 6–12. BKL (11/15/73). PRV (10/73), STE (9/73). 573
Contents: 1. The Variety of Skin Colors. 2. Environment and Adaptation.
Shows how skin color is determined and the role of melanin, genetics and heritage as they affect it. Identifies the evolutionary developments that have led to the multiplication of skin colors, and examines social reactions to color variations.

COLORADO

MONUMENTS TO EROSION (Motion Picture—16mm—Sound). See Erosion

COMIC BOOKS, STRIPS, ETC.

THE DOONESBURY SPECIAL (Motion Picture—16mm—Sound). Pyramid Films, 1977. 16mm color, sound film, animated, 26 min.; ^{3}/$_4$ in. videocassette also available. ($350). Gr. 8–A. BKL (7/15/78), LFR (1978), M&M (1978). 741.5
In an ironic look at contemporary society, Zonker Harris and the other members of Walden Commune come face to face with shifting values and the conclusion of the activist sixties. With its humorous yet uncompromising view of our changing patterns and life-styles, this film shows that transition and process are part of the natural condition.

COMMONER, BARRY

ECOLOGY: BARRY COMMONER'S VIEWPOINT (Motion Picture—16mm—Sound). See Ecology

COMMUNES. See Collective settlements

COMMUNICATION

DADDY, CAN I HEAR THE SUN? (Motion Picture—16mm—Sound). See Deaf

INTRODUCTION TO FEEDBACK (Motion Picture—16mm—Sound). See Feedback (psychology)

IS IT ALWAYS RIGHT TO BE RIGHT? (Motion Picture—16mm—Sound). See Human relations

MAKING WORDS WORK (Filmstrip—Sound). See Vocabulary

MEDIA AND MEANING: HUMAN EXPRESSION AND TECHNOLOGY (Slides/Cassettes). See Mass media

RHETORIC IN EFFECTIVE COMMUNICATION (Kit). See Rhetoric

UNDERSTANDING THE MAIN IDEA AND MAKING INFERENCES (Slides/Cassettes). Center for Humanities, 1976 (How to Get the Most from What You Read). 160 slides in 2 Carousel cartridges; 2 cassettes or discs; Teacher's Guide. ($139.50). Gr. 6–12. M&M (3/76), PRV (1/77), PVB (3/78). 001.54
This program stresses the importance of considering all the facts before making an inference and points out the confusion that results when one fails to understand the main idea in any medium—conversation, book, or film.

USING CLUE WORDS TO UNLOCK MEANING (Slides/Cassettes). Center for Humanities, 1977 (How to Get the Most from What You Read). 160 slides in 2 Carousel cartridges; 2 cassettes or discs; Teacher's Guide; 30 Student Activity Books; Teacher's Desk Copy. ($149.50). Gr. 7–12. BKL (3/1/78), CPR (12/77). 001.54
Since being able to read and understand what one reads depends largely on recognizing how ideas are related, this program teaches students to see how objects, events, and ideas are related in patterns, time order, comparison/contrast, cause-and-effect or simple listing. Clue words are pointed out which will enable students to readily identify each.

WHAT'S GOING ON HERE? (Filmstrip—Sound). See Propaganda

COMMUNISM

COMMUNISM: WHAT YOU SHOULD KNOW ABOUT IT AND WHY (Filmstrip—Captioned). McGraw-Hill Films, 1962. 8 captioned filmstrips, av. 40 fr.; Guide. #643060-X ($65). Gr. 9–12. CFF (1962), STE (1962). 320.532
Contents: 1. Why Study Communism? 2. What Communism Is. 3. The History of

Communism: Marx to Lenin. 4. The History of Communism: Stalin to Khrushchev. 5. Communist Expansion in Asia. 6. Communist Expansion in Europe. 7. Communism as Practiced in the U.S.S.R. 8. Meeting the Challenge of Communism.

These strips present general aspects of communism. They include theory of communism, international Communist movements. political-economic-social aspects of communism, and a review of Russian history.

COMMUNITY LIFE

FOUR FAMILIES (Filmstrip—Sound). *See* Family life

THE POLICE AND THE COMMUNITY (Kit). *See* Police

WHO'S RUNNING THE SHOW? (Filmstrip—Sound). *See* Leadership

COMMUNITY LIFE, COMPARATIVE

FAMILIES AROUND THE WORLD (Filmstrip—Sound). *See* Family life

FAMILIES OF ASIA (Filmstrip—Sound). *See* Asia

FAMILIES OF WEST AFRICA: FARMERS AND FISHERMEN (Filmstrip—Sound). *See* Africa, West

LIVING IN OTHER LANDS (Filmstrip—Sound). Educational Direction/Learning Tree Filmstrips, 1974. 12 color, sound filmstrips, av. 42 fr. ea.; 6 cassettes, av. 8 min. ea.; Teacher's Guide. ($156). Gr. 3–9. CHH (9/76), PRV (4/75), PVB (5/75). 901.9
Contents: 1. A/B Great Britain and the Netherlands. 2. C/D France and Spain. 3. E/F West Germany and Sweden. 4. G/H Australia and New Zealand. 4. I/J Japan and India. 6. K/L Turkey and Iran.

This set is designed to give the student insight into what life is like in key countries around the world from a humanistic, not statistical point of view.

WORKING IN U.S. COMMUNITIES, GROUP ONE (Filmstrip—Sound). *See* Occupations

COMPARATIVE GOVERNMENT

JOURNEY TO DEMOCRACY (Filmstrip—Sound). *See* Political science

COMPARISONS

OUR VISUAL WORLD (Filmstrip—Sound). *See* Art appreciation

COMPETITION

COMPETITIVE VALUES: WINNING AND LOSING (Filmstrip—Sound). *See* Sports

DEAR KURT, (Motion Picture—16mm—Sound). *See* Soap box derbies

THE SOAP BOX DERBY SCANDAL (Motion Picture—16mm—Sound). *See* Soap box derbies

COMPUTERS

A CAREER IN COMPUTERS (Filmstrip—Sound). Pathescope Educational Media, 1973. 2 color, sound filmstrips, 71–83, fr. ea.; 2 cassettes, 13–12 min. ea.; Teacher's Guide. #704, ($50). Gr. 7–12. BKL (5/15/74). 338.4023
Computers have created many new job-systems analysts, programmers, engineers, operators, maintenance technicians, keypunch operators, verifiers, documents and tape librarians. Each job has its own requirements for abilities and qualifications, each its own satisfactions and advancement potential. Shows what training is available.

A COMPUTER GLOSSARY (Motion Picture—16mm—Sound). Charles Eames/EBEC, 1973. 16mm color, sound film, 10 min. #3181 ($150). Gr. 7–A. ATE (1974), BKL (3/15/74), LFR (1975). 621.3819
Computer-related terms such as hardware, mnemonic, and Boolean Logic are defined. The film emphasizes the challenge of computer programming that emerges from its precisely defined procedures—procedures that demand a technical jargon. The live-action prologue takes the viewer along a computer data path; animation is used to describe words related to the computer world.

COMPUTERS AND HUMAN SOCIETY (Filmstrip—Sound). Harper and Row Publishers/Sunburst Communications, 1976. 6 color, sound filmstrips, 77–102 fr.; 6 cassettes or discs, 14–19 min.; Teacher's Guide. #79 ($145). Gr. 9–12. BKL (6/15/77), PRV (12/77). 608
Contents: 1. The Ultimate Servant? 2. The Electronic Brain. 3. The Evolution of Intelligence. 4. The Computer Revolution. 5. The Data Explosion and Central Control. 6. Projections for Man and Machine.

Provides students with a technological and philosophical framework for assessing the computer's impact on humankind. Describing the Computer Age as an evolutionary breakthrough, program details the machine's actual and potential enrichment of human life and speculates on its equally great potential for destruction.

COMPUTERS: FROM PEBBLES TO PROGRAMS (Filmstrip—Sound). Guidance Associates, 1975 (Math Matters). 3 color, sound filmstrips, av. 76 fr.; 3 discs or cassettes, av. 12 min.; Teacher's Guide. #1B–301–422 Discs ($68.50), #1B–301–414 Cassettes ($68.50). Gr. 5–8. ATE. 621.3819

COMPUTERS (cont.)

Outlines the history of calculation from ancient times through the 20th century. Reviews computer development, surveys current applications and suggests future directions for computer technology.

INFORMATION MACHINE (Motion Picture—16mm—Sound). *See* Information storage and retrieval systems

JOBS IN DATA PROCESSING (Filmstrip—Sound). Coronet Instructional Media, 1974. 6 color, sound filmstrips. av. 54 fr.; 3 discs or 6 cassettes, av. 13 min. #M280 Cassettes ($95), #S280 Discs ($95). Gr. 10–C. BKL (1975). 338.0423
Contents: 1. Control Clerk. 2. Keypunch Operator. 3. Computer Operator. 4. Programmer. 5. Custom Engineer. 6. Sales Representative.
Interviews with young people now in computer work tell why they chose their jobs, how they got them, their educational backgrounds and what they do.

VIEW FROM THE PEOPLE WALL (Motion Picture—16mm—Sound). *See* Thought and thinking

CONCERTOS

CONCERTO FOR ORCHESTRA (Phonodiscs). Deutsche Grammophon, 1975. 2 discs. #2530 479 ($8.98). Gr. 4–A. PRV (12/75). 785.6
The orchestral sonorities which the composer achieves are as imaginative as the structure and expressive plan of the work itself. Rafael Kubelik conducts the Boston Symphony Orchestra in this.

PIANO CONCERTO NO. 3 IN D MINOR (Phonodiscs). Philips Records, 1974. 1 disc. #6500–540 ($5.50). Gr. 3–A. PRV (5/75). 785.6
Rafael Orozco, pianist with Edo de Waart conducting the Royal Philarmonic Orchestra in this Rachmaninoff favorite.

CONDUCT OF LIFE. *See* Human behavior

CONDUCTORS (MUSIC)

THE CONDUCTOR (Filmstrip—Sound). Educational Audio Visual, 1973. 2 color, sound filmstrips; 2 cassettes or discs; Guide. Cassettes ($38), Discs ($35). Gr. 7–12. BKL (12/74). 781.635
This set focuses on playing in an orchestra and on what a conductor's responsibilities are in relation to furthering the value of music.

CONFLICT OF GENERATIONS

ADOLESCENT CONFLICT: PARENTS VS. TEENS (Filmstrip—Sound). *See* Adolescence

COPING WITH PARENTS (Motion Picture—16mm—Sound). FilmFair Communications, 1973. 16mm color, sound film, 15$\frac{1}{4}$ min. ($195). Gr. 7–A. FHE (5/74), LFR (2/74), PRV (12/74). 362.8
Demonstrates methods of solving typical parent-teenager conflicts by altering responses and understanding human needs. Three typical conflicts between teenagers and parents are dramatized: a mother nagging; a father regularly criticizing and arguing; a mother "has no time for child." The narrator shows how to change the situations by changing responses and by understanding people's basic needs. Each dramatized action is replayed, using the narrator's suggestions, and shows a positive solution.

LOOKING TOWARDS ADULTHOOD (Filmstrip—Sound). *See* Adolescence

PORTRAIT OF PARENTS (Filmstrip—Sound). *See* Adolescence

PORTRAIT OF TEENAGERS (Filmstrip—Sound). *See* Adolescence

THE STRUGGLE FOR INDEPENDENCE (Filmstrip—Sound). *See* Adolescence

CONFLICT, SOCIAL. *See* Social conflict

CONNELL, RICHARD

A DISCUSSION OF THE HUNT (Motion Picture—16mm—Sound). *See* Motion pictures—History and criticism

THE HUNT (Motion Picture—16mm—Sound). *See* American fiction

CONRAD, JOSEPH

A DISCUSSION OF THE SECRET SHARER (Motion Picture—16mm—Sound). *See* Motion pictures—History and criticism

THE SECRET SHARER (Motion Picture—16mm—Sound). *See* American fiction

CONSERVATION OF NATURAL RESOURCES

COAL, CORN, AND COWS (Filmstrip—Sound). *See* Environment

ECOLOGY: BARRY COMMONER'S VIEWPOINT (Motion Picture—16mm—Sound). *See* Ecology

ENERGY AND OUR ENVIRONMENT (Filmstrip—Sound). *See* Power resources

ENERGY AND THE EARTH (Filmstrip—Sound). Lyceum/Mook & Blanchard, 1973. 2 color, sound filmstrips, 56–69 fr.; 2 cassettes or discs, 11–18 min. #LY35373SC Cassettes ($46), #LY35373SR Discs ($37). Gr. 6–A. IFT (1973), INS (4/74), FLN (2/3/74). 333

Contents: 1. Earth: The Early Years. 2. Earth: The Years of Decision.

This series develops an understanding of the fundamental energy resources of the earth and how they have been and are being expanded. Environmental effects are considered.

JOHN MUIR: CONSERVATION (Filmstrip—Sound). *See* Muir, John

LET NO MAN REGRET (Motion Picture—16mm—Sound). *See* Man—Influence on nature

LIVING WITH A LIMIT (Slides/Cassettes). *See* Power resources

MOUNTAINEERING (Filmstrip—Sound). *See* Mountaineering

UNDERGROUND WILDERNESS (Filmstrip—Sound). *See* Caves

CONSTITUTION. *See* U.S. Constitution

CONSUMER EDUCATION

THE ACTION PROCESS (Filmstrip—Sound). Visual Education/Butterick Publishing, 1977 (Butterick Consumer Education Series). 3 color, sound filmstrips, 73–80 fr.; 3 discs or cassettes, 9–12 min.; Teacher's Guide. #C–528–8E Cassettes ($75), #R–527–XE Discs ($75). Gr. 7–A. BKL (12/15/78), PRV (12/78). 640.73

Contents: 1. Buy It or Do It Yourself. 2. Power to the Purchaser. 3. The Consuming Question.

This set introduces the alternatives to consumer do-it-yourselfers. It examines the trade-offs between spending time and spending money. The historical perspective on the changes in laws protecting the consumers is also shown.

THE AMERICAN CONSUMER (Kit). Films/EBEC, 1976 (Ongoing Issues in American Society). 3 color, sound filmstrips, 64–77 fr.; 3 cassettes, 7–10 min.; 30 Resource Readers; 14 Spirit Masters; 16 Mini-prints; 2 cassettes, 21–30 min.; 1 Guide. #6854K ($133.50). Gr. 8–12. BKL (5/77). 640.73

Contents: 1. The Consumer and the Law. 2. The Consumer and Regulatory Agencies. 3. The Consumer and Special Interest Groups.

The American Consumer examines the three consumer protective devices most commonly used in the United States: laws, regulatory agencies, and private citizens' action groups. Case studies, selected readings, and interviews with a cross-section of participants provoke student discussion and analysis of the success and limitations of each of these devices.

THE BUY LINE (Motion Picture—16mm—Sound). *See* Advertising

CONSUMER CREDIT (Kit). *See* Credit

CONSUMER EDUCATION SERIES (Filmstrip—Sound). McGraw-Hill Films, 1972. 6 color, sound filmstrips, av. 80 fr. ea.; 6 cassettes or discs, 18 min. ea.; Guide. #102342–9 Cassettes ($97), #103439–0 Discs ($97). Gr. 7–A. BKL (11/73), FHE (2/73), PRV (5/74), 640.73

Contents: 1. The Consumer and the Government. 2. Money Is the Medium. 3. Let the Buyer Beware. 4. How to Buy Clothes. 5. How to Buy Food. 6. How to Buy a Used Car.

This set is designed to help students to understand the rights and responsibilities of consumers and to exercise them effectively. Useful tips are offered on where to buy, how to buy, and what to buy.

CONSUMER MATH (Kit). *See* Arithmetic—Study and teaching

THE CONSUMER OFFENSIVE (Motion Picture—16mm—Sound). ABC Television/Benchmark Films, 1976. 16mm color, sound film, 26 min. ($390). Gr. 7–A. BKL (12/15/76), CPR (12/19/77). 640.73

The individual American consumer is discovering how to get better products for less money, how to protect the environment, and how to improve the quality of life—by joining local and national consumer organizations.

CONSUMER'S WORLD: IT'S YOUR DECISION (Filmstrip—Sound). Coronet Instructional Media, 1974. 6 color, sound filmstrips, av. 99 fr. ea.; 6 cassettes or discs, av. 11$\frac{1}{2}$ min. ea. #M706 Cassettes ($105), #S706 Discs ($105). Gr. 10–A. BKL (3/1/75), IDF (1975), M&M (1975). 640.73

Contents: 1. Buying a Car. 2. Buying Clothing. 3. Renting an Apartment. 4. Buying Food. 5. Buying Home Furnishings. 6. Buying Trouble.

An introduction to problems faced when buying items and to concepts that are important in making sensible buying decisions. Warranty, depreciation, comparative shopping, impulse buying, high pressure selling, credit, and ethical retail practices are concepts covered in this set.

FOOD AND NUTRITION: DOLLARS AND SENSE (Kit). *See* Nutrition

JUSTICE IN THE MARKETPLACE (Kit). Changing Times Education Service/EMC, 1974. 2 color, sound filmstrips, 224 fr; 2 cassettes or discs; 10 Linemasters; 1 Reading

CONSUMER EDUCATION (cont.)

and Resources List; 1 Exercise for Review; 1 Guide to Inquiry and Discussion. #2348 Cassettes ($49.50), #2336 Discs ($49.50). Gr. 9–12. IFT (1974). 640.73

Contents: 1. A Play on Wheels. 2. Washday Blues. 3. Where to Now? 4. Your Day in Court. 5. You'll Hear from My Lawyer.

Impresses students that they have a right to be heard when they have been wronged in the marketplace. It guides inquiry and discussion based on five situations involving young consumers seeking redress of grievances.

LET'S GO SHOPPING (Kit). *See* Shopping

PLAY THE SHOPPING GAME (Motion Picture—16mm—Sound). EBEC, 1977. 16mm color, sound film, 20 min. #3548 ($295). Gr. 7–12. BKL (1978), FHE (1978), LFR (1978). 640.73

Based on the TV show that poses the important question, "Do you know how to shop by comparison and avoid consumer fraud?" Six young contestants test their skills as they purchase food and housing, clothing and means of transportation, medical and personal care. Consumer savvy builds through musical lessons on deceptive packaging and labeling, improper weighing of foods, unit vs. quantity pricing, leases and credit contracts.

SOOPERGOOP (Motion Picture—16mm—Sound). *See* Advertising

THERE IS A LAW AGAINST IT (Motion Picture—16mm—Sound). FilmFair Communications, 1972. 16mm color, sound film, 8 min. ($115). Gr. 7–A. BKL (6/73), FHE (9/73), LFR (2/73). 640.73

This film explains and depicts four legal developments in the area of consumer protection. Four familiar consumer problems are introduced; garnishment of wages, unauthorized auto repair work, payment demanded for a debt already paid, and a housewife pressured into signing a purchase contract by a salesman. The film dramatizes how new consumer laws in one state (California) protect the consumer. It explains provisions of the laws, responsibilities of the businesses involved, and the steps the consumer should take.

THIS IS FRAUD! (Motion Picture—16mm—Sound) FilmFair Communications, 1972. 16mm color, sound film, 8¼ min. ($115). Gr. 7–A. BKL (4/1/73), FHE (9/73), PRV (5/74). 640.73

The methods of recourse for common consumer fraud cases are described—e.g., trade association offices, small claims court, licensing bureau, district attorney's office, legal aid society, etc. Using three of the cases dramatized, the film underscores how certain danger signals in each case could have

helped stop the fraud before it happened. Narrator for the film is Herschel Elkins, head of the Consumer Protection Division, State Attorney General's Office, State of California.

TOMMY'S FIRST CAR (BUYING A USED CAR) (Motion Picture—16mm—Sound). A Learning Garden Film/FilmFair Communications, 1972. 16mm color, sound film, 11 min. ($145). Gr. 7–A. BKL (3/73), FHE (9/73), LFR (4/73). 640.73

To show step-by-step how to determine the condition of a used car before buying. At the used car lot, Tommy's father shows him where to look for clues that indicate the car's condition and possible necessary repairs.

CONTINENTAL DRIFT

THE MOVING EARTH: NEW THEORIES OF PLATE TECTONICS (Filmstrip—Sound). *See* Plate tectonics

CONTINENTS

HOW OUR CONTINENT WAS MADE (Filmstrip—Sound). *See* North America

COOKERY

ANIMAL AND PLANT PROTEIN (Kit). Butterick Publishing, 1976 (Look and Cook). 2 color, sound filmstrips, approx. 90 fr.; 2 discs or cassettes, 10–13 min.; 6 Transparencies; 8 Duplicating Masters; 1 Wall Chart; 1 Guide. #C–414–1E Cassettes ($75), #R–413–3E Discs ($75). Gr. 7–12. BKL (9/15/77), JNE (1/77), PRV (12/77). 641.5

An introduction to all the different sources of protein from beef, poultry, fish, eggs, and legumes. What to look for when shopping to get freshness, quality, economy, and nutrition are demonstrated. The principles of protein cookery are explored.

BREADS AND CEREALS (Kit). Butterick Publishing, 1976. 2 color, sound filmstrips, approx. 90 fr.; 2 cassettes or discs, 10–13 min.; 6 Transparencies; 8 Duplicating Masters; 1 Wall Chart; 1 Guide. Cassette #C–420–6E ($75), Disc #R–419–2E ($75). Gr. 7–12. JNE (1/77), PRV (12/77). 641.5

The differences in nutrition and the problems of preparation of a variety of cereals and breads are examined.

EDIBLE ART (Filmstrip—Sound). Paramount Communications, 1975. 6 color, sound filmstrips. 6 cassettes, 8–10 ea.; Teacher's Guide; Recipes. #9106 ($98). Gr. 4–12. BKL (7/76), PRV (10/76), PVB (1977). 641.5

Contents: 1. Modeled Cookies. 2. Candy Clay. 3. Bread Sculpture. 4. Frosting Paint. 5. Graham Cracker Construction. 6. Fruit and Vegetable Assemblage.

This set combines the art elements color, texture, line, and shape and the human inclination to make foods aesthetically pleasing. It provides recipes and instructions for festive foods for many occasions.

FRUITS AND VEGETABLES (Kit). Butterick Publishing, 1976 (Look and Cook). 2 color, sound filmstrips, approx. 90 fr.; 2 cassettes or discs, 10–13 min.; 6 Transparencies; 8 Duplicating Masters; 1 Wall Chart, 1 Guide. Cassette #C–418–4E ($75), Disc #R–417–6E ($75). Gr. 7–12. JNE (1/77), PRV (12/77). 641.5

This set gives useful hints and demonstrates techniques for the selection, storage and preparation of fresh, canned and frozen produce.

KITCHEN EQUIPMENT (Kit). Butterick Publishing, 1976 (Look and Cook). 2 color, sound filmstrips, approx. 90 fr.; 2 cassettes or discs, 10–13 min.; 6 Transparencies; 8 Duplicating Masters; 1 Wall Chart; 1 Guide. #C4–12–5E Cassettes ($75), #411–7E Discs ($75). Gr. 7–12. BKL (9/15/77), JNE (1/77), PRV (12/78). 641.5

This set covers all sizes of equipment from ovens to hand tools. Students learn how they operate, how to use them safely and efficiently, and how to select the right ones for different life-styles.

KITCHEN WITH A MISSION: NUTRITION (Filmstrip—Sound). *See* Nutrition

MILK AND DAIRY PRODUCTS (Kit). Butterick Publishing, 1976 (Look and Cook). 2 color, sound filmstrips, approx. 90 fr.; 2 cassettes or discs, 10–13 min.; 6 Transparencies; 8 Duplicating Masters; 1 Wall Chart; 1 Guide. Cassette #C–416–8E ($75), Disc #R–415–6E ($75). Gr. 7–12. BKL (10/15/77), PRV (12/77). 641.5

Tips on how to select the many forms of milk and milk products are given to help meet the nutritional needs, taste and economy. Cooking and storage methods that are most appropriate for each dairy product, as well as the properties of each, is presented.

NUTRITION BASICS (Kit). Butterick Publishing, 1976 (Look and Cook). 2 color, sound filmstrips, approx. 90 fr.; 2 cassettes or discs, 10–13 min.; 6 Transparencies; 8 Duplicating Masters; 1 Wall Chart; 1 Guide. #C–410–9E Cassettes ($75), #R–409–7E Discs ($75). Gr. 7–12. JNE (1/77), PRV (12/77). 641.5

The quality of our lives, as well as our weight, height, strength, and health, is affected by the food we eat. This fact, along with information about the basics of fats, carbohydrates, proteins, vitamins, and minerals are presented. Students learn about calories and diets and guidelines for meal planning. They are introduced to world food problems through interviews with Francis Lappe, author of *Diet For a Small Planet*, nutritionists and government officials.

COOKERY, BLACK

SOUL FOOD (Kit). Butterick Publishing, 1977 (The American Ethnic Food Series). 2 color, sound filmstrips, 60–72 fr.; 2 discs or cassettes, 8–9 min.; 6 Spirit Masters; 1 Wall Chart; Teacher's Guide. Cassette #C–426–5 ($68), Disc #R–425–7 ($68). Gr. 7–A. BKL (4/15/78), JNE (10/77). 641.59

This kit traces the history and development of soul food to Southern blacks before the Civil War. They created nutritious dishes from a limited variety of foods: namely corn, pork, vegetables, and game. It also shows the step-by-step preparation of a meal of barbecued chicken, greens, and cornbread.

COOKERY, CHINESE

CHINESE FOOD (Kit). Butterick Publishing, 1977 (The American Ethnic Food Series). 2 color, sound filmstrips, 68–72 fr.; 2 cassettes or discs, 9–10 min.; 6 Spirit Masters; 1 Wall Chart; Teacher's Guide. Cassette #C–455–9E ($68), Disc #R–454–0E ($68). Gr. 7–A. BKL (3/15/78), JNE (10/77). 641.59

Demonstrates Chinese cuisine, reflecting one of the world's oldest cultures. Chinese immigrants have preserved this culture and their eating traditions in spite of difficulties and deprivations. A step-by-step program of producing an authentic meal of stir-fried beef and vegetables with rice and tea is shown. Properly slicing vegetables, cooking evenly and quickly, adding the proper color and crunchy texture, measuring and marinating are all presented.

COOKERY FOR INSTITUTIONS. *See* Cookery, Quantity

COOKERY, GERMAN

GERMAN FOOD (Kit). Butterick Publishing, 1977 (The American Ethnic Food Series). 2 color, sound filmstrips, 69–73 fr.; 2 discs or cassettes, 8–11 min.; 6 Spirit Masters; 1 Wall Chart; Teacher's Guide. Cassette #C–463–XE ($68), Disc #R–462–1E ($68). Gr. 7–A. BKL (2/15/79). 641.59

Students discover that dishes usually thought of as American, such as bacon and eggs, frankfurters, sweet pickle relish on hamburger and strawberry gelatin dessert, all come from German recipes. Step-by-step instructions are shown for preparing a German meal of potato soup with cucumber, meatballs with noodles, sweet and sour cabbage, salad with German-style dressing and strawberry Bavarian cream.

COOKERY, ITALIAN

ITALIAN FOOD (Kit). Butterick Publishing, 1977 (The American Ethnic Food Series). 2 color, sound filmstrips, 70–74 fr.; 2 discs or cassettes 8–10 min.; 6 Spirit Masters; 1 Wall Chart; Teacher's Guide. #C–457–5 Cas-

COOKERY, ITALIAN (cont.)

settes ($68), #R–456–7 Discs ($68). Gr. 7–A. BKL (1/15/78), JNE (10/77). 641.59

Traces the history of Italian Americans back to 1800, when millions of farmers began emigrating to America where they maintained contact with their heritage by providing their families with traditional Italian foods. A whole industry grew up to serve the demand for Italian-style food, as the tastes of other Americans were attracted to it. A step-by-step demonstration on how to prepare an Italian meal of lasagne, green salad with Italian dressing, Italian bread and fresh fruit is presented.

COOKERY, MEXICAN

MEXICAN FOOD (Kit). Butterick Publishing, 1977 (The American Ethnic Food Series). 2 color, sound filmstrips, 61–66 fr.; 2 discs or cassettes, 7–8 min.; 6 Spirit Masters; 1 Wall Chart; Teacher's Guide. Cassette #C–59–1E ($68), Disc #R–58–3E ($68). Gr. 7–A. BKL (3/15/78), JNE (10/77). 641.59

Explains the culinary development of Mexican food as a blend of Spanish and Indian foods. Corn, indigenous to the Western Hemisphere, proves the staple starch of Mexican food. Step-by-step instructions for preparing Mexican tacos and guacamole are shown, as well as how to turn a meal into a fiesta.

COOKERY, QUANTITY

CAREERS IN FOOD SERVICE (Filmstrip—Sound). Associated Press/Pathescope Educational Media, 1973 (Careers In). 2 color, sound filmstrips, 84–91 fr.; 2 cassettes, 13–14 min.; Teacher's Manual. #714 ($50). Gr. 9–12. PRV (1/19/75). 642.023

Contents: Part 1. Part 2.

Describes the different jobs available within this field. The set uses interviews to determine necessary training, job satisfactions, and opportunities.

COPLEY, JOHN SINGLETON

AMERICAN CIVILIZATION: 1783–1840 (Filmstrip—Sound). See Art, American

JOHN SINGLETON COPLEY (Motion Picture—16mm—Sound). National Gallery of Art & Visual Images/WETA-TV/EBEC, 1974 (The Art Awareness Collection. Nat'l Gallery of Art). 16mm color, sound film, 6 min. #3538 ($105). Gr. 7–C. EFL (1977), EGT (1978), LFR (1978). 759.13

America's first great painter leaves his country for the richer artistic influence of the old world. There, in 18th century England, Copley nurtures the Rococco style and becomes a forerunner of the Romantic movement.

CORRERAS DE MOTOCROS

MOTOCROSS RACING (Kit). See Spanish language—Reading materials

CORRUPTION IN POLITICS

CORRUPTION IN AMERICA: WHERE DO YOU STAND? (Filmstrip—Sound). Harper and Row Publishers/Sunburst Communications, 1974. 2 color, sound filmstrips, 66–68 fr.; 2 cassettes or discs, 12–13 min.; Teacher's Guide. #80 ($59). Gr. 7–12. PRV (5/76), PVB (1977). 172

Contents: 1. Society's View. 2. The Personal View.

Defining corruption as a "rotting" of personal or social values, this program presents views on the American values system as expressed by political and religious leaders, authors, and private citizens.

COSBY, BILL

BILL COSBY ON PREJUDICE (Motion Picture—16mm—Sound). See Prejudices and antipathies

COSMOLOGY. See Universe

COUNTING

DECIMAL NUMERATION SYSTEM (Filmstrip—Sound). See Arithmetic

NUMBERS: FROM NOTCHES TO NUMERALS (Filmstrip—Sound). Guidance Associates, 1975 (Math Matters). 2 color, sound filmstrips, av. 61 fr.; 2 cassettes or discs, av. 11 min.; Teacher's Guide. #1B–301–364 Discs ($52.50), #1B–301–356 Cassettes ($52.50). Gr. 4–8. ATE. 513

Reviews systems used in Egyptian, Babylonian, Roman, Mayan, and other ancient cultures. Traces development of our Hindu-Arabic number system. It also explains the binary system used in program computers.

COURTS MARTIAL AND COURTS OF INQUIRY

BENEDICT ARNOLD: TRAITOR OR PATRIOT? (Filmstrip—Sound). See Arnold, Benedict

COURTS—U.S.

WITH JUSTICE FOR ALL (Kit). See Justice, Administration of

COURTSHIP. See Dating (social customs)

COVINGTON, LUCY

LUCY COVINGTON: NATIVE AMERICAN INDIAN (Motion Picture—16mm—Sound). Odyssey Productions/EBEC, 1978. 16mm

color, sound film, 16 min. #3567 ($240). Gr.
7-C. BKL (1978), LFR (1978), M&M (1978).
921

As an active leader and spokesperson for the
Colville Indians, Lucy Covington retells the
history of her people as it has been handed
down through oral tradition. The Indian language, ritual music, rare historical photographs, and on-location shots of the Colville
Reservation in northern Washington enrich
the portrayal of these native Americans.

CRANE, STEPHEN

EPISODES FROM CLASSIC NOVELS
(Filmstrip—Sound). *See* Literature—Collections

THE RED BADGE OF COURAGE (Kit). *See*
U.S.—History—1861-1865—Civil War—
Fiction

CREATION

AN INQUIRY INTO THE ORIGIN OF
MAN: SCIENCE AND RELIGION (Slides/
Cassettes). *See* Man—Origin and antiquity

CREATION (LITERARY, ARTISTIC, ETC.)

ADVENTURES IN IMAGINATION SERIES
(Filmstrip—Sound). *See* English language—
Composition and exercises

THE FIRST MOVING PICTURE SHOW
(Motion Picture—16mm—Sound). *See* Motion pictures

WHY MAN CREATES (Motion Picture—
16mm—Sound). Pyramid Films, 1970.
16mm, color, sound film, live action and animation, 25 min.; ³/₄ in. videocassette, also
available. ($325). Gr. 7-A. AAW (1971),
FLN (1971), M&M (1971). 153.3

A series of explorations, episodes, and comments on creativity, by a master of conceptual design. Humor, satire and irony are
combined with serious questions about the
creative process and how it comes into play
for different individuals. A beautiful Academy Award winning film.

CREATIVITY—FICTION

THE BEGINNING (Motion Picture—16mm—
Sound). C. O. Hayward/Bosustow Productions, 1970. 16mm color, sound film, 4¹/₂
min. ($110). Gr. 8-A. CIF (1973), LFR (3/
72), Fic

A butterfly touches a man and inspires him
to fly—a metaphor for any creative effort.
The man's companions ridicule him when he
fails, but he keeps trying and finally succeeds. When the butterfly returns and
touches the man's companions, they have
forgotten their mockery. Now everybody
wants to fly—to duplicate what's been done.

They aren't afraid because someone has
dared and succeeded before them.

A BETTER TRAIN OF THOUGHT (Motion
Picture—16mm—Sound). Pannonia Film
Studios/Bosustow Productions, 1974. 16mm
color, sound film, 9¹/₂ min. ($160). Gr. 8-A.
FLN (12/75), PRV (1975). Fic

An imaginative and humorous parable on
problem solving and the reaction to new
ideas. An old train, whose passengers are in-
dignant that another train passes them try a
variety of traditional methods of speeding up
their train—but no go. They repeatedly re-
ject unconventional solutions proposed by
one passenger. He goes away mad—but
shows up again as designer of a Supertrain.

CREDIT

CONSUMER CREDIT (Kit). Society for Visual Education, 1978. 4 color, sound filmstrips, av. 60 fr. ea.; 4 cassettes, 12 min. ea.;
30 skill extenders; Teacher's Guide. #A618-
SATC ($99). Gr. 9-12. BKL (1/15/79). 332.7

This kit explores the many credit options
available in today's market. The credit terms
and concept as well as the advantages, disadvantages, and responsibilities of using
credit are explained.

CREWELWORK

CREATIVE CREWEL (Filmstrip—Sound).
Warner Educational Productions, 1974. 2
color, sound filmstrips, 70-79 fr.; 1 cassette,
14 min.; Teaching Guide. #520 ($47.50). Gr.
6-A. BKL (1/15/75), PRV (9/75). 746.44

The crewel designs presented are not just
flowers and animals, and the 15 distinctive
stitches and combinations present something
for everyone. Step-by-step instructions present the materials and tools needed as designs
are developed and stitches learned.

CRIME

CRIME: EVERYBODY'S PROBLEM (Filmstrip—Sound). Associated Press/Pathescope
Educational Media, 1976 (Contemporary Issues). 1 color, sound filmstrip, 63 fr.; 1 cassette or disc, 7 min.; Teacher's Guide.
#9504 ($28). Gr. 4-8. PRV (4/78). 364

Crime has grown to such a proportion that it
affects all of us in some manner. A brief look
at how ideas and laws change is presented in
a clear and direct manner. Emphasis is
placed on everyone's responsibility to combat crime.

CRIMINAL JUSTICE: TRIAL AND ERROR
(Filmstrip—Sound). *See* Justice, Administration of

LAW AND SOCIETY: LAW AND CRIME
(Kit). *See* Law enforcement

CRIME (cont.)

PROJECT AWARE (Motion Picture—
16mm—Sound). *See* Juvenile delinquency

SHOPLIFTING, IT'S A CRIME (Motion Picture—16mm—Sound), Filmfair Communications, 1975. 16mm color, sound film, 12 min.
($165). Gr. 4–9. LFR (5/75), LNG (12/75),
PRV (12/75). 364
Typical shoplifting incidents are dramatized
with young people of both elementary and
high school age. The film points out the additional cost to shoppers, the methods of ap-
prehension, the embarrassment and
consequences of being caught.

THE WITNESS (Motion Picture—16mm—
Sound). *See* Values

CRIMEAN WAR, 1853–1856

FLORENCE NIGHTINGALE (Cassette), *See*
Nightingale, Florence

**CRIMINAL JUSTICE, ADMINISTRATION
OF**

LAW AND CRIME (Kit). EBEC, 1977 (Law
and Society). 3 sound filmstrips with cassettes; 2 audiocassettes; Student Resource
Reader (10 copies); Teacher's Guide.
#17012K ($93.50). Gr. 7–12. PRV (12/78).
345
Contents: 1. Crime and Punishment: Changing Views. 2. Criminal Justice: Goals and
Challenges. 3. Criminal Punishment: Two
Alternatives.
In firsthand accounts and written opinions,
law enforcement officials and penologists define the causes of crime, the problems of
criminal justice, and the nature of punishment.

CROCE, JIM

MEN BEHIND THE BRIGHT LIGHTS (Kit).
See Musicians

CRUELTY TO CHILDREN. *See* Child abuse

CRYSTALLOGRAPHY

CRYSTALS AND THEIR STRUCTURES
(Motion Picture—16mm—Sound). Wards
Natural Science Establishment, 1975. 16mm
b/w, sound film, 22 min. ($155). Gr. 9–12.
AAS (9/75). 548
This film is to be used with high-level, general chemistry courses. Phenomenological evidence for crystallinity is explored, followed
by a demonstration of diffraction—first with
x rays, then with water waves. The Bragg
equation is presented and used.

CULTURE

COMPARATIVE CULTURES AND GEOG-
RAPHY, SET ONE (Filmstrip—Sound).
See Civilization

COMPARATIVE CULTURES AND GEOG-
RAPHY, SET TWO (Filmstrip—Sound).
See Civilization

CULTURE AND ENVIRONMENT: LIV-
ING IN THE TROPICS (Filmstrip—
Sound). *See* Tropics

HUMANITIES: A BRIDGE TO OUR-
SELVES (Motion Picture—16mm—Sound).
See Humanities

WHY CULTURES ARE DIFFERENT (Filmstrip—Sound). Filmstrip House/United
Learning, 1974. 6 color, sound filmstrips, av.
55–75 fr. ea.; 3 cassettes or phonodiscs, av.
10–13 min. ea.; Script/Guide. Cassettes
#50202 ($80), Discs #50201 ($80). Gr. 5–8.
BKL (5/75), ESL (10/77), PRV (9/75). 301.2
Contents: 1. Culture: What Is It? 2. Land
and Climate. 3. Government and Economy.
4. Education and Technology. 5. History. 6.
Religion.
An introduction to the concept of different
ways of life which shows clearly the wide
variations between Eskimo and Japanese,
French and Arabian life-styles. The filmstrips present the factors contributing to the
differences, show how they developed, and
illustrate many of them.

CUMMINGS, E. E.

POETS OF THE TWENTIETH CENTURY
(Filmstrip—Sound). *See* American literature—History and criticism

CURIE, MARIE

CHILDHOOD OF FAMOUS WOMEN,
VOLUME THREE (Cassette). *See* Wom-
en—Biography

CURIOSITIES AND WONDERS

MONSTERS AND OTHER SCIENCE MYS-
TERIES (Filmstrip—Sound). Bendick Asso-
ciates/Miller-Brody Productions, 1977. 8
color, sound filmstrips, av. 77–91 fr. ea.; 8
cassettes or discs, av. 18–21 min. ea.
#MSM 100 ($160). Gr. 4–9. BKL (1/78),
NEF (1978), PRV (11/78), 001.9
Contents: 1. The Mystery of the Loch Ness
Monster. 2. The Mystery of the Abominable
Snowman. 3. The Mystery of the Bermuda
Triangle. 4. The Mystery of Life on Other
Worlds. 5. The Mystery of Atlantis. 6. The
Mystery of Witchcraft. 7. The Mystery of
ESP. 8. The Mystery of Astrology.
Eight questions are explored in this pro-
gram. Original art, photography, maps, old

drawings and engravings are used. Each filmstrip presents the facts and theories that should be considered.

CURRIER, NATHANIEL

THE AMERICA OF CURRIER AND IVES (Filmstrip—Sound). *See* Prints, American

CYCLOTRON. *See* Atom smashers

DNA

THE LIVING CELL: DNA (Motion Picture—16mm—Sound). *See* Cells

DANCING

AMERICA DANCES (Phonodisc or Cassette). *See* Folk dancing

DARWIN, CHARLES

CHARLES DARWIN (Cassette). Ivan Berg Associates/Jeffrey Norton Publishers, 1976 (History Makers). 1 cassette, approx. 77 min. #41004 ($11.95). Gr. 10–A. BKL (9/15/77). 921

In an imaginary interview, Darwin tells of his plans to be a preacher, which were altered by the kindling of his interest, during a voyage to South America, for scientific studies. This tape examines the men and ideas that influenced Darwin's thinking.

GALAPAGOS: DARWIN'S WORLD WITH-IN ITSELF (Motion Picture—16mm—Sound). *See* Galapagos Islands

DATING (SOCIAL CUSTOMS)

ADOLESCENCE, LOVE AND DATING (Kit). *See* Adolescence

MATE SELECTION: MAKING THE BEST CHOICE (Filmstrip—Sound). Human Relations Media, 1975. 2 color, sound filmstrips, 71–77 fr.; 2 cassettes or discs, 13 min. ea.; Teacher's Guide. #612 ($60). Gr. 10–C. BKL (4/15/76), MNL (4/76), PVB (4/77), 301.41

Contents: 1. Predicting Success. 2. Beyond Engagement.

Examines factors in mate selection that often influence marital success or failure. Assesses the roles of romantic love, personal values, family and social pressures, and emotional needs during dating and engagement. Stresses the importance of self-knowledge in successful mate selection.

VALUES FOR DATING (Filmstrip—Sound). Sunburst Communications, 1974. 4 color, sound filmstrips, 76–104 fr.; 4 cassettes or discs, 14–15 min.; Teacher's Guide. #206 ($99). Gr. 7–12. NCF (1975), PRV (1/19/76), TSS (3/19/75). 301.42

Contents: 1. Pressures. 2. Traditional Values. 3. Love and Friendships. 4. Two Couples in Love.

Investigates the attitudes of young people regarding two important dating areas—love and sex. Contrasts current dating values with those of the past, suggests that today, widely different approaches to love and sex are possible and that no one approach is necessarily "right." Low key, objective and nonjudgmental.

D'AULAIRE, EDGAR PARIN

CHILDREN OF THE NORTHLIGHTS (Motion Picture—16mm—Sound). *See* Authors, American

D'AULAIRE, INGRI

CHILDREN OF THE NORTHLIGHTS (Motion Picture—16mm—Sound). *See* Authors, American

DEAF

DADDY, CAN I HEAR THE SUN? (Motion Picture—16mm—Sound). Film Company, Adelphi University/Stanfield House, 1978. 16mm color, sound film ($365). Gr. 9–C. IFT (1978). 362.4

About the meaning and advantage of total communications, worthwhile human interaction between the hearing and the nonhearing world, community integration and the disintegration of stereotypes.

MARY (Motion Picture—16mm—Sound). WGBH Educational Foundation/EBEC, 1978 (People You'd Like to Know). 16mm color, sound film, 20 min. #3587 ($185). Gr. 4–8 LFR (1979). 362.4

Born deaf, Mary is still learning how to speak. She attends language class and works individually with a speech therapist. Scenes show the problems that can rise from her condition—and that they have not discouraged her.

DEATH

DEATH: A NATURAL PART OF LIVING (Filmstrip—Sound). Marshfilm, 1977, (Mental Health/Hygiene). 1 color, sound filmstrip, approx. 55 fr.; 1 disc or cassette, approx. 9 min. #1134 ($22). Gr. 5–9. PRV (2/78), PVB (4/78). 128.5

Included are the results of recent research on the five steps of acceptance of death by the dying. An interesting and factual look at death as a natural part of the life cycle of all living organisms, presents ancient and foreign customs, modern day trends and concerns and practical guidance for young people.

DEATH (cont.)

DEATH AND DYING: CLOSING THE CIRCLE (Filmstrip—Sound). Guidance Associates, 1975. 5 color, sound filmstrips, 82–115 fr.; 5 cassettes or discs, 14–20 min. Teacher's Guide. #102549 ($139.50). Gr. 10–A. M&M (1977), PRV (12/76) PVB (4/77). 128.5

In this program the emphasis on youth and the development of technology is shown as suppressing the awareness of death. It is pointed out that there is value in confronting our feelings about death and analyzing the guilt and other emotions experienced by survivors. The five types of ''immortality'' wrought by individuals are explored. Author Doris Lund recreates her son Eric's five-year struggle against leukemia. Her narrative sets the stage for dialogue about loss, courage, hope and finality. New approaches to coping with critical illness are discussed. Positive approaches to talking about death are explored.

ECHOES (Motion Picture—16mm—Sound). *See* Life

GRAMP: A MAN AGES AND DIES (Filmstrip—Sound). *See* Old age

LIVING WITH DYING (Filmstrip—Sound). Sunburst Communications, 1973. 2 color, sound filmstrips, 72–78 fr.; 2 discs or cassettes, 14–15 min.; 2 Student Activity Cards; Teacher's Guide. #110–81 ($59). Gr. 6–C. BKL (8/73), M&M (3/75), PVB (5/74). 128.5

Contents: 1. Acceptance. 2. Immortality.

This set takes the view that children should cope with death's inevitability. A falling leaf, a dead bird, a young woman with a terminal disease create the awareness of death while offsetting the imagery. The five stages of the acceptance of death are presented, including historic flashbacks that relate death customs of the past to the problems of those facing death today.

PERSPECTIVES ON DEATH (Filmstrip—Sound). Sunburst Communications, 1976. 2 color, sound filmstrips, 63–74 fr.; 2 cassettes or discs, 11–12 min.; Teacher's Guide. #228 ($59). Gr. 7–12. MNL (2/77). PRV (2/78), TSS (3/77). 301.322

Contents: 1. Toward an Acceptance. 2. The Right to Die.

Introduces students to different attitudes toward death, investigating various reactions to its inevitability and illustrating how people have used their religious beliefs, philosophy, science, and art to cope with the realization of their own mortality. Discusses the medical, legal, and moral questions concerned with an individual's right to life and death.

THE RIGHT TO DIE? (Filmstrip—Sound). Current Affairs Films, 1977. 1 color, sound filmstrip; 1 cassette; Guide. #570 ($24). Gr. 9–12. PRV (11/78). 128.5

This filmstrip discusses the moral, legal, and practical implications between allowing a person to ''die with dignity'' or sustaining life at all possible costs. The issues are faced from a medical point of view and from the impact on society as a whole.

RONNIE'S TUNE (Motion Picture—16mm—Sound). Wombat Productions, 1977. 16mm color, sound film, 18 min. ($270). Gr. 7–A. AFF (1978), FLN (3/4/78). 301.322

''Why did he do it?'' the question posed by eleven-year-old Julie regarding the death of a teen-age cousin, may be without answer. But the sense of responsibility and guilt of those left behind remains—and must be dealt with. A tasteful, touching film on an increasingly important topic.

WHERE IS DEAD? (Motion Picture—16mm—Sound). Life Style Productions/EBEC, 1975. 16mm color, sound film, 19 min. #3417 ($255). Gr. 4–A. AAS (1976) AFF (1977), LNG (1976). 155.937

Low-keyed drama deals compassionately, yet realistically, with a subject which young children may be forced to face: the death of a loved one. Six-year-old Sarah plays and fights with her nine-year-old brother, David, in a series of recognizable childhood vignettes. David's sudden death rends the family fabric. Sarah's parents attempt to explain what has happened—with tenderness but without sentimentality. Gradually, the little girl is able to cope with her feelings of sadness, confusion and fear . . . and life regains most of its earlier joys.

DECISION MAKING

DE FACTO (Motion Picture—16mm—Sound). Filmbulgaria/EBEC, 1974. 16mm color, sound film, 8 min. #3362 ($125). Gr. 7–A. BKL (1/1/75), LFR (3/75), LNG (1976). 153.8

In this nonnarrated, animated story the ceremony of dedicating a new building is about to begin. The people gather, the officials appear, the drummer beats his drum in salute to the magnificent new building which thereupon collapses! Who is to blame? It ends with an ironic twist.

DECISIONS, DECISIONS (Filmstrip—Sound). Marshfilm, 1976 (Mental Health/Hygiene). 1 color, sound filmstrip, approx. 55 fr.; 1 disc or cassette, approx. 9 min. #1135 ($21). Gr. 5–9. JNE (9/74), PRV (2/78), PVB (4/78). 153.8

Being prepared to make the right decision is what this filmstrip is all about. Value judgment, choices, options and consequences, awareness of alternatives, weighing pros and cons, all are presented with cartooning and photographs.

DIVIDED MAN, COMMITMENT OR COMPROMISE? (Motion Picture—16mm—Sound). Film Polski/Bosustow Productions,

1973. 16mm color, sound, animated film, 5 min. ($120). Gr. 5–10. BKL (11/11/73), CIF (1975), PRV (1973). 153.8

A solitary figure travels a road happily until he comes to a fork. He hesitates, there is no indication what may lie ahead on either path. He makes several false starts, thinks better of it and returns to the original point. Because he is struggling to make the perfect choice, he divides and travels both paths. The two halves reunite at a later crossroads, but the two halves no longer fit together. They have changed on their journeys and cannot be the same man again.

LIFE GOALS: SETTING PERSONAL PRIORITIES (Filmstrip—Sound). *See* Self-realization

LIFESTYLES: OPTIONS FOR LIVING (Kit). *See* Life-styles—U.S.

MAKING JUDGMENTS AND DRAWING CONCLUSIONS (Slides/Cassettes). *See* Reading—Study and teaching

THE PSYCHOLOGY OF JUDGMENT (Filmstrip—Sound). *See* Judgment

DEFOE, DANIEL

EPISODES FROM FAMOUS STORIES (Filmstrip—Sound). *See* Literature—Collections

DEGAS, EDGAR

DEGAS (Motion Picture—16mm—Sound). National Gallery of Art & Visual Images/ EBEC, 1973 (Art Awareness Collection). 16mm color, sound film, 6 min. #3537 ($105). Gr. 7–A. BKL (1978), EFL (1977), LFR (1978). 759.4

Racetracks, circus, dancehalls and ballet challenge the creativity of Edgar Degas. The film presents some of Degas's most vivid work—and compares the fleeting moments he paints, with frames from early motion pictures by Eadweard Muybridge.

DEITCH, GENE

GENE DEITCH: THE PICTURE BOOK ANIMATED (Motion Picture—16mm—Sound). Morton Schindel/Weston Woods Studios, 1977 (Signature Collection). 16mm color, sound film (also available as video-cassette), 25 min. #427 ($305). Gr. 7–A. CFF (1977), FLN (3/77), HOB (8/77). 778.534

Gene Deitch turns his camera upon himself, his colleagues and his production facilities in Prague, Czechoslovakia, to reveal his technique, purpose and art in adapting picture books. In this film, he invites the audience to share with him the excitement and discovery of the unique gifts animation can bring to the meaning of picture books.

DENVER, JOHN

MEN BEHIND THE BRIGHT LIGHTS (Kit). *See* Musicians

DEPRESSIONS

THE TWENTIES AND THIRTIES (Filmstrip—Sound). *See* U.S.—History—1919–1933

DESERTS

DAYBREAK (Motion Picture—16mm—Sound). *See* Nature in poetry

THE DESERT: PROFILE OF AN ARID LAND (Filmstrip—Sound). EBEC, 1974. 5 color, sound filmstrips, av. 101 fr. ea.; 5 discs or cassettes, 15 min. ea.; Teacher's Guide. #6497 Discs ($72.50), #6497K Cassettes ($72.50). Gr. 4–8. CPR (11/75), ESL (1977), PRV (5/76). 551.58

Contents: 1. The Desert Environment: An Overview. 2. Face of the Desert: Patterns of Erosion. 3. Life in the Desert: Plants. 4. Life in the Desert: Animals. 5. The Desert: Environment in Danger.

This series shows how deep canyons are formed by desert streams and sand dunes by violent windstorms; how one plant survives by growing its own shade and how a tiny animal manufactures its own water. The last filmstrip in the series shows how humans are changing the face of the desert and why they pose a threat to this environment.

THE ECOLOGY OF A DESERT: DEATH VALLEY (Filmstrip—Sound). Educational Development, 1973. 2 color, sound filmstrips, 35–40 fr.; 2 cassettes or discs, 11–12 min.; Manual. #406–R Discs ($30), #406–C Cassettes ($32). Gr. 5–8. BKL (3/15/74). 551.58

Contents: 1. Deserts. 2. Ecology.

The set describes the ways in which certain animals have adapted to the Death Valley desert, explains the area's geological structure, shows towns abandoned in the late 1800s because of drought and isolation and examines the brief influx of gold, silver, and borax seekers.

FOUR WHEELS (Filmstrip—Sound). *See* Camping

UNDERSTANDING NATURAL ENVIRONMENTS: SWAMPS AND DESERTS (Slides). *See* Ecology

THE WILD YOUNG DESERT SERIES (Filmstrip—Sound). Lyceum/Mook & Blanchard, 1970. 2 color, sound filmstrips, 50 fr.; 2 cassettes or discs, 10 min. #LY35570SC Cassette ($25), #LY35570SR Disc ($19.50). Gr. K–A. BKL (9/1/70), STE (2/1/71). 551.58

Contents: 1. The Making of a Desert. 2. Life Conquers the Desert.

DESERTS (cont.)

Written and photographed by Ann Atwood. The color photography of this set brings an appreciation of the desert to those who know and love it and to those who have never experienced the desert. This set deepens awareness of the geological background, ecological balance, and the unique beauty of the desert.

DESIGN, DECORATIVE

CHANGING THE FACE OF THINGS (Filmstrip). Visual Publications, 1972. 6 color, silent filmstrips, 22–38 fr.; Handbooks. #CFT/E ($54). Gr. 6–A. PRV (12/73), PVB (5/74). 745.4

Contents: 1. General Introduction. 2. Effects of Pattern and Color. 3. Effects of Reflection and Light. 4. Unification, Emphasis, and Division. 5. Camouflage and Displays. 6. Design and Surface Qualities.

An exploration of the interplay between surface, form, and space using as examples buildings, boats, furniture, clothing, pottery, sculpture, posters, trick photography, face make-up, masks, shop signs and a host of other things.

DESIGN IS A DANDELION (Filmstrip—Sound). BFA Educational Media, 1969. 6 color, sound filmstrips; 6 cassettes or discs. Cassettes #R6000 ($104). Discs #95000 ($104). Gr. 4–8. BKL (7/1/70). 745.4

Contents: 1. Design in Nature. 2. Design in Form. 3. Design in Texture. 4. Design in Balance. 5. Design in Rhythm. 6. Design in Contrast.

Designed to enhance sensitivity to, and appreciation of, the major elements of design as well as the beauty around us. Adapted by George Manitzas from the book by Janice Lovoos.

TEXTILES AND ORNAMENTAL ARTS OF INDIA (Motion Picture—16mm—Sound). *See* India—Handicraft

DEVIL—FICTION

EDGAR ALLAN POE'S THE BLACK CAT (Filmstrip—Sound). Prentice-Hall Media, 1978 (Prentice-Hall Media Classic Adaptations). 1 color, sound filmstrip, 71 fr.; 1 cassette, 14 min.; 1 Guide. #KHC758 ($25). Gr. 7–A. BKL (9/1/78). Fic

The elements of Poe's classic are maintained, but the details are markedly altered. The devil incarnated as a loving black cat shapes the narrator into an instrument of evil.

DICKENS, CHARLES

CHARLES DICKENS (Cassette). Ivan Berg Associates/Jeffrey Norton Publishers, 1976 (History Makers). 1 cassette, approx. 59 min. #41014 ($11.95). Gr. 10–A. BKL (6/15/78). 921

This cassette begins with a biographical sketch. Incidents in Charles Dickens's life are given and dramatizations illustrate how these events were transformed into his works.

A CHRISTMAS CAROL (Phonodisc or Cassette). *See* Christmas—Fiction

A CHRISTMAS CAROL (Kit). *See* Christmas—Fiction

EPISODES FROM CLASSIC NOVELS (Filmstrip—Sound). *See* Literature—Collections

GREAT WRITERS OF THE BRITISH ISLES, SET II (Filmstrip—Sound). *See* Authors, English

LIBRARY 3 (Cassette). *See* Literature—Collections

NOVELISTS AND THEIR TIMES (Kit). *See* Authors

DICKEY, JAMES

JAMES DICKEY: POET (LORD LET ME DIE, BUT NOT DIE OUT) (Motion Picture—16mm—Sound). EBEC, 1970. 16mm color, sound film, 37 min. #47738 ($450). Gr. 9–A. CFF (1971), M&M (1971), MER (1973). 921

A poetry experience in film that does not "teach poetry." The words of the poet and the people he talks to and the camera that follows him for three weeks on a Dickey barnstorming-for-poetry tour project. He is a man trying "to break through this kind of glass wall we all live behind." This film is to "break through" to involve in the actual thoughts and feelings of a man who happens to write fine poetry.

DICKINSON, EMILY

NINETEENTH CENTURY POETS (Filmstrip—Sound). *See* American literature—History and criticism

DICTATORS

ADOLF HITLER (Cassette). *See* Hitler, Adolf

DIET

BEFORE YOU TAKE THAT BITE (Motion Picture—16mm—Sound). *See* Nutrition

EAT, DRINK, AND BE WARY (Motion Picture—16mm—Sound). *See* Nutrition

THE EATING ON THE RUN FILM (Motion Picture—16mm—Sound). *See* Nutrition

FOOD: HEALTH AND DIET (Filmstrip—Sound). *See* Nutrition

NUTRITION: FOOD VS. HEALTH (Filmstrip—Sound). *See* Nutrition

TOO MUCH OF A GOOD THING (Filmstrip—Sound). *See* Nutrition

DINOSAURS

DINOSAURS: THE TERRIBLE LIZARDS (Motion Picture—16mm—Sound). Avatar Productions, 1976 (Wide World of Adventure). 16mm color, sound film, 24 min. #3504 ($320). Gr. 4–12. AVI (1977), LFR (1977), PRV (1978). 568.1

Surveys the variety of life during the Age of Reptiles, explores the methods paleontologists use to excavate an actual dinosaur dig, and shows how a dinosaur skeleton is assembled and reconstructed for museum display.

DISASTERS

EARTHQUAKES: LESSON OF A DISASTER (Motion Picture—16mm—Sound). *See* Earthquakes

DISCOVERIES (IN GEOGRAPHY)—FICTION

THE KING'S FIFTH (Filmstrip—Sound). Miller-Brody Productions, 1976 (Newbery Award Records). 2 color, sound filmstrips, 125–137 fr.; 2 cassettes or discs, 16–17$\frac{1}{2}$ min.; 1 Guide. #NSF-3066C Cassette ($32), #NSF-3066 Disc ($32). Gr. 3–9. BKL (4/15/77). Fic

Scott O'Dell's 1967 Newbery honor book is adapted in this abridged filmstrip version. It is a tale of the Spanish conquistadors.

DISEASES

ENEMIES OF THE BODY (Filmstrip—Sound). Educational Activities, 1977. 4 color, sound filmstrips, 48–53 fr.; 4 cassettes, 10–12 min.; Teacher's Guide. #FSC–487 ($64). Gr. 8–A. INS (2/78), PRV (1/78), SCT (1/78). 616

Contents: 1. Cancer—What Is It? 2. Hypertension—The Quiet Killer. 3. Heart Attack—Sudden Terror. 4. Diabetes—Sugar Gone Away.

This series focuses on the diseases which constitute the major health problems in our society. A comprehensive presentation which indicates how to avoid or conquer these enemies by taking proper precautions, heeding warning signs, and following medical advice.

DISSECTION

BIOLOGICAL DISSECTION (Filmstrip—Sound). Clearvue, 1978. 6 color, sound filmstrips, 40–57 fr.; 6 cassettes, 10$\frac{1}{2}$–11$\frac{1}{2}$ min.: 1 Guide. #CL 420–C ($81.50). Gr. 7–C. BKL (1/1/79). 591.4

Contents: 1. An Introduction to Dissection. 2. The Earthworm. 3. The Crayfish. 4. The Fish. 5. The Frog, Part 1. 6. The Frog, Part 2.

The relationship of environment to the physiology of a species and to the evolutionary adaptations it has made are presented first. The remaining strips show dissection of the four animals—a worm, crayfish, fish and frog. The various tools for dissection are shown and their uses explained. Terminology used is defined.

DISSECTION OF A FROG (Filmstrip—Sound). Library Filmstrip Center, 1977. 2 color, sound filmstrips, 47–63 fr.; 2 cassettes or discs, 13–18$\frac{1}{4}$ min. ($55). Gr. 9–C. PRV (1/79). 591.4

This detailed set illustrates the skin, muscular, nervous, endocrine, and reproductive systems of the frog. It demonstrates dissection methods, scientific observations and experimentations.

DISSENT

WOMEN WRITERS: VOICES OF DISSENT (Filmstrip—Sound). *See* Women authors

DIVORCE

MY PARENTS ARE GETTING A DIVORCE (Filmstrip—Sound). Human Relations Media Center. 2 color, sound filmstrips, 71–80 fr.; 2 discs or cassettes, 9–10 min.; 1 Guide. #618 ($45). Gr. 6–12. BKL (11/15/76), LGB (12/77), M&M (4/77). 301.428

Contents: 1. Separation. 2. Adjusting.

Straightforward answers and mature discussions are what is needed when young people are left hanging in limbo on their parents' separation. The set offers explanations for the unusual behavior parents may exhibit; warns of reactions of other members of the family and of emotions children may feel but not understand.

DOCTOR HEIDEGGER'S EXPERIMENT

A DISCUSSION OF DR. HEIDEGGER'S EXPERIMENT (Motion Picture—16mm—Sound). *See* Motion pictures—History and criticism

DOGS—FICTION

CALL OF THE WILD (Filmstrip—Sound). Listening Library, 2 color, sound filmstrips; 2 cassettes. ($28). Gr. 5–9. LNG (3/76). Fic

The familiar Jack London story, adapted to this two part filmstrip with cartoon illustrations. It contains all the adventure, pathos, and vitality of the original.

DOGS—FICTION (cont.)

THE CALL OF THE WILD (Kit). Current Affairs Films, 197. 1 color, sound filmstrip, 70 fr.; 1 cassette, 15 min.; 1 Book, 1 Guide. #620 ($30). Gr. 6–12. BKL (1/1/79). Fic

In 1897, the Klondike Gold Rush was in full swing, and men were on the lookout for prime sled dogs. This is the story by Jack London of how the dog, Buck, is stolen from his California home and taken to the frozen North, and of what happens to him there. The introductory segment describes the influences of Nietzsche, Marx, and Darwin on London's philosophies and thus his work.

LASSIE COME HOME (Phonodisc or Cassette). Caedmon Records, 1973. 1 cassette or disc, 65 min. #TC1389 Disc ($6.98), #CDL51389 Cassette ($7.95). Gr. 4–8. BKL (2/15/74). Fic

Eric Knight's beloved story which features a resourceful collie whose courage and brightness are tested in each adventurous episode. David McCallum's beautifully controlled reading preserves the suspense and adventure of the story in this carefully abridged adaptation.

DOLLS

THE DOLLMAKER (Filmstrip—Sound). Warner Educational Productions, 1978. 4 color, sound filmstrips, 64–114 fr.; 2 cassettes, 8–15 min.; Teacher's Guide. #165 ($78.50). Gr. 7–A. BKL (2/15/79). 688.722

This program outlines the procedures for making a doll. It describes the construction of a relatively simple Kewpie doll and then moves on to more progressively difficult and complex dolls. The final jointed and elaborately costumed ballerina is demonstrated.

DOLPHINS

BOTTLENOSE DOLPHIN (Film Loop—8mm—Silent). Walt Disney Educational Media, 1966. 8mm color, silent film loop, approx. 4 min. #62-5465L ($30). Gr. K–12. MDU (1/74). 599.53

Presents the birth of a dolphin and shows the herd protecting it. Dolphins play and chase a turtle and shark, while the mother nurses the newborn and guides it to the surface for air.

DOSTOEVSKI, FEODOR M.

THE CROCODILE (Motion Picture—16mm—Sound). See Fantasy—Fiction

A DISCUSSION OF THE CROCODILE (Motion Picture—16mm—Sound). See Motion pictures—History and criticism

NOVELISTS AND THEIR TIMES (Kit). See Authors

DOYLE, ARTHUR CONAN

THE ADVENTURE OF THE SPECKLED BAND BY DOYLE (Cassette). See Mystery and detective stories

EPISODES FROM CLASSIC NOVELS (Filmstrip—Sound). See Literature—Collections

DRAMA

AMERICAN THEATRE (Filmstrip—Sound). See Theater—U.S.

DRAMA—HISTORY AND CRITICISM

MEDIEVAL THEATER: THE PLAY OF ABRAHAM AND ISAAC (Motion Picture—16mm—Sound). See Theater—History

DRAMATISTS

INTERVIEWS WITH PLAYWRIGHTS: SOPHOCLES, SHAKESPEARE, O'NEILL (Cassette). Arthur Meriwether Education Resources, 1975 (Interviews with Immortals Series). 3 cassettes, av. 16 min. ea.; Plastic Folder; Printed Guide. #C–43 ($19.50). Gr. 9–12. BKL (6/76). 920

Dramatized interviews provide insight into each dramatist's life and personality. Included are Sophocles, William Shakespeare, and Eugene O'Neill.

LORRAINE HANSBERRY (Motion Picture—16mm—Sound). See Hansberry, Lorraine

WILLIAM SHAKESPEARE (Cassette). See Shakespeare, William

DRAWING

DRAWING: ARTISTRY WITH A PENCIL (Filmstrip—Sound). Warner Educational Productions, 1978. 4 color sound filmstrips, 81–114 fr.; 2 cassettes 10–16 min.; Teacher's Guide. #820 ($78.50). Gr. 7–A. BKL (3/1/79). 741.2

Basic introduction to drawing with a pencil. Shows techniques to use when working with line, value, and sketches. Demonstrates linear perspective and aerial perspective. Gives a step-by-step analysis of drawings as the artist is working and shows how to draw from photographs, objects, and people.

SEEING AND DRAWING (Filmstrip—Sound). Encore Visual Education, 1977. 4 color, sound filmstrips, 46–54 fr.; 4 cassettes, 6–10 min.; Teacher's Guide. #87 ($59). Gr. 7–A. PRV (1/79). 741.2

Contents: 1. Introduction to Drawing. 2. Drawing Tools, Materials, and Techniques.

3. Contour and Gesture Drawing. 4. Perspective and Fantasy Drawing.

Numerous works of art are presented and the relationships of different types of art. Close-ups are used to demonstrate processes in the various methods shown. The set helps students understand the purposes of drawing and presents a variety of approaches, use of materials, tools and techniques.

DREAMS

THE DREAM (Motion Picture—16mm—Sound). *See* Pantomimes

DRESDEN

OUT OF THE HOLOCAUST (Filmstrip—Sound). *See* Cities and towns

DREYFUS, ALFRED

THE DREYFUS AFFAIR (Motion Picture—16mm—Sound). Texture Films, 1969. 16mm color, sound film, 15 min. ($120). Gr. 8–A. BKL (1971), M&M (1971). 921

Filmed from authentic period graphics, this is the history of Captain Alfred Dreyfus, a French army officer who was falsely accused of treason and whose trials inflamed France and became an international cause celebre.

DRUG ABUSE

ACID (Motion Picture—16mm—Sound). *See* Hallucinogens

ALCOHOL AND ALCOHOLISM: THE DRUG AND THE DISEASE (Filmstrip—Sound). *See* Alcoholism

ALCOHOL, DRUGS, OR ALTERNATIVES (Motion Picture—16mm—Sound). *See* Alcoholism

ALCOHOL: FACTS, MYTHS, AND DECISIONS (Filmstrip—Sound). *See* Alcohol

ALMOST EVERYONE DOES (Motion Picture—16mm—Sound). Wombat Productions, 1970. 16mm color, sound film, 14 min.; Spanish and French available. ($210). Gr. 4–12. CRA (1974), SOC (3/74). 301.47

We live in a time when it's accepted that we take something to change our feelings. But how can we use our feelings—all of them?

PROJECT AWARE (Motion Picture—16mm—Sound). *See* Juvenile delinquency

STORY OF JOE: RECOLLECTIONS OF DRUG ABUSE (Filmstrip—Sound). Richard Bruner Productions/Sunburst Communications, 1970. 6 color, sound filmstrips, 125–163 fr.; 6 cassettes or discs, 15–18 min.; Teacher's Guide. #210 ($145). Gr. 7–12. AFF (1970), PRV (1/76). 301.47

Contents: 1. Early Childhood. 2. Adolescent Rebellion. 3. Initiation into the Drug Subculture. 4. Arrest and Probation. 5. Going Nowhere. 6. New Start.

A true story, narrated in part by Joe, this set examines prevalent attitudes toward causes of drug use, depicts the drug experience itself, and portrays the difficulty of breaking the drug habit.

WEED (MARIJUANA) (Motion Picture—16mm—Sound). *See* Marijuana

WHAT ARE YOU GOING TO DO ABOUT ALCOHOL (Filmstrip—Sound). *See* Alcohol

DRUGS—PHYSIOLOGICAL EFFECT

PSYCHOACTIVE (Motion Picture—16mm—Sound). Pyramid Films, 1976. 16mm color, sound film, 30 min.; $3/4$ in. videocassette, also available. ($350). Gr. 7–12. BKL (11/15/77), LFR (1977), PRV (1977). 613.8

This film informs its viewers about the physiological effects of each of the five classifications of psychoactive drugs on the human body. Combining live-action and animation sequences, it explains how the nine systems of the body function and demonstrates the way in which a drug meant for one body system affects all the other systems as well.

DUBROVIN, VIVIAN

SADDLE UP! (Kit). *See* Horses—Fiction

DURDEN, KENT

TWO FOR ADVENTURE (Kit). *See* Animals—Fiction

DUVOISIN, ROGER

THE HAPPY LION SERIES, SET ONE (Filmstrip—Sound). *See* Lions—Fiction

THE HAPPY LION SERIES, SET TWO (Filmstrip—Sound). *See* Lions—Fiction

DYES AND DYEING

CRAYON BATIK MAGIC (Filmstrip—Sound). *See* Batik

CREATIVE BATIK (Filmstrip—Sound). *See* Batik

CREATIVE TIE/DYE (Filmstrip—Sound). *See* Batik

EAGLES

BIRD OF FREEDOM (Motion Picture—16mm—Sound). Unit One Film Productions,

EAGLES (cont.)

1976. 16mm color, sound film, 13½ min.; Guide. ($225). Gr. 5–10. BKL (10/15/77). 598.9

The majestic spirit of the American bald eagle is shown in the eagle's physical characteristics and range of habitat as well as its symbolic image throughout U.S. history and its interpretation by various artists.

EARTH—CRUST

THE MOVING EARTH: NEW THEORIES OF PLATE TECTONICS (Filmstrip— Sound). *See* Plate tectonics

EARTHQUAKES

EARTHQUAKES: LESSON OF A DISASTER (Motion Picture—16mm—Sound). EBEC, 1971. 16mm color, sound film, 13 min. #3056 ($185). Gr. 5–12. AVI (1975), LFR (1972), SLJ (1972). 551.1

Introduces the phenomena of earthquakes through the case study of two major quakes with on-location footage showing the devastation. Seismologists demonstrate the use of P waves, R waves and other methods to determine an earthquake's occurrence, location, and magnitude.

THE SAN ANDREAS FAULT (Motion Picture—16mm—Sound). EBEC, 1974. 16mm color, sound film, 21 min. #3304 ($275). Gr. 7–A. AAS (1975), CGE (1974), EFL (1975). 551.2

The film shows what the San Andreas fault trace looks like, what it has done to the landscape and to the rocks of California, and how it is being monitored and studied with creep meters, tiltmeters, strain gauges, and seismometers. The use of computers to forecast seismic activity is described.

EASTERN EUROPE

EASTERN EUROPE FROM WITHIN (Filmstrip—Sound). EBEC, 1973. 6 color, sound filmstrips, av. 103 fr. ea.; 6 discs or cassettes, 15 min. ea.; Teacher's Guide. Discs #6470 ($86.95), Cassettes #6740 ($86.95). Gr. 5–9. BKL (11/15/74), ESL (1977), PRV 4/75). 914.3

Contents: 1. Poland: My Country. 2. Czechoslovakia: My Country. 3. Hungary: My Country. 4. Yugoslavia: My Country. 5. Romania: My Country. 6. Bulgaria: My Country.

Shrouded in a veil of half-legend and half-propaganda, the small countries of eastern Europe are difficult for outsiders to know and understand. In this series, nationals of these countries speak out about life in their homelands.

ECOLOGY

BIOLOGICAL CATASTROPHES: WHEN NATURE BECOMES UNBALANCED (Slides/Cassettes). Center for Humanities, 1976. 160 slides in Carousel cartridges; 2 cassettes or discs; Teacher's Guide. #1006 ($139.50). Gr. 6–12. BKL (7/1/76), IDF (1976), SCT (4/77). 574.5

A framework to help understand and evaluate the vast changes created by our technological society is presented as well as the issues involving complex value choices. The program establishes the vital role of natural cycles in maintaining life, documents their delicate balance, and illustrates ways in which people deplete, destroy, disrupt, and overload natural cycles.

CANNERY ROW, LIFE AND DEATH OF AN INDUSTRY (Filmstrip—Sound). *See* Monterey, California—History

ECOLOGICAL COMMUNITIES (Filmstrip—Sound). Coronet Instructional Media, 1971. 6 color, sound filmstrips, av. 53 fr.; 6 cassettes or 3 discs, av. 11 min. #M205 Cassettes ($95), #S205 Discs ($95). Gr. 5–12. BKL, PVB. 574.5

Contents: 1. The Deciduous Forest. 2. Ponds and Lakes. 3. The Streams. 4. The Meadow. 5. The Thicket. 6. The Northern Coniferous Forest.

The photography of six major types of ecological communities. Close-up studies show the interdependence of plant and animal life and adaptations for survival.

ECOLOGY: BALANCE OF NATURE (Filmstrip–Sound). Marshfilm, 1972 (Ecology). 1 color, sound filmstrip, 45 fr.; 1 cassette or disc, 15 min.; Teacher's Guide. #1112 ($21). Gr. 4–9. BKL (4/1/73), ESL (1977), SLJ (4/72). 574.5

Everything on earth affects everything else. Interactions are seen through food chains, changes in natural communities, pyramids of plant and animal mass living in an area, disease, weather, and the cycles of chemicals.

ECOLOGY: BARRY COMMONER'S VIEWPOINT (Motion Picture—16mm—Sound). EBEC, 1977. 16mm color, sound film, 19 min.; Guide. #3519 ($270). Gr. 7–12. BKL (1978), LFR (1978), TED (1979). 574.5

One of America's most outspoken biologists, Dr. Barry Commoner explains the threat of technology to the environment by describing four principles of ecology. These are: "Everything is connected to everything else," "Everything must go somewhere," "Nature knows best," and "There is no such thing as a free lunch." These principles are visualized through a study demonstrating the effects of automobile pollution upon the long-term survival of a National Forest.

THE ECOLOGY OF A DESERT: DEATH VALLEY (Filmstrip—Sound). *See* Deserts

ECOLOGY OF A HOT SPRING: LIFE AT HIGH TEMPERATURE (Motion Picture—16mm—Sound). EBEC, 1972 (Biology). 16mm color, sound film, 15 min.; available in Spanish. #3143 ($185). Gr. 7–12. CGE (1972), LFR (1/19/73), SCT (5/19/73). 574.5

A study of Yellowstone Park's hot springs provides answers to what type of life evolved and survived when the earth was much hotter billions of years ago.

THE ECOLOGY OF A TEMPERATE RAIN FOREST (Filmstrip—Sound). *See* Rain forests

ECOLOGY: SPACESHIP EARTH (Filmstrip—Sound). Marshfilm, 1971 (Ecology). 1 color, sound filmstrip, 45 fr.; 1 cassette or disc, 15 min.; Teacher's Guide. #1111 ($21). Gr. 4–8. BKL (4/1/73), ESL (1976), SLJ (4/72). 574.5

This filmstrip emphasizes the life-support systems of earth, i.e., our energy supply, food, air, water, raw materials, and the living. The problems of pollution, over-population, and consumption are related to the continued survival of our planet.

ECOLOGY: UNDERSTANDING THE CRISIS (Filmstrip—Sound). EBEC, 1971. 6 color, sound filmstrips, av. 82 fr. ea.; 6 cassettes or discs, 9 min. ea.; Teacher's Guide. Discs #6454 ($86.95), Cassettes #6454K ($86.95). Gr. 5–A. BKL (2/15/74), LFR (10/72), PRV (9/72). 574.5

Contents: 1. Environments and Ecosystems. 2. Man in Ecosystems. 3. Human Communities Simple and Complex. 4. Creating Imbalances. 5. Destroying the Future. 6. Creating the Future.

Ecology holds the key to understanding today's environmental crisis—an understanding vital to the survival of our planet. How the crisis has come about is shown with dramatic clarity in this series.

ENERGY AND OUR ENVIRONMENT (Filmstrip—Sound). *See* Power resources

EXPLORING ECOLOGY (Filmstrip—Sound). National Geographic, 1974. 5 color, sound filmstrips, 60–72 fr.; 5 discs or cassettes, 12–14 min. #03758 Discs ($74.50), #03759 Cassettes ($74.50). Gr. 4–12. PRV (5/75), PVB (4/76). 574.5

Contents: 1. The Mountain. 2. The River. 3. The Woodland. 4. The Prairie. 5. The Swamp.

Explores the relationship of living things in five distinct environments. Adaptations of mountain plants and animals and the myriad forms of life in rivers and swamps. The seasonality in the woodlands and the productivity of prairies are observed through the beautiful photography.

FIRE! (Motion Picture—17mm—Sound). *See* Forest fires

FOUR BIOMES (Filmstrip—Sound). Jean-Michel Cousteau/BFA Educational Media, 1976. 4 color, sound filmstrips, av. 55–74 fr.; 4 cassettes or discs, av. 9–13 min. #VHV000 Cassettes ($70), #VHU000 Discs ($70). Gr. 6–9. BKL (7/1/77). 574.5

Contents: 1. The Temperate Forest Biome. 2. The Rain Forest Biome. 3. The Grassland Biome. 4. The Desert Biome.

Climate is the basic factor that determines which organisms live in various parts of the world. However, a host of other factors interact to create the living community.

GRASSLAND ECOLOGY—HABITATS AND CHANGE (Motion Picture—16mm—Sound). *See* Grasslands

LET NO MAN REGRET (Motion Picture—16mm—Sound). *See* Man—Influence on nature

LIFE CYCLE OF COMMON ANIMALS, GROUP 2 (Filmstrip—Sound). *See* Sheep

THE MAYFLY: ECOLOGY OF AN AQUATIC INSECT (Motion Picture—16mm—Sound). *See* Mayflies

THE NEW ALCHEMISTS (Motion Picture—16mm—Sound). *See* Agriculture—Experiments

THE OTHER WORLD (Motion Picture—16mm—Sound). *See* Life (biology)

PARADISE LOST (Motion Picture—16mm—Sound). *See* Man—Influence on nature

THE SALT MARSH: A QUESTION OF VALUES (Motion Picture—16mm—Sound). *See* Marshes

SAND DUNE SUCCESSION (Filmstrip—Sound). Imperial Educational Resources, 1975. 2 color, sound filmstrips; 2 cassettes or discs; Teacher's Notes. #3RG40400 Discs ($30), #3KG40400 Cassette ($33). Gr. 4–8. BKL (11/15/75), PRV (4/76). 574.5

Contents: 1. From Beach to Pine Forest. 2. From Pine Forest to Beech-Maple.

This set begins with the description of environmental communities and their structure, then proceeds to the processes by which these communities are changed through time. The principles of ecological succession are examined through the stages of sand dune succession.

THOREAU ON THE RIVER: PERSPECTIVE ON CHANGE (Filmstrip—Sound). Frances D. Brooks & Fred A. Rowley, 1971. 1 color, sound filmstrip, 104 fr.; 1 cassette or disc, 20 min.; Teacher's Guide. ($21). Gr. 7–A. BKL (4/1/73), PRV (5/73), SLJ (11/72). 574.5

Portrays the contrasts of a changing America. This sound filmstrip juxtaposes selections from Thoreau's *A Week on the*

ECOLOGY (cont.)

Concord and Merrimack Rivers with scenes from the same trip taken recently. Thoreau's descriptions of the river's beauties are sharply contrasted with the ecological disasters of urbanization and industrial waste which are its present reality.

UNDERSTANDING NATURAL ENVIRONMENTS: SWAMPS AND DESERTS (Slides). Science and Mankind, 1977. 2 units with 80 slides ea. with 2 discs and cassettes of the same sound track. Teacher's Guide. #1021 ($140). Gr. 7–C. BKL (11/1/78), PRV (1/79). 574.9

A view of deserts and swamps, presenting plant and animal ecosystems—their organization, how they are stablized and how they are threatened.

A WALK IN THE FOREST (Motion Picture—16mm—Sound). *See* Forests and forestry

WHAT IS ECOLOGY? (SECOND EDITION) (Motion Picture—16mm—Sound). EBEC, 1977. 16mm color, sound film, 21 min. #3542 ($295). Gr. 9–12. LFR (1978), NEF (1978), PRV (1979). 574.5

The film develops an understanding of what an ecosystem is, as well as featuring specific ecosystems: forest, desert, urban setting, tide pool, farm. The film explores the ways in which plant and animal species are related to each other, and how human activity presents a constant danger to ecological balance.

THE WILD YOUNG DESERT SERIES (Filmstrip—Sound). *See* Deserts

WINTER IN THE FOREST (Filmstrip—Sound). *See* Winter

ECONOMICS

THE AMERICAN ECONOMY, SET 1 (Filmstrip—Sound). National Assoc. of Manufacturers/McGraw-Hill Films, 1971. 4 color, sound filmstrips, av. 68 fr. ea.; 4 cassettes or discs, 20 min. ea. #102241–4 Cassettes ($71.75), #101945–6 Discs ($64). Gr. 9–12. PRV (11/72). 330.973

Contents: 1. The Economy and You. 2. Comparative Economic Systems. 3. How to Manage Your Income. 4. Personal Economic Security.

This set shows students how their lives are shaped by economic factors such as private property, freedom of enterprise, self-interest, competition, the right to make private contracts, money management, buying and credit.

THE AMERICAN ECONOMY, SET 2 (Filmstrip—Sound). National Assoc. of Manufacturers/McGraw-Hill Films, 1971. 4 color, sound filmstrips, av. 68 fr, ea.; 4 cassettes or discs, 20 min. ea. #102246–5 Cassettes ($71.75), #101954–5 Discs ($64). Gr. 9–12. PRV (11/72). 330.973

Contents: 1. Prices: Balance Wheel of the Economy. 2. Productivity: The Key to Better Living. 3. Capital: Foundation of the Economy. 4. Wages in a Market Economy.

This set deals with wages and prices and discusses productivity, the consumer, and the factors of supply and demand.

THE AMERICAN ECONOMY, SET 3 (Filmstrip—Sound). National Assoc. of Manufacturers/McGraw-Hill Films, 1971. 4 color, sound filmstrips, av. 68 fr. ea.; 4 cassettes or discs, 20 min. ea. #102251–1 Cassettes ($71.75), #101963–4 Discs ($64). Gr. 9–12. PRV (11/72). 330.973

Contents: 1. Profits: Fuel of the Economy. 2. Money and Banking. 3. Business Cycles. 4. The United States and International Trade.

This set shows the economic functions of profits that influence business, and explains how the economy is affected by banks, business cycles, and international trade.

THE BUSINESS OF MOTION PICTURES (Cassette). *See* Motion pictures—Economic aspects

ECONOMIC ISSUES IN AMERICAN DEMOCRACY (Kit). *See* U.S.—Economic conditions

ECONOMICS (Kit). Sue Beauregard/Cypress Publishing, 1977 (Reading in the Content Area). 2 color, sound filmstrips, 60–65 fr.; 2 cassettes, 10–12 min.; 16 Paperback Books; Teacher's Guide. #051 and #052 ($78). Gr. 5–8. BKL (1/15/79), PRV (11/78). 330

Contents: 1. Getting and Spending. 2. Call It Wishing.

Remedial reading in the content area; vocabulary and conceptual development; learning activities to teach introductory level of economics. Explains consuming and producing and goes on to explore career choices, working to earn money and trying out different jobs.

ECONOMICS AND THE AMERICAN DREAM (Kit). Newsweek Educational Division, 1975. 3 color, sound filmstrips, 98–100 fr.; 3 cassettes or discs, 17–18 min.; Spirit Masters; Visuals for Transparencies; Simulation Game; 2 Case Study Units; Teacher's Guide. #501C Cassettes ($85), #501 Discs ($80). Gr. 8–C. PRV (10/75), PVB (4/76). 330.973

Contents: 1. Our Economic Heritage. 2. Contemporary Capitalism. 3. The Future of Capitalism.

American history, current issues and other factors contributing to the development of capitalism in America are presented while introducing a variety of economic terms and concepts.

ECONOMICS OF THE CITY (Filmstrip—Sound). *See* Cities and towns

HAS BIG BUSINESS GOTTEN TOO BIG?
(Filmstrip—Sound). Educational Enrichment Materials, 1977 (New York Times,
Filmstrip Subscription Series). 1. color,
sound filmstrip, 80 fr.; 1 disc or cassette, 15
min.; Teacher's Guide. #32101 ($22). Gr. 9–
A. PRV (12/78). 330.973

Big business in America today is examined
in terms of its nature and impact on our
lives. The Mobile Corporation is shown
through its various development stages as an
example. The pros and cons of big business
are both presented.

MONEY: FROM BARTER TO BANKING
(Filmstrip—Sound). See Money

PROFIT: A LURE/A RISK (Motion Picture—
16mm—Sound). See Profit

ECONOMICS—HISTORY

ECONOMICS IN AMERICAN HISTORY
(Slides). See U.S.—History

ECOSYSTEMS. See Ecology

EDISON, THOMAS ALVA

THOMAS ALVA EDISON (Cassette). Ivan
Berg Associate/Jeffrey Norton Publishers,
1977 (History Makers). 1 cassette, approx.
69 min. #41015 ($11.95), Gr. 7–A. BKL (6/
15/78). 921

Thomas Alva Edison's biography is presented in narrative form with simulated interviews of Edison and his friends on a variety
of subjects.

EDUCATION

PARENT INVOLVEMENT: A PROGRAM
FOR TEACHERS AND EDUCATORS
(Filmstrip—Sound). See Home and school

EDUCATION—AIMS AND OBJECTIVES

LEARNING: CONDITIONING OR
GROWTH? (Filmstrip—Sound). Harper and
Row Publishers/Sunburst Communications,
1974. 2 filmstrips; 2 cassettes or discs;
Teacher's Guide. #85 ($59). Gr. 7–12. PRV
(5/76), PVB (1977). 370.11
Contents: 1. Society's View. 2. The Personal
View.

Educators, psychologists, anthropologists
and students explore the learning process
and discuss the relationship between education and society's values. Stimulates student
opinion about the personal and social purposes of learning.

EDUCATION, BILINGUAL

MEDIA: RESOURCES FOR DISCOVERY
(Filmstrip—Sound). See Audio-visual materials

OUR LANGUAGE, OUR CULTURE, OUR-
SELVES (Filmstrip—Sound). See Family
life

EDUCATION—CHINA

CHINA: EDUCATION FOR A NEW SO-
CIETY (Motion Picture—16mm—Sound).
See China—Civilization

EDUCATIONAL GUIDANCE

BETTER CHOICE, BETTER CHANCE: SE-
LECTING A HIGH SCHOOL PROGRAM
(Filmstrip—Sound). Guidance Associates,
1975. 2 color, sound filmstrips, av. 75 fr. ea.;
2 cassettes or discs; Guide. #9A–107–258
Cassettes ($52.50), #9A–107–241 Discs
($52.50). Gr. 6–9. LGB (1975). 371.42

These filmstrips explore purposes and requirements of academic, vocational, and
general programs, as well as ways to relate
personal needs and objectives to courses.
Case histories are used to show the importance of careful course selection and the impact of new opportunity and self-discovery
in altering career plans and interest.

WHY AM I STUDYING THIS? (Filmstrip—
Sound). Teaching Resources Films, 1974. 4
color, sound filmstrips, av. 53 fr.; 4 discs or
cassettes, 10–12 min.; Teacher's Guide.
#410710 Discs ($56), #410711 Cassettes
($60). Gr. 4–8. PRV (4/76). 371.42
Contents: 1. Why am I Studying English? 2.
Why am I Studying Mathematics? 3. Why
am I Studying Science? 4. Why am I Studying Social Studies?

This set attempts to show the relationship
between the classroom curriculum and students' daily lives. Portions of the strips deal
with career opportunities in each of the
fields. Cartoon-like visuals are used.

WHY DO WE HAVE TO TAKE SOCIAL
STUDIES (Filmstrip—Sound). Multi-Media
Productions, 1977. 2 color, sound filmstrips,
50–52 fr.; 1 cassette, 10–11 min. #7210C
($19.95). Gr. 8–12. PRV (12/78). 371.42

Reasons for studying the different disciplines
in the social studies curriculum are illustrated. Puts the role of social studies into
perspective.

EGYPT—ANTIQUITIES

ANCIENT EGYPT: LAND/PEOPLE/ART
(Filmstrip—Sound). Educational Dimen-
sions Group, 1976. 2 color, sound filmstrips,
58–61 fr.; 2 cassettes, 14–16 min.; Teacher's
Guide. #820 ($37). Gr. 4–12. BKL (12/1/76),
PRV (4/77), PVB (4/78). 913.32

Life of ancient Egypt is described while
photos of art works and artifacts are shown.
An overview of life and customs of ancient
Egypt is given and information about how
scholars learned about this interesting land
is related by the narrator.

EGYPT—ANTIQUITIES (cont.)

IN THE BEGINNING (Motion Picture—
16mm—Sound). Reader's Digest/Pyramid
Films, 1976). 16mm color, sound film, 27
min.; ³/₄ in. videocassette also available.
($350). Gr. 11–A. BKL (12/1/76), CGE
(1976), LFR (1976). 913.32
Written and narrated by Lord Kenneth
Clark, the noted art historian, as he travels
to Egypt, the home of civilization as we
know it, exploring the remnants of this extraordinary society. Clark traces, as he
stands among the majestic ruins, the ways in
which Egypt flourished until its downfall.

KING TUTANKHAMUN: HIS TOMB AND
HIS TREASURE (Kit). Pathescope Educational Media, 1978. 2 color, sound filmstrips,
75–84 fr.; 10–12¹/₂ min.; Teacher's Manual;
24 in. × 35 in. Color Poster of the King Tut
Gold Mask. #765 ($60). Gr. 7–12. BKL (10/
78). 913.32
This program makes outstanding use of archival stills of the discovery and excavation
of the tomb of Tutankhamun; photos of exhibit objects and of the Valley of the Tombs
of the Kings. Outlines the long search by
Howard Carter and finally his analysis of the
tomb itself. Stirs the imagination and thrills
the students with the magnificent artifacts
uncovered.

TREASURES OF KING TUT (Videocassette). See Tutankhamun

EISELEY, LOREN

CONVERSATION WITH LOREN EISELEY
(Cassette). Cinema Sound Ltd./Jeffrey Norton Publishers, 1976 (Avid Reader). 1 cassette, approx. 55 min. #40218 ($11.95). Gr.
7–A. BKL (7/15/77). 501
Heywood Hale Broun interviews Loren
Eisely, author of All the Strange House
(Scribner, 1975), speculates on the complexity of the universe and the role of the individual within it.

EL MAÑANA ES HOY

LEARNING BEGINS AT HOME (Filmstrip—Sound). See Family life

ELECTIONS

THE ELECTION PROCESS (Filmstrip—
Sound). Westport Communications/Sun-
burst Communications, 1974. 6 color, sound
filmstrips; 6 cassettes or discs; Teacher's
Guide. #42 ($145). Gr. 6–12. BKL (7/15/77),
PRV (5/75), TSS (11/77). 324
Contents: 1. We the People. 2. Where the
Cheering Starts. 3. My Friends and Fellow
Citizens. 4. Prime Time in the Old Town
Tonight. 5. Dangers and Detours. 6. Who
Owns Tomorrow.

Traces the history and functions of suffrage
from the Athens of Aristotle to the Washington of Watergate. Discusses how the right to
vote was established in America and how
various restrictions imposed—property-
ownership, religion, sex, race, age, and
character—were ultimately eliminated. Analyzes the party system in the United States.
Examines how it helps moderate the interests of diverse and conflicting groups. Considers how government in a democracy is
responsible to the people, and explores the
inextricable link between the electoral process and citizenship.

HOW TO GET ELECTED PRESIDENT
(Filmstrip–Sound). Hawkhill Associates,
1977. 1 color, sound filmstrip, 120 fr.; 1 cassette, 18 min,; Teacher's Guide. ($24.50).
Gr. 6–A. PRV (11/78). 324
Using the sound track from Jimmy Carter's
1976 drive to win the Democratic nomi-
nation, this filmstrip follows the rise of a
single Presidential candidate and the back-
home personal hopes and dreams.

ELIOT, GEORGE

GREAT WRITERS OF THE BRITISH
ISLES, SET II (Filmstrip—Sound). See Authors, English

ELIZABETH I, QUEEN OF ENGLAND

ELIZABETH I (Cassette). Ivan Berg Associ-
ates/Jeffrey Norton Publishers, 1976, (History Makers). 1 cassette, approx. 74 min.
#41007 ($11.95). Gr. 10–A. BKL (5/15/78),
PRV (11/19/78). 921
The court of Elizabeth I of England, divided
by jealousies and presided over by a woman
who was vigorous, variable and vain, and
yet a great ruler. Her reign saw the death of
Mary, Queen of Scots, the execution of her
beloved Earl of Essex, the exploits of Drake
and Raleigh and the defeat of the Spanish
Armada, and the brilliance of the playwrights Shakespeare, Marlowe and Ben
Jonson.

ELSA BORN FREE

ELSA BORN FREE (Filmstrip—Sound). See
Lions

EMBROIDERY

CREATIVE CREWEL (Filmstrip—Sound).
See Crewelwork

EMBRYOLOGY

EMBRYOLOGY OF THE CHICK (Film
Loop—8mm—Captioned). BFA Educational Media, 1973 (Animal Behavior). 8mm col-
or, captioned film loop, approx. 4 min.

#481420 ($30). Gr. 4–9. BKL (5/1/76). 501.3

Demonstrates experiments, behavior, and characteristics of chicken embryos.

MIRACLE OF LIFE (Motion Picture— 16mm—Sound). Cine-Science, Tokyo/Pyramid Films, 1977. 16mm color, sound film, 15 min.; ³/₄ in. videocassette, also available. ($275). Gr. 7–A. BKL (1977), LFR (1977), LGB (1977). 612.6

A live-action film shot through microscope photography showing—for the first time— the living processes of fertilization, cell division, and growth and development of the fetus to the first heartbeat. The processes for which animal cells were photographed are the same in all mammals; from the point at which human development is different, the living human fetus is filmed.

PREGNANCY (Filmstrip—Sound). See Pregnancy

EMERGENCIES. See First aid; Lifesaving

EMERSON, RALPH WALDO

THE ROMANTIC AGE (Filmstrip—Sound). See American literature—History and criticism

EMOTIONS

DEALING WITH ANGER (Filmstrip— Sound). Audio Visual Narrative Arts, 1977. 2 color, sound filmstrips, 74–77 fr.; 2 cassettes or discs, 12¹/₂–14 min.; Teacher's Guide. #268C Cassettes ($51.50), #268R Discs ($47.50). Gr. 9–12. PRV (1/79). 152.4

Contents: 1. The Hidden Feeling. 2. The Hidden Emotion.

Anger is a natural and frequently useful emotion, but the results of denied, repressed or misdirected anger can interfere with interpersonal relations, bodily functions, and, in extreme cases, lead to suicide or other types of violence.

THE DEVELOPMENT OF FEELINGS IN CHILDREN (Motion Picture—16mm— Sound). See Child development

MASKS: HOW WE HIDE OUR EMOTIONS (Filmstrip—Sound). Human Relations Media, 1976. 2 color, sound filmstrips, 67–69 fr.; 2 cassettes or discs, approx 13¹/₂ min. ea.; Teacher's Guide. #601 ($72). Gr. 7–12. BKL (5/1/78). M&M (11/77), PRV (12/77). 152.4

Contents: 1. Performing. 2. Understanding Masks.

Shows how healthy human beings present psychological masks and that there are dangers when our masks fail to reflect our true feelings and emotions. It proceeds to analyze the personality hidden behind such masks as body language, speech patterns, dress, and grooming.

EMPLOYMENT

OPPORTUNITY (Kit). See Vocational guidance

ENAMEL AND ENAMELING

ENAMELING: PAINTING WITH GLASS (Filmstrip—Sound). Warner Educational Productions, 1977. 6 color, sound filmstrips, 71–102 fr.; 3 cassettes, 6–13 min.; Teaching Guide. #390 ($105). Gr. 7–A. PRV (11/19/78). 738.4

These strips teach the history of European and Oriental enameling up to the present artisans. They show how to do modern enameling, step-by-step, and discuss the choosing of tools, designs for enamels, painting with glass, and the choosing of materials and colors.

ENCYCLOPAEDIA BRITANNICA

FIVE THOUSAND BRAINS (Motion Picture—16mm—Sound). See Encyclopedias

ENCYCLOPEDIAS

FIVE THOUSAND BRAINS (Motion Picture—16mm—Sound). EBEC, 1973. 16mm color, sound film, 27 min. #3345 ($195). Gr. 4–A. CGE (1975), FLN (1974). 031

Control of the knowledge explosion focuses new attention on the encyclopedia. This film reports the inside story of how Encyclopaedia Britannica developed a revolutionary new three-part reference work capable of storing all human knowledge and making it equally available to students, researchers, and home-users as a means of contending with the flood of new information and technology.

ENDANGERED SPECIES. See Rare animals; Rare birds; Rare plants; Rare reptiles

ENERGY. See Force and energy; Power resources

ENGINEERING

CAREERS IN ENGINEERING (Filmstrip— Sound). Pathescope Educational Media, 1974. 2 color, sound filmstrips, 89–90 fr. ea.; 2 cassettes, 14–18 min. ea.; Teacher's Guide. #712 ($50). Gr. 7–12. PRV (11/74). 620.023

Engineering is shown as a bridge between the theoretical world of the dreamers, thinkers, and inventors and the real world of the constructors and the manufacturers. Five engineering work areas are covered: civil, mechanical, electrical, chemical, and industrial. Interviews highlight some of the duties and responsibilities of engineers in specific jobs.

ENGLAND—HISTORY—BIOGRAPHY

ELIZABETH I (Cassette). *See* Elizabeth I, Queen of England

ENGLAND—HISTORY—FICTION

A CHRISTMAS CAROL (Kit). *See* Christmas—Fiction

A CHRISTMAS CAROL (Phonodisc or Cassette). *See* Christmas—Fiction

ENGLISH AS A SECOND LANGUAGE

PASSPORT TO FRANCE (Filmstrip—Sound). *See* France

ENGLISH COMPOSITION. *See* English language—Composition and exercises

ENGLISH DRAMA

HIS INFINITE VARIETY: A SHAKE-SPEAREAN ANTHOLOGY (Cassette). Miller–Brody Productions, 1974 (The Margaret Webster Cassette Library). 4 cassettes, 30 min. ea. MW1201/4 ($31.80). Gr. 7–12. BKL (7/15/75), NYT (12/22/74). 822.33

Margaret Webster, a grande dame of the English theater, interprets scenes from Shakespeare. She presents his versatility and genius as a lyricist and playwright. His philosophy on the theater and his early concern with the audience's responsiveness are discussed. Finally, she deals with Shakespeare's treatment of women.

NO COWARD SOUL: A PORTRAIT OF THE BRONTES (Cassette). Miller–Brody, 1974 (Margaret Webster Cassette Library). 4 cassettes, 30 min. ea. #MW1401–1404 ($31.80). Gr. 9–A. BKL (7/15/75), NYT (12/22/74). 822

Margaret Webster, a grande dame of the English theater, interprets the three Bronte sisters, Charlotte, Emily, and Anne. She uses bits of letters, diaries and poems to provide insight into their background.

THE SEVEN AGES OF GEORGE BERNARD SHAW (Cassette). Miller–Brody Productions, 1974 (Margaret Webster Cassette Library). 4 cassettes, 30 min. ea. #MW1301/4 ($31.80). Gr. 7–12. NYT (12/22/74). 822

Margaret Webster, a grande dame of English theater, couches the views of Shaw in terms of the seven "ages" of his development as a comedian, satirist, iconoclast, social reformer, crusader, philosopher, and critic.

ENGLISH FICTION

THE HOBBIT (Kit). *See* Fantasy—Fiction

ENGLISH LANGUAGE—BUSINESS ENGLISH

CAREER ENGLISH: COMMUNICATING ON THE JOB (Slides/Cassettes). Center for Humanities, 1978. 160 slides in 2 Carousel cartridges; 2 cassettes or discs, 40 min.; Guide. #327 ($139.50). Gr. 9–A. BKL (9/1/78). 808

Contents: 1. Business Letters. 2. Requests for Information.

Based on a series of at work scenarios, this program provides instruction in basic business communication. Learning is reinforced by a series of exercises related to the vignettes. Included is information on job application letters, sales letters, collection and adjustment letters, and letters of inquiry. Part two discusses requests for information, progress reports, proposals designed to persuade, and short cover memos that highlight longer reports.

ENGLISH LANGUAGE—COMPOSITION AND EXERCISES

ADVENTURES IN IMAGINATION SERIES (Filmstrip—Sound). McGraw–Hill Films, 1974. 4 color, sound filmstrips, av. 25 fr. ea.; 4 cassettes or discs, $4^1/2$ min. ea.; Teacher's Guide. ($68). Gr. 4–12. BKL (4/15/75), PRV (11/75). 372.6

Contents: 1. Being Alone. 2. Being with Others. 3. Getting Caught and Going Free. 4. Something's Going to Happen—What?

Each of the four sound filmstrips explores a different theme, using bold visual images and an original musical score to encourage students to draw from their own experiences and fantasies and to start ideas flowing for stories, poems, essays, and other endeavors in creative communication.

COMPOSITION POWER (Kit). *See* English language—Study and teaching

PARAGRAPH POWER (Kit). *See* English language—Study and teaching

SENTENCE PATTERNS (Filmstrip—Sound). Coronet Instructional Media, 1969. 8 color, sound filmstrips, av. 49 fr. ea.; 4 cassettes or discs, av. $8^1/2$ min. #M176 Cassettes ($120), #S176 Discs ($120). Gr. 7–12. BKL (1971), SLJ (1971). 375.4

Contents: 1. Introduction. 2. Subject-verb. 3. Subject-verb-modifier. 4. Subject-verb-predicate noun. 5. Subject-verb-direct object. 5. Subject-verb-direct object-modifier. 7. Subject-verb-direct object-object complement. 8. Subject-verb-indirect object-direct object.

Simple structural linguistics learned with a little help from fun-filled zoos, blueprints, and more! All structural elements-determiners, modifiers and prepositional groups—are emphasized in artwork and photography.

SENTENCE PROBLEMS TWO (Filmstrip—Sound). Filmstrip House, 1972. 4 color, sound filmstrips, 49–55 fr.; 2 cassettes or discs, 11 min.; Script Book. ($45). Gr. 5–10. PRV (12/73), PVB (5/74). 372.61

Contents: 1. Choppy Sentences. 2. Run-on Sentences. 3. Dangling Modifiers. 4. Wordiness—Excess Baggage.

The various forms of each problem are explored in detail, then summarized. Photographs give effective visual support to ideas in example sentences.

WRITING SKILLS—THE FINAL TOUCH: EDITING, REWRITING & POLISHING (Slides/Cassettes). Center for Humanities, 1977. 3 units with 80 slides ea.; 3 discs and cassettes; Teacher's Guide. #0324 ($139). Gr. 9–C. PRV (12/78). 375.4

This unit reviews structure, style, and development guidelines for writing. It outlines a basic framework for a "final touch" at the completion of a writing project.

WRITING THE EXPOSITORY ESSAY (Kit). *See* Essay

ENGLISH LANGUAGE—DICTIONARIES

DICTIONARY SKILL BOX (Cassette). Troll Associates, 1977. 6 cassette tapes, av. 7 min. ea.; 24 Spirit Masters; Teacher's Guide. ($56). Gr. 3–8. BKL (1/15/78). 028.7

Contents: 1. What Is a Dictionary? 2. How Do You Find It? 3. How Do You Say It? 4. What Does It Mean? 5. More about the Entry. 6. Front and Back Matter.

Introducing the middle-grade student to the dictionary, this program shows how the tool aids in spelling, definition, pronunciation, and providing other sources of information.

ENGLISH LANGUAGE—GRAMMAR

COMMUNICATION SKILLS: WHO'S AFRAID OF GRAMMAR? (Slides/Cassettes). Center for Humanities, 1976. 240 slides in 3 Carousel cartridges; 3 cassettes; 3 discs; Teacher's Guide. #2090 ($179.50). Gr. 9–C. BKL (3/1/78), M&M (5/77), PRV (11/77). 425

Contents: 1. What Is "Good" English? 2. Sentences Plain. 3. Sentences Fancy.

The program is designed to motivate students to learn grammar as a tool of effective communication.

GRAMMAR (Slides/Cassettes). Center for Humanities, 1978. 160 color slides in 2 Carousel trays; 2 cassettes or discs, av. 56 min.; Guide. #306 ($145). Gr. 9–12. BKL (9/1/78). 425

Visual metaphors help students understand and remember crucial grammatical principles. Run-on sentences, comma splices,

fragments, sentence consistency, are all studied in this humorous presentation.

GRAMMAR: EVERYTHING YOU WANTED TO KNOW ABOUT USAGE (Slides/Cassettes). Center for Humanities, 1977. 240 color slides in 3 Carousel cartridges; 3 cassettes or discs; Teacher's Guide. ($179.50). Gr. 9–C. PRV (5/78). 425

This program presents the sentence as the basic building block of standard English, subject/verb agreement, pronoun/antecedent agreement, and common usage errors. Part one illustrates how different language is used for different purposes. Two kinds of sentence problems are presented; the sentence fragment and the run-on sentence. Part two concentrates on two problem areas: the agreement of subject and verb and of pronoun and antecedent. Part three focuses on common usage problems such as double subjects, verbs, negatives, and comparisons as well as common subject/verb errors.

ON YOUR MARKS: PART I (Motion Picture—16mm—Sound). *See* Punctuation

ON YOUR MARKS: PART II (Motion Picture—16mm—Sound). *See* Punctuation

PUNCTUATION PROBLEMS (Filmstrip—Sound). *See* Punctuation

SENTENCE PATTERNS (Filmstrip—Sound). *See* English language—Composition and exercises

SENTENCE PROBLEMS TWO (Filmstrip—Sound). *See* English language—Composition and exercises

SPEAKING OF GRAMMAR (Filmstrip—Sound). Guidance Associates, 1975 (Language Skills Series). 2 color, sound filmstrips, 60–74 fr.; 2 cassettes or discs, 11–14 min. #301–992 Cassettes ($52.50), #301–984 Discs ($52.50). Gr. 5–8. BKL (2/15/76), LGB (1975), PRV (1975). 372.61

Contents: 1. The Parts of Speech. 2. A Language About Language.

An overview of grammatical terms and techniques, helping students define the parts of speech and learn how words, phrases, and clauses combine to make sentences. Identifies simple sentence patterns as well as compound and complex sentences.

THE STRUCTURE OF LANGUAGE (Filmstrip—Sound). Multi-Media Productions, 1976. 4 color, sound filmstrips, 41–46 fr.; 4 cassettes, 7–9 min.; Available separately. #5009C ($50). Gr. 4–12. PRV (4/78). 372.61

Contents: 1. End Punctuation. 2. Sentences. 3. Capitalization and Clauses. 4. Those Crazy Hooks . . . and Friends.

This group of four sound filmstrips covers punctuation rules and marks, the use of capitalization, sentence structure, and clauses. Each item is defined and illustrated. Color

ENGLISH LANGUAGE—GRAMMAR (cont.)

photographs of catchy billboards frequently serve as visuals.

WATCH YOUR LANGUAGE: USAGE (AND ABUSAGE) (Slides/Cassettes). *See* English language—Usage

ENGLISH LANGUAGE—HISTORY

LANGUAGE—THE MIRROR OF MAN'S GROWTH (Filmstrip—Sound). Centron Films, 1971. 5 color, sound filmstrips; 5 discs or cassettes. Discs ($59.50), Cassettes ($69.50). Gr. 4–8. BKL (5/15/71), ELE (11/72), STE (9/72). 420.9

Contents: 1. Language and Its Mysteries. 2. What Age Has Done to English. 3. How Is it that an Englishman Speaks English? 4. Languages are Born, Sometimes They Die. 5. The American Language—Or When the King's English Came to America.

These five artwork filmstrips trace the development of language in general and American English in particular.

ENGLISH LANGUAGE—SPELLING

SPEAKING OF SPELLING (Filmstrip—Sound). Guidance Associates, 1975 (Language Skills Series). 2 color, sound filmstrips, av. 61 fr.; 2 discs or cassettes, av. 11 min. #1B-301968 ($52.50), #1B-301976 ($52.50). Gr. 5–8. BKL (2/15/76), LGB (1975), PRV (1975). 421.52

Contents: 1. The System. 2. Using the System.

Illustrates "how to" tips for competent spelling: learning basic patterns, dealing with complex words and devices to help with irregular spellings.

ENGLISH LANGUAGE—STUDY AND TEACHING

THE BUILDING BLOCKS OF LANGUAGE (Filmstrip—Sound). Multi-Media Productions, 1976. 4 color, sound filmstrips, 41–45 fr; 4 cassettes, 8–9 min. ea.; Teacher's Guide. #5014C ($50). Gr. 9–A. PRV (5/78). 375.4

Contents: 1. Before and After Prefixes and Suffixes. 2. The Nym Family—Synonyms/Antonyms and Acronyms. 3. Soundly Confusing—Homonyms and Homophones. 4. Avoiding Pitfalls in Spelling.

Provides students with building blocks to improve competency in the use of the English language. This set strives to put the student at ease with some of the inconsistencies in the language.

COMPOSITION POWER (Kit). United Learning, 1971. 4 color, sound filmstrips; 2 cassettes or discs; 16 Spirit Masters; 1 Script/Guide. #79502 Cassette ($55), #79501 Disc ($55). Gr. 4–12. BKL (2/72). 372.6

Contents: 1. Outline Power. 2. Opening. 3. Meat of the Sandwich. 4. Closing and Revising.

Presents the fundamentals of composition and revising so that students will be able to write intelligently and clearly.

PARAGRAPH POWER (Kit). United Learning, 1971. 4 color, sound filmstrips; 4 cassettes or discs; 16 Spirit Masters; 1 Script/Guide book. #79402 Cassettes ($55), #79401 Discs ($55). Gr. 4–12. BKL (1/72). 372.6

Contents: 1. Thinking in Paragraphs. 2. Topic Sentence Power. 3. Paragraph Unity Power. 4. Paragraph Development.

Illustrates how to organize ideas into paragraphs for clear, logical flow, how to construct topic sentences, and how to unify sentences into paragraphs.

THE STRUCTURE OF LANGUAGE (Filmstrip—Sound). *See* English language—Grammar

WRITING: FROM ASSIGNMENT TO COMPOSITION (Filmstrip—Sound). Guidance Associates, 1974. 2 color, sound filmstrips, av. 78 fr. ea.; 2 discs or cassettes, av. 9–12 min. ea.; Teacher's Guide. #4E-503-357 Cassettes ($52.50), #E-503-340 Discs ($52.50). Gr. 4–8. BKL (1/76), PRV (12/75), PVB (4/76). 375.4

The step-by-step process for writing an interesting, coherent composition. "How to" tips for selecting a topic, organizing an outline, doing a first draft and doing a rewrite.

ENGLISH LANGUAGE—USAGE

AN ALMANAC OF WORDS AT PLAY (Cassette). Cinema Sound Ltd./Jeffrey Norton Publishers, 1976 (Avid Reader). 1 cassette, 55 min. #40223 ($11.95). Gr. 10–A. BKL (6/15/77). 428

In this radio interview, Heywood Hale Broun and Willard R. Espy, author of the book *An Almanac of Words At Play* (Potter, 1976), engage in an interchange on some aspects of the English language: its special pronunciation, rhymes, use of foreign words, double entendres, slang, punctuation, and palindromes—words meaning the same read both forwards and backwards.

EXPANDING YOUR VOCABULARY (Kit). Society for Visual Education, 1978. 4 filmstrips; 4 cassettes; 1 Guide; 25 worksheets, #327-SATC ($99). Gr. 9–C. BKL (9/15/78). 428

Contents: 1. Introduction to Vocabulary Building. 2. Word Pairs. 3. The Right Word. 4. Figures of Speech.

Explores use of word histories, synonyms, antonyms, homonyms, connotation, denotation, and figures of speech as aids in vocabulary-building. Emphasizes importance of vocabulary in effective communication.

WATCH YOUR LANGUAGE: USAGE
(AND ABUSAGE) (Slides/Cassettes). Cen-
ter for Humanities, 1977. 160 slides in 2
Carousel cartridges; 2 cassettes or discs;
Teacher's Guide. ($139.50). Gr. 9-C. BKL
(3/1/78). 425

An introduction to the basic consideration of
correct usage, this set alerts students to
the language they speak, hear, read and
write. It helps them to understand the dis-
tinctions between standard and nonstandard
English, and to know why some words and
phrases are acceptable in writing or formal
speaking. Part two focuses on abuses of language, and demonstrates what happens
when words are used carelessly, deceptively
or maliciously. Jargon, loaded words and
other aspects of the topic are presented.

ENGLISH LITERATURE

TALES FROM SHAKESPEARE (Phonodiscs). Caedmon Records, 1976. 1 disc TC-
1469 ($6.98). Gr. 9-12. PRV (4/76). Fic

Julie Harris reads these rewritings of Shakespeare's plays. "The Tempest" and "A
Midsummer Night's Dream" are spoken at
an easily understandable rate.

ENGLISH POETRY

COLERIDGE: THE FOUNTAIN AND THE
CAVE (Motion Picture—16mm—Sound).
Bayley Silleck/Pyramid Films, 1974. 16mm
color, sound film, 32 min.; ¾" videocassette
also available. ($375). Gr. 9-A. BKL (1974),
EGT (1974), FLN (1974). 821

The life and poetry of Samuel Taylor Coleridge, a British romantic poet, filmed on location in the Lake Country and London,
England. The examination of the major influences on the poet's life and work are aug-
mented by commentary from contemporary
scholars.

ENGRAVING

ETCHING AND ENGRAVING (Filmstrip—
Sound). See Etching

ENTERTAINERS

WOMEN BEHIND THE BRIGHT LIGHTS
(Kit). EMC, 1975, (Behind the Bright
Lights). 4 paperbacks; 4 read-along cas-
settes; activities; Teacher's Guide.
#ELC2300000 ($65). Gr. 5-12. BKL (1/76),
PRV (2/76), SLJ (12/75). 920

Contents: 1. Cher: Simply Cher. 2. Valerie
Harper: The Unforgettable Snowflake. 3.
Roberta Flack: Sound of Velvet Melting. 4.
Olivia Newton John: Sunshine Supergirl.

Four popular female personalities are presented. Their struggles to reach the top
should act as motivation in these high-interest, low vocabulary materials.

ENVIRONMENT

ATMOSPHERE IN MOTION (Motion Picture—16mm—Sound). See Atmosphere

COAL, CORN, AND COWS (Filmstrip—
Sound). Hawkhill Associates, 1977. 1 color,
sound filmstrip, 96 fr.; 1 cassette, 18½
min.; Teacher's Guide. ($26.50). Gr. 6-C.
BKL (12/15/77), SCT (1978). 301.31

An examination of the land use dilemma in
Illinois, where strip mining of coal comes into conflict with corn growing on some of the
richest prairie soil in the world. While the
situation is addressed specifically in Illinois,
it is easily generalized to other areas and to
other environmental problems.

See also Ecology; Man—Influence of environment

ENVIRONMENT—LAW AND LEGISLATION

LAW AND THE ENVIRONMENT (Kit).
EBEC, 1977 (Law and Society). 2 sound
filmstrips with cassettes; audio cassette; Student Resource Reader (10 copies); Teacher's
Guide. #17009K ($67). Gr. 7-12. BKL (3/
79). 346.044

Contents: 1. The Reserve Mining Case: A
Legal Conflict. 2. The Silver Bay Dilemma:
Health v. Livelihood.

Citing recent landmark decisions, materials
explain how law and the courts resolve conflicts between differing value-systems. Case
in point: a Minnesota iron-ore refinery is
charged with polluting Lake Superior. Students follow the legal proceedings, step by
step, as the case unfolds. An unusual twist to
the documentary approach is the use of original TV art from the trial coverage by
WCCO-TV, Minneapolis.

ERDOES, RICHARD

SUN DANCE PEOPLE: THE PLAINS IN-
DIANS (Filmstrip—Sound). See Indians of
North America—Plains

ERIE CANAL—SONGS

THE ERIE CANAL (Filmstrip—Sound).
Weston Woods Studios, 1974. 1 color, sound
filmstrip, 30 fr.; 1 cassette, 6 min.; 1 script.
#157C ($12.75). Gr. 4-9. PRV (1975).
784.756

Based on the book by Peter Spier, it tells the
story of the boats that carried people and
freight on the canal in the 19th century.
Spier's colorful and accurate pictures add
much to the production.

EROSION

EROSION: WIND AND RAIN (Slides). Educational Dimensions Group, 1977. 20 slides;

EROSION (cont.)

Teacher's Guide. #9123 ($25). Gr. 9–12. PRV (12/19/78). 551.3

These slides illustrate the erosion caused by wind and rain. Photographs help present the development of ventifacts, sand dunes, blowouts, wind shadows and water erosion.

MONUMENTS TO EROSION (Motion Picture—16mm—Sound). EBEC, 1974. 16mm color, sound film, 11 min. #3340 ($150). Gr. 4–12. PRV (5/76), SCT (4/76). 551.3

This film presents the red rock country of the Colorado Plateau as it has never been seen before. The camera glides over towering stone arches and delicate spires, revealing a unique natural mosaic of suggestive shapes and colors. The narration approaches the land from several viewpoints: as an inspiring reflection of the patterns and rhythms of nature; as the product of the erosion process that shaped it; and as a brief moment in the geologic lifetime of an ancient planet.

ERRORS

THE SNAKE: VILLAIN OR VICTIM? (Motion Picture—16mm—Sound). *See* Snakes

ESKIMOS

HOW TO BUILD AN IGLOO—SLIDE SET (Slides). *See* Igloos

ESKIMOS—FICTION

JULIE OF THE WOLVES (Phonodisc or Cassette). Caedmon Records, 1973. 1 cassette or disc, 40 min.; Jacket Notes. #TC1534 Disc ($6.98), #CDL51534 Cassette ($7.95). Gr. 4–8. BKL (5/1/74). Fic

From the Newbery winner by the same name, Irene Worth presents this dramatization which captures the unique and thrilling experiences of Julie (Miyax) the Eskimo girl.

ESPY, WILLARD R.

AN ALMANAC OF WORDS AT PLAY (Cassette). *See* English language—Usage

ESSAY

WRITING THE EXPOSITORY ESSAY (Kit). Society for Visual Education, 1978. 4 filmstrips; 4 cassettes; worksheets; Guide. #322–SATC ($99). Gr. 9–C. BKL (10/15/78). 808.4

Contents: 1. Shaping the Expository Paragraph. 2. Thinking for the Expository Paragraph. 3. Unifying the Expository Paragraph. 4. The Expository Essay.

Covers planning, approaches to writing, assembling paragraphs, organizing paragraphs, unifying the essay.

ESTES, ELEANOR

ELEANOR ESTES (Filmstrip—Sound). *See* Authors

ETCHING

ETCHING AND ENGRAVING (Filmstrip—Sound). Educational Audio Visual, 1977. 2 color, sound filmstrips, 86–104 fr.; 2 cassettes or discs, 11–16 min.; Teacher's Guide. Cassette #P7KF ($48), Disc #P7RF ($44). Gr. 7–A. BKL (11/15/77), PRV (2/78). 767

Contents: 1. The Intaglio Print. 2. Techniques.

A learning package on the history and techniques involved in the graphic processes of etching and engraving, this is a broad introduction with in-depth technical details. Starts with the intaglio prints of the artists of the 15th century through Durer, Breugel, Callot, Rembrandt, and contemporary artists, Bruce Bleach and Donald Mckay.

ETHICS

I THINK (Motion Picture—16mm—Sound). *See* Values

JOSEPH SCHULTZ (Motion Picture—16mm—Sound). Wombat Productions, 1973. 16mm color, sound film, 13 min. ($195). Gr. 7–A. AFF (1974), AIF (1973), SOC (5/74). 170

How morally responsible is each one of us? Does moral responsibility end in war time—or other times of crisis?

VALUES (Filmstrip–Sound). *See* Values

ETHICS, AMERICAN

MODERN MORALITY: OLD VALUES IN NEW SETTINGS (Filmstrip—Sound). Current Affairs Films, 1976. 1 color, sound filmstrip, 74 fr.; 1 cassette, 15 min.; Teacher's Guide. #525 ($24). Gr. 7–A. PRV (4/77). 170.973

Deals with the moral re-examination in American society. Some observers say the "New Morality" had no real meaning and the only thing that has changed is that we are now more honest. Others disagree, and argue that there have been real and serious changes. This filmstrip examines the two points of view, as they apply to different areas of our society, and analyzes possible future trends.

UNIVERSAL VALUES IN AMERICAN HISTORY (Kit). *See* U.S.—History

ETHNIC GROUPS

CHINESE FOOD (Kit). *See* Cookery, Chinese

GERMAN FOOD (Kit). *See* Cookery, German

IMMIGRANT AMERICA (Filmstrip—Sound). *See* Immigration and emigration

ITALIAN FOOD (Kit). *See* Cookery, Italian

JEWISH IMMIGRANTS TO AMERICA (Filmstrip—Sound). *See* Jews in the U.S.

MEXICAN FOOD (Kit). *See* Cookery, Mexican

SOUL FOOD (Kit). *See* Cookery, Black

See also Minorities

EULENSPIEGEL, TILL—FICTION

A TALE OF TILL (Motion Picture—16mm—Sound). *See* Folklore—Germany

EUROPE

EUROPE: DIVERSE CONTINENT (Filmstrip—Sound). EBEC, 1977 (Europe). 6 color, sound filmstrips, av. 108 fr.; 6 discs or cassettes, 14 min ea.; #17004K Cassettes ($86.95), #17004 Discs ($86.95). Gr. 7–12. BKL (3/78), LFR (5/78), PRV (11/78). 914
Contents: 1. Peninsular Continent. 2. What Unites and What Divides Europe. 3. New Towns and Villages. 4. The Common Market. 5. World Model of Ideas and Technology. 6. Living in a Welfare State.
The pressures for and against a "United States of Europe" are carefully considered. In documentary style, Europeans discuss their diverse social, cultural, and linguistic backgrounds. From the planned communities of England to the planned economy of Sweden, from the medieval village market to the modern Common Market, we follow the rise of European cooperation and economic interdependence.

SOUTHERN EUROPE: MEDITERRANEAN LANDS (Filmstrip—Sound). EBEC, 1977. 4 color, sound filmstrips, av. 112 fr. ea.; 4 discs or cassettes, av. 15 min. ea. #17005K Cassettes ($57.95), #17005 Discs ($57.95). Gr. 7–12. BKL (5/15/78), MDU (8/78). 914
Contents: 1. Spain: The Past Holds Back the Future. 2. Italy: Perpetual Crises. 3. Greece: Old Values, New Ways. 4. The Mediterranean: Highway of Trade and Culture.
This set tells of three Mediterranean lands and the sea that links them to the rest of the world: Spain, a country held together by common traditions, but held back by regional differences; Italy, inequities between the rural south and the industrial north are accentuated by unstable government; Greece, the cradle of democracy, rocks to a 19th-century rhythm. The sea, itself, is an economic lifeline for contemporary Mediterranean people.

WESTERN EUROPE, GROUP ONE (Filmstrip—Sound). Society for Visual Education, 1975. 6 color, sound filmstrips, 90–100 fr.; 3 discs or cassettes, av. 20 min.; Teacher's Guide. #293-SAR Discs ($83) #293-SATC Cassettes ($83). Gr. 4–10. BKL (3/15/76), PRV (11/76). 914
Contents: 1. Benelux: A United European Community. 2. Austria: Modern Nation, Ancient Crafts. 3. France: The Basques in the 20th Century. 4. Andorra and Liechtenstein: Two Mini-Countries. 5. Monaco: Skyward and Seaward. 6. Switzerland: The Banking Nation.
People, places, customs, and contemporary problems are discussed. Emphasis is placed on the idea that each country retains its own culture and character even though European economy is interdependent.

WESTERN EUROPE, GROUP TWO (Filmstrip—Sound). Society For Visual Education, 1975. 6 color, sound filmstrips, 90–100 fr.; 3 discs or cassettes, av. 20 min.; Teacher's Guide. #293-SBR Discs ($83), #293-SBTC Cassettes ($83). Gr. 4–10. BKL (3/15/76), PRV (11/76). 914
Contents: 1. Italy: Venice, the Modern Atlantis. 2. Malta: Mediterranean Melting Pot. 3. Portugal: People of the Land and Sea. 4. San Marino: The Postage Stamp Republic. 5. Spain: A Nation Goes Forward. 6. The Vatican: A Nation Within a City.
People, places, customs, and contemporary problems are discussed. Emphasis is placed on the idea that each country retains its own culture and character even though European economy is interdependent.

EUROPE—HISTORY—476–1492

MEDIEVAL EUROPE (Filmstrip—Sound). *See* Middle Ages

EUROPE—HISTORY—1492–1789

THE REFORMATION: AGE OF REVOLT (Motion Picture—16mm—Sound). *See* Reformation

SPIRIT OF THE RENAISSANCE (Motion Picture—16mm—Sound). *See* Renaissance

EVOLUTION

CHARLES DARWIN (Cassette). *See* Darwin, Charles

GALAPAGOS: DARWIN'S WORLD WITHIN ITSELF (Motion Picture—16mm—Sound). *See* Galapagos Islands

EXERCISE

THE PHYSIOLOGY OF EXERCISE (Filmstrip—Sound). Sunburst Communications, 1976. 2 color, sound filmstrips, 70–74 fr.; 2 cassettes or discs, 13 min. ea.; Teacher's Guide. #232 ($59). Gr. 9–A. BKL (5/1/77), PRV (1/78). 613.7

EXERCISE (cont.)

Contents: 1. The Physiology of Exercise. 2. Your Exercise Program.

Investigates how the body is improved and strengthened by proper exercise and stamina conditioning. Shows how heart disease, the number one killer of Americans, can be avoided by steady, rhythmic, moderate exercise of the heart, lungs and circulatory system through regular hiking, swimming, jogging, brisk walking or other aerobic exercises.

EXPLORERS

AGE OF EXPLORATION AND DISCOVERY (Filmstrip—Sound). Coronet Instructional Media, 1975. 6 color, sound filmstrips, 46–50 fr.; 6 cassettes or 3 discs, 13–15 min. ($95). Gr. 5–8. BKL (6/15/75), PRV. 910.92

Contents: 1. Marco Polo. 2. Prince Henry and the Portuguese Navigators. 3. Christopher Columbus. 4. Ferdinand Magellan. 5. Sir Francis Drake. 6. Search for the Northern Passages.

Capturing the historic significance of oceanic voyages and experiences of famous discoverers, this set traces important exploration and our increasing knowledge of the world, from the 13th to the 20th century.

DISCOVERY AND EXPLORATION (Filmstrip—Sound). *See* America—Exploration

JOHN WESLEY POWELL: DISCOVERY (Filmstrip—Sound). *See* Powell, John Wesley

EXPLORERS—FICTION

THE KING'S FIFTH (Filmstrip—Sound). *See* Discoveries (in geography)—Fiction

EXPLOSIVES

ALFRED NOBEL—THE MERCHANT OF DEATH? (Motion Picture—16mm—Sound). *See* Nobel, Alfred

EYE

HOW WE SEE (Filmstrip—Captioned). *See* Vision

EYEGLASSES

GLASSES FOR SUSAN (Motion Picture—16mm—Sound). *See* Vision

FABLES

BIRDS OF A FEATHER (Motion Picture—16mm—Sound). *See* Birds—Fiction

THE GRIZZLY AND THE GADGETS AND FURTHER FABLES FOR OUR TIME (Phonodisc or Cassette). Caedmon Records, 1972. 1 cassette or disc, approx. 55 min. #TC1412 Disc ($6.98), #CDL51412 Cassette ($7.95). Gr. 4–A. BKL (12/15/72), LBJ, JOR. 398.2

Here is the story of a grizzly bear who finds the antics of his relatives, the gadgets of the twentieth century and the complacency of his wife intolerable, and decides to take matters into his own paws to set his world to rights. The other animal heroes of these humorous fables reveal once again James Thurber's marvelous inventiveness and magical insight into human character.

THE HOARDER (Motion Picture—16mm—Sound). National Film Board of Canada/Benchmark Films, 1971. 16mm color, sound, animated film, 6 min. ($140). Gr. 2–A. AFF (1971). 398.2

A witty story about greed and sharing. Evelyn Lambert's colorfully animated blue jay carries away and hides all it can see—including the sun.

THE OWL WHO MARRIED A GOOSE (Motion Picture—16mm—Sound). *See* Folklore, Eskimo

FADIMAN, CLIFTON

A DISCUSSION OF THE CROCODILE (Motion Picture—16mm—Sound). *See* Motion pictures—History and criticism

FAIRY TALES

THE GREAT QUILLOW (Phonodisc or Cassette). *See* Giants—Fiction

THE HAPPY PRINCE AND OTHER OSCAR WILDE FAIRY TALES (Phonodisc or Cassette). Caedmon Records, 1956. 1 cassette or disc, approx. 50 min. #TC1044 Disc ($6.98), #CDL51044 Cassette ($7.95). Gr. 4–A. ESL, NCT, NYR. 398.2

Contents: 1. The Happy Prince. 2. The Selfish Giant. 3. The Nightingale and the Rose.

Among the immortal stories written for children are these three by Oscar Wilde and performed by Basil Rathbone. They are stamped with their creator's personal genius and nothing he ever wrote is likely to be remembered longer than these.

IRISH FAIRY TALES (Phonodisc or Cassette). *See* Folklore—Ireland

THE LITTLE LAME PRINCE (Phonodisc or Cassette). Caedmon Records, 1975. 1 cassette or disc, approx. 60 min. #TC1293 Disc ($6.98), #CDL51293 Cassette ($7.95). Gr. 4–8. LBJ (1975), NYR. Fic

This story tells of a little prince who escaped from the prison tower of his paralysis and on the way conquered the hearts of his countrymen. This slightly abridged version of the story is read by Cathleen Nesbitt.

THE LITTLE MATCH GIRL AND OTHER TALES (Phonodisc or Cassette). Caedmon Records, 1961. 1 cassette or disc, approx. 50 min. #TC1117 Disc ($6.98), #CDL51117 Cassette ($7.95). Gr. K–8. ESL, HOB, NYR. 398.2

Contents: 1. The Swineherd. 2. The Top and the Ball. 3. The Red Shoes. 4. Thumbelina. 5. The Little Match Girl.

Five of Hans Christian Andersen's fairy tales are presented by Boris Karloff. There is a warmth, characteristic of no other writer, and there is sympathy for all the world's creatures which matches that of every little child. These tales were first translated into English in 1846.

THE LITTLE MERMAID (Phonodisc or Cassette). Caedmon Records, 1967. 1 cassette or disc, approx. 55 min. #CDL51230 Cassette ($7.95), #TC1230 Disc ($6.98), Gr. K–8. ESL, HOB, LBJ. 398.2

The particular attraction of this fairy tale is that it offers children the opportunity to weep for themselves when they weep for the lovely little mermaid who longs for what she cannot have.

FAMILY

ECHOES (Motion Picture—16mm—Sound). *See* Life

FAMILIES IN CRISIS (Filmstrip—Sound). *See* Social problems

THE FAMILY IN CHANGING SOCIETY (Filmstrip—Sound). EBEC, 1978. 6 color, sound filmstrips, av. 103 fr. ea; 6 discs or cassettes, 16 min. ea.; Teacher's Guide. #17048K Cassettes ($99.50), #17048 Discs ($99.50). Gr. 7–A. BKL (11/78), MDU (8/78). 301.427

Contents: 1. Marriage: An Overview. 2. Marriage: Case Studies. 3. Parenthood: An Overview. 4. Parenthood: Case Studies. 5. Working Parents: An Overview. 6. Working Parents: Case Studies.

A combination of historical perspective and contemporary case studies, this filmstrip set examines the problems facing the family today in marriage, parenthood, and work. Each topic is treated in two paired filmstrips: the first—an overview—uses historical prints, photos and cartoons; the second offers case studies revealing candid interviews with people coping with the problems and decision-making in their chosen life-styles.

SALLY GARCIA AND FAMILY (Motion Picture—16mm—Sound). Educational Development Center, 1977 (Role of Women in American Society). 16mm color, sound film, 35 min. ($425). Gr. 7–A. AFF (1978), BKL (6/15/78), LFR (5/78). 301.427

Sally Garcia and Family is an unscripted documentary about a growing phenomenon of our time: a woman returning to work as she enters middle age, despite her satisfaction with home and family. While the Garcias are a close family, they are experiencing many of the changes and tensions which have affected, and often disrupted, other families.

TAKING RESPONSIBILITY (Filmstrip—Sound). *See* Self realization

TODAY'S FAMILY: A CHANGING CONCEPT (Filmstrip—Sound). Current Affairs Films, 1978. 1 color, sound filmstrip, 74 fr.; 1 cassette, 16 min.; Guide. #605 ($24). Gr. 8–12. BKL (4/1/78). 301.427

This filmstrip deals with the personal and social effects taking place as we move from the extended family to the nuclear family and single parent family.

FAMILY LIFE

ADOLESCENT RESPONSIBILITIES: CRAIG AND MARK (Motion Picture—16mm—Sound). *See* Adolescence

BECOMING A PARENT: THE EMOTIONAL IMPACT (Filmstrip—Sound). *See* Youth as parents

BUILDING A FUTURE (Filmstrip—Sound). *See* Youth as parents

COPING WITH PARENTS (Motion Picture—16mm—Sound). *See* Conflict of generations

DEALING WITH PRACTICAL PROBLEMS OF PARENTHOOD (Filmstrip—Sound). *See* Youth as parents

FAMILIES AROUND THE WORLD (Filmstrip—Sound). Science Research Associates, 1977. 8 color, sound filmstrips; 4 cassettes; Teacher's Guide. ($165). Gr. 4–12. MDU (1977). 301.42

Contents: 1. Family in China. 2. Family in Japan. 3. Family in Bangladesh. 4. Family on the Ivory Coast. 5. Family in Israel. 6. Family in Lebanon. 7. Family in Germany. 8. Family in Mexico.

Each family is shown carrying on its daily activities—working, eating, educating children, enjoying each others' company and expressing hope for the future. They show students' differences and likenesses of people around the world. Although the family structure may differ in different countries and cultures the basic functions of the family are the same.

FAMILIES OF ASIA (Filmstrip—Sound). *See* Asia

THE FAMILY IN TRANSITION (Kit). Butterick Publishing, 1976 (Family Life). 6 color, sound filmstrips, 65–75 fr.; 6 cassettes or discs, 10–12 min.; 11 Spirit Masters; Teacher's Guide. Cassette #C-607-1E ($125), Disc #R-606-3E ($125). Gr. 7–A. PRV (9/77). 301.42

FAMILY LIFE (cont.)

Contents: 1. & 2. Divorce. 3. & 4. Aging Families. 5. & 6. Future Family Life.

This set explores family problems and future family relationships.

THE FAMILY (Kit). Scholastic Book Services, 1975 (Family Living Program). 3 color, sound filmstrips, 92–102 fr.; 3 cassettes or discs, 12–15 min.; 9 Ditto Masters; Student Worksheets; Teacher's Guide. #1154 Cassette ($69.50), #1144 Disc ($69.50). Gr. 8–A. BKL (10/1/79). 301.427

Contents: 1. Parenthood: Why and When. 2. Family Dynamics. 3. A Family Album.

This program has three sets of parents who discuss both when and why to have children, as well as how this event has shaped their lives. Three other families talk about chores, appearances, school, drinking, smoking, and curfews. Finally, one family is explored in depth. Each member's ideas is offered.

FOUR FAMILIES (Filmstrip—Sound). Educational Direction/Learning Tree Filmstrips, 1975. 4 color, sound filmstrips, av. 53 fr. ea.; 2 cassettes, av. 6 min. ea.; Teacher's Guide. #LT553 ($52). Gr. 5–8. BKL (1/76), ESL (1976), PRV (11/75). 301.34

Contents: 1. We Live in New York City. 2. We Live in Springfield. 3. We Live in Middletown. 4. We Live in Clinton Corners.

This set presents the differences that are caused by living in a megalopolis, an average-sized city, a town, and a farm community.

FROM HOME TO SCHOOL (Filmstrip—Sound). Parents' Magazine Films, 1978 (El Manana Es Hoy). 5 color, sound filmstrips, 45–60 fr.; 1 disc or 3 cassettes, 8 min.; English and Spanish Scripts; Teacher's Guide. ($65). Gr. 10–A. PRV (2/19/79). 301.42

Emphasizes the importance of Hispanic parents helping to prepare their children for the intellectual, emotional, and social challenges they will face in school and as they move from a Spanish-speaking environment to an English-speaking one. Spanish sound track.

LEARNING AWAY FROM HOME (Kit). Parents' Magazine Films, 1976 (The Effective Parent). 5 color, sound filmstrips, 56–64 fr. ea.; 1 disc or 3 cassettes; 5 Audio Script Booklets and 1 Guide. ($65). Gr. 10–A. BKL (10/1/77), PRV (2/79). 301.42

Contents: 1. Family Excursions. 2. Going to the Supermarket. 3. A Trip to the Laundromat. 4. The Waiting Game. 5. A Nature Walk.

Three steps which parents can follow to make neighborhood trips into learning experiences are presented.

LEARNING BEGINS AT HOME (Filmstrip—Sound). Parents' Magazine Films, 1978 (El Manana Es Hoy). 5 color, sound filmstrips, 45–60 fr.; 1 disc or 3 cassettes, 8 min.; English and Spanish Scripts; Teacher's Guide. ($65). Gr. 10–A. PRV (3/19/79). 301.42

Stresses that parents are their children's first and most important teachers and offers advice on how Hispanic parents can help prepare their children for school. Spanish sound track.

LEARNING IN THE HOME (Kit). Parents' Magazine Films, 1976 (The Effective Parent). 5 color, sound filmstrips, 56–64 fr. ea.; 1 disc or 3 cassettes; 5 Audio Script Booklets; 1 Guide. ($65). Gr. 10–A. BKL (10/1/77), PRV (2/79). 601.42

Contents: 1. The Teachable Moments. 2. Cooking. 3. Reading. 4. Number Concepts. 5. Using TV Wisely.

This set illustrates how everyday routines provide opportunities for parents to introduce new words, concepts, and ideas. Also discussed is how to introduce such ideas as numerical order, weight and measure through games and experience.

LEARNING THROUGH PLAY (Kit). Parents' Magazine Films, 1976 (The Effective Parent). 5 color, sound filmstrips 56–64 fr. ea.; 1 disc or 3 cassettes; 5 Audio Script Booklets; 1 Guide. ($65). Gr. 10–A. BKL (10/1/77), PRV (2/79). 301.42

Contents: 1. Choosing Toys. 2. Creative Play. 3. Toys You Can Make. 4. Simple Learning Aids. 5. Games Inside Your Head.

Parents need to evaluate playthings for their usefulness as learning tools as well as for fun. Toys need not be "store bought" and viewers will learn how to make playthings from household items.

MARRIAGE AND PARENTHOOD (Kit). Butterick Publishing, 1976. 6 color, sound filmstrips, 65–75 fr.; 6 cassettes or discs, 10–12 min.; 11 Spirit Masters; Teacher's Guide. Cassette #C–605–5E ($125), Disc #R–604–7E ($125). Gr. 7–A. PRV (9/77). 301.42

Contents: 1. & 2. Getting Married. 3. & 4. Newly Married Couples. 5. & 6. Parenthood.

This set gives students the input needed to think about whether and how marriage fits into their future. It also views the day-to-day adjustments marriage requires. And, finally, it explains the changes in life-styles and values that have created chaotic situations for many parents and children.

OUR LANGUAGE, OUR CULTURE, OURSELVES (Filmstrip—Sound). Parents' Magazine Films, 1978 (El Manana Es Hoy). 5 color, sound filmstrips, 45–60 fr.; 1 disc, or 3 cassettes, 8 min. ea.; English and Spanish Scripts; Teacher's Guide. ($65). Gr. 10–A. PRV (2/79). 301.42

This set discusses many ways in which Spanish-speaking parents can contribute to their child's positive self-image realizing that language is the key to communication and culture the basis of identity.

THE PARENT AS A TEACHER (Kit). Parents' Magazine Films, 1976 (The Effective Parent). 5 color, sound filmstrips, 56–64 fr. ea.; 1 disc or 3 cassettes; 5 Audio Script Booklets; 1 Guide. ($65). Gr. 10–A. BKL (10/1/77), PRV (2/79). 301.42

Contents: 1. Learning Foundations. 2. Self Concept. 3. Developing Independence. 4. Effective Discipline. 5. Language Development.

Viewers are urged to make the child feel important and valuable by talking to the child, responding to his or her needs and by holding the child. Positive reinforcement for teaching self-discipline and how to express critical feelings most effectively are explained.

PARENT-SCHOOL RELATIONSHIPS (Filmstrip—Sound). Parents' Magazine Films, 1978 (El Manana Es Hoy). 5 color, sound filmstrips, 45–60 fr.; 1 disc or 3 cassettes, 8 min. ea.; English and Spanish Scripts; Teacher's Guide. ($65). Gr. 10–A. PRV (2/79). 301.42

This set discusses both the transitional and permanent bilingual education programs and stresses the importance for Hispanic parents to work with the school system to ensure that their child's needs are being met.

PREPARATION FOR PARENTHOOD (Filmstrip—Sound). *See* Parent and Child

RIGHTS AND OPPORTUNITIES (Filmstrip—Sound). *See* Youth as parents

FAMILY LIFE—AFRICA

FAMILIES OF WEST AFRICA: FARMERS AND FISHERMEN (Filmstrip—Sound). *See* Africa, West

FOUR FAMILIES OF KENYA (Filmstrip—Sound). *See* Kenya

FAMILY LIFE—EDUCATION

A LIFE BEGINS . . . LIFE CHANGES . . . THE SCHOOL-AGE PARENT (Filmstrip—Sound). *See* Pregnancy

FAMILY LIFE—FICTION

LITTLE WOMEN (Phonodisc or Cassette). Caedmon Records, 1975. 1 cassette or disc. #CDL51470 Cassette ($7.95), #TC1470 Disc ($6.98). Gr. 4–8. PRV (4/76). Fic

Julie Harris reads three chapters of this popular book by Louisa May Alcott. They are ''A Merry Christmas,'' ''Gossip,'' and ''The First Wedding.'' Young listeners will enjoy the reading and should be motivated to read the book.

LITTLE WOMEN (Phonodisc or Cassette). Listening Library, 1974. 2 phonodiscs or 2 cassette tapes, 105 min. #CX-3110 Cas-

settes ($15.90), #AA-33110 Discs ($12.95). Gr. 5–A. ESL (1977), LTP (4/75), PRV (11/75). Fic

The March girls grow up under the care and guidance of a watchful and loving mother. Two of them find happiness in marriage and a family tragedy takes Beth. This is a story of family affection at the turn of the century and has long been a classic favorite with children.

THE PETERKIN PAPERS (Phonodisc or Cassette). Caedmon Records, 1 disc or cassette. #TC1443 Disc ($6.98), #CDL51443 Cassette ($7.95). Gr. 4–A. INS. Fic

Contents: 1. The Lady Who Puts Salt in Her Coffee. 2. About Elizabeth Eliza's Piano. 3. The Peterkins Try to Be Wise. 4. The Peterkins at Home. 5. The Peterkins Snowed-up. 6. The Peterkins' Picnic.

These delightful stories by Lucretia Hale laugh at human folly.

FAMILY LIFE—JAPAN

JAPAN: SPIRIT OF IEMOTO (Filmstrip—Sound). *See* Japan—Social life and customs.

FAMILY LIFE—MIDDLE EAST

FAMILIES OF THE DRY MUSLIM WORLD (Filmstrip—Sound). *See* Middle East—Social life and customs

FANTASTIC FICTION

AROUND THE WORLD IN EIGHTY DAYS (Phonodisc or Cassette). Caedmon Records, 1977. 1 cassette or disc, 60 min., with notes. #CDL51553 Cassettes ($7.98), #TC1553 Discs ($7.98). Gr. 7–A. BKL (1/15/79). Fic

An abridged reading of the novel by Jules Verne and read by Christopher Plummer. He takes us on the 1873 around-the-world journey of Phileas Fogg and his valet, Passepartout. Several adventures are omitted, but the story flows smoothly.

DOCTOR JEKYLL AND MR. HYDE (Filmstrip—Sound). Listening Library, 1977. 2 color, sound filmstrips, 61–63 fr.; 2 cassettes, approx. 15 min. ea.; Teacher's Guide. YL58CFX ($35). Gr. 7–12. PRV (5/78). Fic

A presentation of the classic by Robert Louis Stevenson delineating the spiritual and baser nature of people. Watercolor drawings are used to illustrate the story.

FANTASY

THE DREAM (Motion Picture—16mm—Sound). *See* Pantomimes

HORSE FLICKERS (Motion Picture—16mm—Sound). *See* Collage

FANTASY (cont.)

IF TREES CAN FLY (Motion Picture— 16mm—Sound). *See* Imagination

JOHN HENRY AND JOE MAGARAC (Phonodisc or Cassette). *See* Folklore—U.S.

FANTASY—FICTION

ALICE IN WONDERLAND (Kit). Jabberwocky Cassette Classics, 1972. 1 cassette, 60 min.; 6 Read-along Scripts. #1051 ($15). Gr. 5–A. BKL (4/74), PRV (11/73). Fic
In this combined reading and dramatization of Lewis Carroll's classic story, Alice, the White Rabbit, the Caterpillar, the Cheshire Cat, the Mad Hatter, and the Queen of Hearts are aptly characterized. This is an adaptation and not a straight reading of the story.

ANIMAL FARM BY GEORGE ORWELL (Filmstrip—Sound). Current Affairs Films, 1978. 1 color, sound filmstrip; 1 cassette; Teacher's Edition of the Book & Testing Materials; Discussion Guide. #LP-617 ($30). Gr. 7–A. BKL (12/78). Fic
Orwell's parable pointing up the dangers of revolutionary corruption to the extent of restoring, in another form, the very tyranny that caused the revolution. It parallels directly the Russian Revolution and its aftermath, but is relevant to national revolutions in general.

CLOSED MONDAYS (Motion Picture— 16mm—Sound). *See* Art—Exhibitions—Fiction

THE CROCODILE (Motion Picture— 16mm—Sound) EBEC, 1973 (Short Story #47787 ($360). Gr. 7–A. BKL (1973), EFL (1974), PRV (1973). Fic
The crocodile stretches its jaws and swallows Ivan Matveyevitch. What's more, Ivan refuses to come out. "This is utopia!" he cries from within his cozy spot in the creature's belly. "I always thought this would happen to him," remarks Ivan's superior. A satire of 19th-century Russian petty officials? A pessimistic statement on the progressive society? Draw your own conclusions.

A DISCUSSION OF THE CROCODILE (Motion Picture—16mm—Sound). *See* Motion pictures—History and criticism

A DISCUSSION OF THE LADY OR THE TIGER? (Motion Picture—16mm—Sound). *See* Motion pictures—History and criticism

THE GAMMAGE CUP (Filmstrip—Sound). Miller-Brody Productions, 1976 (Newbery Award Sound Filmstrip Library). 2 color, sound filmstrips, 131–132 fr.; 1 disc or 2 cassettes, 22–23 min.; Teacher's Guide. #NSF3073 ($32). Gr. 4–8. PRV (4/78). Fic
The Minnipins, a mountain valley society, exile a group of inhabitants for questioning the wisdom of the leading family, the Periods. The exiles establish a new home for themselves and then discover that the Minnipins' ancient enemies are preparing an attack. This fantasy world story is based on the book by Carol Kendall.

HEROES OF THE ILIAD (Cassette). *See* Mythology, Classical

THE HOBBIT (Filmstrip—Sound). Current Affairs Films, 1978. 1 color, sound filmstrip, 66 fr.; 1 cassette, 16 min.; 1 Book, Teacher's Guide. #629 ($30). Gr. 7–12. PRV (12/19/78). Fic
An introduction to *The Hobbit* and Tolkien's Middle Earth world. Fantasy is explained and the characters are analyzed.

HOW A PICTURE BOOK IS MADE (Filmstrip—Sound). *See* Books

THE LADY OR THE TIGER? (Motion Picture—16mm—Sound). *See* American fiction

MINDSCAPE (Motion Picture—16mm—Sound). *See* Painters—Fiction

THE PRINCE AND THE PAUPER (Phonodisc or Cassette). Caedmon Records, 1977. 1 cassette or disc, approx. 60 min. #TC1541-1 Disc ($6.98), #CDL51541-1 Cassette ($7.95). Gr. 5–9. AUR (1978). Fic
Mark Twain wrote this great yarn of two boys in medieval England. The story has all of Twain's love of a tale for a tale's sake. Ian Richardson provides the many voices and sharp characterizations.

TCHOU TCHOU (Motion Picture—16mm—Sound). National Film Board of Canada/ EBEC, 1972. 16 mm color, sound film, 15 min. Teacher's Guide. #3336 ($185). Gr. K–C. LAM (2/76), LFR (10/74), LNG (12/75). Fic
This imaginative, nonnarrated film shows toy blocks forming a child's city. All animation is within the confines of the six sides of each block. One block is a bird fluttering wings, another a ladybug moving legs. It takes three blocks each to form a boy and a girl who move as though self-propelled. A chain of blocks make a dragon which topples buildings to warning sounds in music and drumrolls. While the dragon sleeps, the children outwit him in a surprise ending.

THROUGH THE LOOKING GLASS (Phonodiscs). CMS Records. 3 discs; Complete Illustrated Text. #673/3L ($20.94). Gr. 4–A. PRV (4/76). Fic
This Lewis Carroll work is told by George Rose with Sarah Jane Gwillim as Alice. Carroll's witty writing and George Rose's brilliant characterizations (he plays everyone but Alice) and Sarah Jane Gwillim (charming, well-mannered and awfully British) recreate this classic.

THE VELVETEEN RABBIT (Filmstrip—Sound). Miller-Brody productions, 1976. 2

color, sound filmstrips, av. 60 fr. ea.; 2 cassettes or discs, av. 12 min. ea.; Guide. Cassettes #L-512-FC ($32), Discs #L-512-FR ($32). Gr. 1-A. BKL (1/1/77), LGB (12/76), PVB (4/78). Fic

The narration conveys the mood of the fantasy of Margery Williams's book about the lifegiving power of love in a small boy's nursery.

FARM LIFE—U.S.

THE FARMER IN A CHANGING AMERICA (Motion Picture—16mm—Sound). *See* Agriculture—U.S.

HARVEST (Motion Picture—16mm—Sound). *See* Agriculture—U.S.

FASHION

CAREERS IN THE FASHION INDUSTRY (Kit). *See* Clothing trade

FATIO, LOUISE

THE HAPPY LION SERIES, SET ONE (Filmstrip—Sound). *See* Lions—Fiction

THE HAPPY LION SERIES, SET TWO (Filmstrip—Sound). *See* Lions—Fiction

FAULKNER, WILLIAM

FIVE MODERN NOVELISTS (Filmstrip—Sound). *See* American literature—History and criticism

FAUVISM

GEORGES ROUAULT (Motion Picture—16mm—Sound). *See* Rouault, Georges

FEEDBACK CONTROL SYSTEMS

INTRODUCTION TO FEEDBACK (Motion Picture—16mm—Sound). *See* Feedback (psychology)

FEEDBACK (PSYCHOLOGY)

INTRODUCTION TO FEEDBACK (Motion Picture—16mm—Sound). Eames Film, 1973, 16mm color, sound film, 11 min. #3183 ($150). Gr. 7-A. BLK (3/15/74), LFR (9/19/75). 153.1

Feedback—the cycle of measuring a performance, evaluating it, and correcting future performances—occurs all around us, even inside us. The film illustrates a wide range of activities, both human and mechanical, that defines and describes the feedback process and its important component—oscillation.

FICTION

UNDERSTANDING FICTION (Filmstrip—Sound). Educational Direction/Xerox Educational Publications, 1977. 5 color, sound filmstrips, av. 62 fr. ea.; cassettes, av. 9 min. ea.; Teacher's Guide. #SC030 ($105). Gr. 5-12. LAM (10/78). 807

Contents: 1. The Track Star. 2. The Track Star—A Closer Look. 3. The Dream. 4. The Dream—A Closer Look. 5. Stories to Finish.

Through audiovisual lessons, the basic elements of fiction are taught. The elements of plot, setting, imagery, mood, and style are discussed. The last strip presents "story-starters" that are intended to encourage or inspire writing.

FIEDLER, ARTHUR

CLASSICAL MUSIC FOR PEOPLE WHO HATE CLASSICAL MUSIC (Phonodiscs). *See* Orchestral music

FINANCE, PERSONAL

MEET MARGIE (Motion Picture—16mm—Sound). FilmFair Communications, 1976, 16mm color, sound film, 10 min. ($140). Gr. 7-A. ACI (3/76). 332.024

It does not take a great deal of money to lead a rewarding life. Margie, although living at a dollar-figure poverty level, uses energy, imagination, and creativity to gain a rich life. She travels by bike, belongs to a food coop, trades chores with her neighbors, and barters her skills for someone else's. While showing how she manages to get a "lot of fun out of life" in low-income circumstances, the film celebrates a simple, non-money-oriented life-style and the value of human interdependence. The film stresses that success in life comes from using time, energy, health, friendship, and intelligence.

MONEY MANAGEMENT (Kit). Society for Visual Education, 1978. 4 filmstrips; 4 cassettes; 30 Worksheets; Guide. #619-SATC ($99). Gr. 9-A. BKL (11/15/78). 332.024

Contents: 1. Setting Financial Goals. 2. Saving for Security. 3. Managing a Checking Account. 4. Using Other Financial Services.

Examines personal money management and the utilization of financial institutions to reach financial goals. Emphasizes planning, handling records, paying bills, saving, and wise use of credit.

FIRE PREVENTION

ALL ABOUT FIRE (Motion Picture—16mm—Sound). Farmhouse Films, National Safety Council/Pyramid Films, 1976. 16mm color, sound film, 10 min. #0153 ($175). Gr. K-8. LNG (2/78). 614.8

Colorful animation and a cynical cat bring the message on fire prevention. Accented

FIRE PREVENTION (cont.)

are the hazards in the average family home as pointed out by the fire captain.

FIRE EXTINGUISHERS AND SMOKE DE-TECTORS (Motion Picture—16mm—Sound). Handel Film, 1977 (Health and Safety). 16mm color, sound film, 18 min. ($270). Gr. 4–12. BKL (12/1/77), LFR (11/12/77), PRV (5/78). 614.8

The various types of smoke detectors and heat detectors are examined. Described are the four classes of fires, ''A, B, C, and D,'' and the corresponding fire extinguishers. Also discussed are the four keys to fire safety.

TIGER, TIGER, BURNING BRIGHT (Filmstrip—Sound). Marshfilm, 1976 (Safety). 1 color, sound filmstrip, 50 fr.; 1 cassette or disc, 15 min.; Teacher's Guide. #1129 ($21). Gr. 4–8. INS (1/78), NYF (1976). 614.8

Dramatizes the benefits as well as the hazards of fire through familiar situations; shows how to prevent fires in the home and outdoors; how to get help in an emergency and how to give first aid treatment for burns and shock.

FIRST AID

C P R: TO SAVE A LIFE (Motion Picture—16mm—Sound). EBEC, 1977. 16mm color, sound film, 14 min. #3544 ($200). Gr. 7–A. AAS (1978), BKL (2/1/78), LFR (1/2/78). 614.8

Simulated rescue scenes demonstrate the basic emergency techniques to be used in the event of heart arrest. The film's illustrations show that, with the proper training, the steps for cardiopulmonary resuscitation can be used effectively to save the lives of both adults and children. Each step of the procedure—reinforcing the airway, rescue breathing, cardiac compression—is simply and vividly demonstrated by paramedics and reinforced with art illustrations.

CHOKING: TO SAVE A LIFE (Motion Picture—16mm—Sound). EBEC, 1977. 16mm color, sound film, 12 min. #3545 ($180). Gr. 5–A. AAS (1978), AVI (1977), BKL (1978). 614.8

This film clearly explains choking rescue techniques to apply to others and to oneself. Trained paramedics demonstrate the back blow, the abdominal thrust and the finger probe. The film also presents ways to avoid choking situations.

EMERGENCY FIRST AID (Slides/Cassettes). Film Communicators, 1977. 129 slides; 1 cassette, 19 min.; Guide. ($125). Gr. 9–A. BKL (9/1/78). 614.8

A presentation of the basics of first aid, demonstrating mouth-to-mouth resuscitation, and what to do for severe bleeding, choking, poisoning, and shock.

FIRST AID: NEWEST TECHNIQUES (Filmstrip—Sound). Sunburst Communications, 1975. 8 color, sound filmstrips, 42–53 fr.; 8 cassettes or discs, 6–6½ min.; Teacher's Guide. #215 ($149). Gr. 7–12. BKL (4/1/75), LGB (12/75), PRV (12/75). 614.8

Contents: 1. Artificial Respiration. 2. Bleeding. 3. Poison. 4. Shock. 5. Burns. 6. Fractures. 7. Rescue and Transfer. 8. First Aid Review.

True-to-life first aid emergencies with a series of alternative actions for each. Carefully explains the nationally approved action that should be taken in each case, and discusses some common mistakes. Covers the life sustaining techniques everyone should know in seven major areas of first aid.

FISHERIES

CANNERY ROW, LIFE AND DEATH OF AN INDUSTRY (Filmstrip—Sound). *See* Monterey, California—History

FISHES

THE BROWN TROUT (Film Loop—8mm—Captioned). *See* Trout

SALMON RUN (Film Loop—8mm—Silent). *See* Salmon

See also Names of individual fish, e.g., Trout, etc.

FITZGERALD, F. SCOTT

FIVE MODERN NOVELISTS (Filmstrip—Sound). *See* American literature—History and criticism

FLACK, ROBERTA

THE LEGEND OF JOHN HENRY (Motion Picture—16mm—Sound). *See* Folklore—U.S.

FLAGS—U.S.

FOLK SONGS IN AMERICAN HISTORY: THE AMERICAN FLAG (Filmstrip—Sound). *See* Folk songs—U.S.

FLIGHT

QUEST FOR FLIGHT (Motion Picture—16mm—Sound). Avatar Productions/EBEC, 1976 (Wide World of Adventure). 16mm color, sound film, 23 min. #3374 ($320). Gr. 5–A. AAS (9/77), BKL (5/15/77), EFL (6/77). 629.13

Traces the history of humans' attempts to imitate the bird—from Icarus's feathered wings to the Supersonic Transport system.

FLIGHT TO THE MOON. *See* Space flight to the moon

FOG

FOG (Motion Picture—16mm—Sound). *See* Art and nature

FOLK ART

NATIVE ARTS (Filmstrip—Sound). Encore Visual Education, 1975. 4 color, sound filmstrips, 35–38 fr.; 4 cassettes, 9–11 min. ($55). Gr. 7–12. BKL (12/15/75), PRV (4/77). 709.011

Contents: 1. Introduction to Native Arts. 2. Northwest Coat Indian Arts. 3. Arts of the South Seas. 4. Mask and Figures around the World.

Describes the basic elements of primitive art as it surveys native artifacts in terms of materials, colorful designs, emotional expression, and craftsmanship. It views the totem poles, masks, baskets, food containers, slate carvings, and musical instruments of various areas.

FOLK ART, AMERICAN

AMERICAN FOLK ARTS (Filmstrip—Sound). Troll Associates, 1975. 8 color, sound filmstrips, av. 55 fr. ea.; 8 cassettes, av. 10 min. ea.; Teacher's guide. Each title ($32), Series ($128). Gr. 4–8. BKL (12/1/75), LGB (1975/76), PRV (3/76). 745.44

Contents: 1. Weathervanes and Whirligigs. 2. Scrimshaw. 3. Sculpture. 4. Art Technique. 5. Household Crafts. 6. Pottery. 7. Oil Painting. 8. Documentary Art.

Set displays the work of weavers, sculptors, artisans, and craftspeople, who using only the most basic materials, produced a wide variety of folk art forms—the authentic art of the young United States.

FOLK DANCING

AMERICA DANCES (Phonodisc or Cassette). Activity Records/Educational Activities. 1 disc or cassette, 30 min.; Teacher's Guide. #AC57 Cassette ($8.95), #AR57 Disc ($7.95). Gr. 2–8. PRV (9/75). 793.31

Contents: 1. Hornpipe. 2. Mowrah Cawkah. 3. Cherkessia. 4. Hokey Pokey. 5. Cissy. 6. Come Dance with Me. 7. Rabbit and the Fox. 8. Bingo.

A variety of American folk dances reflecting the spirit, variety, and multicultural makings of our country.

FOLK DANCING, SWISS

SONGS AND DANCES OF SWITZERLAND (Phonodiscs). Folkways Records, 1954. 1 disc, 10 in. 33^1/$_3$ rpm; Guide. #6807 ($5.00). Gr. 6–12. MDU (1978). 793.3

Includes yodeling, Alpine horn, and the bell-tree.

FOLK SONGS

AMERICA SINGS (Phonodisc or Cassette). Activity Records/Educational Activities. 1 disc or cassette, 30 min. Teacher's Guide. #AR56 Disc ($6.95), #AC56 Cassette ($7.95). Gr. 2–8. PRV (9/75). 784.4

Contents: 1. America the Beautiful. 2. Arkansas Traveler. 3. Blow Ye Winds. 4. Casey Jones. 5. Down the Stream. 6. Get Along Little Doggies. 7. Old Joe Clark. 8. Pop Goes the Weasel. 9. Yankee Doodle, etc..

A collection of American folk songs especially chosen for their animation and singability. The songs are representative of different strands of American Folk Music Heritage.

HAYWIRE MAC (Phonodiscs). Folkways Records. 1 disc. #FD5272 ($6.98). Gr. 4–A. PRV (12/75). 784.7

An important documentary of the music of W W I, performed by Harry K. McClintock. He tells stories, sings songs and spreads his message. Informative and entertaining. Strong in musical content and sociological significance.

FOLK SONGS, AMERICAN. *See* Folk songs—U.S.

FOLK SONGS, INDIAN

AS LONG AS THE GRASS SHALL GROW (Phonodiscs). Folkways Records, 1963. 1 disc, 12 in. 33^1/$_3$ rpm; Guide #2532 ($5.95). Gr. 6–A. MDU (1978). 784.75

Peter La Farge sings of the Indians and as a member of the Narraganset Tribe, La Farge fulfills a long dream of interpreting his people in music.

PETER LA FARGE ON THE WARPATH (Phonodiscs). Folkways Records, 1965. 1 disc, 12 in. 33^1/$_3$ rpm; Guide. #2535 ($5.95). Gr. 6–A. MDU (1978). 784.75

Bridging the worlds of white man and red man are these contemporary songs.

WAR WHOOPS AND MEDICINE SONGS (Phonodiscs). Folkways Records, 1964. 1 disc, 12 in. 33^1/$_3$ rpm; Guide. #4381 ($8.95). Gr. 3–A. MDU (1978). 784.75

Collected at Upper Dells, Wisconsin, during the annual Stand Rock Indian Ceremonials.

FOLK SONGS—U.S.

AN AUDIO VISUAL HISTORY OF AMERICAN FOLK MUSIC (Filmstrip—Sound). Educational Audio Visual, 1975. 5 color, sound film strips, 44–74 fr.; 5 discs or cassettes, 16–20 min.; Teacher's Guide. #9KF-026 Cassettes ($96), #9RF-025 Discs ($85). Gr. 7–12. PRV (2/76), PVB (4/76). 784.75

Contents: 1. Roots of American Folk Music. 2. Country Music in America. 3. Black Folk

FOLK SONGS—U.S. (cont.)

Music in America. 4. Folk Music in American History, Part I. 5. Folk Music in American History, Part II.

This set traces the history of American folk music from its roots to the present. Antique photographs and early oil paintings are used for the visuals.

THE ERIE CANAL (Filmstrip—Sound). *See* Erie Canal—Songs

FOLK SONGS IN AMERICAN HISTORY, SET ONE: 1700-1864 (Filmstrip—Sound). Warren Schloat Productions/Prentice-Hall Media, 1967. 6 color, sound filmstrips; 6 cassettes or discs. #HAC501 Cassettes ($138), #HAC501 Discs ($138). Gr. 5–A. MDU (1978), SLJ. 784.75

Contents: 1. Early Colonial Days. 2. The Revolutionary War. 3. Workers of America. 4. In Search of Gold. 5. The South. 6. The Civil War.

The spirit and folklore of America's heritage reflected through traditional folk songs. Performers include Joan Baez; Peter, Paul, and Mary; Pete Seeger; Burl Ives; and Josh White. The focus is on paintings, documentary photographs, and authentic illustrations.

FOLK SONGS IN AMERICAN HISTORY, SET TWO: 1865-1967 (Filmstrip—Sound). Warren Schloat Productions/Prentice-Hall Media, 1969. 6 color, sound filmstrips; 6 cassettes or discs. #HAC503 Cassettes ($138), #HAR503 Discs ($138). Gr. 5–A. MDU (1978), STE. 784.75

Contents: 1. Reconstruction and the West. 2. Immigration and Industrialization. 3. World War I. 4. 1920s and the Depression. 5. World War II. 6. The Post War Years.

America's heritage is reflected through traditional folk songs performed by such famous artists as Joan Baez; Peter, Paul, and Mary; Pete Seeger; Burl Ives; and Josh White. The lyrics appear on the screen and important period painting, documentary photographs, and authentic illustrations are also used.

FOLK SONGS IN AMERICAN HISTORY: THE AMERICAN FLAG (Filmstrip—Sound). Warren Schloat Productions/Prentice-Hall Media, 1970. 2 color, sound filmstrips; 2 cassettes or discs. #HAC cassettes ($48), #HAC discs ($48). Gr. 5–A. AFF. 784.75

This two-part set on the American flag reflects the spirit and folklore of America's heritage through traditional folk songs performed by famous artists.

FOLK SONGS IN AMERICA'S HISTORY SINCE 1865 (Filmstrip—Sound). Globe Filmstrips/Coronet Instructional Media, 1977. 6 color, sound filmstrips; 6 cassettes. #M716 ($99). Gr. 4–9. MDU (1978). 784.75

Contents: 1. Songs of Labor. 2. Songs of Happiness. 3. Songs of Protest. 4. Songs of Sorrow. 5. Songs of Injustice. 6. Songs of Heroism.

Here is U.S. history in drawings, photographs, and song. Some of the songs included in this set are "Hallelujah, I'm a Bum" and "This Land Is Your Land."

FOLKLORE

AN AUDIO VISUAL HISTORY OF AMERICAN FOLK MUSIC (Filmstrip—Sound). *See* Folk songs—U.S.

THE GREAT QUILLOW (Phonodisc or Cassette). *See* Giants—Fiction

WORLD MYTHS AND FOLKTALES (Filmstrip—Sound). Coronet Instructional Media, 1974. 8 color, sound filmstrips, av. 52 fr.; 8 discs or cassettes, av. 11½ min. #M701 Cassettes ($129), #S701 Discs ($129). Gr. 5–12. BKL (7/15/74). 398.2

Contents: 1. Hiawatha and the Iroquois Nation (U.S.A.). 2. Petit Jean (Canada). 3. Quetzalcoatl, Aztec God (Mexico). 4. Pele, the Fire Goddess (Hawaii). 5. The Excellent Archer (China). 6. Siegfried and the Jealous Queen (Germany). 7. Paris and the Golden Apple (Greece). 8. The Day the World Went Dark (Nigeria).

Each tale is retold in an appropriate dialect and in a suitable ethnic context.

FOLKLORE—AFRICA

ANANSI THE SPIDER (Motion Picture—16mm—Sound). Texture Films, 1969. 16mm color, sound film, 10 min. ($165). Gr. 1–A. BKL (10/1/76), CGE (1970), FLN (9/69). 398.2

An animated film which graphically relates the adventures of the cunning spider Anansi, trickster hero of the Ashanti people of Ghana's West Africa.

THE MAGIC TREE (Motion Picture—16mm—Sound). Texture Films, 1970. 16mm color, sound film, 10 min. ($165). Gr. 4–A. BKL (5/1/73), CGE (1971), FLN (1971). 398.2

This tale from the Congo tells of a homely, unloved boy who leaves his family and finds a secret paradise. But he loses it all when he breaks his vow of secrecy and reveals its mystery.

MEN BEHIND THE BRIGHT LIGHTS (Kit). *See* Musicians

A MISERABLE MERRY CHRISTMAS (Motion Picture—16mm—Sound). *See* Christmas

FOLKLORE, BLACK

FOLKTALES OF BLACK AMERICA (Filmstrip—Sound). International Film Bureau,

1977. 4 color, sound filmstrips, av. 56 fr.; 4 cassettes, av. 7 min. ($62.50). Gr. 5–A. BKL (11/15/77), PRV (4/78). 398.2

Contents: 1. A Tale of Tell: Introducing Folk Poetry. 2. The Signifyin' Monkey. 3. Dolomite. 4. The Titanic.

These are warm, relaxed, and exaggerated tales of the streets where these tales are usually told. Pin-the-tail-on-the-donkey, jump rope songs, cat calls, the Bermuda Triangle, and Bigfoot are part of the ''living folklore'' covered. An excellent example of the rhythm, dialect, and tone of black oral literature is provided by the narrator.

FOLKLORE—CHINA

THE SUPERLATIVE HORSE (Motion Picture—16mm—Sound). Phoenix Films, 1975. 16mm color, sound film, 36 min. #0160 ($450). Gr. 4–12. PRV (11/77). 398.2

Based on Jean Merrill's allegorical children's story, a tale from ancient China. It proves that people should not judge things, animals or even other people by their outer appearances, but rather by their inner qualities.

FOLKLORE, ESKIMO

THE OWL WHO MARRIED A GOOSE (Motion Picture—16mm—Sound). National Film Board of Canada/Stephen Bosustow Productions, 1975. 16mm color, sound film, 7 1/2 min. ($145). Gr. K–A. BKL (1/1/77), AFF (1977), LGB (1977). 398.2

In the solitude of the arctic, a goose captures the fancy of an owl. While the owl is hopelessly in love, he can never hope to keep up with the goose and her goslings. The owl pursues his dream of sharing her life, even though it means his destruction.

FOLKLORE—GERMANY

A TALE OF TILL (Motion Picture—16mm—Sound). FilmFair Communications, 1975. 16mm color, sound film, 11 1/4 min. ($160). Gr. 4–12. BKL (6/19/75), EGT (1976), LGB (1974). 398.2

After briefly characterizing the Dark Ages with visuals setting the scene in Schleswig-Holstein, Germany, the film introduces Till Eulenspiegel and his place in Germany literature. The remainder of the film is then devoted to an outdoor puppet show of one of Till's stories.

FOLKLORE, INDIAN

ARROW TO THE SUN (Motion Picture—16mm—Sound). *See* Indians of North America—Legends

NATIVE AMERICAN MYTHS (Filmstrip—Sound). *See* Indians of North America—Legends

FOLKLORE—IRELAND

IRISH FAIRY TALES (Phonodisc or Cassette). Caedmon Records, 1972. 1 cassette or disc, approx. 50 min. #TC1349 Disc ($6.98), #CDL51349 Cassette ($7.95). Gr. 4–8. LBJ. 398.2

Contents: 1. Cucullin and the Legend of Knockmany. 2. Guleesh.

These two Celtic fairy tales, gathered from Celtic oral literature collected during the nineteenth century from storytellers whose native language was Irish or Scottish-Gaelic, is performed here by Cyril Cusack.

FOLKLORE—JAPAN

REFLECTIONS: A JAPANESE FOLK TALE (Motion Picture—16mm—Sound). EBEC, 1975. 16mm color, sound film, 19 min. #3473 ($255). Gr. 5–9. BKL (12/15/75), LFR (3/4/75), LGB (1975/76). 398.2

A tale designed to probe the many faces of perception. The old Japanese tale in which a son believes that he has found his dead father inside a small box in a gift shop. Unfamiliar with mirrors he takes the reflection for that of his father. His wife, curious about the box and as unfamiliar with mirrors as her husband, is enraged when she looks in the box and thinks that he is hiding a young woman there. They take the box to a holy woman, but she discovers an old woman inside.

FOLKLORE—MEXICO

GOLDEN LIZARD: A FOLK TALE FROM MEXICO (Motion Picture—16mm—Sound). EBEC, 1976. 16mm color, sound film, 19 min. #3493 ($255). Gr. 4–8. LFR (3/19/77), LGB (11/19/77). 398.2

This dramatization of a Mexican folk tale presents a symbolic lesson involving a young man, a wandering stranger, and a lizard turned to gold. Transcending its Mexican setting, the film explores a simple story of wealth and generosity, of despair and new-found faith.

MEXICAN INDIAN LEGENDS (Motion Picture—16mm—Sound). BFA Educational Media, 1976. 16mm color, sound film, 16 min. #11668 ($230). Gr. 4–10. BKL (1/15/77). 398.2

Tales from the Toltec, Aztec, and Mayan are retold by a narrator with a Spanish accent as actors pantomime the legends. The legend behind the founding of Mexico City and the interpretation of such natural phenomena as the sun, moon, and creation of the mountains are discussed.

FOLKLORE—POLAND

ZLATEH THE GOAT (Motion Picture—16mm—Sound). Morton Schindel/Weston Woods Studios, 1966. 16mm color, sound film, 20 min. #419 ($325). Gr. K–9. CFF (1974), IFT (1974), PRV (2/77). 398.2

This tender folk tale by the great Yiddish author, Isaac Bashevis Singer, is a deceptively simple story in which an unforgettable parable unfolds to reveal the ecological balance that exists throughout nature and which links humans to all living things.

FOLKLORE—U.S.

FOLKTALES OF BLACK AMERICA (Filmstrip—Sound). *See* Folklore, Black

JOHN HENRY AND JOE MAGARAC (Phonodisc or Cassette). Caedmon Records, 1972. 1 cassette or disc, approx. 50 min. #TC1318 Disc ($6.98), #CDL51318 Cassette ($7.95). Gr. 3–8. BKL (1972), LBJ (1972), NYR. 398.2

Ed Begley reads two folk tales of legendary American heroes: powerful, black John Henry was the hero of the rock tunnel gangs who blasted out the tunnels for the trains; Joe Magarac was the name legend makers bestowed on the man they considered the greatest steel maker of them all.

JOHNNY APPLESEED AND PAUL BUNYAN (Phonodisc or Cassette). Caedmon Records, 1972. 1 cassette or disc, approx. 60 min. #TC1321 Disc ($6.98), #CDL51321 Cassette ($7.95). Gr. 3–8. BKL, LBJ. 398.2

Prophet, fanatic, saint-like, or whatever the real man was, Johnny Appleseed is a permanent part of American folk literature. Paul Bunyan, the other hero described on this fourth volume of American tall tales recorded by Caedmon, was not only a symbol of brawn but of American ingenuity in overcoming obstacles.

THE LEGEND OF JOHN HENRY (Motion Picture—16mm—Sound). Bosustow Productions, 1974 (American Folktales). 16mm color, sound, animated film, 11 min.; Reader's Guide. ($160). Gr. 5–A. CGE (1974), CFF (1974), WLB (6/74). 398.2

This is the legend of the great steel driving man. With a deadline of tunneling through Big Ben mountain before winter, the railroad bosses decide to replace John Henry and his crew with a steam drill. John Henry declares "Before I give in and let the steam drill win, I'll hammer myself to death." The race pits the power of one mighty and determined human against the steady relentless machine. After days of exhausting hammering, John Henry wins the race. He beats the steam drill, but it costs him his life. It is sung by Roberta Flack.

THE LEGEND OF PAUL BUNYAN (Motion Picture—16mm—Sound). Bosustow Productions, 1973 (American Folktales). 16mm color, sound, animated film, 13 min.; Reader's Guide. ($180). Gr. 2–A. BKL (1/1/74), CGE (1974), PRV (11/74). 398.2

The legend of the giant woodsman, Paul Bunyan, tells of Babe, the blue ox, and Paul's prosperous logging business. A favorite in American folklore, Paul Bunyan was known for his gentle temperament and fair dealing as well as his size and strength. But one day, Hels Helsun, "The Bull of the Woods," pushes him a little too far by giving poor Babe a dose of poison sumac. They have a battle, and the force of their blows creates the Grand Canyon, the Great Lakes, the Mississippi River, and Niagara Falls.

LEGEND OF SLEEPY HOLLOW AND OTHER STORIES (Cassette). *See* Legends—U.S.

THE LEGEND OF SLEEPY HOLLOW (Motion Picture—16mm—Sound). *See* Legends—U.S.

FONDA, PETER

NOT SO EASY (MOTORCYCLE SAFETY) (Motion Picture—16mm—Sound). *See* Motorcycles

FOOD SUPPLY

HARVEST (Motion Picture—16mm—Sound). *See* Agriculture—U.S.

THE HUNGRY PLANT (Filmstrip—Sound). *See* Plants—Nutrition

THE PEOPLE PROBLEM: FEEDING THE WORLD FAMILY (Filmstrip—Sound). Associated Press/Pathescope Educational Media, 1977 (Let's Find Out). 1 color, sound filmstrip, 63 fr.; 1 disc or cassette, 7$^{1/2}$ min.; Spirit Masters; Teacher's Guide. #9510 Disc ($25), #9510C Cassette ($25). Gr. 4–8. PRV (4/78). 338.1

Designed to give overall insight into a problem affecting the entire world.

SOYBEANS—THE MAGIC BEANSTALK (Motion Picture—16mm—Sound). *See* Soybeans

FOOTBALL

CENTER FOR A FIELD GOAL (Film Loop—8mm—Silent). Athletic Institute (Football). 8mm color, silent film loop, approx. 4 min.; Guide. #I-7 ($22.95). Gr. 3–12. AUR (1978). 796.33

A single concept loop demonstrating the basic technique.

CENTER SNAP FOR PUNT (Film Loop—8mm—Silent). Athletic Institute (Football). 8mm color, silent film loop, approx. 4 min.; Guide. #I-6 ($22.95). Gr. 3–12. AUR (1978). 796.33

A single concept loop demonstrating the basic technique.

CENTER TO QUARTERBACK EX-
CHANGE (Film Loop—8mm—Silent). Athletic Institute (Football). 8mm, color, silent
film loop, approx. 4 min.; Guide. #I–5
($22.95). Gr. 3–12. AUR (1978). 796.33
A single concept loop demonstrating the basic technique.

FIELD GOAL AND EXTRA POINTS (Film
Loop—8mm—Silent). Athletic Institute
(Football). 8mm color, silent film loop, ap-
prox. 4 min.; Guide. #I–9 ($22.95). Gr. 3–12.
AUR (1978). 796.33
A single concept loop demonstrates the bas-
ic technique.

FIELD GOAL AND KICKOFF (SOCCER
STYLE) (Film Loop—8mm—Silent). Athletic Institute (Football). 8mm color, silent
film loop, approx. 4 min.; Guide. #I–15
($22.95). Gr. 3–12. AUR (1978). 796.33
A single concept loop demonstrates the basic technique.

HAND OFF (Film Loop—8mm—Silent). Athletic Institute (Football). 8mm color, silent
film loop, approx. 4 min.; Guide. #I–2
($22.95). Gr. 3–12. AUR (1978). 796.33
A single concept loop demonstrates the basic technique.

KICKOFF AND ONSIDE KICK (Film
Loop—8mm—Silent). Athletic Institue
(Football). 8mm color, silent film loop, approx. 4 min.; Guide. #I–14 ($22.95). Gr. 3–
12. AUR (1978). 796.33
A single concept loop demonstrates the basic technique.

MIDDLE GUARD PLAY (Film Loop—
8mm—Silent). Athletic Institute (Football).
8mm color, silent film loop, approx. 4 min.;
Guide #I–10 ($22.95). Gr. 3–12. AUR (1978).
796.33
A single concept loop demonstrates the basic technique.

OFFENSIVE BACKS (Film Loop—8mm—Silent). Athletic Institute (Football). 8mm col-
or, silent film loop. approx. 4 min.; Guide.
#I–1 ($22.95). Gr. 3–12. AUR (1978). 796.33
A single concept loop demonstrates the basic technique.

OFFENSIVE LINE BLOCKING (Film
Loop—8mm—Silent). Athletic Institute
(Football). 8mm color, silent film loop, approx. 4 min.; Guide. #I–12 ($22.95). Gr. 3–
12. AUR (1978). 796.33
A single concept loop demonstrates the basic technique.

PASS PROTECTION (Film Loop—8mm—Si-
lent). Athletic Institute (Football). 8 mm col-
or, silent film loop, approx. 4 min.; Guide.
#I–13 ($22.95). Gr. 3–12. AUR (1978).
796.33
A single concept loop demonstrates the basic technique.

PASSING SKILLS, PART 1 (Film Loop—
8mm—Silent). Athletic Institute (Football).
8mm color, silent film loop, approx. 4 min.;
Guide. #I–3 ($22.95). Gr. 3–12. AUR (1978).
796.33
A single concept loop demonstrates the basic technique.

PASSING SKILLS, PART 2 (Film Loop—
8mm—Silent). Athletic Institute (Football).
8mm color, silent film loop, approx. 4 min.;
Guide. #I–4 ($22.95). Gr. 3–12. AUR (1978).
796.33
A single concept loop demonstrates the basic technique.

PUNTING (Film Loop—8mm—Silent). Athletic Institute (Football). 8 mm color, silent
film loop, approx. 4 min.; Guide. #I–8
($22.95). Gr. 3–12. AUR (1978). 796.33
A single concept loop demonstrates the basic technique.

FORCE AND ENERGY

FORCES MAKE FORMS (Motion Picture—
16mm—Sound). FilmFair Communications,
1974. 16mm color, sound film, 12^1/2 min.
#D272 ($165). Gr. 3–8. BKL (1/15/75), CFF
(1975), PRV (9/75). 531
Unusual and everyday objects and events
are used to encourage critical observation
and logical association between forces in-op-
eration and their resultant forms.

FORECASTING

DIMENSIONS OF CHANGE (Filmstrip—
Sound). *See* Social change

FORECASTING THE FUTURE: CAN WE
MAKE TOMORROW WORK? (Filmstrip—
Sound). Harper and Row Publishers/Sun-
burst Communications, 1976. 5 color, sound
filmstrips, 83–102 fr.; 5 cassettes or discs,
18–20^1/2 min.; Teacher's Guide. #77 ($125).
Gr. 7–C. BKL (7/1/77), LGB (12/76), TSS (5/
77). 301.24
Contents: 1. Choosing Tomorrow's World.
2. Energy and Human Values. 3. Science,
Technology, and the Year 2000. 4. The Future of Work. 5. The Family in Transition.
Uses exciting visuals from the futuristic
worlds of Woody Allen's *Sleeper* and Isaac
Asimov's *Profession* to turn classroom into
a think tank about the future of human values. Presents likely changes in the fabric of
our working habits that might be necessi-
tated by new economic and social conditions. Looks at the variety of new family
forms that forecasters anticipate for the future. Emphasizes that the future cannot be
predicted but can be invented—and will be
strongly influenced by choices made today.

REDESIGNING MAN: SCIENCE AND HU-
MAN VALUES (Filmstrip—Sound). Harper and Row Publishers/Sunburst

FORECASTING (cont.)

Communications, 1974. 6 color, sound film-strips, 72–90 fr.; 6 cassettes of discs, 14–16^1/$_2$ min.; Teacher's Guide. #70 ($145). Gr. 9–C. ABT (11/77), M&M (9/75), MDM (11/77). 218

Contents: 1. Corrections and Carbon Copies. 2. Breeding Tomorrow's Man. 3. Transplants and Implants. 4. Exploring Man's Mind. 5. The World Unborn. 6. The Search for Immortality.

Presents a startling look at the future of the human species. New and fascinating bio-medical developments have created a Pandora's box—knowing the possible consequences, should we open it? Poses value questions about topics such as cloning, genetic engineering, eugenics, and test tube babies.

A TIME OF CHANGES UNIT THREE (Motion Picture—16mm—Sound). *See* Occupations

FOREST FIRES

FIRE! (Motion Picture—16mm—Sound). Film Polski/Encyclopaedia Britannica Educational, 1977. 16mm color, sound film 9 min. #3562 ($145). Gr. K–A. NEA (1978). 634.9

Unusual animation technique (oil painting on glass slides) is combined with dramatic music to carry this nonnarrated film. The forest is at first alive with birds singing and animals playing. Suddenly the animals and birds leave and a raging fire approaches, destroying everything in its path. Then rain falls, the fire smolders and dies. Soon a single plant and bud emerge and the life cycle begins anew.

FORESTS AND FORESTRY

AMAZON JUNGLE (Film Loop—8mm—Silent). Walt Disney Educational Media, 1966. 8mm color, silent film loop, approx. 4 min. #62-5315L ($30). Gr. 4–12. MDU (1974). 634.9

The Amazon jungle is surveyed in this motion picture. Views of impenetrable rain forest, rivers, uncharted streams, and extraordinary vegetation, as well as some of the animal life are presented.

THE GODS WERE TALL AND GREEN (Filmstrip—Sound). Lyceum/Mook & Blanchard, 1972. 2 color, sound filmstrips, 56 fr.; 2 cassettes or discs, 13^3/$_4$ min. #LY35272SC Cassettes ($46), #LY35272SR Discs ($46). Gr. 6–A. BKL (9/15/72), IFT (1972), PVB (5/73). 574.981

Contents: 1. Trees: An Ancient Kinship. 2. The Kingdom of the Forest.

Ann Atwood takes us on a photographic art journey into nature. She crosses the latitudes of the topics, the temperate zone, California's redwoods, and the Olympic Rain Forest. She helps us understand the ecology of the forest.

A WALK IN THE FOREST (Motion Picture—16mm—Sound). Pyramid Films, 1976. 16mm color, sound film, 28 min.; ^{3}/$_4$ in. videocassette, also available. ($375). Gr. 7–A. AAS (1976), IDF (1976), LGB (1976). 574.5

Details the delicate balance and intricate harmony of plants and creatures that make the forest a living whole. Time-lapse photography chronicles the varying moods of the forest through the change of seasons, and the catastrophic effects of a raging forest fire underline the message that we must learn to replace what we take from the earth.

WINTER IN THE FOREST (Filmstrip—Sound). *See* Winter

FOSSILS

DINOSAURS: THE TERRIBLE LIZARDS (Motion Picture—16mm—Sound). *See* Dinosaurs

FOSSILS (Filmstrip—Sound). Educational Dimensions Group, 1976. 2 color, sound filmstrips, 57–60 fr.; 2 cassettes, 15–17 min.; Teacher's Guide. #520 ($37). Gr. 4–9. PRV (11/76), PVB (1977). 560

A fossil is any trace of life which lived through ancient times. The fossil record begins roughly 600 million years ago. Tracing the twelve geological periods through the three eras, students see ice caps and oceans' ebb and flow, plants and animals briefly appear and then disappear, remaining only as fossils.

FOSSILS: EXPLORING THE PAST (Motion Picture—16mm—Sound). EBEC, 1978. 16mm color, sound film, 16 min. #3600 ($425). Gr. 7–A. BKL (1/1/79), PRV (1978). 560

The film offers viewers a close look at paleontologists working in the field and laboratory, collecting, examining, and interpreting many types of fossils. The film highlights processes of fossil formation.

FRAGONARD, JEAN HONORE

FRAGONARD (Motion Picture—16mm—Sound). National Gallery of Art & Visual Images/WETA-TV/EBEC, 1973 (The Art Awareness Collection, National Gallery of Art). 16mm color, sound film 7 min. #3531 ($105). Gr. 7–C. EGT (1978), LFR (1978). 759.4

Fragonard paints the luxurious world of the French aristocrat: airy gardens, frolicking aristocrats, etc.

FRANCE

PASSPORT TO FRANCE (Filmstrip—Sound). EMC, 1977. 5 color, sound filmstrips, 91–107 fr.; 5 cassettes, 15–22 min.; Teacher's Guide. FRC 111000 ($98). Gr. 4–12. BKL (1/1/79), PRV (1/79). 914.4

Contents: 1. Preparation and Flight. 2. At the Bank, Drugstore, Post Office, Filling Station, Newsstand, Sidewalk Cafe. 3. Breakfast, at the Tourist Office, Transportation, the Seine. 4. At the Fleamarket, Pastry Shop, Delicatessen, Department Store. 5. Coping with the Metric System.

A cultural program in English and French, designed for exploratory study, travelers, adult education or French language students. It will acquaint students with day-to-day experiences of traveling in France.

WESTERN EUROPE: FRANCE (Filmstrip—Sound). Milan Herzog & Assoc./EBEC, 1977. 3 color, sound filmstrips, 95 fr. ea.; 3 discs or cassettes, 14 min. ea. #6957K Cassettes ($43.50), #6957 Discs ($43.50). Gr. 7–12. LFR (10/77). 914.4

Contents: 1. The Market Comes to the Farmer. 2. Land and People. 3. The Buffeteau Family.

This set examines French responses to changing world conditions. It details France's booming agricultural growth—including its contributions to the Common Market—and introduces us to a typical French family, still the cornerstone of French life.

FRANCE—HISTORY

THE DREYFUS AFFAIR (Motion Picture—16mm—Sound). See Dreyfus, Alfred

FRANCE—HISTORY—BIOGRAPHY

CATHERINE DE'MEDICI (Cassette). See Catherine de'Medici

JOAN OF ARC (Cassette). See Joan of Arc

NAPOLEON BONAPARTE (Cassette). See Napoleon I

FRANK, ANNE

THE DIARY OF ANNE FRANK (Filmstrip—Sound). See Autobiographies

FRANKLIN, BENJAMIN

BENJAMIN FRANKLIN—SCIENTIST, STATESMAN, SCHOLAR, AND SAGE (Motion Picture—16mm—Sound). Handel Film, 1970 (Americana Series #5). 16mm color, sound film, 30 min.; Film Guide. ($360). Gr. 4–A. FLN (1970), LFR (2/70). 921

Traces the life of Benjamin Franklin. Includes his career as printer, publisher, scientists, inventor, philosopher, and statesman; his work with the development of the Constitution and the postal system; and his role in the founding of the first subscription library and the first charity hospital in America.

THE FOUNDING FATHERS IN PERSON (Cassette). See U.S.—History—Biography

FREEDOM

FREEDOM RIVER (Motion Picture—16mm—Sound). Bosustow Productions, 1971. 16mm color, sound film, 8 min. ($150). Gr. 7–A. AVI (4/19/72), NEF (1972), PRV (9/19/72). 323.44

An animated film illustrating the parable that the life or death of the "River of Freedom" notion is in our hands. A comparison with our nation is obvious and the responsibility is on each and every citizen.

I SHALL MOULDER BEFORE I SHALL BE TAKEN (Motion Picture—16mm—Sound). See Surinam—History

FREEDOM OF THE PRESS

FREE PRESS: A NEED TO KNOW THE NEWS (Kit). Pathescope Educational Media, 1977 (Let's Find Out). 1 color, sound filmstrip, 62 fr.; 1 cassette, 9 min.; 6 Spirit Masters; Teacher's Guide. #9514 ($28). Gr. 4–8. BKL (12/77), PRV (4/78). 323.44

This explores the need for a free press in a democracy as well as the limitations and responsibilities, an introductory look at the news media and their role in society.

THE JOHN PETER ZENGER TRIAL (Filmstrip—Sound). Current Affairs Films, 1977. 1 color, sound filmstrip, 74 fr.; 2 cassettes—"Pro-and-Con," 16 min. ea.; Teacher's Guide. #584 ($30). Gr. 7–A. PRV (3/78). 323.44

Before the thirteen English colonies became the United States, John Peter Zenger, a New York newspaper publisher, printed articles criticizing the royal governor. In 1735, Zenger was brought to trial for seditious libel. This trial not only set the precedent of freedom of the press in America, but also caused the first step to be taken to adjust existing British laws to fit the new circumstances and characteristics of American life.

YOUR NEWSPAPER (Filmstrip—Sound). See Newspapers

FRENCH DRAMA

DIRECTING A FILM (IONESCO'S: THE NEW TENANT) (Motion Picture—16mm—Sound). See Motion pictures—History and criticism

THE NEW TENANT (Motion Picture—16mm—Sound). EBEC, 1975. 16mm color,

FRENCH DRAMA (cont.)

sound film, 31 min. #47816 ($445). Gr. 11–A. EFL (1978), INS (1977), PRV (1976). 842

Almost a vaudeville act, this play from the Theater of the Absurd expresses Ionesco's sense of the ambiguity and incoherence of much of modern life. The scene opens in an empty room at the top of a six-story walk-up. Soon the new tenant arrives—and begins filling the room with furniture. Finally the man is crowded into a tiny square—isolated, lost, barricaded by four walls of his own furniture. In this peculiar way, he has cut himself off from the world; he has created for himself a kind of death.

FRENCH FICTION

LA GRANDE BRETECHE (Motion Picture— 16mm—Sound). *See* Horror—Fiction

VOLTAIRE PRESENTS CANDIDE: AN INTRODUCTION TO THE AGE OF ENLIGHTENMENT (Motion Picture— 16mm—Sound). EBEC, 1976. 16mm color, sound film, 34 min. #47826 ($475). Gr. 9–A. EFL (1978), LFR (1977), PRV (5/77). Fic

This dramatization of Voltaire's satiric masterpiece provides a look at the history and culture of 18th-century Western civilization. The irascible and witty Voltaire acts as our host: slyly questioning the "Age of Enlightenment" and its doctrinaire optimism, introducing the naive characters from *Candide,* and commenting on their disastrous adventures.

FRENCH LANGUAGE—READING MATERIALS

A. J. MILLER'S WEST: THE PLAINS INDIAN—1837—SLIDE SET (Slides). *See* Indians of North America—Paintings

FRENCH LANGUAGE—STUDY AND TEACHING

PASSPORT TO FRANCE (Filmstrip— Sound). *See* France

FRESH-WATER BIOLOGY

THE MAYFLY: ECOLOGY OF AN AQUATIC INSECT (Motion Picture— 16mm—Sound). *See* Mayflies

FRIENDSHIP

PARTNERS (Kit). *See* Love

TEENAGE RELATIONSHIPS: VENESSA AND HER FRIENDS (Motion Picture— 16mm—Sound). *See* Adolescence

VALUES FOR DATING (Filmstrip—Sound). *See* Dating (social customs)

FRIENDSHIP—FICTION

ANGEL AND BIG JOE (Motion Picture— 16mm—Sound). *See* Values—Fiction

THE GIVING TREE (Motion Picture— 16mm—Sound). Bosustow Productions, 1973. 16mm color, sound, animated film, 10 min.; Reader's Guide. ($175). Gr. 2–A. CFF (1974), INS (6/7/73), NEA (1973). Fic

This is the story of a boy and a tree. The tree unselfishly offers everything it has for the boy's comfort. He never gives anything in return, but the tree is always happy to see him and happy to offer its resources. The boy grows and his life takes him away from his beloved tree, but old age brings him back again. They enjoy a peaceful and heartening reunion. Based on the book by Shel Silverstein and narrated by him.

PHILIP HALL LIKES ME: I RECKON MAYBE (Phonodisc or Cassette). Miller-Brody Productions, 1976. 1 cassette or disc, 58 min.; Notes. #NAR–3085 Disc ($6.95), #NAC–3085 Cassette ($7.95). Gr. 6–10. BKL (9/15/77). Fic

Narrated by Ruby Dee, this recording is an abridgment of the Newbery Honor Award book by the same name. The trials of growing up plus a test of friendship arising from competition at a country fair in Arkansas are related.

THE PIGMAN (Phonodisc or Cassette). Miller-Brody Productions, 1976 (Young Adult Recordings). 1 cassette or disc, 40 min.; Teacher's Notes. #YA404C Cassette ($7.95), #YA404 Disc ($6.95). Gr. 7–12. BKL (4/15/78), CRC (10/19/76), MMB (3/19/77). Fic

Paul Zindel's book has been condensed and dramatized, but the recording brings out the novel's themes of alienation and loneliness. The two teenagers' practical joke on Angelo Pignots, a lonely old man, fizzles when they realize they like him and that he accepts them for what they are. They become friends until a tragedy leads to the old man's death.

SUMMER OF MY GERMAN SOLDIER (Phonodisc or Cassette). *See* World War, 1939-1945—Prisoners & prisons—Fiction

FROGS

DISSECTION OF A FROG (Filmstrip— Sound). *See* Dissection

FRONTIER AND PIONEER LIFE

GREAT GRANDMOTHER (Motion Picture— 16mm—Sound). *See* Canada—History

PIONEER SKILLS—SLIDE SET (Slides). National Film Board of Canada/Donars Productions, 1974. 20 cardboard mounted, color slides in plastic storage pages. ($15). Gr. 4-9. BKL (2/15/75). 973

Illustrates skills and crafts of 100 years ago, such as spinning, weaving, making butter, candles, cheese, bread, splitting shingles, pit sawing, making ropes, and blacksmithing.

SETTLERS OF NORTH AMERICA (Filmstrip—Sound). United Learning, 1973. 5 color, sound filmstrips, av. 49–59 fr. ea.; 5 cassettes, av. 12–15 min. ea.; Teacher's Guide. #16 ($75). Gr. K–8. PRV (3/75). 973
Contents: 1. Transportation. 2. Commerce. 3. Furniture and Household Goods. 4. The Making of a Farm. 5. Community Life.
Introduces students to pioneer life with realistic reenactment of the daily routines, difficulties, and pleasures of pioneer living.

FROST, ROBERT

POETS OF THE TWENTIETH CENTURY (Filmstrip—Sound). *See* American literature—History and criticism

FUEL

BATE'S CAR (Motion Picture—16mm—Sound). National Film Board of Canada/Arthur Mokin Productions, 1975. 16mm color, sound film, 15¹/₂ min. ($245). Gr. 7–A. AAS (12/76), BKL (4/76), PRV (2/76). 662
Mr. Harold Bate, a vital and creative man in his seventies, lives in rural England. Long ago he discovered that methane gas, generated by animal waste, would run his car once it was fed into the engine. Mr. Bate explains how he turns ripe manure into potent fuel.

ENERGY: A MATTER OF CHOICES (Motion Picture—16mm—Sound). *See* Power resources

ENERGY: CRISIS AND RESOLUTION (Filmstrip—Sound). *See* Power resources

ENERGY FOR THE FUTURE (Motion Picture—16mm—Sound). *See* Power resources

LEARNING ABOUT HEAT: SECOND EDITION (Motion Picture—16mm—Sound). *See* Heat

FULLER, BUCKMINSTER

AMERICANS WHO CHANGED THINGS (Filmstrip—Sound). *See* U.S.—Civilization—Biography

FURNITURE

HOUSING AND HOME FURNISHINGS (Kit). *See* Houses

HOUSING AND HOME FURNISHINGS: YOUR PERSONAL ENVIRONMENT (Kit). *See* Houses

FUTURE. *See* Forecasting

FUTURISM

FUTURISM (Motion Picture—16mm—Sound). *See* Painters, Italian

GALAPAGOS ISLANDS

GALAPAGOS: DARWIN'S WORLD WITHIN ITSELF (Motion Picture—16mm—Sound). EBEC, 1971. 16mm color, sound film, 20 min. #3098 ($255). Gr. 7–12. BKL (12/15/72), CGE (1975), PRV (1975). 918.665
Discloses the diverse and unique biology of the volcanically formed Galapagos Islands which intrigued Charles Darwin and provided the basis for his theory of evolution.

GANGS. *See* Juvenile delinquency

GARDENING

PLANT A SEED (Motion Picture—16mm—Sound). *See* Cities and towns

GENERATION GAP. *See* Conflict of generations

GENETICS

THE GROWING TRIP (Filmstrip—Sound). *See* Reproduction

THE LIVING CELL: DNA (Motion Picture—16mm—Sound). *See* Cells

GENIUS

IDIOT . . . GENIUS? (Motion Picture—16mm—Sound). *See* Intellect

GEOGRAPHICAL DISTRIBUTION OF ANIMALS AND PLANTS

FOUR BIOMES (Filmstrip—Sound). *See* Ecology

GEOLOGY

EARTH AND UNIVERSE SERIES, SET 1 (Filmstrip—Sound). University Films/McGraw-Hill Films, 1972. 3 color, sound filmstrips, 76–84 fr.; 3 cassettes, 19–20 min.; Teacher's Guide. #102694–0 ($60). Gr. 4–9. PRV 11/76. 551
Contents: 1. How the Earth's Surface Is Worn Down. 2. How the Earth's Surface Is Built Up. 3. The Air Around Us.
A comprehensive study of the subject. Concepts are simply presented. The effects of erosion, weather, wind, plant growth, and human activity are discussed. Properties of unseen air as weight, volume, and wind movement are also presented.

EARTH AND UNIVERSE SERIES, SET 2 (Filmstrip—Sound). University Films/McGraw-Hill Films, 1976. 3 color, sound filmstrips, av. 76 fr.; 2 cassettes, av. 15–18

GEOLOGY (cont.)

min. #10269-3 ($54). Gr. 4-8. BKL (9/15/77). 551

Contents: 1. The Oceans. 2. The Water Cycle. 3. Water and Its Properties.

Clear photographs, diagrams, and maps offer a blend of earth science and social studies to demonstrate concepts that have contributed to our transportation, recreation, food production, and mineral extraction.

EARTH SCIENCE LOOPS SERIES, SET ONE (Film Loop—8mm—Silent). McGraw-Hill Films, 1972. 7 silent, 8mm technicolor loops, 3-4 min. ea. #101731-3 ($154). Gr. 5-9. PRV (2/73). 551

Contents: 1. Coastal Processes. 2. Deep Ocean Sediments. 3. Fronts. 4. Glaciation. 5. Hurricanes. 6. Igneous Processes. 7. Metamorphism and Coal Formation.

Featuring time-lapse photography and the animation techniques, this set is designed to develop interest and understanding. Each loop explains a fundamental concept and illustrates natural processes.

EARTH SCIENCE LOOPS SERIES, SET TWO (Film Loop—8mm—Silent). McGraw-Hill Films, 1970. 7 8mm color, silent film loops, av. 3-4 min. ea.; Teacher's Guide. #101739-9 ($154). Gr 7-12. PRV (2/73). 551

Contents: 1. Crustal Evolution. 2. Ocean Basin Topography. 3. Ocean Currents. 4. Sedimentation and Sedimentary Rocks. 5. Stream Action. 6. Thunderstorms. 7. Wind Erosion and Deposition.

This set is designed to develop understanding of fundamental concepts illustrated through time-lapse photography and animation.

THE EARTH (Filmstrip—Sound). Scholastic Book Services, 1977 (Adventures in Science). 4 color, sound filmstrips, av. 53-78 fr.; 4 cassettes or discs, av. 9-14 min. Discs #8625 ($69.50). Cassettes #8626 ($69.50). Gr. 4-8. BKL (12/15/77). 551.1

Contents: 1. Continental Drift. 2. Mountain Building. 3. Earthquakes and Tsunamis. 4. Volcanoes.

This set treats geology on the basis of theories. For example, Charles Darwin's discovery of seashells in the Andes Mountains provides the basis for theories on the formation of mountains.

THE EARTH'S RESOURCES (Filmstrip—Captioned). *See* Natural resources

EROSION: WIND AND RAIN (Slides). *See* Erosion

FACE OF THE EARTH (Motion Picture—16mm—Sound). National Film Board of Canada, 1976. 16mm color, sound film, 17 min. ($275). Gr. 5-A. AAS (1977), AFF (1977), BKL (12/1/76). 551.3

A substantial amount of scientific information about the origin of the earth's crust is provided while viewers watch spectacular views of mountains, volcanoes, geysers, and hot springs. Interspersed throughout the film are animated scenes which explain the causes of folding and buckling surfaces that create mountains and valleys. The underground heat and pressure causing geysers and volcanoes are discussed and illustrated. The wind and rain erode the earth's surface, building new formations, and the earth's crust continues to change in a never-ending cycle.

FIELDSTRIPS: EXPLORING EARTH SCIENCE (Filmstrip—Sound). EBEC, 1977 (Fieldstrips). 5 color, sound filmstrips, av. 101 fr. ea.; 5 discs or cassettes, 15 min. ea.; Teacher's Guide. #17027K Cassettes ($72.50), #17027 Discs ($72.50). Gr. 7-C. LFR (6/78), MCU (8/78), PRV (1/79). 551.1

Contents: 1. Investigating Soils. 2. The Grand Canyon. 3. A Weather Station. 4. A Volcano. 5. The San Andreas Fault.

Five expeditions explore geologic phenomena, using scientific methods and instruments. Each trip is led by a different expert. First-person camera work and location sound effects create a "you-are-there" feeling.

FOSSILS (Filmstrip—Sound). *See* Fossils

GEOLOGY: OUR DYNAMIC EARTH (Filmstrip—Sound). National Geographic Educational Services, 1977. 4 color, sound filmstrips; 4 cassettes or discs, av. 11-14 min. ea. #03248 Cassettes ($62.50), #03247 Discs ($62.50). Gr. 5-12. BKL (1978). 551

Contents: 1. The Face of the Earth. 2. The Changing Land. 3. The Restless Earth. 4. Geology and You.

The planet and its constantly changing surface are presented in this set.

GLACIER ON THE MOVE (Motion Picture—16mm—Sound). *See* Glaciers

HOW OUR CONTINENT WAS MADE (Filmstrip—Sound). *See* North America

INTRODUCTION TO EARTH SCIENCES (Filmstrip—Sound). Educational Dimensions Group, 1977. 6 color, sound filmstrips, approx. 40 fr. ea.; 6 cassettes, av. 10 min. ea.; Teacher's Guide. #541 ($95). Gr. 7-12. PRV (12/19/78). 551.1

Contents: 1. Volcanoes. 2. Earthquakes. 3. Plate Tectonics. 4. Glaciers. 5. Erosion: Wind and Thermal. 6. Erosion: Water.

Photos and illustrations are used in this set to illustrate the dynamic forces of nature. All are examined and analyzed.

MONUMENTS TO EROSION (Motion Picture—16mm—Sound). *See* Erosion

OUR CHANGING EARTH (Filmstrip—Sound). Coronet Instructional Media, 1974. 6 color, sound filmstrips, av. 50 fr.; 3 discs or

6 cassettes, av. 12½ min. #S129 Discs ($70), #M129 Cassettes ($78). Gr. 6–12. PRV (11/15), PVB (4/76), TEA (1975). 551.1

Contents: 1. How We Study It. 2. Water and Its Work. 3. Wind, Weathering, and Wasting. 4. Pressure and Change Beneath the Earth's Surface. 5. Thermal Activity and Igneous Formations. 6. Man and His Geological Environment.

A thorough and comprehensive overview of the earth. Photography and diagrams clearly define a great many complicated concepts.

POWERS OF NATURE (Filmstrip—Sound). National Geographic Educational Services, 1973. 5 color, sound filmstrips, 54–70 fr. ea.; 5 cassettes or discs, av. 12–13 min. ea. #03731 Cassettes ($74.50), #03730 Discs ($74.50). Gr. 5–12. BKL (1/1/74). 551

Contents: 1. Weather and Man. 2. Floods. 3. Forest Fires. 4. Earthquakes. 5. Volcanoes.

Shows the efforts to understand, predict, and control the upheavals of nature.

ROCKS AND MINERALS (Filmstrip—Sound). Imperial Educational Resources, 1974. 4 color, sound filmstrips; 4 cassettes. #3KG41000 ($62). Gr. 4–8. INS (10/74), STE (10/74). 552

Contents: 1. Igneous Rocks. 2. Sedimentary Rocks. 3. Metamorphic Rocks. 4. Minerals.

An introduction to basic concepts about rocks and minerals. Shows the actual location on the North American continent where various kinds of rock are located and makes clear their relationship to specific geological formations.

THE SAN ANDREAS FAULT (Motion Picture—16mm—Sound). See Earthquakes

GEORGE, JEAN CRAIGHEAD

JEAN CRAIGHEAD GEORGE (Filmstrip—Sound). See Authors

JULIE OF THE WOLVES (Phonodisc or Cassette). See Eskimos—Fiction

GEOTHERMAL RESOURCES

ECOLOGY OF A HOT SPRING: LIFE AT HIGH TEMPERATURE (Motion Picture—16mm—Sound). See Ecology

See also Names of geothermal resources, e.g., Geysers, etc.

GERMANS IN THE U.S.

GERMAN FOOD (Kit). See Cookery, German

GERMANY

WESTERN EUROPE: GERMANY (Filmstrip—Sound). Milan Herzog & Assoc./EBEC, 1977. 3 color, sound filmstrips, 105 fr. ea.; 3 discs or cassettes, 15 min. ea.

#6958K Cassettes ($43.50), #6958 Discs ($43.50). Gr. 7–12. BKL (2/78), LFR (10/77), PRV (9/78). 914.3087

Contents: 1. Keystone of the Continent. 2. Industrial Heartland. 3. The Alberts Family.

Despite regional differences and political division, Germany has achieved an "economic miracle." This series looks for the reasons in the roots of German history and tradition. It examines the German economic role in Europe, and shares the daily routine of a typical German family.

GERSHWIN, GEORGE

AMERICANS WHO CHANGED THINGS (Filmstrip—Sound). See U.S.—Civilization—Biography

GEYSERS

GEYSER VALLEY (Motion Picture—16mm—Sound). EBEC, 1972. 16mm color, nonnarrated, sound film, 9 min. #3142 ($115). Gr. 4–12. AIF (1972), BKL (12/15/72), CIF (1972). 551.2

Yellowstone's geysers and other thermal phenomena are shown and heard in this nonnarrated film. Viewers travel through valleys, over deep caverns, and rocky precipices. Unusual forms, movements, and striking colors suggest many possibilities for writing, drawing, painting, and storytelling.

GHETTO LIFE

COMMITMENT TO CHANGE: REBIRTH OF A CITY (Filmstrip—Sound). See Urban renewal

GHOST TOWNS. See Cities and towns—Ruined, extinct, etc.

GIANTS—FICTION

THE GREAT QUILLOW (Phonodisc or Cassette). Caedmon Records, 1972. 1 cassette or disc, 41 min.; Jacket Notes. #TC1411 Disc ($6.98), #CDL51411 Cassette ($7.95). Gr. 3–A. BKL (5/1/73), LBJ, STE (1973). Fic

James Thurber's tale of the creative toymaker who drives a giant out of his mind and out of town is related by Peter Ustinov. A modern fable set in the ageless time of folk tales.

GIRAFFES

GIRAFFE (Motion Picture—16mm—Sound). EBEC, 1971 (Silent Safari). 16mm color, sound, nonnarrated film, 10 min.; Teacher's Guide. #3124 ($150). Gr. K–9. BKL (10/1/76). 599.7

The giraffe, the tallest mammal, is shown at home in Tanzania. Close-ups examine its unusual shape and size, the markings which camouflage it in bush country, how inde-

GIRAFFES (cont.)

pendent family units follow simple daily patterns for foraging, carrying tick birds, and drinking.

GIRLS

THERE'S A NEW YOU COMIN' FOR GIRLS (Filmstrip—Sound). Marshfilm, 1974 (The Human Growth Series). 1 color, sound filmstrip, 50 fr.; 1 cassette or disc, 15 min.; Teacher's Guide. Gr. 5–8. ESL (1976), PRV (12/74), PVB (5/75). 613.6

Tailored to the special needs of the growing girl, a basic study of female anatomy and functions is included.

GIRLS—FICTION

GIRL STUFF (Kit). EMC, 1973. 4 Paperback Books; 4 Read-along Cassettes; Teacher's Guide. #ELC–216000 ($55). Gr 4–10. INS (1/74), STE (11/73). Fic

Contents: 1. Stand Off. 2. Break In. 3. Stray. 4. Hot Shot.

Difficult situations are dealt with in an entertaining and resourceful way. Designed to improve reluctant readers' abilities, comprehension, and vocabulary skills.

REALLY ME (Kit). EMC, 1974. 4 cassettes, av. 30 min.; 4 Paperback Books; 1 Teacher's Guide. ($53.10). Gr. 6–8. BKL (6/75), CPR (2/77), SLJ (4/75). Fic

Contents: 1. Will the Really Jeannie Murphy Please Stand Up. 2. Checkmate Julie. 3. Everyone's Watching Tammy. 4. A Candle, a Feather, a Wooden Spoon.

High-interest, low vocabulary stories about the problems of growing up as teenagers.

GLACIERS

GLACIER ON THE MOVE (Motion Picture—16mm—Sound). EBEC, 1973. 16mm color, sound film, 11 min. #3177 ($150). Gr. 7–C. LFR (1973), PRV (1975), SCT (1974). 551.3

Four years of unique cinematic study led to this unusual film of a glacier in action. Using a time-lapse camera and still photographs, the film shows the rhythm of glacial movement in summer and in winter, while dramatizing the role of the camera as a scientific tool. The narration describes the effects of gravity, pressure, and temperature on the large masses of ice. Two laboratory experiments demonstrate the deformation of ice under high pressure.

GLASGOW, ELLEN

WOMEN WRITERS: VOICES OF DISSENT (Filmstrip—Sound). *See* Women authors

GLASS PAINTING AND STAINING

ENAMELING: PAINTING WITH GLASS (Filmstrip—Sound). *See* Enamel and enameling

STAINED GLASS: THE AGELESS ART (Filmstrip—Sound). Warner Educational Productions, 1975. 4 color, sound filmstrips, 72–95 fr.; 2 cassettes, 8–17 min.; Teaching Guide and Project Patterns. #490 ($78.50). Gr. 7–A. BKL (2/1/76), PRV (12/76). 748.5

Introduces the craft of stained glass through its history, industrial production and actual doing. It traces the major innovations and development in style and techniques. Detailed directions are given and shown for making a terrarium, box, and window. Also demonstrates how glass is made by hand and by machine.

GLIDING AND SOARING

HANG GLIDING: RIDING THE WIND (Kit). Troll Associates, 1976 (Troll Reading Program). 1 cassette, 10 Soft Cover Books, 1 Library Edition, 4 Duplicating Masters, 1 Teachers Guide. ($48). Gr. 4–9. BKL (2/15/77). 797.5

The filmstrip visuals are the pictures in the book, and the cassette is a word-for-word reading of the text. The kit provides information about the sport of hang gliding.

GLOBE THEATRE

THEATER IN SHAKESPEARE'S ENGLAND (Filmstrip—Sound). *See* Theater—England

GLOBES

HOW TO USE MAPS AND GLOBES (Filmstrip—Sound). *See* Maps

GNU

THE YEAR OF THE WILDEBEEST (Motion Picture—16mm—Sound). Benchmark Films, 1976. 16mm, sound film, 55 min. ($675). Gr. 7–A. CPR (3/19/78), INS (1/19/77), SCT (4/19/77). 599.735

This is an extraordinary documentary of a herd of over one million wildebeest on their annual, epic 2,000 mile migration in Kenya in search of grass and water during the dry summer months. Immense herds of antelope, zebra, giraffe, and buffalo accompany them.

GOLD RUSH. *See* California—Gold discoveries; Klondike gold fields

GOLDING, WILLIAM

LORD OF THE FLIES BY WILLIAM GOLDING (Filmstrip—Sound). *See* Human behavior—Fiction

GOLDSMITH, OLIVER

GREAT WRITERS OF THE BRITISH ISLES, SET I (Filmstrip—Sound). *See* Authors, English

GOOD AND EVIL—FICTION

VOLTAIRE PRESENTS CANDIDE: AN INTRODUCTION TO THE AGE OF ENLIGHTENMENT (Motion Picture—16mm—Sound). *See* French fiction

GOVERNMENT. *See* Political science; State governments; U.S.—Politics and government

GOVERNMENT AND THE PRESS

FREE PRESS: A NEED TO KNOW THE NEWS (Kit). *See* Freedom of the press

GOYA, FRANCISCO

GOYA (Motion Picture—16mm—Sound). National Gallery of Art & Visual Images/ WETA-TV/EBEC, 1973 (The Art Awareness Collection, National Gallery of Art). 16mm color, sound film, 7 min. #3535 ($125). Gr. 7–C. EFL (1977), EGT (1978), LFR (1978). 759.6

The masquerade lived by men and women is brilliantly revealed by Goya's portraits of Spanish nobility. For he unmasks them even as he pleases them. And when the horrors of war fill his head with nightmare visions, these, too, he sets down in haunting and powerful sequences.

GRAMMAR. *See* English language—Grammar

GRAPHIC ARTS

CAREERS IN GRAPHIC ARTS (Filmstrip— Sound). Pathescope Educational Media, 1973 (Careers In). 2 color, sound filmstrips, 83–93 fr. ea.; 2 cassettes, 11–15 min. ea.; Teacher's Guide. #715 ($50). Gr. 7–12. PRV (4/75). 760.023

Explores possible areas of employment in the graphic arts field. Interviews with graphic artists in different fields tell about the training possibilities and emphasizes the competition for employment.

THE GRAPHIC ARTS: AN INTRODUCTION (Filmstrip—Sound). Educational Audio Visual, 1977. 4 color, sound filmstrips, 77–98 fr. ea.; 4 cassettes or discs, av. 11 min. ea. #P7KF-0049 Cassettes ($86), #P7RF-0049 Discs ($78). Gr. 9–A. BKL (10/ 15/77). 760

Contents: 1. Typography and Design. 2. Illustrations. 3. Photography. 4. Fine Art Prints.

Graphic art refers to something that is reproducible and occurs more than once. Each strip deals with one style of graphic art as the narrator highlights the techniques, history, and details of each form. These are supplemented by remarks from a practitioner in the field.

GRASSLANDS

GRASSLAND ECOLOGY—HABITATS AND CHANGE (Motion Picture—16mm— Sound). Centron Films, 1970. 16mm color, sound film, 13 min.; Reader's Guide. ($205). Gr. 2–8. BKL (12/1/71), CGE (1971), STE (10/71). 574.5

This takes an historical approach to the ecology of the American prairies. Unique wildlife photography shows how bison, prairie dogs, etc., affected our nation's grasslands. It also examines the human role in changing the prairies through agriculture, construction, and use of chemicals.

GRAVITATION

BLACK HOLES OF GRAVITY (Motion Picture—16mm—Sound). *See* Astronomy

GRAY, GENEVIEVE

GIRL STUFF (Kit). *See* Girls—Fiction

GREAT EXPECTATIONS

EPISODES FROM CLASSIC NOVELS (Filmstrip—Sound). *See* Literature—Collections

GREECE

SOUTHERN EUROPE: MEDITERRANEAN LANDS (Filmstrip—Sound). *See* Europe

GREECE—HISTORY

ANCIENT GREECE (Filmstrip—Sound). Coronet Instructional Media, 1973. 4 color, sound filmstrips, av. 56 fr. ea.; 4 cassettes or 2 discs, av. 15 min. ea. #M268 Cassettes ($65), #S268 Discs ($65). Gr. 7–12. BKL (1/ 1/73), PRV (1974). 938

Contents: 1. Aegean Civilizations. 2. Athens and Sparta. 3. Athens' Golden Age. 4. City States at War.

Photographs of antiquities and artifacts display the culture whose democracy, philosophy, and art still influence our life today.

GREEK POETRY

HOMER'S MYTHOLOGY: TRACING A TRADITION (Filmstrip—Sound). Guidance Associates, 1977. 3 color, sound filmstrips, 81–116 fr. ea.; 3 cassettes or discs, 13–18 min. ea.; Guide. #9A–500–718 Cassettes ($72.50), #9A–500–700 Discs ($72.50). Gr. 9–C. BKL (12/1/77). 883

GREEK POETRY (cont.)

Contents: 1. Homer's World. 2. The Iliad. 3. The Odyssey.

Built around on-location photography, fine art and dramatic reading, this set explores the appeal and significance of Homer's two famous epic poems as they relate to Greek belief and thought. The substantive script relates the story line of both poems while analyzing their literary structure, importance, and their enduring universal themes.

GREENE, BETTE

PHILIP HALL LIKES ME: I RECKON MAYBE (Phonodisc or Cassette). *See* Friendship—Fiction

SUMMER OF MY GERMAN SOLDIER (Phonodisc or Cassette). *See* World War, 1939–1945—Prisoners & prisons—Fiction

GREENE, NATHANAEL

FAMOUS PATRIOTS OF THE AMERICAN REVOLUTION (Filmstrip—Sound). *See* U.S.—History—Biography

GROFE, FERDE

AMERICAN SCENES (Phonodiscs). *See* Orchestral music

GROOMING, PERSONAL

PERSON POWER (Filmstrip—Sound). Marshfilm, 1977 (Mental Health/Hygiene). 1 color, sound filmstrip, approx. 56 fr.; 1 cassette or disc, approx 9½ min. #1133 ($21). Gr. 5–9. PRV (2/78), PVB (4/78). 646.7

The filmstrip includes the basics of good grooming and reasons why personal hygiene is not only pleasant but necessary.

GROWTH

ADOLESCENCE TO ADULTHOOD: RITES OF PASSAGE (Filmstrip—Sound). *See* Adolescence

ANIMAL LIFE SERIES, SET ONE (Film Loop—8mm—Silent). McGraw-Hill Films, 1969. 7 silent, 8mm technicolor loops, 3–4 min. ea. #668550-0 ($154). Gr. 5–12. INS (2/72). 591.3

Contents: 1. Life Cycle of a Frog. 2. Life Cycle of a Fish. 3. Life Cycle of an Alligator. 4. Life Cycle of a Chicken. 5. Life Cycle of a Dog. 6. Life Cycle of a Mantis. 7. Life Cycle of a Butterfly.

The loops in this set show the important stages in animal growth; many similarities and differences in growth patterns of major animal groups can be observed.

BECOMING AN ADULT: THE PSYCHOLOGICAL TASKS OF ADOLESCENCE (Filmstrip—Sound). *See* Adolescence

GROWTH (PLANTS)

ANGIOSPERMS: THE LIFE CYCLE OF A STRAWBERRY (Film Loop—8mm—Captioned). BFA Educational Media, 1973 (Plant Behavior). 8mm color, captioned film loop, approx. 4 min. #481425 ($30). Gr. K–9. BKL (5/1/76). 581.3

Demonstrates experiments, phenomena, and behavior in plants.

GWILLIM, SARAH JANE

THROUGH THE LOOKING GLASS (Phonodiscs). *See* Fantasy—Fiction

GYMNASTICS

WOMEN'S GYMNASTICS, BEGINNING LEVEL (Motion Picture—16mm—Sound). Athletic Institute, 1975. 16mm color, sound film, 20 min.; super 8mm with cassette also available. #NC–1R ($220). Gr. 7–12. BKL (6/1/76). 796.4

Contains complete routine, including floor exercises, balance beam, uneven parallel bars, and vault.

HAIKU

HAIKU: THE HIDDEN GLIMMERING (Filmstrip—Sound). Lyceum/Mook & Blanchard, 1973. 1 color, sound filmstrip, 60 fr.; 1 cassette or disc, 14½ min.; Teacher's Guide. #LY35173C Cassettes ($25), #LY35173R Discs ($19.50). Gr. 5–A. BKL (6/1/73), NEA (1973), PVB (5/74). 895.61

Discover with Ann Atwood, author/photographer, in the lives of three Japanese poets—Basho, Issa, and Buson—their world views and values, which helps to develop an appreciation for haiku and eastern culture.

HAIKU: THE MOOD OF EARTH (Filmstrip—Sound). Lyceum/Mook & Blanchard, 1971. 2 color, sound filmstrips, 55 fr.; 2 cassettes or discs, 14 min. #LY35371SC Cassettes ($46), #LY35371SR Discs ($37). Gr. 6–12. EGT (12/73), M&M (2/74), NEF (1972). 895.61

Contents: 1. The Heart of Haiku. 2. Haiku: A Photographic Interpretation.

Ann Atwood, through this series, conveys the concept or sense of Haiku with thoughtful consideration of its Japanese origin. Helps to understand, appreciate, and compose Haiku poetry. Also available with the book.

HALE, LUCRETIA P.

THE PETERKIN PAPERS (Phonodisc or Cassette). *See* Family life—Fiction

HALEY, ALEX

ROOTS WITH ALEX HALEY (Cassette). Cinema Sound Ltd./Jeffrey Norton Publishers, 1976 (Avid Reader). 1 cassette, approx. 55 min. #40253 ($11.95). Gr. 10–A. BKL (7/15/78). 921

In this conversation with Heywood Hale Broun, author Alex Haley talks about his book *Roots* (Doubleday, 1976) from the standpoint of family history, black experience, and self-identity. The discussion is enhanced by Broun's questions that go beyond the book and probe Haley's views on the relevance of ethnic heritage and family studies to the lives of present-day black Americans.

HALEY, GAIL E.

GAIL E. HALEY: WOOD AND LINOLEUM ILLUSTRATION (Filmstrip—Sound). Weston Woods Studios, 1978. 1 color, filmstrip, 72 fr.; 1 cassette, 17 min. SF 456 C ($25). Gr. 3–A. BKL. 921

Gail E. Haley discusses her technique of using wood and linoleum blocks to create a story's mood through illustration. She traces the historical development of printmaking, demonstrates with masks and dolls the source of some of her ideas, and details her research into different cultures.

HALLUCINOGENS

ACID (Motion Picture—16mm—Sound). Concept Films, 1971 (Drug Abuse Education). 16mm color, sound film, 26 min. #3049 ($360). Gr. 7–C. BKL (10/72), CGE (1975), NTS (2/72). 615.788

From the death of a boy on LSD to the success of LSD treatments in curbing alcoholism, *Acid* explores the unpredictable power of this chemical tiger. Relationships of LSD to creativity, to love, and to the ego are examined objectively.

HAMILTON, VIRGINIA

VIRGINIA HAMILTON (Filmstrip—Sound). Miller-Brody Productions, 1976 (Meet the Newbery Author). 1 color, sound filmstrip, 101 fr.; 1 disc or cassette, 15 min. #MNA–1007C Cassette ($32), #MNA–1007 Disc ($32). Gr. 5–12. BKL (12/1/76). 921

Many relationships between the author and her works, as well as biographical information, notes on her family life, her writing habits, and her feelings about her occupation, are related in a conversational tone by the voice-over narrator and by Hamilton herself. Excerpts from Miller-Brody's sound filmstrip production of *M. C. Higgins, the Great* are used in the early portion of this filmstrip, as are old snapshots selected from the author's family album. The majority of the visuals consist of photos taken of Hamilton with her husband and her children.

HANDICAPPED

GET IT TOGETHER (Motion Picture—16mm—Sound). Pyramid Films, 1977. 16mm color, sound film, 20 min.; $^{3}/_{4}$ in. videocassette also available. ($300). Gr. 7–A. BKL (11/15/77), EFL (1977), LNG (1977). 362

This film tells the true story of a remarkable man and his relations with those around him. After losing the use of his legs in an automobile accident, Jeff Minnebraker has achieved a happy life, a successful marriage, a fulfilling career and a meaningful role in society.

HANDICAPPED—RECREATION

A MATTER OF INCONVENIENCE (Motion Picture—16mm—Sound). Stanfield House, 1974. 16mm color, sound film, 10 min. ($170). Gr. 5–A. BKL (11/15/74). 796.019

This film illustrates how amputee skiers negotiate the slopes using specially designed poles and how blind skiers judge distances and place turns by trusting the familiar voices of their instructors. In some on-screen interviews, some disabled skiers explain they "don't feel handicapped, just inconvenienced."

HANDICAPPED CHILDREN

ELIZABETH (Motion Picture—16mm—Sound). WGBH Educational Foundation/EBEC, 1978 (People You'd Like to Know). 16mm color, sound film, 10 min.; Teacher's Guide. #3593 ($185). Gr. 5–A. BKL (1979). 362.7

This film is designed to create acceptance and understanding of young people with disabilities. It captures moments in the life of Elizabeth and it tells us of her feelings and attitudes toward her disability—the importance of self acceptance, her need to become independent, and her desire to actively participate in school and social activities.

I'M JUST LIKE YOU: MAINSTREAMING THE HANDICAPPED (Filmstrip—Sound). Victoria Productions/Sunburst Communications, 1977. 2 color, sound filmstrips, 113 fr.; 2 cassettes or discs, 21 min.; Teacher's Guide. #245–81 ($59). Gr. 5–A. BKL (5/1/78), MDI (4/78), PRV (1/79). 362.7

Contents: 1. Taking It in Stride. 2. Dealing with the Problem.

Viewers evaluate mainstreaming by observing case histories of children mainstreamed to different extents. Basic communication skills and techniques to overcome discrimination and embarrassment are presented.

MY SON, KEVIN (Motion Picture—16mm—Sound). Granada International Television/Wombat Productions, Inc., 1974. 16mm color, sound film, 24 min. ($360). Gr. 12–A. AFF (1975), BKL (1975). 362.7

There are over 400 children in England born malformed because of the drug thalidomide.

HANDICAPPED CHILDREN (cont.)

This film is about such a child, as seen by his mother: a woman who communicates her strength to 11-year-old Kevin and the rest of the family.

HANDICAPPED CHILDREN—EDUCATION

MARK (Motion Picture—16mm—Sound). WGBH Educational Foundation/EBEC, 1978 (People You'd Like to Know). 16mm color, sound film, 20 min. #3586 ($185). Gr. 4–12. BKL (1979). 362.7
Fourteen-year-old Mark has had a reading problem all his life. He understands that this problem is but one aspect in the total picture of what he is and what he has to offer. His positive outlook is reflected in the tutoring help he gives an elementary grade student who also has a reading problem.

MARY (Motion Picture—16mm—Sound). *See* Deaf

THE MUSIC CHILD (Motion Picture—16mm—Sound). Benchmark Films, 1976. 16mm b/w, sound film, 45 min. ($495). Gr. 11–A. AFF (1977), IFT (1977), SPN (1977). 371.9
Contents: 1. Individual Therapy. 2. Group Therapy.
A moving demonstration of the use of improvised music by four widely respected music therapists to achieve communication individually and in group sessions with 30 nonverbal children who are either autistic, emotionally disturbed, mentally retarded, or cerebral palsied.

PAIGE (Motion Picture—16mm—Sound). *See* Mentally handicapped

HANDICRAFT

CRAYON BATIK MAGIC (Filmstrip—Sound). *See* Batik

CREATIVE BATIK (Filmstrip—Sound). *See* Batik

CREATIVE CREWEL (Filmstrip—Sound). *See* Crewelwork

CREATIVE MACRAME (Filmstrip—Sound). *See* Macrame

CREATIVE TIE/DYE (Filmstrip—Sound). *See* Batik

THE DOLLMAKER (Filmstrip—Sound). *See* Dolls

DRAWING: ARTISTRY WITH A PENCIL (Filmstrip—Sound). *See* Drawing

EDIBLE ART (Filmstrip—Sound). *See* Cookery

ENAMELING: PAINTING WITH GLASS (Filmstrip—Sound). *See* Enamel and enameling

HOW TO DO: CARDBOARD SCULPTURE (Filmstrip—Sound). *See* Sculpture

IN PRAISE OF HANDS (Motion Picture—16mm—Sound). National Film Board of Canada, 1974. 16mm color, sound film, 27¹/₂ min. ($375). Gr. 7–A. EFL (1976), LFR (9/10/76), PRV (1976). 745.5
The hands of craftsmen living in Japan, Finland, Mexico, Poland, Nigeria, India, and the Canadian Arctic shape their work as the film takes the viewer around the world to visit the fantastic artisans. The viewer sees the artist at work throwing pots on a potter's wheel, weaving rugs, applying enamel to metal birds and jewlery boxes, tie-dying fabric, carving sculptures, making puppets, hammering cloth and decorating houses with floral patterns. A narrationless film conveys its story.

JEWELRY: THE FINE ART OF ADORNMENT, FABRICATION METHOD (Filmstrip—Sound). *See* Jewelry

JEWELERY: THE FINE ART OF ADORNMENT, LOST WAX METHOD (Filmstrip—Sound). *See* Jewelry

JUNK ECOLOGY (Filmstrip—Sound). Troll Associates, 1975. 6 color, sound filmstrips, av. 50 fr. ea.; 3 cassettes av. 10 min. ea.; Teacher Guide. ($78). Gr. 3–8. PRV (4/76), TEA (2/76). 745.5
Contents: 1. Recycling Paper and Cardboard/Wood and Sticks. 2. Recycling Plastic Throw-Aways. 3. Tin Cans and Bottle Caps. 4. Recycling Jars and Bottles. 5. Fix Up, Clean Up. 6. Special Crafts.
Gets students involved in crafts and ecology by showing how different types of discarded junk materials can be collected and turned into something interesting and useful.

THE LEATHER-CRAFTER (Filmstrip—Sound). *See* Leather work

LOOMLESS WEAVING (Filmstrip—Sound). *See* Weaving

MOBILES: ARTISTRY IN MOTION (Filmstrip—Sound). *See* Mobiles (sculpture)

PIONEER SKILLS—SLIDE SET (Slides). *See* Frontier and pioneer life

POTTERY: ARTISTRY WITH EARTH (Filmstrip—Sound). *See* Pottery

THE PRINT-MAKER (Filmstrip—Sound). *See* Prints

PUPPETRY: MINIATURES FOR THEATRE (Filmstrip—Sound). *See* Puppets and puppet plays

REDISCOVERY: ART MEDIA SERIES, SET 1 (Filmstrip—Sound). ACI Media/Paramount Communications, 1974. 4 color, sound filmstrips, 37–59 fr.; 4 cassettes, 8 min. ea.; Teacher's Guide. #9561 ($78). Gr. 4–12. BKL (6/74). 745.5

Contents: 1. Leather. 2. Macrame. 3. Stitchery. 4. Weaving.

Demonstrates the creation of utilitarian and aesthetically pleasing objects.

REDISCOVERY: ART MEDIA SERIES, SET 2 (Filmstrip—Sound). ACI Media/Paramount Communications, 1974. 4 color, sound filmstrips, 57–74 fr.; 4 cassettes, 8 min.; Teacher's Guide. #9562 ($78). Gr. 4–A. BKL (6/15/74), PRV (1/75). 745.5

Contents: 1. Collage. 2. Crayon. 3. Posters. 4. Prints.

This set offers four techniques using very basic material—paper.

REDISCOVERY: ART MEDIA SERIES, SET 3 (Filmstrip—Sound). ACI Media/Paramount Communications, 1974. 4 color, sound filmstrips; 4 cassettes, 8 min. ea.; Teacher's Guide. #9563 ($78). Gr. 4–12. BKL (6/15/74). 745.5

Contents: 1. Clay. 2. Paper Construction. 3. Paper Mache. 4. Puppets.

Places emphasis on the processes themselves in these four art media. Basic tools, raw materials, and the simple techniques are cited.

REDISCOVERY: ART MEDIA SERIES, SET 4 (Filmstrip—Sound). ACI Media/Paramount Communications, 1974. 4 color, sound filmstrips, av. 44 fr.; 4 cassettes, 8 min. ea.; Teacher's Guide. #9564 ($78). Gr. 4–12. BKL (6/15/74), PRV (5/75). 745.5

Contents: 1. Batik. 2. Enameling. 3. Silkscreen. 4. Watercolor.

Brief overviews include techniques and show completed works.

RUG HOOKING: A MODERN APPROACH (Filmstrip—Sound). See Rugs, Hooked

SCULPTURE WITH STRING (Filmstrip—Sound). See String art

STAINED GLASS: THE AGELESS ART (Filmstrip—Sound). See Glass painting and staining

HANSBERRY, LORRAINE

LORRAINE HANSBERRY (Motion Picture—16mm—Sound). Films for the Humanities, 1976. 16mm color, sound film, 35 min. $3/4$ in. videocassette also available. FFH128 ($395). Gr. 7–A. BKL (1976), CGE (1976), WLB (1976). 921

This film traces Mrs. Hansberry's life, largely in her own words and voice, from her early childhood in Chicago to her premature death at age 34. Presents the life and work of America's leading black woman playwright, showing how she used the obstacles that confronted her and overcame them creatively. Included are excerpts from *A Raisin in the Sun*, *The Sign in Sidney Brustein's Window*, and *Les Blaves*.

HARDY, THOMAS

GREAT WRITERS OF THE BRITISH ISLES, SET II (Filmstrip—Sound). See Authors, English

HARPSICHORD

HARPSICHORD BUILDER (Motion Picture—16mm—Sound). Labyrinth Films/Wombat Productions, 1977. 16mm color, sound film, 28 min. ($395). Gr. 7–A. BKL (3/1/78). 681.816

It takes W. S. Kater $2^1/2$ months to make a harpsichord. As we watch a beautiful instrument come into being, we also discover the vital connection between a dedicated builder and the work that captures his total loving attention.

HARVEY PHYLLIS

SQUARE PEGS—ROUND HOLES (Motion Picture—16mm—Sound). See Individuality

HAWAII

FIRE IN THE SEA (Motion Picture—16mm—Sound). See Volcanoes

HAWAII: THE FIFTIETH STATE (Filmstrip—Sound). EBEC, 1974. 4 color, sound filmstrips, av. 87 fr. ea.; 4 discs or cassettes, 16 min. ea. Discs #6493 ($66.50), Cassettes #6493K ($66.50). Gr. 5–A. PRV (4/76), PVB (5/75). 919.69

Contents: 1. Hawaii's Origins: Its First People. 2. Hawaii's History: From Kingdom to Statehood. 3. Hawaii's Economy: Growth and the Future. 4. Hawaii's People: Islands of Contrasts.

Ancient songs and legends, firsthand accounts, drawings and artifacts help to recreate the story of our fiftieth state. The series identifies the customs and traditions of Hawaii's first society, then traces Hawaii's transformation from an agricultural kingdom to a modern industrial state. Visits to three Hawaiian families representing a rich cultural mix of backgrounds and life-styles are revealing.

HAWTHORNE, NATHANIEL

A DISCUSSION OF DR. HEIDEGGER'S EXPERIMENT (Motion Picture—16mm—Sound). See Motion pictures—History and criticism

DOCTOR HEIDEGGER'S EXPERIMENT (Motion Picture—16mm—Sound). See American fiction

THE ROMANTIC AGE (Filmstrip—Sound). See American literature—History and criticism

HEALTH EDUCATION

BUYING HEALTH CARE (Kit). *See* Insurance, Health

NUTRITION AND GOOD HEALTH (Kit). *See* Nutrition

PHYSICAL FITNESS: IT CAN SAVE YOUR LIFE (Motion Picture—16mm—Sound). *See* Physical education and training

HEART

THE HEART AND THE CIRCULATORY SYSTEM (Motion Picture—16mm—Sound). *See* Blood—Circulation

HEAT

LEARNING ABOUT HEAT: SECOND EDITION (Motion Picture—16mm—Sound). EBEC, 1974 (Introduction to Physical Science). 16mm color, sound film, 15 min. #3365 ($220). Gr. 4-9. LFR (4/75), PRV (4/19/76). 536

Any understanding of the energy crisis presupposes an understanding of the nature of heat as the basic energy force. Through laboratory demonstrations interwoven with animated episodes, this film discusses the molecular basis of heat, friction, changes of state in gases, solids, and liquids, the movement of heat through radiation, convection and conduction.

HEMINGWAY, ERNEST

A DISCUSSION OF MY OLD MAN (Motion Picture—16mm—Sound). *See* Motion—History and criticism

MY OLD MAN (Motion Picture—16mm—Sound). *See* American fiction

SCOURBY READS HEMINGWAY (Phonodisc or Cassette). *See* American fiction

HENDIN, HERBERT

THE AGE OF SENSATION (Cassette). *See* Youth—Attitudes

HENRY, JOHN

THE LEGEND OF JOHN HENRY (Motion Picture—16mm—Sound). *See* Folklore—U.S.

HENRY, PATRICK

FAMOUS PATRIOTS OF THE AMERICAN REVOLUTION (Filmstrip—Sound). *See* U.S.—History—Biography

HEROES AND HEROINES

HEROIC ADVENTURES (Filmstrip—Sound). *See* Legends

HIBERNATION OF ANIMALS. *See* Animals—Hibernation

HICKS, MARK

GRAVITY IS MY ENEMY (Motion Picture—16mm—Sound). Churchill Films, 1977. 16mm color, sound film, 26 min. ($370). Gr. 7-A. AAW (1977), NEF (1977). 921

Mark Hicks, essentially without nerve or muscle function below the neck, has become a superlative artist. He sees his skill, his images, his life and others' perception of him with a clear vision.

HIKING

BACKPACKING: REVISED (Filmstrip—Sound). *See* Backpacking

HINE, LEWIS

IMMIGRANT AMERICA (Filmstrip—Sound). *See* Immigration and emigration

HINTON, S. E.

THE OUTSIDERS (Phonodisc or Cassette). *See* Juvenile delinquency—Fiction

HIROSHIMA

OUT OF THE HOLOCAUST (Filmstrip—Sound). *See* Cities and towns

HISTORY, ANCIENT

ANCIENT GREECE (Filmstrip—Sound). *See* Greece—History

HISTORY—PHILOSOPHY

THE GOOD OLD DAYS—THEY WERE TERRIBLE (Cassette). Current Affairs Films/Jeffrey Norton Publishers, 1975 (Avid Reader). 1 cassette, approx. 55 min. #40095 ($11.95). Gr. 10-A. BKL (3/15/77). 901

Otto Bettman, of the Bettman Archive, and Viola Scott Thomas, historian and curator of the Museum of Immigration in New York, discuss Bettman's book, *The Good Old Days-They Were Terrible!* (Random, 1974). Thomas argues that today is worse than yesterday. Bettman suggests that a constant fixture of our imagination is the bucolic past. Heywood Hale Broun brings cohesiveness to the discussion and all conclude that the quality of life is a little better now than yesterday.

HITCHHIKING

THUMBS DOWN (HITCHHIKING) (Motion Picture—16mm—Sound). Sanders/Rose/ Swerdloff Film/FilmFair Communications, 1974. 16mm color, sound film, 17 min. ($235). Gr. 5–A. BKL (1/19/75), LFR (2/19/ 75), WLB (1974). 614.8

Hitchhiking is analogous to Russian roulette: eventually there's a victim. Using dramatizations and interviews with real victims of hitchhiking-related crimes and accidents, the film demonstrates the variety of potential dangers to both the hitchhiker and the driver.

HITLER, ADOLF

ADOLF HITLER (Cassette). Cinema Sound Ltd./Jeffrey Norton Publishers, 1977 (Avid Reader). 1 cassette, 55 min. #40254 ($11.95). Gr. 10–A. PRV (5/78). 921

Authors of newly published books examine their respective topics from varying points of view. John Toland, author of *Adolf Hitler*, and Telford Taylor, former chief counsel for the War Crimes Commission, ponder the charismatic yet human side of Hitler's personality, the devotion and/or fear he inspired in his associates, and analysis of his madness. The cassette is a tape of a former radio broadcast; an interesting exposure to current authors and issues.

HOBOES. *See* Tramps

HOLIDAYS

HOLIDAYS: SET ONE (FIlmstrip—Sound). Random House, 1977. 4 color, sound filmstrips, 66–80 fr.; 4 cassettes or discs, 9–12 min.; Teacher's Guide. #05072 Discs ($72), #05073 Cassettes ($72). Gr. K–8. LGB (1977), PRV (2/78). 394.2

Contents: 1. Halloween. 2. Christmas/Hanukkah. 3. Easter/Passover. 4. Independence Day.

The origins, history, and practices of major holidays are presented in a mixture of original and archival photography and appropriate music and effects.

HOLOGRAPHY

AN INTRODUCTION TO HOLOGRAPHY (Motion Picture—16mm—Sound). EBEC, 1972. 16mm color, sound film, 17 min. #3149 ($220). Gr. 9–A. AAS (1975), NEF (1973), PRV (1973). 774

Through a progression of optical demonstrations and laboratory experiments rarely possible in the classroom, the film identifies the principle of holography. It demonstrates types of holograms, explains holographic interferometry, and illustrates the hologram's three dimensionality and data storage and multi-channel capacities. The film shows how holograms are made for display and research and explains the nature of water interference and laser coherence.

HOME AND SCHOOL

FROM HOME TO SCHOOL (Filmstrip— Sound). *See* Family life

PARENT INVOLVEMENT: A PROGRAM FOR TEACHERS AND EDUCATORS (Filmstrip—Sound). Parents' Magazine Films, 1978. 5 color, sound filmstrips; 3 cassettes or 1 disc; Script; Teacher's Guide. ($275). Gr. 12–A. CPR (11/19/78, LNG (3/19/ 79). 371.103

Contents: 1. The Importance of Parent Involvement. 2. Working in the Classroom. 3. Parents and Policy-making. 4. Parent-Teacher Communication. 5. Viewpoints on Parent Participation.

Parent involvement is meant to support the premise that the most effective programs of early childhood education are those which involve parents. This series outlines the advantages of parent involvement programs to teachers, parents, and children. Discusses some of the problems in a successful program. Guidelines are provided to help teachers plan and implement parent participation in the classroom and beyond. Parents and educators discuss parent involvement and offer suggestions on how they feel they can best contribute to children's education.

PARENT-SCHOOL RELATIONSHIPS (Filmstrip—Sound). *See* Family life

HOME ECONOMICS

THE BUTTERICK INTERIOR DESIGN SERIES (Kit). *See* Interior decoration

CAREERS IN HOME ECONOMICS (Filmstrip—Sound). University Films/McGraw-Hill Films, 1975. 6 color, sound filmstrips, 78–136 fr. ea.; 6 cassettes or discs, 13–21 min. ea.; Teacher's Guide. #102630–4 Cassettes ($119), #106283–1 Discs ($119). Gr. 7–A. FHE (5/6/75), PRV (5/76). 640.23

Contents: 1. Careers in Child Care. 2. Careers in Clothing and Textiles. 3. Careers in Consumer Education. 4. Careers in Dietetics and Food Service. 5. Careers in Food and Nutrition. 6. Careers in Housing and Home Environment.

A set to acquaint students with the personal and educational requirements necessary to success in a variety of home economics occupations. They will meet some people who have found success in this field and are doing jobs they love.

CLOTHING (Kit). *See* Clothing and dress

HOME ECONOMICS (cont.)

CONSUMER EDUCATION SERIES (Filmstrip—Sound). *See* Consumer education

FOOD AND NUTRITION: DOLLARS AND SENSE (Kit). *See* Nutrition

HEALTH AND SAFETY: KEEPING FIT (Kit). *See* Hygiene

HOME DECORATION SERIES, SET ONE (Filmstrip—Sound). *See* Interior decoration

HOME DECORATION SERIES, SET TWO (Filmstrip—Sound). *See* Interior decoration

HOUSING AND HOME FURNISHINGS (Kit). *See* Houses

HOUSING AND HOME FURNISHINGS: YOUR PERSONAL ENVIRONMENT (Kit). *See* Houses

LIFESTYLES: OPTIONS FOR LIVING (Kit). *See* Life styles—U.S.

HOMER

HEROES OF THE ILIAD (Cassette). *See* Mythology, Classical

HOMER'S MYTHOLOGY: TRACING A TRADITION (Filmstrip—Sound). *See* Greek poetry

HONEY

BEEKEEPING (Filmstrip—Sound). *See* Bees

HOPI

HOPIS—GUARDIANS OF THE LAND (Motion Picture—16mm—Sound). FilmFair Communications, 1971. 16mm color, sound film, 9³/₄ min. ($135). Gr. 4–10. LFR (12/71), PRV (11/73). 970.3
This film explores the traditional Hopi way of life and the threat that hangs over them. The Hopi philosophy of unity with the land and the way their land is being threatened are presented.

HORROR—FICTION

A DISCUSSION OF THE FALL OF THE HOUSE OF USHER (Motion Picture—16mm—Sound). *See* Motion pictures—History and criticism

A DISCUSSION OF THE HUNT (Motion Picture—16mm—Sound). *See* Motion pictures—History and criticism

EDGAR ALLAN POE'S THE BLACK CAT (Filmstrip—Sound). *See* Devil—Fiction

THE FALL OF THE HOUSE OF USHER (Motion Picture—16mm—Sound). *See* American literature—Study and teaching

LA GRANDE BRETECHE (Motion Picture—16mm—Sound). EBEC, 1973 (Orson Welles Great Mysteries). 16mm color, sound film, 24 min.; Guide. #3422 ($325). Gr. 7–A. BKL (12/1/75), EGT (1976), PRV (1975). Fic
This tale of revenge, set in France during the Napoleonic wars, begins as a love story involving a French countess, a Spanish prisoner of war and a jealous husband. On a visit to his wife's bedroom, the count finds signs of a visitor. Suspecting a lover hiding in the closet, the count tricks his devout wife into swearing the closet is empty. Pretending to believe her, he calmly proceeds to have the closet sealed up forever behind an impenetrable brick wall—while his wife looks on in horror.

THE HUNT (Motion Picture—16mm—Sound). *See* American fiction

I COULDN'T PUT IT DOWN: HOOKED ON READING (Kit). *See* Literature—Collections

THE MANY FACES OF TERROR (Cassette). Arthur Meriwether Education Resources, 1977 (Literary Analysis Series). 3 cassettes, 16 min. ea.; Guide. ($19.50) Gr. 9–12. BKL (7/15/78). Fic
Contents: 1. "The Cask of Amontillado" by Edgar Allan Poe. 2. "The Guest" by Albert Camus. 3. "Miss Brill" by Katherine Mansfield. 4. "The Lottery" by Shirley Jackson. 5. "Cat's Cradle" by Kurt Vonnegut.
Brief analyses provide a framework for study of Poe's "The Cask of Amontillado," Camus's "The Guest," Jackson's "The Lottery," and Vonnegut's "Cat's Cradle." The second side of the third cassette is discussion and questions for the five stories stressing setting, irony, characterization, and reader reaction. The critiques and questions are narrated.

TEN TALES OF MYSTERY AND TERROR (Cassette). *See* Literature—Collections

HORSES

HORSE FLICKERS (Motion Picture—16mm—Sound). *See* Collage

HORSES . . . TO CARE IS TO LOVE (Motion Picture—16mm—Sound). AIMS Instructional Media Service, 1976. 16mm color, sound film, 12 min. #9447 ($190). Gr. 4–10. BKL (2/15/77). 636.1
This film shows that owning a horse places continuing demands on the owner to care for it. It provides valuable detail on the housing, hygiene, grooming, and feeding of a horse. How to purchase a horse is covered as well as the tremendous cost of owning one.

HORSES—FICTION

SADDLE UP! (Kit). EMC, 1975. 4 Paperback Books; 4 Read-along Cassettes; Teacher's

Guide. #ELC127000 ($55). Gr. 4-9. BKL (5/76), SLJ (3/76). Fic

Contents: 1. A Better Bit and Bridle. 2. A Chance to Win. 3. Trailering Troubles. 4. Open the Gate.

These four books revolve around early teenage girls who meet the difficulties and demands of horses and learn some hard truths about themselves as well. The illustrations help relate the world of horses, their care, and competition.

THE SUPERLATIVE HORSE (Motion Picture—16mm—Sound). *See* Folklore—China

HOUSEHOLD EQUIPMENT AND SUPPLIES

KITCHEN EQUIPMENT (Kit). *See* Cookery

HOUSES

HOUSING AND HOME FURNISHINGS: YOUR PERSONAL ENVIRONMENT (Kit). Butterick Publishing, 1975 (Independent Living Series). 4 color, sound filmstrips, 65-75 fr. ea.; 4 cassettes or discs, 8-9 min. ea.; 12 Spirit Masters; Teacher's Guide. #C-501-6E Cassettes ($85), #R-500-8E Discs ($85). Gr. 9-12. PRV (9/19/76). 643

Contents: 1. Different People, Different Homes. 2. You and Your Living Space. 3. Housekeeping Upkeep. 4. Spending Your Housing Dollar.

This kit explores ways to make living quarters into a home. Information on home design, maintenance, financing, interior decorating are all presented.

HUMAN BEHAVIOR

CHANGING HUMAN BEHAVIOR (Filmstrip—Sound). *See* Behavior modification

DEALING WITH STRESS (Filmstrip—Sound). *See* Stress (psychology)

DIVIDED MAN, COMMITMENT OR COMPROMISE? (Motion Picture—16mm—Sound). *See* Decision making

EYE OF THE STORM (Motion Picture—16mm—Sound). *See* Prejudices and antipathies

JOSEPH SCHULTZ (Motion Picture—16mm—Sound). *See* Ethics

LIVING TOGETHER AS AMERICANS (Filmstrip—Sound). *See* Human relations

THE MENTAL/SOCIAL ME (Kit). *See* Anatomy

PROJECT AWARE (Motion Picture—16mm—Sound). *See* Juvenile delinquency

PSYCHOLOGICAL DEFENSES: SERIES A (Filmstrip—Sound). Human Relations Media, 1975. 3 color, sound filmstrips, 71-96 fr.; 3 cassettes or discs, 13-16 min.; Teacher's

Guide. #614 ($90). Gr. 9-C. BKL (4/15/76), PRV (1/77), TSS (1/76). 158.2

Contents: 1. The Unconscious Mind/Repression. 2. Avoidance, Denial, Undoing. 3. Fantasy, Regression, Dreams.

Demonstrates the way people handle everyday stresses through unconscious psychological defenses. Illustrates some common defense mechanisms. Helps students weigh helpful and harmful influences of these defenses on behavior, enabling them to evaluate better their own behavior and that of others.

PSYCHOLOGICAL DEFENSES: SERIES B (Filmstrip—Sound). Human Relations Media, 1975. 3 color, sound filmstrips, 71-89 fr.; 3 cassettes or discs, 13-15 min.; Teacher's Guide. #615 ($90). Gr. 7-12. BKL (4/15/76), PRV (1/77), TSS (1/76). 158.2

Contents: 1. Projection/Rationalization. 2. Identification/Displacement. 3. Reaction Formation/Sublimation.

Examines some of the processes that help people cope with the frustrations of living in a complex society. Elucidates the defense mechanisms which help us preserve self-esteem by avoiding memories, impulses, and actions that make us feel threatened.

SPINNOLIO (Motion Picture—16mm—Sound). National Film Board of Canada/Bosustow Productions, 1977. 16mm color, sound film, 10 min. ($160). Gr. 10-A. BKL (1/15/79). 158.1

A bitingly humorous parody of Pinocchio, with a touch of Dorian Gray. Spinnolio goes through life with a smile painted on his mouth (he's a wooden puppet), even as he moves from school hallways, corporate byways, counterculture jails and alleys to his final release by a fairy godmother.

VALUES (Filmstrip—Sound). *See* Values

WHY WE DO WHAT WE DO! HUMAN MOTIVATION (Filmstrip—Sound). *See* Motivation (psychology)

HUMAN BEHAVIOR—FICTION

BARTLEBY (Motion Picture—16mm—Sound). *See* American fiction

A DISCUSSION OF BARTLEBY (Motion Picture—16mm—Sound). *See* Motion pictures—History and criticism

LORD OF THE FLIES BY WILLIAM GOLDING (Filmstrip—Sound). Current Affairs Films, 1978. 1 color, sound filmstrip, 1 cassette; Teacher's Edition of the Book; Discussion Guide and Testing Materials. #632 ($30). Gr. 9-12. BKL (12/15/78). Fic

Whether readers agree or disagree with Golding's view of human nature, this story of the descent of a group of ''proper'' English schoolboys into savagery is a powerful piece of writing—and one which is sure to provoke lively discussion.

HUMAN BODY. *See* Anatomy; Physiology

HUMAN ECOLOGY

THE CITY AT THE END OF THE CEN-
TURY (Motion Picture—16mm—Sound).
See Cities and towns—U.S.

ON POPULATION (Filmstrip—Sound). *See*
Population

PLANT A SEED (Motion Picture—16mm—
Sound). *See* Cities and towns

THE POPULATION DEBATE (Filmstrip—
Sound). *See* Population

POPULATION: THE PEOPLE PROBLEM
(Filmstrip—Sound). *See* Population

HUMAN RELATIONS

ALL MY FAMILIES (Kit). *See* Humanities

THE ARTIST INSIDE ME (Kit). *See* Humanities

COPING WITH LIFE: THE ROLE OF
SELF/CONTROL (Filmstrip—Sound). *See*
Self-control

THE HUMAN-I-TIES OF LANGUAGE
(Kit). *See* Humanities

HUMAN RIGHTS: WHO SPEAKS FOR
MAN? (Filmstrip—Sound). Current Affairs
Films, 1978. 1 color, sound filmstrip, 74 fr.; 1
cassette, 16 min.; Teacher's Guide. #608
($24). Gr. 8–A. BKL (9/1/78), PRV (2/79).
341.76
This set states that despite a 1948 U.N.
"Universal Declaration of Human Rights"
and even against the laws of individual countries, serious abuses of human rights
abound. A close look is taken at the overall
issue of human rights and what can be done
internationally to safeguard them.

THE I IN IDENTITY (Kit). *See* Humanities

IS IT ALWAYS RIGHT TO BE RIGHT? (Motion Picture—16mm—Sound). Bosustow
Productions, 1970 (Parables for the Present).
16mm color, sound, animated film, 8 min.;
Teacher's Guide. ($150). Gr. 6–A. AFF
(1971), BKL (6/1/71), NEA (1971). 158.2
This fast-moving parable interlaces live action and animation to describe a land where
everyone "is always right." The resulting
conflict and lack of communication lead to
total cessation of activity. Old, young,
black, white, rich, and poor refuse to deal
with any other group. Finally someone has
the courage to admit "I may be wrong," to
which an opposing person answers "No,
you may be right." Discussion resumes and
the film ends on a note of hope and challenge. It is narrated by Orson Welles.

LIVING TOGETHER AS AMERICANS
(Filmstrip—Sound). Eye Gate Media, 1976.

6 color, sound filmstrips; 3 cassettes.
#TH748 ($74.70). Gr. 4–9. MDU (4/76). 158
Contents: 1. Our Minorities—Just Who Is
An American? 2. Indians. 3. Asian Ameri-
cans. 4. Mexican Americans. 5. Blacks. 6.
Middle Eastern Cultures.
The overall theme of this set is that a real
American is a person who has learned to live
together with his fellow Americans respect-
ing their customs and traditions.

MAN ALONE AND LONELINESS: THE
DILEMMA OF MODERN SOCIETY
(Slides/Cassettes). The Center for Humanities, 1974. 160 slides in 2 Carousel cartridges; 2 cassettes; 2 discs also available;
Teacher's Guide. #0235 ($139.50). Gr. 9–C.
M&M (4/74), PVB (4/75). 301.1
The program explores the many faces of
loneliness, showing how the vision of artists,
innovators, and explorers often sets them
apart from the rest of society. The content
includes a discussion of the change in American attitudes about loneliness, from the chosen solitude of the early settlers to the
compelling group-mindedness of today.

MASKS: HOW WE HIDE OUR EMOTIONS
(Filmstrip—Sound). *See* Emotions

RELATING: THE ART OF HUMAN IN-
TERACTION (Filmstrip—Sound). Human
Relations Media, 1977. 2 color, sound filmstrips, 73–74 fr.; 2 cassettes or discs, 10^{1}/$_{2}$–
11 min.; Teacher's Guide with Script. #624
($60). Gr. 7–12. BKL (7/15/78), M&M (4/78),
NCF (1978). 158.2
Contents: 1. Impressions. 2. Communicating.
Encourages students to critically evaluate
the ways in which they relate to others.
Demonstrates how we give and receive impressions. Pinpoints characteristics and effects of manipulative behavior.

THEY (Motion Picture—16mm—Sound). Centron Films, 1972 (Trilogy). 16mm color,
sound film, 16 min.; Leader's Guide. ($250).
Gr. 2–A. CFF. STE (2/74). 301.45
Through symbolism, the film examines the
human society in microcosm. A young boy
meets the "we" people and the "they"
people. When he crosses the river to meet
the "they" people, he raises questions about
human relationships, which makes this film
outstanding for class discussion and individual contemplation.

WALLS AND WALLS (Motion Picture—
16mm—Sound). *See* Prejudices and antipathies

THE WORLD OUTSIDE ME (Kit). *See* Humanities

YOU AND THE GROUP (Filmstrip—Sound).
Hum, 1977. 2 color, sound filmstrips, 65–77
fr.; 2 cassettes or discs, 11 min. ea.; Teacher's Guide. #620 ($60). Gr. 8–12. BKL (4/15/
78), CPR (5/78), FHE (5/78). 301.45

Contents: 1. Belonging. 2. Conformity and Individuality.

Probes the complex workings and dynamics of group interaction to help students assess their own groups. Stresses the importance as well as the problems of belonging to groups. Identifies the most influential adolescent groups (clubs, gangs, cliques, classes, teams) and analyzes their structure, leadership and power, conflicts, and climates of trust.

HUMAN RELATIONS—FICTION

ANGEL AND BIG JOE (Motion Picture— 16mm—Sound). *See* Values—Fiction

TO KILL A MOCKINGBIRD (Kit). *See* Southern states—Fiction

HUMANITIES

ADVENTURE AND SUSPENSE (Filmstrip—Sound). *See* Literature—Study and teaching

ALL MY FAMILIES (Kit). EBEC, 1975 (Discovery-Journey in the Humanities). 4 color, sound filmstrips, av. 58–93 fr. ea.; 3 cassettes, av. 5–13 min. ea.; 8 Study Prints; 10 Student Books; 16 Spirit Masters; Cassette; Teacher's Guide. #66302 ($108.50). Gr. 7– 12. BKL (9/15/76). 001.9

Contents: 1. Everyone's Family. 2. Her Story. 3. Family Life Cycle. 4. Dreams of a Place Just Right.

This set focuses on the family and its impact on you. Materials from literature, music, art, history, philosophy and theater are used to help the student understand the place of the family in his or her development.

THE ARTIST INSIDE ME (Kit). EBEC, 1975 (Discovery-Journey in the Humanities). 3 color, sound filmstrips, av. 58–93 fr. ea.; 3 cassettes, 5–13 min. ea.; 8 Study Prints; 16 Spirit Masters; Cassette; Teacher's Guide. #66304 ($92.50). Gr. 7–12. BKL (9/15/76). 001.9

Contents: 1. Discovering Nature. 2. Discovering Self. 3. Artists.

Delving deeper to obtain insights about yourself, you discover the influence of the artist in you. This set uses material from literature, music, art, history, philosophy, and theater to help understand the artist in you.

COURAGE (Filmstrip—Sound). *See* Literature—Study and teaching

A DISCUSSION OF THE CROCODILE (Motion Picture—16mm—Sound). *See* Motion pictures—History and criticism

DRAMA (Filmstrip—Sound). *See* Literature—Study and teaching

FOLKLORE AND FABLE (Filmstrip—Sound). *See* Literature—Study and teaching

THE FUTURE (Filmstrip—Sound). *See* Literature—Study and teaching

HUMANITIES: A BRIDGE TO OURSELVES (Motion Picture—16mm—Sound). EBEC, 1974. 16mm color, sound film, 29 min. #47797 ($395). Gr. 7–C. LFR (1975). 001.3

Art . . . music . . . drama . . . the dance— are they relevant? This film attempts an answer by presenting a brief cavalcade of human cultural history. Certain constant themes link divergent times and places and begins a transformation that ends as "Who—and what—am I?"

THE HUMAN-I-TIES OF LANGUAGE (Kit). EBEC, 1975 (Discovery-Journey in the Humanities). 3 color, sound filmstrips, av. 58–93 fr. ea.; 3 cassettes, av. 5–13 min. ea.; 10 Student Books; 8 Study Prints, 16 Spirit Masters, Cassette, Teacher's Guide. #66305 ($92.50). Gr. 7–12. BKL (9/15/76). 001.9

Contents: 1. Message Beginnings. 2. A Universe of Messages. 3. My Scripts.

This set uses materials from literature, music, art, history, philosophy, and theater to help students discover insights about themselves.

HUMOR AND SATIRE (Filmstrip—Sound). *See* Literature—Study and teaching

THE I IN IDENTITY (Kit). EBEC, 1975 (Discovery-Journey in the Humanities). 4 color, sound filmstrips, av. 58–93 fr. ea.; 4 cassettes, av. 5–13 min. ea.; 10 Student Books; 8 Study Prints; 16 Spirit Masters; Cassette; Teacher's Guide. #66301 ($108.50). Gr. 7–12. BKL (6/15/76). 001.9

Contents: 1. Humanities. 2. Identity Man. 3. Fads and Fashions. 4. Making Worlds to See.

This set helps you have deeper insights about yourself. Materials from literature, music, art, history, philosophy, and theater are used to assist with this discovery.

MYTHOLOGY (Filmstrip—Sound). *See* Literature—Study and teaching

SHORT STORY (Filmstrip—Sound). *See* Literature—Study and teaching

THE WORLD OUTSIDE ME (Kit). EBEC, 1975 (Discovery-Journey in the Humanities). 2 color, sound filmstrips, 58–93 fr. ea.; 2 cassettes, 5–13 min. ea.; 10 Student Books; 8 Study Prints; 16 Spirit Masters; Cassette and Teacher's Guide. #66303 ($76.50). Gr. 7–12. BKL (9/15/76). 001.9

Contents: 1. Journeys. 2. Museum Field Trip.

This set in the Discovery unit focuses on outer worlds we all live in and then explores that world as it relates to you. Materials from literature, music, art, history, philosophy, and theater are integrated into this program.

HUMANITIES AND SCIENCE. *See* Science and the humanities

HUMMINGBIRDS

CHUPAROSAS: THE HUMMINGBIRD TWINS (Filmstrip—Sound). Lyceum/Mook & Blanchard. 1 color, sound filmstrip, 42 fr.; 1 cassette or disc, 9¼ min. #LY351070C Cassette ($25), #LY351070R Disc ($19.50). Gr. K-12. BKL (1/1/72), FLN (12/71). 598.8
The life and growth of hummingbirds, smallest of the vertebrates are presented. Photographed in a high desert habitat, the hummingbird twins are followed from egg to first flight.

HUMPERDINCK, ENGELBERT

STORIES IN BALLET AND OPERA (Phonodiscs). *See* Music—Analysis, Appreciation

HUMPHREY, HUBERT H.

THE UNITED STATES CONGRESS: OF, BY, AND FOR THE PEOPLE (Motion Picture—16mm—Sound). *See* U.S. Congress

HUNT, IRENE

ACROSS FIVE APRILS (Filmstrip—Sound). *See* U.S.—History—1861-1865—Civil War—Fiction

HUNTING—FICTION

THE HUNT (Motion Picture—16mm—Sound). *See* American fiction

HYGIENE

HEALTH AND SAFETY: KEEPING FIT (Kit). Butterick Publishing, 1976 (Independent Living). 4 color, sound filmstrips, 65-75 fr. ea.; 4 discs or cassettes, av. 8 min. ea.; 6 Spirit Masters; Teacher's Guide. #C-515-6E Cassettes ($85), #R-514-8E Discs ($85). Gr. 7-12. FHE (9/76). 613
This program concentrates on the positive practices that assure our well-being. It demonstrates how good health and safety come about as a result of choice, not by chance. Presents the basic rules for staying healthy, continuing with sensible information about drugs, available public health services and safety and accident prevention.

NOISE AND ITS EFFECTS ON HEALTH (Motion Picture—16mm—Sound). *See* Noise pollution

THERE'S A NEW YOU COMIN' FOR GIRLS (Filmstrip—Sound). *See* Girls

ICE HOCKEY

HOCKEY HEROES (Kit). *See* Athletes

IDENTITY. *See* Individuality; Personality; Self

IGLOOS

HOW TO BUILD AN IGLOO—SLIDE SET (Slides). National Film Board of Canada/Donars Productions, 1971. 10 cardboard mounted, color slides in plastic storage page. Notes in English and Spanish. ($10). Gr. 4-A. PRV (12/74), PVB (5/75). 970.3
With only a saw and a knife, an Eskimo can build an Igloo in less than half an hour. This slide set and the accompanying guide show you how to do it.

ILLINOIS

COAL, CORN, AND COWS (Filmstrip—Sound). *See* Environment

ILLUSTRATION OF BOOKS

GAIL E. HALEY: WOOD AND LINOLEUM ILLUSTRATION (Filmstrip—Sound). *See* Haley, Gail E.

ILLUSTRATORS

CHILDREN OF THE NORTHLIGHTS (Motion Picture—16mm—Sound). *See* Authors, American

EZRA JACK KEATS (Motion Picture—16mm—Sound). *See* Keats, Ezra Jack

MR. SHEPARD AND MR. MILNE (Motion Picture—16mm—Sound). *See* Authors, English

IMAGINATION

ADVENTURES IN IMAGINATION SERIES (Filmstrip—Sound). *See* English language—Composition and exercises

THE BEGINNING (Motion Picture—16mm—Sound). *See* Creativity—Fiction

A BETTER TRAIN OF THOUGHT (Motion Picture—16mm—Sound). *See* Creativity—Fiction

IF TREES CAN FLY (Motion Picture—16mm—Sound). Phoenix Films, 1976. 16mm color, sound film, 12 min. #0175 ($180). Gr. 5-A. BKL (4/1/77). 153.3
Scenes of seagulls, eagles, and other birds fill this production's visuals with a sense of freedom. An introspective, poetic film which captures both children's and adults' attention.

JOHN HENRY AND JOE MAGARAC (Phonodisc or Cassette). *See* Folklore—U.S.

IMMIGRATION AND EMIGRATION

IMMIGRANT AMERICA (Filmstrip—Sound). Sunburst Communications, 1974. 2

sound filmstrips (1 color, 1 b/w), 74–109 fr.; 2 cassettes or discs, 11–18 min.; Teacher's Guide. #209 ($59). Gr. 6–12. BKL (4/1/75), CPR (2/76), PRV (12/75). 301.45

Contents: 1. The Immigrants Arrive. 2. An Ethnic Community Today.

Presents a profile of immigrant life in New York City and the coal fields of Pennsylvania in the 1900s, and of the Polish and Chicano residents in Chicago today. Uses prize-winning b/w documentary photographs of Lewis Hine to illustrate the middle European emigrants who came to America between 1903–1913. Contributions they made and hardships faced are included. Demonstrates how changing conditions have affected attitudes of different ethnic groups.

IMMIGRATION AND MIGRATION (Motion Picture—16mm—Sound). *See* U.S.—History

THEY CHOSE AMERICA: VOLUME ONE (Cassette). Visual Education 1975, (Conversations with Immigrants). 6 cassettes, 30–40 min. ea.; Listener's Guide. #5301H ($67). Gr. 7–A. BKL (11/15/76), M&M (2/77), PRV (11/75). 325

Contents: 1. Chinese Immigrants. 2. Irish Immigrants. 3. Italian Immigrants. 4. Jewish Immigrants. 5. Mexican Immigrants. 6. Polish Immigrants.

A group of Americans who immigrated to the United States tell of their trials in getting to America. Twenty-three people tell their fascinating experiences.

THEY CHOSE AMERICA: VOLUME TWO (Cassette). Visual Education, 1975 (Conversations with Immigrants). 6 cassettes, 40–60 min. ea.; Listener's Guide. #5302H ($67). Gr. 7–A. BKL (11/15/76), PRV (10/76). 325

Contents: 1. Cuban Immigrants. 2. German Immigrants. 3. Greek Immigrants. 4. Hungarian Immigrants. 5. Japanese Immigrants. 6. Scandinavian Immigrants.

As in Volume One, voices of people who have come from other countries to become American citizens are presented. Especially interesting are the Cuban refugees and the Japanese-Americans who were interned in their own country during World War II.

IMMORTALITY

DEATH AND DYING: CLOSING THE CIRCLE (Filmstrip—Sound). *See* Death

LIVING WITH DYING (Filmstrip—Sound). *See* Death

IMPEACHMENTS

ANDREW JOHNSON COMES TO TRIAL (Filmstrip—Sound). *See* U.S.—History—1865–1898

IMPRESSIONISM (ART)

DEGAS (Motion Picture—16mm—Sound). *See* Degas, Edgar

THE IMPRESSIONIST EPOCH (Filmstrip—Sound). Miller-Brody Productions, 1975. 4 color, sound filmstrips, 50–60 fr.; 4 cassettes or discs, 13–20 min.; Poster; Teacher's Guide. #MB-801/4C Cassettes ($100), #MB-801/4 Discs ($100). Gr. 7–12. MLJ (4/77), M&M (4/77), PRV (5/77). 759.06

Contents: 1. Events That Led to Impressionism. 2. Techniques of Impressionistic Painting. 3. A Survey of Major American Impressionistic Paintings. 4. Later Works Strongly Influenced by Impressionism.

A landmark exhibition, called "The Impressionist Epoch," held at the Metropolitan Museum of Art in New York, is commemorated in these filmstrips. Nearly 300 frames were drawn from the collections of the Louvre (Galerie du Jeu de Paume), the Metropolitan Museum, and public and private collections in Europe and the United States.

INDEXES

INDEXES (Filmstrip—Sound). Library Filmstrip Center, 1976. 1 color, sound filmstrip, 61 fr.; 1 disc or cassette, 22 min. ($34). Gr. 9–A. PRV (2/79). 029.9

Explains the best-known indexes available in the areas of art, business language arts, music, science and social studies. General and specialized indexes are included.

INDIA

AGRICULTURAL REFORM IN INDIA: A CASE STUDY (Filmstrip—Sound). *See* Agriculture—India

INDIA: TRADITION AND CHANGE (Filmstrip—Sound). Society for Visual Education, 1977. 5 color, sound filmstrips, av. 79 fr. ea.; 5 cassettes of 33¹/₃ rpm discs, av. 13¹/₂ min. ea.; Teacher's Guide. #A387–SATC Cassette ($85), #A387–SAR Discs ($85). Gr. 5–9. BKL (1/1/78). 915.4

Contents: 1. India's Land. 2. India's People. 3. Patterns of Family Life. 4. Village Life. 5. City Life.

While presenting much basic information on India's geography and culture, the filmstrips highlight some of the key problems and challenges facing the nation's people, and provide a balanced perspective on the nation's cultural heritage and development goals.

SOUTH ASIA: REGION IN TRANSITION (Filmstrip—Sound). *See* Asia

SOUTH ASIA: THE INDIAN SUBCONTINENT (Filmstrip—Sound). EBEC, 1976. 5 color, sound filmstrips, av. 81 fr. ea.; 5 discs or cassettes, av. 9 min. ea.; Teacher's Guide. #6923 Discs ($72.50), #6923K Cas-

INDIA (cont.)

settes ($72.50). Gr. 5–12. ESL (1977), PRV (12/77). 915.4

Contents: 1. The Physical Base. 2. The Economic Base. 3. Religions and Cultures. 4. The Villages. 5. The Cities.

Presents a pictorial investigation of India and the small countries surrounding it. Although characterized by great diversity of culture, language and religious tradition, these six basically agricultural economies share the common burdens of over-population, inadequate food supply, and lack of capital.

INDIA—FICTION

KIM (Phonodisc or Cassette). Caedmon Records, 1976. 1–12 in. LP disc or 1 cassette, 60 min. #CDL 51480 Cassette ($7.95), #TC1480 Disc ($6.95). Gr. 6–8. BKL (12/15/77). Fic

Excerpts are from the first, third, and fourth chapters of Kipling's tale in which Kim, an English boy who is brought up as a native of colonial India, meets the Lama with whom he will travel over the Indian countryside. Anthony Quayle's reading of this masterpiece gives it color, power, and life.

THE WORLD OF JUST SO STORIES, SET ONE (Kit). *See* Animals—Fiction

INDIA—HANDICRAFT

TEXTILES AND ORNAMENTAL ARTS OF INDIA (Motion Picture—16mm—Sound). EBEC, 1973. 16mm color, sound film, 11 min. #3189 ($150). Gr. 7–A. BKL (1974), LFR (1975). 745.44

Color, sound, texture, poetry, and rhythm come together in this interpretation of an exhibition at the Museum of Modern Art. As the narration discusses the significance of color in Indian textile design, the camera moves over the articles on exhibit. The film presents the flavor of Indian tradition and philosophy and explores the spiritual significance that relates art to everyday life.

INDIANS OF CENTRAL AMERICA

CULTURE AND ENVIRONMENT: LIVING IN THE TROPICS (Filmstrip—Sound). *See* Tropics

INDIANS OF NORTH AMERICA

ADOBE OVEN BUILDING (Film Loop—8 mm—Silent). Thorne Films/Prentice-Hall Media, 1972 (Southwest Indians). 8 mm color, silent film loop, approx. 4 min. #HAT 290 ($28). Gr. 4–A. AUR (1978). 970.1

Indians of the Southwest are shown building an adobe oven.

AMERICAN INDIANS OF THE NORTH PACIFIC COAST (Filmstrip—Sound). Coronet Instructional Media, 1971. 6 color, sound filmstrips, av. 51 fr.; 6 cassettes or 3 discs, av. 12 min. #M194 Cassettes ($95), #S194 Discs ($95). Gr. 4–8. PRV (3/73), PVB (5/73), SLJ (1973). 970.1

Contents: 1. Lands and Tribes. 2. How They Lived. 3. Arts and Crafts. 4. Myths and Ceremonies. 5. How They Changed. 6. Their Life Today.

The Indians' dependence on land and nature and their culture were disrupted by the white settlers. Modern photographs, as well as rare historic photographs from the Oregon Historical Society, add depth to this history lesson.

AMERICAN INDIANS OF THE NORTH-EAST (Filmstrip—Sound). Coronet Instructional Media, 1971. 6 color, sound filmstrips, av. 48 fr.; 6 cassettes or discs, av. 11½ min. #M214 Cassettes ($95), #S214 Disc ($95). Gr. 4–8. BKL (12/15/76), INS. 970.1

Contents: 1. Who They Are. 2. Their History. 3. How They Lived. 4. Their Religions. 5. Their Handicrafts. 6. Their Life Today.

These filmstrips study the rise and fall of the Algonquin and Iroquois Indian empires, migrants from Asia to the northeastern United States and southern Canada. Their life today in a modern society is explained through photography, original art work, and period drawings. The set traces ancient history, culture, and Indian reliance on nature and religious ceremonies.

CANYON DE CHELLY CLIFF DWELLINGS (Film Loop—8mm—Silent). *See* Cliff dwellers and cliff dwellings

THE DAWN HORSE (Motion Picture—16mm—Sound). Coronet Instructional Media, 1976. 4 color, sound filmstrips, 47–55 fr.; 4 cassettes or discs, 10–12 min.; 1 Guide. #M257 Cassettes ($65), #S257 Discs ($65). Gr. 4–10. BKL (5/15/77). 598.9

Imparts traditional American Indian attitudes toward nature. Captures the beauty of native earth as explored in Indian poetry, art, and music.

THE FIRST AMERICA: CULTURE PATTERNS (Filmstrip—Sound). Prentice-Hall Media, 1974. 4 color, sound filmstrips, 84–114 fr.; 4 cassettes or discs, 15–19 min.; Teacher's Guide. ($76). Gr. 6–A. PRV (5/75), PVB (4/76). 970.1

Contents: 1. The Paleo—Indians. 2. The Arctic. 3. The Southwest. 4. The Mound Builders.

The primary purpose of this series is to expel the myth that the Pilgrims/Puritans from England were ''the first Americans,'' and to acquaint students with the ancient history of the western hemisphere. Many of the color photographs were provided by American museums.

HORIZONTAL BELT LOOM (Film Loop—8mm—Silent). *See* Looms

INDIAN LIFE IN NORTH AMERICA (Filmstrip—Sound). Imperial Educational Resources, 1973. 4 color, sound filmstrips; 4 cassettes or discs. #3RG50200 Discs ($56), #3KG50200 Cassette ($62). Gr. 4–8. STE (2/74). 970.1

Contents: 1. The Havasupai of the Grand Canyon—Part I. 2. The Havasupai of the Grand Canyon—Part II. 3. The Pueblo Indians of the Southwest—Part I. 4. The Pueblo Indians of the Southwest—Part II.

This set will help you to understand contemporary Indian life, its cultural heritage, and the challenges it faces.

INDIANS OF NORTH AMERICA (Filmstrip—Sound). Society For Visual Education, 1976. 6 color, sound filmstrips, 36–47 fr.; 6 cassettes or discs, 8–11 min. #A477-SATC Cassettes ($90), #A477-SAR Discs ($90). Gr. 4–8. BKL (5/1/77). 970.1

Contents: 1. Indians of the Northeast. 2. Indians of the Southeast. 3. Indians of the Plains. 4. Indians of the Northwest Coast. 5. Indians of the Southwest. 6. Indians of the Far North.

The life-styles and cultures of North American Indians in six geographic areas are presented. Aspects covered include diet, dress, ceremonies, artifacts, spiritual beliefs, social structure, and inter-tribal relationships.

THE INDIANS OF NORTH AMERICA (Filmstrip—Sound). National Geographic, 1973. 5 color, sound filmstrips, 55–61 fr.; 5 discs or cassettes, 13–14 min.; Teacher's Guide. #03736 Discs ($74.50), #03737 Cassettes ($74.50). Gr. 5–12. PRV (10/74), PVB (5/75). 970.1

Contents: 1. The First American. 2. The Eastern Woodlands. 3. The Plains. 4. West of the Shining Mountains. 5. Indians Today.

The subject content is authentic, noncontroversial, unbiased, and up-to-date and excellent photography of current personalities adds to the set.

MESA VERDE (Study Print). *See* Cliff dwellers and cliff dwellings

NATIVE AMERICANS: YESTERDAY AND TODAY (Filmstrip—Sound). BFA Educational Media, 1976. 4 color, sound filmstrips, 4 discs or cassettes. #VGS000 Cassettes ($70), #VGR000 Discs ($70). Gr. 9–12. INS (4/77), M&M (4/77). 970.1

Contents: 1. The Hopi Way. 2. An Iroquois Way of Life. 3. Cherokee Land, White Nation. 4. A Navajo Reservation.

This set helps to define, compare and contrast the cultural and religious values of our representative Indian groups.

THE NATIVE LAND (Motion Picture—16mm—Sound). *See* Cherokee

POTTERY DECORATION (Film Loop—8mm—Silent). *See* Pottery, Indian

POTTERY FIRING (Film Loop—8mm—Silent). *See* Pottery, Indian

POTTERY MAKING (Film Loop—8mm—Silent). *See* Pottery, Indian

PUYE COMMUNITY HOUSES (Film Loop—8mm—Silent). Thorne Films/Prentice-Hall Media, 1972 (Southwest Indians). 8mm color, silent film loop, approx. 4 min. #HAT 272 ($26). Gr. 4–A. AUR (1978). 970.4

This loop illustrates the community houses used by the Southwest Indians.

PUYE DE CHELLY CLIFF DWELLINGS (Film Loop—8mm—Silent). *See* Cliff dwellers and cliff dwellings

SPINNING WOOL (Film Loop—8mm—Silent). Thorne Films/Prentice-Hall Media, 1972 (Southwest Indians). 8 mm color, silent film loop, approx. 4 min. #HAT 284 ($28). Gr. 4–A. AUR (1978). 970.1

This film loop illustrates the spinning of wool by the Indians of the Southwest.

UPRIGHT LOOM (Film Loop—8mm—Silent). *See* Looms

INDIANS OF NORTH AMERICA—ART

INDIAN ARTISTS OF THE SOUTHWEST (Motion Picture—16mm—Sound). EBEC, 1972. 16mm color, sound film, 15 min. #3151 ($185). Gr. 5–12. BKL (1972), LFR (1972), PRV (1972). 709.01

Modern Southwest Indians have retained ancient art and craft techniques as part of their cultural heritage. In this film, three Pueblo Indian tribes, the Zuni, Hopi, and Navajo, introduce four of their major art forms: stone and silverwork, pottery making, weaving, and kachina carving. As the Indians demonstrate their crafts, viewers see that the Indians' life-style and breathtaking surroundings are reflected in their work.

NAVAJO SAND PAINTING CEREMONY (Film Loop—8mm—Silent). Thorne Films/Prentice-Hall Media, 1972 (Southwest Indians). 8mm color, silent film loop, approx. 4 min. #HAT 291 ($26). Gr. 4–A. MDU (1978). 709.01

The Navajo sand painting ceremony is demonstrated by the Southwest Indians.

PETROGLYPHS: ANCIENT ART OF THE MOJAVE (Filmstrip—Sound). Lyceum/Mook & Blanchard, 1971. 1 color, sound filmstrip, 51 fr.; 1 cassette or disc, 11½ min. #LY351270C Cassette ($25), #LY351270R Disc ($19.50). Gr. 4–12. BKL (3/15/72). 709.01

This filmstrip explores petroglyphs and their possible meanings. It stimulates interest in the Indians, their history and culture. These

INDIANS OF NORTH AMERICA—ART (cont.)

petroglyphs picture the Indian's life; hunting with spear and bow, ceremonies attendant to the hunt, and animals that provided them with food, shelter, and clothing.

INDIANS OF NORTH AMERICA—BIOGRAPHY

LUCY COVINGTON: NATIVE AMERICAN INDIAN (Motion Picture—16mm—Sound). *See* Covington, Lucy

INDIANS OF NORTH AMERICA—CANADA

PAUL KANE GOES WEST (Motion Picture—16mm—Sound). National Film Board of Canada/Encyclopaedia Britannica Educational Corporation, 1972. 16mm color, sound film, 15 min. #3337 ($220). Gr. 4–12. BKL (10/74), CFF (10/75), PRV (5/75). 970.1
In the mid-19th century, when the Indians of the Canadian West were proud and powerful, when the scenery was magnificent and natural, artist Paul Kane traveled thousands of miles between Canada's east and west. This film, accompanied by a narration from his diary, is a vivid record of Kane's adventures as seen through his forceful sketches and paintings. The Indians posed gladly, believing he was a medicine man, an impression Kane encouraged.

INDIANS OF NORTH AMERICA—FICTION

THE LIGHT IN THE FOREST (Phonodisc or Cassette). Caedmon Records, 1974. 1 cassette or disc. #TC1428 Disc ($6.98), #CDL51428 Cassette ($7.95). Gr. 5–A. BKL (5/15/74), JOR (1974). Fic
A skillfully abridged recording of Conrad Richter's novel by the same name. Read by E. G. Marshall and others, it presents the masterful story of True Son, adopted son of Cuyloga, an Indian chief, but really John Butler, captured eleven years earlier by the Indians. When a treaty is made, True Son is returned to his real family and he experiences a confrontation between the two cultures as he matures and must choose between them.

SING DOWN THE MOON (Filmstrip—Sound). *See* Navajo—Fiction

INDIANS OF NORTH AMERICA—HISTORY

NATIVE AMERICAN HERITAGE (Filmstrip—Sound). EBEC, 1977. 5 color, sound filmstrips, av. 92 fr. ea.; 5 cassettes or discs, 15 min. ea.; Teacher's Guide. #6961K Cassettes ($72.50), #6961 Discs ($72.50). Gr. 5–12. BKL (2/78), INS (1/78), LFR (11/78). 970.1

Contents: 1. The Old Ways. 2. Art: Reflections of a Culture. 3. The Fight for Survival. 4. Beliefs and Ceremonies. 5. The New Ways.
This series uses artwork and photography backed by illuminating narration to banish stereotypes as it traces a richly diverse culture from pre-Columbian days to the present. Common elements are explored, particularly the universal themes found in art and religion. It shows how Indians today are finding new strength and a new sense of identity as they cope with common problems.

INDIANS OF NORTH AMERICA—LEGENDS

ARROW TO THE SUN (Motion Picture—16mm—Sound). Texture Films, 1973. 16mm, color, sound film, 12 min. ($180). Gr. 1–A. CGE (1974), FLN (1974), PRV (4/75). 398.2
This uses animation to tell the tale of the Acoma Pueblo Indians. It is about a boy's search for his father, entailing his voyage on an arrow to the sun and the trials he undergoes in the sky village until he is recognized by his father, the Lord of the Sun, and his return to Earth to spread the Sun's warm delights.

HOW BEAVER STOLE FIRE (Motion Picture—16mm—Sound). Paramount Communications, 1972. 16mm color film, 12 min., Teacher's Guide. #7117 ($210). Gr. K–A. CGE (1972), FLN (12/73), MER (4/73). 398.2
Caroline Leaf's special sand animation technique recreates an American Indian myth on the origin of fire.

THE LEGEND OF THE MAGIC KNIVES (Motion Picture—16mm—Sound). EBEC, 1971. 16mm color, sound film, 11 min. #2999 ($150). Gr. 5–12. CGE (1976), PRV (1978), SLJ (1976). 398.2
A totem village in the Pacific Northwest provides the setting for this dramatic portrayal of an ancient Indian legend. The legend is recounted by means of the figures on a totem and authentic Indian masks.

THE LOON'S NECKLACE (Motion Picture—16mm—Sound). Crawley Films/EBEC, 1949. 16mm color, sound film, 11 min. #423 ($150). Gr. 5–A. FLN (1978), PRV (1978). 398.2
A charming Indian legend of how the loon, a water bird, received his distinguishing neckband. Authentic ceremonial masks, carved by Indians of British Columbia, portray the Indian's sensitivity to the moods of nature.

NATIVE AMERICAN MYTHS (Filmstrip—Sound). EBEC, 1978. 4 color, sound filmstrips, av. 64 fr. ea.; 4 cassettes or discs, 6 min. ea.; Teacher's Guide. #1706K Cassettes ($66.50), #17067 Discs ($66.50). Gr. 4–9. BKL (11/78), PRV (9/78). 398.2

Contents: 1. Seneca: Sky Woman. 2. Haida: The Raven Gave Daylight Unto Man. 3. Klamath: How Coyote Stole Fire. 4. Hopi: How the People Came Out of the Underground.

Colorful, original artwork brings to life myths from four different native American tribes—Seneca, Haida, Klamath, and Hopi. Viewers explore the rich cultural, spiritual, and artistic heritage preserved through the storytelling of these tribes.

INDIANS OF NORTH AMERICA— PAINTINGS

A. J. MILLER'S WEST: THE PLAINS IN-DIAN—1837—SLIDE SET (Slides). National Film Board of Canada/Donars Productions, 1975. 40 cardboard mounted, color slides in a plastic storage page. Teacher's Guide ($30). Gr. 5–A. PRV (9/76), PVB (4/77). 759.13

Photographs of A. J. Miller's paintings of the Plains Indians as he saw them. The drawings for these paintings were made in the wild west region that now comprises the states of Kansas, Nebraska, Missouri, and Wyoming. The script, in English and French, was written by the artist.

INDIANS OF NORTH AMERICA—PLAINS

SUN DANCE PEOPLE: THE PLAINS IN-DIANS (Filmstrip—Sound). Random House, 1973 (Indians: the Southwest & the Plains Indians, Part 2). 2 color, sound filmstrips, 80–85 fr.; 2 cassettes or discs, 15–20 min.; Discussion Guide. #12622–X Cassettes ($39), #12621–1 Discs ($39). Gr. 5–A. BKL (4/1/74). PRV (4/74), PVB (5/1/74), 970.4

This set describes the North American Plains Indians life-style—from blanket courting to horse racing—revealing important facets of the culture. The visuals include paintings, drawings, and historical and contemporary photographs.

INDIANS OF NORTH AMERICA—POETRY

AMERICAN INDIAN POETRY (Filmstrip—Sound). Educational Dimensions Group, 1976. 4 color, sound filmstrips; 4 cassettes, 15–18 min. #737 ($74). Gr. 4–9. MDU (4/77), PRV (10/76). 897

Contents: 1. Eastern Indian Poetry. 2. Great Plains Indians. 3. Southwestern Indians. 4. Eskimo and Pacific Northwestern Indians.

This material increases the awareness and understanding of the Native American people: "I feel this material would help the Native American Indian identify with and be proud of his/her heritage," says Ron Hatch.

INDIANS OF NORTH AMERICA—SOCIAL LIFE AND CUSTOMS

BLESSINGWAY: TALES OF A NAVAJO FAMILY (Kit). *See* Navajo

HOPIS—GUARDIANS OF THE LAND (Motion Picture—16mm—Sound). *See* Hopi

INDIVIDUALITY

JERRY'S RESTAURANT (Motion Picture—16mm—Sound). EBEC, 1977. 16mm color, sound film, 12 min. #3539 ($180). Gr. 7–A. BKL (4/1/78), EGT (1978), M&M (1978). 158.1

Anyone venturing into Jerry's Restaurant knows he or she is in a unique establishment. Filmed during the frantic and noisy lunch hour, this character study challenges thinking about unconventionality and business success.

LIVING IN THE FUTURE: NOW—INDIVIDUAL CHOICES (Filmstrip—Sound). *See* Self-realization

MAN ALONE AND LONELINESS: THE DILEMMA OF MODERN SOCIETY (Slides/Cassettes). *See* Human relations

SQUARE PEGS—ROUND HOLES (Motion Picture—16mm—Sound). FilmFair Communications, 1974. 16mm color, sound film, 7 1/2 min. ($100). Gr. 5–12. BKL (5/75), EFL (1974), LFR (9/74). 155.2

This animated film begins with the birth of a cube. As it grows, it finds it just isn't like its fellow shapes, the cones, spheres, etc., who easily fall into grooves made for them. The influences on him to fit neatly into a hole are to no avail. He tries to fit, but all the holes are filled. He finally realizes that each is unique and he digs his own hole, jumps in. Based on the original story by Phyllis Harvey.

TAKING RESPONSIBILITY (Filmstrip—Sound). *See* Self-realization

YOU AND THE GROUP (Filmstrip—Sound). *See* Human relations

INDIVIDUALITY—FICTION

THE BAT POET (Phonodisc or Cassette). Caedmon Records, 1971. 1, 12 in. disc or cassette; #CDL51364 Cassettes ($7.95), #TC1364 Discs ($6.98). Gr. 3–A. HOB, LBJ, NYR. Fic

The Bat Poet, according to Jarrell, was a story "half for children and half for grown-ups." It is about the personal world one finds and how one is treated by the conventional world when one is different.

A DISCUSSION OF MY OLD MAN (Motion Picture—16mm—Sound). *See* Motion pictures—History and criticism

A DISCUSSION OF THE LOTTERY (Motion Picture—16mm—Sound). *See* Motion pictures—History and criticism

THE LOTTERY BY SHIRLEY JACKSON (Motion Picture—16mm—Sound). *See* American fiction

INDIVIDUALITY (cont.)

MY OLD MAN (Motion Picture—16mm—Sound). *See* American fiction

UP IS DOWN (Motion Picture—16mm—Sound). Pyramid Films, 1970. 16mm color, animated, sound film, 6 min.; ³/₄ in. videocassette, also available. ($100). Gr. 7–A. AFF (1971), BKL (1971), M&M (1971). Fic
An animated tale about a boy who walks on his hands; consequently his perspective differs from those around him. All efforts to make him conform to accepted practice fail. He says, "If you want me to stand on my feet, you'll have to make some big changes first."

INDUSTRIAL ARTS EDUCATION

AIR CONDITIONING SERVICEMAN, SET 7 (Film Loop—8mm—Silent). *See* Automobiles—Maintenance and repair

AIR CONDITIONING SERVICEMAN, SET 8 (Film Loop—8mm—Silent). *See* Automobiles—Maintenance and repair

AUTO-BODY SHEET METAL MAN, SET ONE (Film Loop—8mm—Silent). *See* Automobiles—Maintenance and repair

AUTO-BODY SHEET METAL MAN, SET TWO (Film Loop—8mm—Silent). *See* Automobiles—Maintenance and repair

AUTO PAINTER HELPER, SET 3 (Film Loop—8mm—Silent). *See* Automobiles—Maintenance and repair

AUTO PAINTER, SET 4 (Film Loop—8mm—Silent). *See* Automobiles—Maintenance and repair

AUTOMOBILE GLASS MAN, SET NINE (Film Loop—8mm—Silent). *See* Automobiles—Maintenance and repair

AUTOMOBILE GLASS MAN, SET TEN (Film Loop—8mm—Silent). *See* Automobiles—Maintenance and repair

AUTOMOBILE UPHOLSTERY REPAIRMAN, SET ELEVEN (Film Loop—8mm—Silent). *See* Automobiles—Maintenance and repair

CIRCULAR SAW, SET FOUR (Film Loop—8mm—Silent). *See* Woodwork

FIBERGLASS REPAIRMAN, SET FIVE (Film Loop—8mm—Silent). *See* Automobiles—Maintenance and repair

FIBERGLASS REPAIRMAN, SET SIX (Film Loop—8mm—Silent). *See* Automobiles—Maintenance and repair

HAND TOOL OPERATIONS, SET 1 (Film Loop—8mm—Silent). *See* Woodwork

HAND TOOL OPERATIONS, SET 2 (Film Loop—8mm—Silent). *See* Woodwork

HAND TOOL OPERATIONS, SET 3 (Film Loop—8mm—Silent). *See* Woodwork

LATHE (Film Loop—8mm—Silent). *See* Woodwork

INFANTS—CARE AND HYGIENE

INFANT CARE AND UNDERSTANDING (Kit). Society for Visual Education, 1977. 4 filmstrips; 4 cassettes; 30 Worksheets; Guide. #578–SATC ($99). Gr. 9–A. BKL (11/15/78). 649
Contents: 1. Prenatal Care and the New Baby. 2. Infant's First Months. 3. Infant's First Year. 4. Infant's Emotional and Intellectual Growth.
Practical prenatal and infant care are clearly presented. Emphasis is on commitment to parenthood, child's need for love, and financial considerations of parenthood.

INFORMATION STORAGE AND RETRIEVAL SYSTEMS

INFORMATION MACHINE (Motion Picture—16mm—Sound). Eames Films/EBEC, 1973. 16mm color, sound film, 10 min. #3182 ($150). Gr. 5–C. BKL (3/15/74), LFR (9/19/75). 621.3819
This film places the computer in historical perspective, showing it to be the culmination of the abstractions and measuring tools that we have been developing since primitive times. Using animation, the film reveals ways in which people use the computer to help define and solve problems.

JOBS IN DATA PROCESSING (Filmstrip—Sound). *See* Computers

INJURIES. *See* First aid

INSECTIVOROUS PLANTS

CARNIVOROUS PLANTS (Motion Picture—16mm—Sound). National Geographic Educational Services, 1974 (Bio-Science Series). 16mm color, sound film, 12 min.; ³/₄ in videocassette also available. ($160). Gr. 7–12. SCT (4/75). 581.53
Nature's exceptions are plants that devour animals. The film contains phenomena of these plants that can only be seen through photography. The film emphasizes scientific principles and methods of investigation.

INSECTS

THE HONEY BEE (Film Loop—8mm—Captioned). *See* Bees

INSECT COMMUNITIES (Film Loop—8mm—Silent). Troll Associates, 4 8mm color, silent film loops, approx. 4 min. Available separately. ($99.80). Gr. K–8. AUR (1978). 595.7

Contents: 1. Ants. 2. Bees. 3. Wasps. 4. Termites.

These loops record in detail the lives and habits of insects whose lives are spent in a "community" arrangement.

INSECT LIFE CYCLE (PERIODICAL CICADA) (Motion Picture—16mm—Sound). *See* Cicadas

LIFE OF A WORKER BEE (Film Loop—8mm—Silent). *See* Bees

INSURANCE

CAREERS IN BANKING AND INSURANCE (Filmstrip—Sound). *See* Banks and banking

INSURANCE, HEALTH

BUYING HEALTH CARE (Kit). Changing Times Education Service/EMC, 1975. 2 color, sound filmstrips, 218 fr.; 2 discs or cassettes; 10 Linemasters; 1 Reading List; 1 Guide. Cassette #2347 ($49.50), Disc #2335 ($49.50). Gr. 9–12. IFT (1975). 368.3
This set is organized around five units: shopping for a doctor, choosing a hospital, saving on prescriptions, paying for health care, it's your health.

INTELLECT

IDIOT . . . GENIUS? (Motion Picture—16mm—Sound). Aesop Films/EBEC, 1977. 16mm color, sound film, 6 min. #3561 ($105). Gr. 5–A. EFL (1979), EGT (1978). 153.4
A baby grows . . . becomes well-educated and understood by his peers. As his intelligence grows, he begins to speak incoherently and his seemingly eccentric babbling alienates him—people don't understand him anymore. This nonnarrated allegory considers human responses to the learning person and the learned person.

INTERIOR DECORATION

THE BUTTERICK INTERIOR DESIGN SERIES (Kit). Chrome Yellow Films/Butterick Publishing, 1977. 8 color, sound filmstrips, 68–86 fr.; 8 cassettes or discs, 8–10 min. ea.; Teacher's Guide; Wall chart, Transparencies. Cassettes #R-423-0 ($198), Discs #C-424-9 ($198). Gr. 7–12. BKL (10/1/77), PRV (2/79). 747
Contents: 1. A Living Environment. 2. Working with Space. 3. Working with Backgrounds. 4. Creating Personal Space.
An introduction to the basics of interior design. Provides information about types of houses and what makes a home; organizing space, including use of furniture; concepts in color, texture, light; and suggestions for personalizing a room on a budget.

HOME DECORATION SERIES, SET ONE (Filmstrip—Sound). University Films/McGraw-Hill Films, 1971. 4 color, sound filmstrips, av. 70 fr. ea.; 4 cassettes or discs, 8 min. ea.; Guide. #102190-6 Cassettes ($68), #101770-4 Discs ($68). Gr. 7–A. BKL (4/15/73), PRV (3/73). 747
Contents: 1. Elements and Principles of Design. 2. Selecting Furniture, Part 1. 3. Selecting Furniture, Part 2. 4. Arranging Furniture.
Various aspects of home decoration are explored—from principles of good design to stretching the budget. Numerous examples of room arrangements are used to point out decorative modes, decorating problems and their solutions.

HOME DECORATION SERIES, SET TWO (Filmstrip—Sound). University Films/McGraw-Hill Films, 1971. 4 color, sound filmstrips, av. 70 fr. ea.; 4 cassettes or discs, 8 min. ea.; Guide. #102195-7 ($68), #101779-8 ($68). Gr. 7–A. BKL (4/15/73), PRV (3/73). 747
Contents: 1. Selecting Tableware. 2. Selecting Fabrics. 3. Lighting. 4. Decorating: An Individual Approach.
Various aspects of home decoration are explored—from principles of good design to stretching the budget. Numerous examples of room arrangements are used to point out decorative modes, decorating problems and their solutions.

HOUSING AND HOME FURNISHINGS: YOUR PERSONAL ENVIRONMENT (Kit). *See* Houses

INTERPERSONAL RELATIONS. *See* Human relations

INVENTIONS—HISTORY

THE STORY OF GREAT AMERICAN INVENTORS (Filmstrip—Sound). *See* Inventors

INVENTORS

THE STORY OF GREAT AMERICAN INVENTORS (Filmstrip—Sound). Bill Boal Prod./Eye Gate Media, 1978. 6 color, sound filmstrips, 47–59 fr.; 3 cassettes, 11–14 min.; 1 guide. #E829 ($82.50). Gr. 7–11. BKL (9/1/78). 609
Contents: 1. Inventors of New Machine. 2. Inventors in Agriculture. 3. Inventors of New Communications. 4. Inventors of New Safety Devices. 5. Inventors in Electricity. 6. Inventors of Automobiles.
Brief biographies with interesting anecdotes highlight several of America's most ingenious and successful inventors, such as Whitney, Singer, McCormick, Morse, Westinghouse, Edison, Olds, and Ford.

INVENTORS (cont.)

THOMAS ALVA EDISON (Cassette). *See* Edison, Thomas Alva

INVERTEBRATES

THE HIGHER INVERTEBRATES (Filmstrip—Sound). Milan Herzog & Assoc./EBEC, 1976. 5 color, sound filmstrips, av. 84 fr. ea.; 5 cassettes or discs, 18 min. ea. #6935K Cassettes ($72.50), #6935 Discs ($72.50). Gr. 9–12. BKL (7/15/78), LFR (2/78). 592

Contents: 1. Echinoderms. 2. Annelids. 3. Arthropods: Crustaceans and Their Allies. 4. Insects: Survey of Representative Orders. 5. Social Insects.

Emphasizes the structural and functional characteristics of arthropods. Traces the nervous, digestive, and reproductive systems of the annelid worm and shows how its body plan represents an important evolutionary development.

IONESCO, EUGENE

DIRECTING A FILM (IONESCO'S: THE NEW TENANT) (Motion Picture—16mm—Sound). *See* Motion pictures—History and criticism

THE NEW TENANT (Motion Picture—16mm—Sound). *See* French drama

IONESCO'S THE NEW TENANT

DIRECTING A FILM (IONESCO'S: THE NEW TENANT) (Motion Picture—16mm—Sound). *See* Motion pictures—History and criticism

IRAN

IRAN (Motion Picture—16mm—Sound). Pyramid Films, 1974. 16mm color, sound film, 18 min.; ³/₄ in. videocassette also available. ($260). Gr. K–A. BKL (1975), PRV (1975), WLB (1975). 915.5

Non-narrated documentary examining the myriad elements of Iranian society and culture. Visual emphasis on people in all walks of life. Lelouch lets the contrasts—through brief, objective, beautiful shots—tell the story of Iran's past, present, and future.

IRAN—ANTIQUITIES

ART OF PERSEPOLIS (Slides). Educational Dimensions Group, 1977. 3 units with 20 slides ea. cardboard mounts; Teacher's Guide. Units available separately. #664 ($75). Gr. 4–A. PRV (4/78). 935

Contents: 1. Its Architecture. 2. Architectural Detail. 3. The Stairways.

Through photography, this set demonstrates how the lives of the earliest Persians affected their arts—paintings, sculpture, crafts, and architecture.

IRAN—FICTION

FLOWER STORM (Motion Picture—16mm—Sound). ACI Media/Paramount Communications, 1974. 16mm color, sound film, 12 min.; Teacher's Guide. ($175). Gr. K–A. CFF (1975), PRV (11/76). Fic

An animated film from Iran, in the style of Persian miniatures. Two emirs quarrel over which one shot a bird during a hunt. They decide on war, but the younger citizens don't want a war and they rig the cannon balls to release birds for the one side and flowers for the other. The war is called off.

IRELAND

THE BRITISH ISLES: SCOTLAND AND IRELAND (Filmstrip—Sound). *See* Scotland

EUROPEAN GEOGRAPHY, SET FOUR (Filmstrip—Sound). *See* Scotland

IRISH DRAMA

THE WELL OF THE SAINTS (Motion Picture—16mm—Sound). EBEC, 1975. 16mm color, sound film, 40 min. #47820 ($550). Gr. 9–C. BKL (1976), EFL (1977), EGT (1976). 822

This traditional Irish folk drama fuses rhythmic, musical speech with live characters. The satiric theme—that sometimes it is better to preserve one's illusions than to face reality—is expressed through the episode of miraculously restored sight. Through his simple superstitious characters, Synge expresses a sense of life's comedy, richness and pathos.

IRVING, WASHINGTON

EPISODES FROM FAMOUS STORIES (Filmstrip—Sound). *See* Literature—Collections

THE LEGEND OF SLEEPY HOLLOW (Motion Picture—16mm—Sound). *See* Legends—U.S.

THE LEGEND OF SLEEPY HOLLOW AND ICHABOD CRANE (Phonodisc or Cassette). *See* Legends—U.S.

ISLAM

ISLAM: THE PROPHET AND THE PEOPLE (Motion Picture—16mm—Sound). Texture Films, 1975. 16mm color, sound film, 34 min. ($435). Gr. 9–A. AFF (1975), CFF (1975), CGE (1978). 297

The cultural background of the Middle East is conveyed by tracing the development of its major religion, Islam. The film delineates the force and power of the Muslim world in both its practical and spiritual aspects. It is a biography of Muhammad the Prophet, a history of Islam from its sixth-century origins to present day, and an explanation of the basic beliefs of Muslims today.

ISLAM—HISTORY

THE ARAB CIVILIZATION (Filmstrip—Sound). *See* Arab countries—Civilization

ISRAEL

ISRAELI BOY: LIFE ON A KIBBUTZ (Motion Picture—16mm—Sound). EBEC, 1973. 16mm color, sound film, 17 min. #3178 ($240). Gr. 3–9. LNG (5/6/76), PRV (3/75), TEA (5/6/74). 915.694

Israeli society is viewed through the eyes of Nir, a young boy born and raised on a kibbutz. Scenes of milking, farming, schooling, and socializing illustrate the busy goings-on of communal life, and show why the kibbutz succeeds as a unique social, political, and economic entity.

ISRAEL—ANTIQUITIES

THE BIG DIG (Motion Picture—16mm—Sound). Tele-Visual Productions/EBEC, 1973. 16mm color, sound film, 22 min. #3363 ($290). Gr. 4–C. BKL (7/15/75), LFR (9/10/75), PRV (1/76). 913.33

A huge mound in the Israeli desert is all that remains of Gezer, a city that saw more than 140 successive generations of inhabitants. This film records some of the events in a season of excavation at Tel Gezer, as an international team of archeologists probes through strata of earth to uncover traces of an ancient civilization. Students are shown the meticulous care and back-breaking labor of professional archeologists, and are introduced to the specialized techniques and equipment required for excavation.

ITALIANS IN THE U.S.

ITALIAN FOOD (Kit). *See* Cookery, Italian

IVES, BURL

FOLK SONGS IN AMERICAN HISTORY, SET ONE: 1700–1864 (Filmstrip—Sound). *See* Folk songs—U.S.

FOLK SONGS IN AMERICAN HISTORY, SET TWO: 1865–1967 (Filmstrip—Sound). *See* Folk songs—U.S.

IVES, JAMES MERRITT

THE AMERICA OF CURRIER AND IVES (Filmstrip—Sound). *See* Prints, American

IVORY COAST

NIGERIA AND THE IVORY COAST: ENTERING THE 21ST CENTURY (Filmstrip—Sound). *See* Africa

JACKSON, ANDREW

PRESIDENTS AND PRECEDENTS (Kit). *See* Presidents—U.S.

JACKSON, SHIRLEY

A DISCUSSION OF THE LOTTERY (Motion Picture—16mm—Sound). *See* Motion pictures—History and criticism

THE LOTTERY BY SHIRLEY JACKSON (Motion Picture—16mm—Sound). *See* American fiction

JACOBS, LINDA

REALLY ME (Kit). *See* Girls—Fiction

WOMEN WHO WIN, SET 1 (Kit). *See* Athletes

WOMEN WHO WIN, SET 2 (Kit). *See* Athletes

WOMEN WHO WIN, SET 3 (Kit). *See* Athletes

WOMEN WHO WIN, SET 4 (Kit). *See* Athletes

JAGUARS

JAGUAR: MOTHER AND CUBS (Film Loop—8mm—Silent). Walt Disney Educational Media, 1966. 8mm color, silent film loop, approx. 4 min. #62-5307L ($30). Gr. K–12. MDU (1974). 599.7

The female's gentleness and patience with her kittens are demonstrated. The jaguar's fine ability to swim is also presented.

JAPAN

JAPAN: ECONOMIC MIRACLE (Filmstrip—Sound). EBEC, 1974. 5 color, sound filmstrips, av. 77 fr. ea.; cassettes or discs, 8 min. ea.; Teacher's Guide. #6907K Cassettes ($72.50), #6907 Discs ($72.50). Gr. 5–10. ESL (1976/77). 915.2

Contents: 1. The Physical Base. 2. Revolution in Food Supply. 3. The Industrial Revolution. 4. Industry and Trade. 5. The Urban Explosion.

The miracle that is modern Japan is explored through its geography, its history, and the temperament and character of its people.

THE NEW JAPAN (Filmstrip—Sound). Society for Visual Education, 1973. 6 color, sound filmstrips, 94–101 fr.; 3 discs or cassettes, av. 19$\frac{1}{2}$ min. #298-SAR Discs ($93),

JAPAN (cont.)

#298-SATC Cassettes ($83). Gr. 4–9. BKL (1/74), PRV (4/74). 915.2

Contents: 1. Tokyo: World's Largest City. 2. A Traditional Japanese Family. 3. Japan's Life from the Sea. 4. Silk Farming at Takatoya. 5. Nagasaki and Her Shipbuilders. 6. Okinawa: Keystone of the Pacific.

Photography, instructional maps, authentic art and music, along with the narration, picture Japanese people as they live and work in an organized and complex society. Japan is portrayed as both an industrial, progressive urban society and as a simple agricultural country deeply committed to tradition.

JAPAN—FICTION

REFLECTIONS: A JAPANESE FOLK TALE (Motion Picture—16mm—Sound). *See* Folklore—Japan

JAPAN—SOCIAL LIFE AND CUSTOMS

JAPAN: SPIRIT OF IEMOTO (Filmstrip—Sound). EBEC, 1974. 5 color, sound filmstrips, av. 79 fr. ea.; 5 discs or cassettes, 11 min. ea. #6908 Discs ($72.50), #6908K Cassettes ($72.50). Gr. 5–10. ESL (1976/77). 915.2

Contents: 1. The Spirit of Japan. 2. Iemoto. 3. The Japanese Family. 4. The Japanese at Work. 5. The Japanese at Play.

The unique blend of personality traits, temperament, and respect for tradition typified by the word *Japanese* and the life-style it evokes are the subjects of these filmstrips. The series illustrates the many seeming contradictions in the Japanese character and manages to unearth a few constants—a love of nature, respect for superiors, and subordination of the individual to the welfare of the group.

RITUAL (Motion Picture—16mm—Sound). Psychomedia Productions/Wombat Productions, 1977. 16mm color, sound film, 30 min. ($430). Gr. 7–A. BKL (11/1/78). 915.2

A provocative, visually compelling documentary about the ways the Japanese have survived as a people, linked to their traditions and each other, despite the demands of a highly industrialized society.

JAPANESE IN THE U.S.

PREJUDICE IN AMERICA—THE JAPANESE AMERICANS (Filmstrip—Sound). Multi-Media Productions, 1971. 4 color, sound filmstrips, 39–52 fr.; 4 cassettes or discs, 11–15 min.; Teacher's Manual. #7011R Discs ($29.90), #7011C Cassettes ($33). Gr. 8–12. PVB (9/72). 301.45

Covers the early contacts and attitude formations among Japanese and Westerners, the early immigration to the United States, the development of open and legal discrimination, the experiences of the detention camps, and the effect of this totality of experience on the Japanese American community today.

JARRELL, RANDAL

THE BAT POET (Phonodisc or Cassette). *See* Individuality—Fiction

JAZZ MUSIC

ART OF DAVE BRUBECK: THE FANTASY YEARS (Phonodiscs). Atlantic Records. 2 discs. #SD2-317 ($7.98). Gr. 3–A. PRV (10/75). 785.42

These recordings come from two early concerts. This is all pretime signature improvisation. The 11 selections include "Laura," "Stardust," and "Lullaby in Rhythm."

JAZZ MILESTONES (Filmstrip—Sound). Educational Enrichment Materials, 1976. 6 color, sound filmstrips, 52–66 fr.; 6 cassettes or discs, 11–13 min.; Teacher's Guide. #51006 C Discs or Cassettes ($108). Gr. 7–C. PVB (3/77). 781.57

Contents: 1. Storyville. 2. Chicago. 3. The Rent Party. 4. Swing. 5. Bop. 6. Soul Jazz.

Examines the stylistic variations and historical development of jazz, the great American form of musical expression. Musical excerpts help communicate a feeling for jazz music and the culture in which it thrived.

JEFFERSON, THOMAS

THE PICTORIAL LIFE-STORY OF THOMAS JEFFERSON (Kit). Davco Publishers, 1977 (Life Stories of Great Presidents). 4 color, sound filmstrips, av. 68 fr.; 4 cassettes, av. 13 min.; 1 Book; 1 Teacher's Guide. ($89). Gr. 5–11. BKL (12/15/77). 921

The filmstrips begin with the American Revolution and an effective reading of the Declaration of Independence. Through flashbacks, Jefferson's life is outlined.

THOMAS JEFFERSON (Motion Picture—16mm—Sound). Handel Film, 1966 (Americana Series # Three). 16mm color, sound film, 28 min.; Film Guide. ($360). Gr. 4–A. FLN (1967), LAA (1967), LFR (3/67). 921

Covers Jefferson as a farmer, student, scientist, inventor, architect, politician, and president. Highlights the accomplishments and versatility of this great American leader.

JENNINGS, BLANCHE

MORE SILVER PENNIES (Phonodisc or Cassette). *See* Poetry—Collections

JEWELRY

JEWELRY: THE FINE ART OF ADORN-
MENT, FABRICATION METHOD (Filmstrip—Sound). Warner Educational
Productions, 1976. 6 color, sound filmstrips,
69–107 fr.; 3 cassettes, 8–20 min.; Teaching
Guide. #370 ($105). Gr. 7–A. PRV (3/19/78).
739.27

The history of jewelry, its styles, and how to
do the fabrication method in creating beautiful places are presented in these strips.

JEWELRY: THE FINE ART OF ADORN-
MENT, LOST WAX METHOD (Filmstrip—Sound). Warner Educational
Productions, 1976. 6 color, sound filmstrips,
80–100 fr.; 3 cassettes, 14–19 min; Teaching
Guide. #335 ($105.50). Gr. 7–A, PRV (10/19/
77). 739.27

These strips teach the history of the role of
jewelry, its styles, and about contemporary
artisans. They show how to do wax castings,
make wax models, and other techniques associated with the art of the lost wax method.

JEWLERY—HISTORY

THE ART OF JEWELRY (Art Prints). Educational Dimensions Group, 1974. 20 color
slides; Teacher's Guide and Lecture Notes.
C944 ($30). Gr. 4–C. PRV (5/75), PVB (4/76).
739.27

Ancient and modern treasures in bone,
feather, gems, and gold are demonstrated as
the viewer examines jewelry produced by
many cultures.

JEWS—BIOGRAPHY

THE DIARY OF ANNE FRANK (Filmstrip—
Sound). *See* Autobiographies

JEWS IN THE U.S.

JEWISH IMMIGRANTS TO AMERICA
(Filmstrip—Sound). Sunburst Communica-
tions, 1974. 2 color, sound filmstrips, 73–74
fr.; 2 cassettes or discs, 13 min.; Teacher's
Guide. #207 ($59). Gr. 7–12. BKL (3/15/75),
INS (12/76), PRV (12/75). 301.45.
Contents: 1. A Difficult Task. 2. Contribu-
tions to America.
Examines the causes of Jewish emigration to
America and traces the effect on both the
emigrants themselves and society in general.

JEWS IN THE U.S.—FICTION

SUMMER OF MY GERMAN SOLDIER
(Phonodisc or Cassette). *See* World War,
1939–1945—Prisoners & prisons—Fiction

JOAN OF ARC

JOAN OF ARC (Cassette). Ivan Berg Associ-
ates/Jeffrey Norton Publishers, 1976 (His-

tory Makers). 1 cassette, approx. 56. min.
#41012 ($11.95). Gr. 10–A. BKL (1/15/78),
PRV (5/19/78). 921

Joan of Arc's brief life, her spectacular end,
and her unshakable belief in her ''voices''
have no parallel outside folklore.

JOBS. *See* Occupations; Vocational guidance

JOHN, ELTON

MEN BEHIND THE BRIGHT LIGHTS (Kit).
See Musicians

JOHNNY APPLESEED. *See* Chapman, John—
Fiction

JOHNSON, ANDREW

ANDREW JOHNSON COMES TO TRIAL
(Filmstrip—Sound). *See* U.S.—History—
1865–1898

JOHNSON, ANNABEL

THE GRIZZLY (Kit). *See* Bears—Fiction

JOHNSON, EDGAR

THE GRIZZLY (Kit). *See* Bears—Fiction

JONES, JOHN PAUL

FAMOUS PATRIOTS OF THE AMERICAN
REVOLUTION (Filmstrip—Sound). *See*
U.S. History—Biography

JOURNALISM

FREE PRESS: A NEED TO KNOW THE
NEWS (Kit). *See* Freedom of the press

NEWSPAPER STORY (SECOND EDITION)
(Motion Picture—16mm—Sound). *See*
Newspapers

SATIRICAL JOURNALISM (Cassette). *See*
Cartoons and caricatures

TV NEWS: BEHIND THE SCENES (Motion
Picture—16mm—Sound). *See* Television—
Production and direction

WHAT IS JOURNALISM? (Filmstrip—
Sound). Guidance Associates, 1977. 2 color,
sound filmstrips, av. 89 fr. ea.; 2 cassettes or
discs, av. 19 min. ea. #2B-500 411 Discs
($54.95), #2B-500 403 Cassettes ($54.95).
Gr. 7–A. PRV (1978), PVB (4/78). 070
Contents: 1. News Stories. 2. Print and
Broadcast Journalism.
Analyzes actual news stories, explaining
''straight'' news, specialized news, editorials, and features. Edwin Newman compares
the functions and problems of print and
broadcast journalism.

YOUR NEWSPAPER (Filmstrip—Sound).
See Newspapers

JUDGMENT

THE PSYCHOLOGY OF JUDGMENT (Filmstrip—Sound). Multi-Media Productions, 1977. 1 color, sound filmstrip, 62 fr.; 1 cassette, 16 min.; Teacher's Manual, Script. #7202C ($12.95). Gr. 9–12. MDU (1978). 153.46

Presents the mental process of report-inference-judgment through three example situations. Illustrates how the process is employed and the consequences of not using it to the fullest extent.

JUSTICE, ADMINISTRATION OF

CRIMINAL JUSTICE: TRIAL AND ERROR (Filmstrip—Sound). Current Affairs Films, 1978. 1 color, sound filmstrip, 74 fr.; 1 cassette, 16 min.; Guide. #611 ($24). Gr. 9–A. BKL (10/1/78). 347.371

A narrator outlines the inequities of America's criminal justice system. It is pointed out that the poor must be satisfied with the limited efforts of an overloaded public defender, but the rich can afford to hire an experienced attorney, and that more emphasis is placed on clearing the court calendar and getting through the case loads than on the effects of the proceedings upon the accused person's life. Does the fault lie in the system or in the nature of American society? How many of the charges against the system are warranted? Such questions are examined from the average citizen's standpoint.

WITH JUSTICE FOR ALL (Kit). EMC, 1976. 4 color, sound filmstrips, 96–103 fr.; 4 cassettes, 16–21 min.; Student Activities, Wall Chart, Teacher's Guide. #SS-217000 ($84). Gr. 10–12. BKL (6/15/77), CPR (2/77), PVB (5/77). 347.371

Contents: 1. What Is American Justice? 2. Does the System Work? 3. Decisions Affect Everyone. 4. Who Decides?

An examination of the system of courts and justice in the United States from historical development, present role, and to future alternatives. Mini-case studies explore the functions of the court and the far-reaching impact of judicial decisions.

JUVENILE DELINQUENCY

PROJECT AWARE (Motion Picture—16mm—Sound). See'n Eye Productions/Perennial Films, 1978. 16mm color, sound film, 29 min. ($555). Gr. 4–A. AUR (1978), NEF (5/78). 364.2

This film is a powerful statement on the prevention of juvenile delinquency. Using dramatic slides of the harsh realities of life inside prison walls, former convict David Crawford illustrates the end effects of petty thievery, bad associations, drugs, and alcohol. Workshops (with the film included) for juveniles, parents, and adult groups or agencies are available.

SHOPLIFTING, IT'S A CRIME (Motion Picture—16mm—Sound). See Crime

YOUTH AND THE LAW (Filmstrip—Sound). See Youth—Law and legislation

JUVENILE DELINQUENCY—FICTION

THE OUTSIDERS (Phonodisc or Cassette). Miller-Brody Productions, 1976 (Young Adult Recordings). 1 cassette or disc, 40 min.; Teacher's Notes. #YA403C Cassette ($7.95), #YA403 Disc ($6.95). Gr. 7–12. BKL (7/15/77), CRC (10/19/77), WLB (3/19/77). Fic

A recorded adaptation of S. E. Hinton's *The Outsiders* (Viking, 1976), a story of the inside violence, heroism, and hope of youth. It revolves around the rivalry of two teenage gangs and shows the loyalty, friendships, and hatred of their members.

KANE, PAUL

PAUL KANE GOES WEST (Motion Picture—16mm—Sound). See Indians of North America—Canada

KEATS, EZRA JACK

EZRA JACK KEATS (Motion Picture—16mm—Sound). Morton Schindel/Weston Woods Studios, 1970 (Signature Collecton). 16mm color, sound film, 17 min. #410 (245). Gr. 7–C. BKL (6/15/71), FLN (10/71). 921

In this informal film interview, Keats discusses the experiences that have influenced his work. He demonstrates how to make one type of collage material, talks about his early work, and conducts an informal tour of his New York neighborhood. The film concludes with complete motion pictures of *A Letter to Amy*.

KEATS, JOHN

JOHN KEATS: POET (Motion Picture—16mm—Sound). EBEC, 1973. 16mm color, sound film, 31 min. #47791 ($390). Gr. 9–C. BKL (1975), EFL (1974), EGT (1975). 921

Written by Archibald MacLeish, this film dramatizes the life of Keats from his early years in England until his death in Rome at the age of 26. Vividly portrayed are the tragic family deaths, the savage criticism, the money worries, the beautiful love affair, and the failing health that Keats endured and rose above in his devotion to his art. Excerpts from Keats's letters and poetry reveal the integrity of the man and the genius of the poet.

KELLOGG, STEVEN

HOW A PICTURE BOOK IS MADE (Filmstrip—Sound). See Books

THE ISLAND OF THE SKOG (Filmstrip—Sound). *See* Mice—Fiction

KENDALL, CAROL

THE GAMMAGE CUP (Filmstrip—Sound). *See* Fantasy—Fiction

KENNEDY (FAMILY)

THE KENNEDYS (Videocassette). Videotape Network/Jeffrey Norton Publishers, 1977. Videocassette, 3/4 in. U-Matic, 1/2 in. Betamax, or 1/2 in. EIAJ (b/w); 60 min. #71017 ($480). Gr. 9–A. BKL (4/1/78). 920

This program traces with original footage and family home movies the myths and realities of the Kennedy family from the beginnings of Joe Kennedy's career through Jack and Bobby to Teddy today. The activity and hope that the Kennedys generated are clearly shown, but so are the human qualities that made them unique figures in American history. Cliff Robertson narrates.

KENNEDY, JOHN FITZGERALD

PRESIDENTS AND PRECEDENTS (Kit). *See* Presidents—U.S.

KENYA

FOUR FAMILIES OF KENYA (Filmstrip—Sound). BFA Educational Media, 1974. 4 color, sound filmstrips; cassettes or discs. #VZ4000 Cassettes ($70), #VZ3000 Discs ($70). Gr. 4–8. BKL (12/1/74). 916.762

Contents: 1. Traditional Farming Family. 2. Tea Estate Family. 3. Nairobi Family. 4. Swahili Coast Family.

Shows that Kenya, like the rest of Africa, is composed of many established native tribal cultures meeting with foreign ones imported at various times in the past.

THE YEAR OF THE WILDEBEEST (Motion Picture—16mm—Sound). *See* Gnu

KEYES, DANIEL

FLOWERS FOR ALGERNON BY DANIEL KEYES (Filmstrip—Sound). *See* Mentally handicapped—Fiction

KIBBUTZ. *See* Israel

KIDNEYS

WORK OF THE KIDNEYS (SECOND EDITION) (Motion Picture—16mm—Sound). EBEC, 1972. 16mm color, sound film, 20 min. #3163 ($255). Gr. 9–C. PRV (1973). 612.4

Demonstrates the essential work of the kidneys in maintaining the fluid environment body cells must have.

KING, MARTIN LUTHER

KING: A FILMED RECORD, MONTGOMERY TO MEMPHIS (Videocassette). Videotape Network/Jeffrey Norton Publishers, 1970. Videocassette, b/w, 103 min. ($450). Gr. 9–A. BKL (7/1/78). 323.1

A documentary of the civil rights movement from the bus boycott in Montgomery, Alabama, that ended with an assassin's bullet in Memphis, Tennessee. Newsreel and television accounts of the events are used. Because there is no commentary, each action speaks for itself, while King's eloquent sermons clearly express the mood of the times. Bible readings and the poetry excerpts add flavor and successfully hold the events together.

MARTIN LUTHER KING JR.—THE ASSASSIN YEARS (Motion Picture—16mm—Sound). Chatsworth Film Distributors/Centron Films, 1978 (The Prizewinners). 16mm color, sound film, 26 min. #77530 ($440). Gr. 7–A. CFF (1978), IDF (1978). 921

The film blends dramatized sequences photographed in Montgomery, Alabama, with historical newsreel footage. It concentrates on the Sunday morning in 1955 when Martin Luther King, Jr. delivers a sermon that assures that America will never be quite the same again; Dr. King's leadership of the civil rights movement; the victories in the form of court decisions and new laws; and Dr. King's winning of the Nobel prize for peace. Four years later he falls to an assassin's bullet.

MARTIN LUTHER KING: THE MAN AND HIS MEANING (Filmstrip—Sound). Martin Luther King Jr. Foundation/Sunburst Communications, 1974. 1 color, sound filmstrip, approx. 70 fr.; 1 cassette or disc, approx. 12 min.; Teacher's Guide. #220 ($29). Gr. 6–12. CPR (2/77). 921

A historical document of Martin Luther King's leadership in the civil rights movement of the 1950s and 1960s. Through James Earl Jones's moving narration, King's inspiring speeches, and Coretta Scott King's explanation of her husband's nonviolent philosophy, the program vividly depicts the blacks' struggle to achieve their constitutional rights.

KIPLING, RUDYARD

KIM (Phonodisc or Cassette). *See* India—Fiction

THE WORLD OF JUNGLE BOOKS, SET ONE (Kit). *See* Animals—Fictions

THE WORLD OF JUST SO STORIES, SET ONE (Kit). *See* Animals—Fiction

KITCHEN UTENSILS. *See* Cookery

KLIBAN, B.

B. KLIBAN (Cassette). Tapes for Readers, 1978. 1 cassette, 23 min. #ART-005A ($10.95). Gr. 11–A. BKL (1/15/79). 741.5
Kliban, the popular cartoonist known best for his cat drawings, talks about cartooning as an art form, the work of Saul Steinberg, the artist, and cartooning as a career.

BARTLEBY (Motion Picture—16mm—Sound). *See* American fiction

KLONDIKE GOLD FIELDS—FICTION

CALL OF THE WILD (Filmstrip—Sound). *See* Dogs—Fiction

THE CALL OF THE WILD (Kit). *See* Dogs—Fiction

KNIEVEL, EVEL

NOT SO EASY (MOTORCYCLE SAFETY) (Motion Picture—16mm—Sound). *See* Motorcycles

KNIGHT, ERIC

LASSIE COME HOME (Phonodisc or Cassette). *See* Dogs—Fiction

KNOWLEDGE, THEORY OF

IDIOT . . . GENIUS? (Motion Picture—16mm—Sound). *See* Intellect

KOALA BEARS

KOALA BEAR (Film Loop—8mm—Silent). Walt Disney Educational Media, 1966. 8mm color, silent film loop, approx. 4 min. #62-5252L ($30). Gr. K–12. MDU (1974). 599.7
These marsupials are seen climbing and swinging through eucalyptus trees. Interesting close-up shots of the animals' faces are presented.

KÜBLER-ROSS, ELISABETH

AN AMERICAN SAMPLER (Filmstrip—Sound). *See* U.S.—Social life and customs

LIVING WITH DYING (Filmstrip—Sound). *See* Death

LSD. *See* Hallucinogens

LA FARGE, PETER

AS LONG AS THE GRASS SHALL GROW (Phonodiscs). *See* Folk songs, Indian

PETER LA FARGE ON THE WARPATH (Phonodiscs). *See* Folk songs, Indian

LABOR UNIONS

LABOR UNIONS: POWER TO THE PEOPLE (Filmstrip—Sound). Educational Enrichment Materials, 1977. 1 color, sound filmstrip, 80 fr.; 1 disc or cassette; Spirit Master; Teacher's Guide. #32102 ($25). Gr. 7–12. PRV (12/78). 331.88
The history and future of labor unions are examined in this filmstrip. Historical photographs are used and representatives of labor and management state their respective positions. Attention is given to technological changes, demographic changes, and multinational corporations on the union structure.

LABOR UNIONS: WHAT YOU SHOULD KNOW (Filmstrip—Sound). Guidance Associates, 1977. 2 color, sound filmstrips, av. 82 fr. ea.; 2 cassettes or discs, av. 15 min. ea.; Guide. #9A-102-770 Cassettes ($52.50), #9A-102-762 Discs ($52.50) Gr. 9–A. BKL (1/15/78), PRV (1977). 331.8
Contents: 1. Unions and Their Members. 2. Collective Bargaining.
Suggests practical information on apprenticeship, open and closed shops, grievance channels, arbitration, strike, boycott, and dues. Outlines primary union goals, roles of shop stewards, and other officers. Reviews levels of rank-and-file participation. Takes the viewer through collective bargaining procedures and presents issues from both the union's and management's points of view.

UNIONS AND YOU (Kit). Changing Times Education Service/EMC, 1976 (Career Directions). 2 color, sound filmstrips; 30 Booklets; 2 Spirit Masters; 2 Transparencies; Guide. #6045 ($89.50). Gr. 9–12. IFT (1977)
Contents: 1. Why Unions. 2. Unions at Work.
Helps to familiarize student with trade unionism, both as it affects society and the individual worker. Concepts developed include historical and current developments, union rights, jurisdictional disputes, unions and political power, alternatives to traditional unionism, worker participation in management, union membership, elections, dues, seniority, and grievance procedures.

LAMB, CHARLES

TALES FROM SHAKESPEARE (Phonodiscs). *See* English literature

LAMBERT, EVELYN

THE HOARDER (Motion Picture—16mm—Sound). *See* Fables

PARADISE LOST (Motion Picture—16mm—Sound). *See* Man—Influence on nature

LANGUAGE AND LANGUAGES

LANGUAGE—THE MIRROR OF MAN'S GROWTH (Filmstrip—Sound). *See* English language—History

SPEAKING OF LANGUAGE (Filmstrip—Sound). Guidance Associates, 1971 (Language Skills Series). 2 color, sound filmstrips, 75–85 fr.; 2 cassettes or discs, 12–13 min. #512-275 Discs ($52.50), #512-283 Cassettes ($52.50). Gr. 5–8. BKL (2/15/76), PRV (9/72), PVB (5/7?). 401

Contents: 1. What Is It? How Does It Work?

Builds conceptual grasp of linguistics. Program examines the universality of speech, history of language, components of grammar, relationship of language to thought.

LANGUAGE ARTS

COURAGE (Filmstrip—Sound). See Literature—Study and teaching

DAYBREAK (Motion Picture—16mm—Sound). See Nature in poetry

DRAMA (Filmstrip—Sound). See Literature—Study and teaching

FIRE! (Motion Picture—16mm—Sound). See Forest fires

FIRE MOUNTAIN (Motion Picture—16mm—Sound). See Volcanoes

FOLKLORE AND FABLE (Filmstrip—Sound). See Literature—Study and teaching

THE FUTURE (Filmstrip—Sound). See Literature—Study and teaching

HUMOR AND SATIRE (Filmstrip—Sound). See Literature—Study and teaching

MYTHOLOGY (Filmstrip—Sound). See Literature—Study and teaching

SHORT STORY (Filmstrip—Sound). See Literature—Study and teaching

UNDERSTANDING FICTION (Filmstrip—Sound). See Fiction

See also Creation (literary, artistic, etc.); English language; and its various divisions; Literature and its various forms, e.g., Poetry; Reading

LASERS

INTRODUCTION TO LASERS (Motion Picture—16mm—Sound). EBEC, 1973. 16mm color, sound film, 17 min. #3175 ($220). Gr. 9–A. AAS (1975), BKL (1974), SCT (1974). 535.5

The laser—its development, structure, and uses—is examined in conversation with three prominent scientists responsible for its discovery. Their explanations are illustrated by graphic laboratory demonstrations and animated sequences. Some of the many uses for lasers—eye surgery, measuring the distance to the moon, metal cutting, communications, even the laser as a typewriter eraser—are illustrated.

LATIN AMERICA

INTRODUCTION TO LATIN AMERICA (FIlmstrip—Sound). EBEC, 1973. 5 color, sound filmstrips, av. 78 fr. ea.; 5 cassettes or discs, 11 min. ea.; Teacher's Guide. Available in Spanish-English cassettes (6492K). #6474K Cassettes ($72.50). Gr. 3–9. AVI (11/75), PRV (5/74). 918

Contents: 1. Latin America: Its Land. 2. Latin America: Its History. 3. Latin America: Its People. 4. Latin America: Its Agriculture. 5. Latin America: Its Industry.

Why Latin America is still underdeveloped and how tradition affects the social, political, and economic lives of today's Latin Americans are illustrated in this set. Problems these people face and what their hopes are for the future are explained. A view that ranges from Mexico and Central America to the Caribbean Islands and all of South America is presented. The diversity of these lands stands out as students examine them from many vantage points: geographic, historical, cultural, and economic.

LATIN AMERICANS—BIOGRAPHY

HISPANIC HEROES OF THE U.S.A. (Kit). See Biography—Collections

LAW

JUSTICE AND THE LAW (Filmstrip—Sound). Pathescope Educational Media, 1973. 6 color, sound filmstrips, 50–63 fr.; 6 cassettes, 15–19 min.; Teacher's Guide. #513 ($100). Gr. 7–12. PRV (4/74), TSS (3/73). 347.9

Contents: 1. Freedom of Expression. 2. Freedom of Belief. 3. Freedom of the Press. 4. The Right to a Fair Trial; Due Process of Law. 5. Equal Protection of the Law. 6. Equal Opportunity under the Law.

Involves students in the legal process and demonstrates the constant quest for justice. Cases of historical importance are shown to indicate their relevancy today. The trial of Socrates, the Peter Zenger trial, and the Pentagon Papers trial are presented. Arguments for both plaintiff and defendant are given for cases without prejudice. Decisions are withheld so the student may decide the verdict.

LAW AND CRIME (Kit). See Criminal justice, Administration of

LAW AND JUSTICE: MAKING VALUE DECISIONS (Filmstrip—Sound). Pathescope Educational Media, 1973. 3 color, sound filmstrips; 3 cassettes; Teacher's Manual. #308 ($60). Gr. 4–8. STE (4/5/74). 340.1

Contents: 1. The Case of the Blue and White Whistle. 2. The Case of the Stolen Hub Cap. 3. The Case of the Boss' Son.

This set presents law as ideals, values, and fair play rather than a collection of rules, and

LAW (cont.)

involves students emotionally and intellectually in making value decisions.

LAW AND LAWMAKERS (Kit). EBEC, 1977 (Law and Society). 3 sound filmstrips; 2 cassettes; 10 Student Resource Readers; Teacher's Guide. #17024K ($93.50). Gr. 7–12. PRV (9/78). 340

Contents: 1. Law and Social Order. 2. Foundations of Law. 3. How Law Is Made.

These materials take the abstract concept we call "laws" and reduces it to terms young people can readily grasp. What is law? How does it affect our lives? By exploring questions such as these, students develop a useful understanding of law's origins, as well as its everyday applications. Two audiocassettes and related readings expand on basic concepts highlighted in the filmstrips.

LAW AND THE ENVIRONMENT (Kit). *See* Environment—Law and legislation

LAW AND SOCIETY

LAW AND CRIME (Kit). *See* Criminal justice, Administration of

LAW AND LAWMAKERS (Kit). *See* Law

LAW AND THE ENVIRONMENT (Kit). *See* Environment—Law and legislation

LAW ENFORCEMENT

CAREERS IN LAW ENFORCEMENT (Filmstrip—Sound). Associated Press/Pathescope Educational Media, 1973 (Careers In). 2 color, sound filmstrips, 71–76 fr.; 2 cassettes, 13 min. ea.; Teacher's Manual. #708 ($50). Gr. 9–12. NVG (1974), PRV (4/19/75). 363.2

Contents: Part 1. Part 2.

This program examines the challenges, dangers, functions, satisfactions, and drawbacks of the law enforcement profession. It describes the different jobs available in the field of law enforcement using interviews to determine necessary training, job satisfactions, and opportunities.

LAW AND SOCIETY: LAW AND CRIME (Kit). EBEC, 1977. 3 color, sound filmstrips, 96–105 fr.; 3 cassettes, 17–18 min.; 10 Student Books: *Reading in Law;* Teacher's Guide. #17012K ($93.50). Gr. 9–12. PRV (12/78). 363.2

Contents: 1. Crime and Punishment: Changing Views. 2. Criminal Justice: Goals and Challenges. 3. Criminal Punishment: Two Alternatives.

Various viewpoints of the criminal justice system are discussed. Individual cases are used to dramatize a point following through sometimes to the rehabilitation program.

YOUTH AND THE LAW (Filmstrip—Sound). *See* Youth—Law and legislation

LEACOCK, STEPHEN

GERTRUDE, THE GOVERNESS AND OTHER WORKS (Phonodisc or Cassette). *See* Canadian fiction

LEADERSHIP

THE ART OF MOTIVATION (Motion Picture—16mm—Sound). *See* Motivation (psychology)

WHO'S RUNNING THE SHOW? (Filmstrip—Sound). Lyceum/Mook & Blanchard, 1973. 1 color, sound filmstrip, 51 fr.; 1 cassette or disc, 6¹/₂ min.; Teacher's Guide. #LY35572C Cassette ($25), #LY35572R Disc ($19.50). Gr. 6–9. FLN (1973). 158

The importance of good leadership qualities is stressed, presenting examples of what our society could be like if there were no leaders to make decisions and think ahead. Among the leaders cited are judges, parents, and student council presidents.

LEAF, CAROLINE

HOW BEAVER STOLE FIRE (Motion Picture—16mm—Sound). *See* Indians of North America—Legends

LEARNING, PSYCHOLOGY OF

LEARNING AWAY FROM HOME (Kit). *See* Family life

LEARNING BEGINS AT HOME (Filmstrip—Sound). *See* Family life

LEARNING: CONDITIONING OR GROWTH? (Filmstrip—Sound). *See* Education—Aims and objectives

LEARNING IN THE HOME (kit). *See* Family life

LEARNING THROUGH PLAY (Kit). *See* Family life

THE PARENT AS A TEACHER (Kit). *See* Family life

LEATHER WORK

THE LEATHER-CRAFTER (Filmstrip—Sound). Warner Educational Productions, 1978. 4 color, sound filmstrips, 86–119 fr.; 2 cassettes, 12–15 min.; Teaching Guide. #933 ($78.50). Gr. 7–A. BKL (3/15/79). 745.53

From the selection of leather to the design and pattern making, we watch five projects develop step by step. Examples are shown as well as tips given on tools and supplies needed.

LEE, HARPER

TO KILL A MOCKINGBIRD (Kit). *See* Southern states—Fiction

LEGENDS

HEROIC ADVENTURES (Filmstrip—
Sound). Acorn Films, 1976. 4 color, sound
filmstrips, 51–75 fr.; 4 cassettes, 9–13 min.;
Guide. ($65). Gr. 3–8. BKL (4/15/77). 398.2
Contents: 1. Gilgamesh and the Monster in
the Woods. 2. Hercules and the Golden
Apple. 3. Arthur and the Magic Sword. 4.
Robin Hood and the Poor Knight.
Exciting excerpts from the adventures of
four legendary heroes.

THE LEGEND OF THE MAGIC KNIVES
(Motion Picture—16mm—Sound). See Indians of North America—Legends

LEGENDS—GREAT BRITAIN

BEOWULF AND THE MONSTERS (Cassette). See Beowulf

LEGENDS—U.S.

THE LEGEND OF SLEEPY HOLLOW (Motion Picture—16mm—Sound). Bosustow
Productions, 1972 (American Folktales).
16mm color, sound, animated film, 14 min.;
Reader's Guide. ($18). Gr. 4–A. BKL (7/1/
72), NEA (1973), PRV (10/72). 398.2
Presents the legend of Ichabod Crane and his
courtship of Katrina Van Tassel, the only
daughter of a prosperous Dutch farmer. Ichabod's rival, Brom Bones, exploits Ichabod's belief in ghosts. Returning from a
gala at the farmer's home. Ichabod encounters the dreaded Headless Horseman. In a
wild chase, the ghostly rider hurls his head at
poor Ichabod, who mysteriously disappears
and is never seen again in Sleepy Hollow.
Narrated by John Carradine.

THE LEGEND OF SLEEPY HOLLOW
AND ICHABOD CRANE (Phonodisc or
Cassette). Caedmon Records, 1967. 1 cassette or disc, approx. 60 min. #TC1242 Disc
($6.98), #CDL51242 Cassette ($7.95). Gr. 5–
A. EGT, NYR, SLJ. Fic
Contents: 1. The Legend of Sleepy Hollow.
2. Ichabod Crane and the Headless Horseman.
The Legend of Sleepy Hollow, written by
Washington Irving, is his attempt to receive
some degree of comfort and delight from his
youthful scenes. This story is a combination
of Dutch folklore and personal boyhood experiences and remembrances. It is narrated
by Ed Begley.

LEGEND OF SLEEPY HOLLOW AND
OTHER STORIES (Cassette). Jabberwocky
Cassette Classics, 1972. 1 cassette, 60 min.
#1111 ($7.98). Gr. 5–12. BKL (3/15/74), JOR
(11/74). 398.2
Contents: 1. Legend of Sleepy Hollow. 2.
Celebrated Jumping Frog. 3. The Bride
Comes to Yellow Sky.

This dramatization faithfully captures the
spirit of a trio of famous American short stories. Without changing the structure or quality of the original story, each has been
condensed. Sound effects and brief musical
interludes add to the dramatization.

THE LOON'S NECKLACE (Motion Picture—16mm—Sound). See Indians of North
America—Legends

LEGISLATION

FOR ALL THE PEOPLE (Kit). See U.S. Congress

STATE GOVERNMENT—RESURGENCE
OF POWER (Motion Picture—16mm—
Sound). See State governments

LEISURE

LEISURE (Motion Picture—16mm—Sound).
See Man

LEMMINGS

LEMMING MIGRATION (Film Loop—
8mm—Silent). Walt Disney Educational Media, 1966. 8mm color, silent film loop, approx. 4 min. #62-5367L ($30). Gr. K–12.
MDU (1974). 599.3
Thousands of lemmings are observed during
their disastrous migration toward the sea.
The migration is climaxed as the lemmings
cast themselves over the edge of a cliff

LENSES

A LIGHT BEAM NAMED RAY (Motion Picture—16mm—Sound). See Color

A LETTER TO AMY

EZRA JACK KEATS (Motion Picture—
16mm—Sound). See Keats, Ezra Jack

LEVINE, DAVID

SATIRICAL JOURNALISM (Cassette). See
Cartoons and caricatures

LEVINE, JACK

JACK LEVINE (Motion Picture—16mm—
Sound). Texture Films, 1966. 16mm color,
sound film, 23 min. ($290). Gr. 10–A. AFF
(1967). 759.13
An at-work profile of one of America's leading artists. Painter Jack Levine combines
powerful social and political content with
classic painterly craft.

LEWIS, SINCLAIR

FIVE MODERN NOVELISTS (Filmstrip—Sound). *See* American literature—History and criticism

LIBERTY. *See* Freedom

LIBRARIES—REFERENCE SERVICE

THE ACTION PROCESS (Filmstrip—Sound). *See* Consumer education

LIBRARIES—U.S.

LIBRARY OF CONGRESS (Motion Picture—16mm—Sound). *See* Library of Congress

LIBRARY OF CONGRESS

LIBRARY OF CONGRESS (Motion Picture—16mm—Sound). EBEC, 1969. 16mm color, sound film, 23 min. #2775 b/w ($150), #2776 color ($290). Gr. 9–A. BKL (1973), PRV (1979), SLJ (1971). 027.5753

A working library for government; a storehouse of historical documents; a cultural institution that commissions chamber music and plays it in concert; a living, breathing agency that affects every school and public library in the country.

LIBRARY SKILLS

BIOGRAPHY: BACKGROUND FOR INSPIRATION (Filmstrip—Sound). *See* Biography

DEWEY DECIMAL CLASSIFICATION (Filmstrip—Sound). *See* Classification, Dewey Decimal

INDEXES (Filmstrip—Sound). *See* Indexes

LIBRARY SKILL BOX (Cassette). Troll Associates, 1975. 10 cassette tapes, av. 13 min. ea.; 50 Spirit Masters, Teacher's Guide. ($96). Gr. 4–10. BKL (9/76). 028.7

Contents: 1. How to Use Your Library. 2. What Is a Book? 3. How to Find Reference Material. 4. How to Use a Dictionary. 5. How to Use an Encyclopedia. 6. How to Use an Atlas. 7. How to Use Magazines and Newspapers. 8. How to Research and Write a Report. 9. How to Study. 10. How to Give a Report.

Tapes especially valuable for individual student use; this set covers all facets of library use. The duplicating masters are for reviewing the tapes and for expanding the student's awareness through visual reinforcement of supplementary information.

MEDIA: RESOURCES FOR DISCOVERY (Filmstrip—Sound). *See* Audiovisual materials

LIFE

ECHOES (Motion Picture—16mm—Sound). Guidance Associates/Xerox Educational Publications, 1974. 16mm color, sound film; Teaching Guide. #7704 Film ($210), #7704 Videocassette ($140). Gr. 7–9. CGE (1974), CIF (1974). 128

A sensitive film about life and death—growing up. As 11-year-old Ellen learns about an 11-year-old who died in 1883, she realizes that she is connected to her ancestors and to the lives of those who will follow her

LIFE TIMES NINE (Motion Picture—16mm—Sound). Insight Media Programs/Pyramid Films, 1974. 16mm color, sound film, 15 min. ($225). Gr. 5–A. AFF (1974), BKL (7/1/74). 128

A compilation of brief "commercials for life" made by nine youngsters between the ages of 11 and 16. The directional control of each "commercial" was assumed by the author who created each idea. Two professionals assisted the productions, but final responsibility was with the individual youngsters.

SPINNOLIO (Motion Picture—16mm—Sound). *See* Human behavior

LIFE (BIOLOGY)

ANIMAL LIFE SERIES, SET ONE (Film Loop—8mm—Silent). *See* Growth

THE OTHER WORLD (Motion Picture—16mm—Sound). Arthur Mokin Productions, 1977. 16mm color, sound film, 19 min. ($315). Gr. 4–A. AAS (12/78), BKL (4/15/78), PRV (5/78). 574

Underwater photography takes the viewer up the biological ladder from microscopic plants and animals to the otter, a mammalian bridge between land and water. The lesson is that in a balanced ecology the materials of life and death are constantly recycled. There is no waste and no one creature is more important than another.

LIFE ON OTHER PLANETS

THE SEARCH FOR EXTRATERRESTRIAL LIFE: IS ANYBODY OUT THERE? (Filmstrip—Sound). Harper and Row Publishers/Sunburst Communications, 1976. 6 color, sound filmstrips, 83–93 fr. ea.; 6 cassettes or discs, 14–16 min. ea.; Teacher's Guide. #78 ($145). Gr. 7–C. BKL (6/1/77), LGB (1977), TSS (5/77). 574.999

Contents: 1. The Endless Quest. 2. Life: Rare or Commonplace? 3. Finding Habitable Planets. 4. Have We Been Visited? 5. The Odds of Intelligence. 6. Will Anybody Answer?

Studies pivotal theories and experiments of astronomers, chemists, biologists, and geologists regarding the conditions necessary for life. Considers the possibility that these conditions may exist elsewhere in the universe.

LIFE-STYLES—U.S.

LIFESTYLES: OPTIONS FOR LIVING (Kit). Butterick Publishing, 1976 (Independent Living Series). 4 color, sound filmstrips, 67–88 fr.; 4 cassettes or discs, 8–11 min; 12 Duplicating Masters; Teacher's Guide. #C-510-5B Cassettes ($85), #R-509-5B Discs ($85). Gr. 9–12. BKL (12/15/76), FHE (9/76). 301.44

Contents: 1. Life-styles. 2. Values, Goals, and Decisions. 3. Changing Lifestyles. 4. Lifestyles of Bygone Eras.

This set stresses the parental, peer, and experiential influences that act on the individual in the process of making decisions on what to be, what to do, and how best to achieve personal goals. It stresses the need to "know yourself" before selecting a lifestyle.

THE QUALITY OF LIFE IN THE UNITED STATES (Kit). *See* U.S.—Social life and customs

LIFESAVING

CPR: TO SAVE A LIFE (Motion Picture—16mm—Sound). *See* First aid

CHOKING: TO SAVE A LIFE (Motion Picture—16mm—Sound). *See* First aid

RESCUE SQUAD (Motion Picture—16mm—Sound). *See* Rescue work

LIGHT

LEARNING ABOUT LIGHT: SECOND EDITION (Motion Picture—16mm—Sound). EBEC, 1976 (Introduction to Physical Science). 16mm color, sound film, 15 min. #3385 ($220). Gr. 4–9. BKL (12/1/76), INS (12/77), LFR (9/10/76). 538

A series of engrossing demonstrations reveals properties of light that account for human ability to harness it to our needs. Students discover how light travels; why light rays bend; and why the same primary colors produce white when light waves are mixed, but near black when paint pigments are blended.

A LIGHT BEAM NAMED RAY (Motion Picture—16mm—Sound). *See* Color

LINCOLN, ABRAHAM

PRESIDENTS AND PRECEDENTS (Kit). *See* Presidents—U.S.

LIONS

ELSA BORN FREE (Filmstrip—Sound). Benchmark Films, 1978 (Joy Adamson's Af-

rica). 3 color, sound filmstrips, 82–104 fr.; 3 cassettes, 13–16 min. ($80). Gr. 4–A. BKL (7/1/78), MED (5/78), PRV (9/78). 599.7

True story of friendship that unfolded between Joy and George Adamson and lioness Elsa from the time she was a cub.

LION (Motion Picture—16mm—Sound). EBEC, 1971 (Silent Safari Series). 16mm color, sound, nonnarrated film, 11 min.; Teacher's Guide. #3123 ($150). Gr. 2–9. BKL (10/1/76). 599.7

Close-up photography shows the lion as he eats, sleeps, and plays with female and cubs in his natural habitat.

LION: MOTHER AND CUBS (Film Loop—8mm—Silent). Walt Disney Educational Media, 1966. 8mm color, silent film loop, approx. 4 min. #62-5202L ($30). Gr. K–12. MDU (1974). 599.7

The lion cubs are shown nursing, playing, and being carried about by their mother. Close-up of mother carrying cub by the scruff of the neck, and cubs viewed in mock battles.

PRIDE OF LIONS (Film Loop—8mm—Silent). Walt Disney Educational Media, 1966. 8mm color, silent film loop, approx. 4 min. #62-5201L ($30). Gr. K–12. MDU (1974). 599.7

An African lion family is shown as it actually lives in its natural habitat. Close-up view of them climbing trees, yawning, scratching, and relaxing in the shade.

LIONS—FICTION

THE HAPPY LION SERIES, SET ONE (Filmstrip—Sound). University Films/McGraw-Hill Films, 1971. 4 color, sound filmstrips, av. 40 fr. ea.; 4 cassettes or discs, 11 min. ea.; Teacher's Guide. #102180 Cassettes ($65), #103058 Discs ($65). Gr. 3–A. PRV (11/1/73). Fic

Contents: 1. The Happy Lion. 2. The Happy Lion's Vacation. 3. The Happy Lion in Africa. 4. The Happy Lion Roars.

Roger Duvoisin's illustrations are used to bring to life the adventures of the enchanting French lion. Narrated by Charles Duval. Based on the series of stories by Louise Fatio.

THE HAPPY LION SERIES, SET TWO (Filmstrip—Sound). University Films/McGraw-Hill Films, 1971. 4 color, sound filmstrips, av. 40 fr. ea.; 4 discs or cassettes, 11 min. ea.; Teacher's Guide. #103067 Discs ($65), #102185 Cassettes ($65). Gr. 3–A. BKL (11/1/73), PRV (9/10/73). Fic

Contents: 1. The Happy Lion's Quest. 2. The Three Happy Lions. 3. The Happy Lion and the Bear. 4. The Happy Lion's Treasure.

Roger Duvoisin's illustrations are used to bring to life the adventures of the enchanting

LIONS—FICTION (cont.)

French lion. Narrated by Charles Duval. Based on the series of stories by Louise Fatio.

LIPSYTE, ROBERT

THE CONTENDER BY ROBERT LIPSYTE (Filmstrip—Sound). *See* Boxing—Fiction

LISTENING

LISTENING BETWEEN THE LINES (Motion Picture—16mm—Sound). Alfred Higgins Productions, 1975. 16mm color, sound film, 16 min.; Guide. ($240). Gr. 7–12. LFR (9/10/75), PRV (3/76). 152.1

This treatment of listening skills presents situations that challenge the ability to listen effectively. It explores listening obstacles, such as distractions and personal prejudices. It also examines issues of interpretation and evaluation.

SUPER THINK PROGRAM (Cassette). *See* Sports—Biography

LITERATURE

AN AUDIO VISUAL HISTORY OF EUROPEAN LITERATURE (Filmstrip—Sound). Educational Audio Visual, 1978. 6 color, sound filmstrips, 64–70 fr. ea.; 6 cassettes or discs, 13–16 min. ea.; Guide. #T7KF-0037 Cassettes ($128), #T7RF-0037 Discs ($128). Gr. 10–A. BKL (10/15/79). 809

Contents: 1. The Ancient World. 2. The Middle Ages. 3. The Renaissance. 4. Reason and Enlightenment. 5. Romanticism and Realism. 6. The Beginnings of Modern Literature.

This set tries to answer perplexing questions that are woven through European literature, but can be traced through history. Art masterpieces are used to illustrate the philosophy and events of the times discussed as well as presenting appropriate writings that follow the selected themes.

LITERATURE—COLLECTIONS

EPISODES FROM CLASSIC NOVELS (Filmstrip—Sound). EBEC, 1976. 6 color filmstrips, av. 75 fr. ea.; 6 discs or cassettes, 18 min. ea.; #6959K Cassettes ($86.95), #6959 Discs ($86.95). Gr. 5–12. BKL (6/1/77), EGT (1/77), RTE (2/78). 807

Contents: 1. Scenes from *Moby Dick* by Herman Melville. 2. Scenes from *The Red Badge of Courage* by Stephen Crane. 3. Scenes from *Pride and Prejudice* by Jane Austen. 4. Scenes from *Adventures of Sherlock Holmes* by A. Conan Doyle. 5. Scenes from *O Pioneers!* by Willa Cather. 6. Scenes from *Great Expectations* by Charles Dickens.

Emotions are the driving force of great literature. The people, the situations, the times that evoke these emotions are portrayed in these filmstrips as each dramatizes just enough of its story to create suspense and arouse interest. The filmstrips are designed to motivate students to read more classic literature.

EPISODES FROM FAMOUS STORIES (Filmstrip—Sound). EBEC, 1973. 6 color, sound filmstrips, av. 72 fr. ea.; 6 discs or cassettes, 13 min. ea.; Teacher's Guide. #6477 Discs ($86.95), #6477K Cassettes ($86.95). Gr. 3–10. PRV (9/74). 808.83

Contents: 1. Highlights from *Robinson Crusoe*. 2. Highlights from *The Adventures of Tom Sawyer*. 3. Highlights from *Treasure Island*. 4. Highlights from *The Legend of Sleepy Hollow*. 5. Highlights from *Heidi*. 6. Highlights from *Alice's Adventures in Wonderland*.

Colorful illustrations, faithful-to-original-text narrations, near-perfect voice characterizations, and exciting sound effects introduce children and young teenagers to the high points in the plots of six outstanding classics in literature.

I COULDN'T PUT IT DOWN: HOOKED ON READING (Kit). Center for Humanities, 1977. 2 units with 154 slides; 2 discs or cassettes; Teacher's Guide; 4 Activity Books; 4 Paperbacks; 4 Desk Copies. #0335 ($139.50). Gr. 5–9. BKL (9/1/78). PRV (1/79). 808.83

A reading motivation kit with four "monster" stories re-created in comic-style format. Selections from *Frankenstein* by Mary Shelley and *The Strange Case of Dr. Jekyll and Mr. Hyde* by Robert Louis Stevenson are in Part I. In Part II the focus is on *Dracula* by Bram Stoker and *The Invisible Man* by H. G. Wells. Enough of each story is given so that students will want to read the entire story.

LIBRARY 3 (Cassette). Jabberwocky Cassette Classics, 1975. 6 cassettes; 6 Scripts; 1 Guide. #1003C ($60). Gr. 3–8. BKL (10/15/75), PRV (9/75), TEA (5/6/75). 808.83

Contents: 1. *Aladdin; or the Wonderful Lamp*. 2. *Heidi* by Johanna Spyri. 3. *The Merry Adventures of Robin Hood* by Howard Pyle. 4. *Oliver Twist* by Charles Dickens.

Adaptations of original stories are acted out as dramas with a full cast of characters, music, and sound effects.

MEDIA CLASSIC ADAPTATIONS (Filmstrip—Sound). Prentice-Hall Media, 1978. 5 color, sound filmstrips, 60–72 fr.; 5 cassettes, 14–15 min.; Teacher's Guide. ($115). Gr. 7–12. PRV (2/70). 808.83

Contents: 1. "The Black Cat," Edgar Allan Poe. 2. "The Lodger," Marie Belloc-Lownde. 3. "The Man without a Country," Edward Everett Hale. 4. "Bartleby the

LITERATURE—COLLECTIONS (cont.)

Scrivener,`` Herman Melville. 5. ``Salomy Jane`s Kiss,`` Bret Harte.

Original artwork is used with 40-year-old radio play recordings of these short stories.

TEN TALES OF MYSTERY AND TERROR (Cassette). Troll Associates, 1973. 10 cassette tapes. ($65). Gr. 5–12. PRV (1/76). 808.83

Contents: 1. *Fall of the House of Usher*. 2. *Dr. Jekyll and Mr. Hyde*. 3. *Hound of the Baskervilles*. 4. *Journey to the Center of the Earth*. 5. *Lost World*. 6. *Mysterious Island*. 7. *The Pit and the Pendulum*. 8. *Time Machine*. 9. *20,000 Leagues under the Sea*. 10. *War of the Worlds*.

These cassettes paraphrase but retain the mood and plots of Poe, Verne, and Doyle, stimulating interest in reading the original.

LITERATURE—STUDY AND TEACHING

ADVENTURE AND SUSPENSE (Filmstrip—Sound). Scholastic Book Services, 1976 (Scholastic Literature Filmstrips). 4 color, sound filmstrips, av. 94–111 fr. ea.; 4 cassettes or discs, av. 12–15 min. ea.; guide. #05334 Cassettes ($79.50), #05324 Discs ($79.50). Gr. 7–12. M&M (4/76). 807

Contents: 1. ``The Adventurers.`` 2. ``The Case of the Feathersbury Diamond.`` 3. ``The Tell-Tale Heart.`` 4. ``The Whole Town's Sleeping.``

A theme that runs through the history of literature and has endless attraction to readers is adventure and suspense. This set provides the basis for understanding the major kinds of adventure/suspense fiction.

THE CALL OF THE WILD (Kit). *See* Dogs—Fiction

CANNERY ROW, LIFE AND DEATH OF AN INDUSTRY (Filmstrip—Sound). *See* Monterey, California—History

COURAGE (Filmstrip—Sound). Scholastic Book Services, 1976 (Scholastic Literature Filmstrips). 4 color, sound filmstrips, 94–111 fr. ea.; 4 cassettes or discs, 12–15 min. ea.; Guide. #05349 Cassettes ($79.50), #0539 Discs ($79.50). Gr. 9–12. M&M (4/76). 807

Contents: 1. *Voices of Courage*. 2. *Red Badge of Courage*. 3. *Escape to Freedom*. 4. *Almost Brave*.

These strips dramatize both courage in real life and famous fictional interpretations.

DRAMA (Filmstrip—Sound). Scholastic Book Services, 1976 (Scholastic Literature Filmstrips). 4 color, sound filmstrips, av. 94–111 fr. ea.; 4 cassettes or discs, 12–16 min. ea.; Guide. #95274 Cassettes ($79.50), #05264 Discs ($79.50). Gr. 9–12. M&M (4/76), PRV (12/76). 807

Contents: 1. The Playwright. 2. The Actor. 3. The Designer. 4. The Director and Opening Night.

These filmstrips encourage students to participate in the production of a play—from script to opening night.

EPISODES FROM CLASSIC NOVELS (Filmstrip—Sound). *See* Literature—Collections

FOLKLORE AND FABLE (Filmstrip—Sound). Scholastic Book Services, 1976 (Scholastic Literature Filmstrips). 4 color, sound filmstrips, av. 94–111 fr. ea.; 4 cassettes or discs, 12–16 min. ea.; Guide. #05274 Cassettes ($79.50), #5264 Discs ($79.50). Gr. 9–12. M&M (4/76), PRV (12/76). 807

Contents: 1. Magic and Superstition. 2. Pecos Bill. 3. Dracula. 4. Folk Music.

This set acquaints students with such folk forms as stories, tales, magic, superstitions, and songs.

THE FUTURE (Filmstrip—Sound). Scholastic Book Services, 1976 (Scholastic Literature Filmstrips). 4 color, sound filmstrips, av. 94–111 fr. ea.; 4 cassettes or discs, av. 12–15 min. ea.; Guide. #05319 Cassettes ($79.50), #05309 Discs ($79.50). Gr. 9–12. M&M (4/76). 807

Contents: 1. Things to Come. 2. Science Fiction. 3. A Martian Chronicle. 4. Brave New World.

Science fiction writers of the past and present introduce the world of tomorrow. Imaginative themes, inventive style, and fantastic visuals combine to stimulate students to create their own science fiction stories as well as to enjoy those already published.

THE HOBBIT (Filmstrip—Sound). *See* Fantasy—Fiction

HUMOR AND SATIRE (Filmstrip—Sound). Scholastic Book Services, 1976 (Scholastic Literature Filmstrips). 4 color, sound filmstrips, av. 94–111 fr. ea.; 4 cassettes or discs, 12–15 min. ea.; Guide. #05259 Cassettes ($79.50), #05249 Discs ($79.50). Gr. 7–12. M&M (4/76). 807

Contents: 1. What Makes Us Laugh? 2. Mark Twain. 3. Interview with Jules Feiffer. 4. The Mechanization of Man.

Explores the many uses and forms of humor and satire in life and literature.

MYTHOLOGY (Filmstrip—Sound). Scholastic Book Services, 1976 (Scholastic Literature Filmstrips). 4 color, sound filmstrips, 94–111 fr. ea.; 4 discs or cassettes, 15 min. ea.; Teacher's Guide. #05244 Cassettes ($79.50), #05234 Discs ($79.50). Gr. 9–12. M&M (4/76), PRV (12/76). 807

Contents: 1. The New People. 2. The Ballad of King Arthur. 3. Mount Olympus. 4. Orpheus and the Underworld.

These filmstrips provide four new points of view on an old subject. Students will discov-

LITERATURE—STUDY AND TEACHING (cont.)

er how myths evolve from the lives of real people, understand some of the forms the mythic experience may take, and create some myths of their own.

READING SCIENCE FICTION (Filmstrip— Sound). *See* Science fiction—History and criticism

SHORT STORY (Filmstrip—Sound). Scholastic Book Services, 1976 (Scholastic Literature Filmstrips). 4 color sound filmstrips, 94–111 fr. ea.; 4 cassettes or discs, 12–15 min. ea.; Guide. #05304 Cassettes ($79.50), #05294 Discs ($79.50). Gr. 7–12. M&M (4/76). 807

Contents: 1. The Short Story. 2. Character. 3. Plot. 4. Mood.

The styles of O. Henry, Faulkner, Poe, Hawthorne, Shirley Jackson, and others are dramatized to make the essential elements of short story writing exciting and easy to understand.

UNDERSTANDING FICTION (Filmstrip— Sound). *See* Fiction

LOBBYING

LOBBYING: A CASE HISTORY (2ND EDITION) (Motion Picture—16mm—Sound). EBEC, 1977. 16mm Color, sound film, 18 min. #3543 ($255). Gr. 7–12. LFR (1978). 328

Viewers see lobbying in action as the interest groups involved mobilize support for their cause and attempt to influence Washington legislators. An ERA rally and an interview with John Gardner, founder of Common Cause, a consumer lobbying agency, illustrate another vital form of lobbying—citizen action.

LOGIC

LET`S LOOK AT LOGIC (Filmstrip— Sound). Guidance Associates, 1977. 2 color, sound filmstrips, 60–78 fr.; 2 cassettes or discs, av. 15 min. ea.; Teacher`s Guide, Activity Cards. #1B-304-665 Discs ($59.50), #1B-304-673 Cassettes ($59.50). Gr. 4–8. PRV (1/78), PVB (4/78). 130

Introduces simple aspects of formal logic. Relates common-sense problem-solving techniques to all elementary subjects. A trip to ``Logicland`` sets the stage for learning about Venn diagrams; helps discover precise meanings of *true, valid, possible, any, some, all,* and *follow from.*

LONDON

THE ISLAND NATION CAPITALS (Filmstrip—Sound). *See* Cities and towns

LONDON, JACK

CALL OF THE WILD (Filmstrip—Sound). *See* Dogs—Fiction

THE CALL OF THE WILD (Kit). *See* Dogs—Fiction

JACK LONDON CASSETTE LIBRARY (Cassette). *See* Short stories

LONGFELLOW, HENRY WADSWORTH

NINETEENTH CENTURY POETS (Filmstrip—Sound). *See* American literature—History and criticism

PAUL REVERE`S RIDE AND HIS OWN STORY (Cassette). *See* American poetry

LOOMS

HORIZONTAL BELT LOOM (Film Loop—8mm—Silent). Thorne Films/Prentice-Hall Media, 1972 (Southwest Indians). 8mm color, silent film loop, approx. 4 min. #HAT 286 ($26). Gr. 4–A. MDU (1978). 746.1

The horizontal belt loom used by Southwest Indians is demonstrated.

UPRIGHT LOOM (Film Loop—8mm—Silent). Thorne Films/Prentice-Hall Media (Southwest Indians). 8mm color, silent film loop, approx. 4 min. #HAT 285 ($28). Gr. 4–A. AUR (1978). 746.1

This film loop illustrates the use of the upright loom by Indians of the Southwest.

LOPEZ, CLAUDE-ANN

THE FOUNDING FATHERS IN PERSON (Cassette). *See* U.S.—History—Biography

LOVE

PARTNERS (Kit). Scholastic Book Services, 1975 (Family Living Program). 3 color, sound filmstrips, 84–102 fr.; 2 cassettes or discs, 10–13 min.; 9 Ditto Masters; Student Worksheets; Teacher`s Guide. #1124 Cassettes ($69.50), #1114 Discs ($69.50). Gr. 10–A. BKL (10/1/75). 301.42

Contents: 1. Love, All You Need Is. . . . 2. A Friend Is. . . . 3. Marriage. Five Views.

The subject is partners, dealing with lovers, friends, or spouses. Added to interviews are such elements as quotations by Cicero, Lincoln, and others.

VALUES FOR DATING (Filmstrip—Sound). *See* Dating (social customs)

LOVOOS, JANICE

DESIGN IS A DANDELION (Filmstrip— Sound). *See* Design, Decorative

LUNAR GEOLOGY

THE MOON: A GIANT STEP IN GEOLOGY (Motion Picture—16mm—Sound). Media-Four Productions/EBEC, 1976. 16mm color, sound film, 24 min. #3376 ($325). Gr. 7–12. AAS (1977), NEF (1977). 559.9

This film is a documentation of the first lunar landing and rock samples brought back by astronauts. Scientists at Houston's Lunar Receiving Laboratory are shown analyzing clues to a once-secret past. Animated segments illustrate the origins of lunar topography. Other segments show how moon rocks provide clues to the origins of the earth.

LUTHER, MARTIN

THE REFORMATION: AGE OF REVOLT (Motion Picture—16mm—Sound). *See* Reformation

MCCLINTOCK, HARRY K.

HAYWIRE MAC (Phonodiscs). *See* Folk songs

MCDERMOTT, GERALD

ANANSI THE SPIDER (Motion Picture—16mm—Sound). *See* Folklore—Africa

ARROW TO THE SUN (Motion Picture—16mm—Sound). *See* Indians of North America—Legends

THE MAGIC TREE (Motion Picture—16mm—Sound). *See* Folklore—Africa

MACDERMOTT, GERALD. *See* McDermott, Gerald

MACK, KAREN

INTERN: A LONG YEAR (Motion Picture—16mm—Sound). EBEC, 1972. 16mm color, sound film, 20 min. #3115 ($225). Gr. 5–C. BKL (1974), M&M (1975), PRV (1975). 921

Philadelphia General Hospital is understaffed and underequipped. The film is a candid look at a dedicated intern whose long work weeks are not long enough. Dr. Mack's patients are society's dispossessed: the homeless, the jobless, and those suffering from malnutrition, cancer, heart disease, and alcoholism. Viewers witness the full scope of her activities at the hospital and at home, from treating a drug overdose to cooking to delivering a baby.

MACLEISH, ARCHIBALD

JOHN KEATS: POET (Motion Picture—16mm—Sound). *See* Keats, John

MACRAME

CREATIVE MACRAME (Filmstrip—Sound). Warner Educational Productions, 1972. 2 color, sound filmstrips, 58–72 fr.; 1 cassette, 14 min.; Teaching Guide. #275 ($47.50). Gr. 7–A. BKL (5/1/74), PRV (9/75). 746.43

A practical approach to learning this skill. Nine popular knots are shown, along with a description of materials and supplies needed to plan and construct projects. Such projects as hanging flower pots and wall hangings demonstrate the finished works.

MAGNETISM

LEARNING ABOUT MAGNETISM: SECOND EDITION (Motion Picture—16mm—Sound). EBEC, 1975 (Introduction to Physical Science). 16mm color, sound film, 14 min. #3396 ($220). Gr. 4–9. BKL (2/76), PRV (1/76). 538

Magnetism, its mysteries, and its practical applications are clearly explored. The subject of magnetic fields is dramatized with scenes of an 83-ton electromagnet and a magnetic field so huge that its entire expanse can be viewed only from a helicopter.

MAKEUP, THEATRICAL

DICK SMITH, MAKE-UP ARTIST (Motion Picture—16mm—Sound). Texture Films, 1976. 16mm color, sound film, 18 min. ($250). Gr. 9–A. BKL (2/1/78), CGE (1978), IFT (1978). 791.43

Dick Smith has an exceptional career, and he loves it. A top Hollywood makeup man, he is a specialist. He transformed Dustin Hoffman into a 90-year-old man in *Little Big Man*, made up Marlon Brando for *The Godfather*, has created faces for *The Exorcist*, and many others in his long career.

MAMMALS

BATS (Film Loop—8mm—Silent). *See* Bats

MAMMALS (Film Loop—8mm—Captioned). *See* Animals—Infancy

POLAR BEAR: MOTHER AND CUBS (Film Loop—8mm—Silent). *See* Bears

WHALES SURFACING (Film Loop—8mm—Silent). *See* Whales

WOLF FAMILY (Film Loop—8mm—Silent). *See* Wolves

THE WOLVES OF ISLE ROYALE (Kit). *See* Wolves

ZEBRA (Motion Picture—16mm—Sound). *See* Zebras

See also names of mammals, e.g., Lions, etc.

MAN

FROM CAVE TO CITY (Motion Picture—
16mm—Sound). *See* Civilization

LEISURE (Motion Picture—16mm—Sound).
Pyramid Films, 1976. 16mm, color, animated, sound film 14 min.; ³/₄ in. videocassette also available. ($250). Gr. 10–A. LGB
(1977), M&M (1977). 301.2
An animated history of human development
from caveman to modern people with emphasis on the development of leisure and
presenting the idea that what people do with
their leisure time may become the determining factor in society's system of values.

REDESIGNING MAN: SCIENCE AND HU-
MAN VALUES (Filmstrip—Sound). *See*
Forecasting

WHAT DOES IT MEAN TO BE HUMAN?
(Slides/Cassettes). *See* Anthropology

MAN—INFLUENCE OF ENVIRONMENT

CULTURE AND ENVIRONMENT: LIV-
ING IN THE TROPICS (Filmstrip—
Sound). *See* Tropics

NOISE AND ITS EFFECTS ON HEALTH
(Motion Picture—16mm—Sound). *See*
Noise pollution

MAN—INFLUENCE ON NATURE

BIOLOGICAL CATASTROPHES: WHEN
NATURE BECOMES UNBALANCED
(Slides/Cassettes). *See* Ecology

BIRDS OF PREY (Filmstrip—Sound). *See*
Birds of prey

BOOMSVILLE (Motion Picture—16mm—
Sound). *See* Cities and towns

THE DESERT: PROFILE OF AN ARID
LAND (Filmstrip—Sound). *See* Deserts

ECOLOGY AND THE ROLE OF MAN
(Filmstrip—Sound). *See* Wildlife conservation

ECOLOGY: UNDERSTANDING THE
CRISIS (Filmstrip—Sound). *See* Ecology

LET NO MAN REGRET (Motion Picture—
16mm—Sound). Alfred Higgins Productions, 1973. 16mm color, sound film, 11 min.
($160). Gr. 7–A. BKL (3/1/74), CRA (1974),
EFL (2/77). 574.5
This film reminds us of our personal responsibility to preserve the natural beauty of our
recreational area.

PARADISE LOST (Motion Picture—16mm—
Sound). National Film Board of Canada/
Benchmark Films, 16mm color, animated
film, 4 min. ($95). Gr. 3–A. LBJ, NEF. 574.5
Evelyn Lambert's colorful animation sym-
bolizes the threat to all living creatures

posed by the great despoilers, people.
Manufactured pesticides drift into the ecological balance in the fields and forests.

THE SALT MARSH: A QUESTION OF
VALUES (Motion Picture—16mm—
Sound). *See* Marshes

WHAT IS ECOLOGY? (SECOND EDITION)
(Motion Picture—16mm—Sound). *See* Ecology

MAN—ORIGIN AND ANTIQUITY

AN INQUIRY INTO THE ORIGIN OF
MAN: SCIENCE AND RELIGION (Slides/
Cassettes). Center for Humanities, 1974. 2
Kodak Carousel cartridges, each with 80 col-
or, b/w slides; 2 cassettes or discs; Teacher's
Guide. ($139.50) Gr. 9–12. PRV (4/75), SOC
(3/75). 573.2
An examination of Eastern and Western religious beliefs about the creation of the universe. The scientific hypotheses of Baron
George Cuvier, Sir Charles Lyell, Charles
Darwin, Thomas Malthus, Alfred Wallace,
and Thomas Huxley are then presented. The
program shows how science defines our limits and religion explains our purpose.

MAN, NONLITERATE

THE TASADAY: STONE AGE PEOPLE IN
A SPACE AGE WORLD (Filmstrip—
Sound). Pathescope Educational Media,
1975. 2 color, sound filmstrips, 91–100 fr.
ea.; 2 cassettes, 14–15 min. ea.; Teacher's
Manual. #761 ($50). Gr. 7–12. LGB (1976),
M&M (4/76). 301.2
Contents: 1. The Cave People. 2. Civilization: Curse of Blessing?
The discovery of what is probably the last
tribe of Stone Age people in the world today
was made known in early 1970. This series
shows them in their actual surroundings, living as their ancestors did, as well as their
reaction to modern men and women.

MANAGEMENT

CAREERS IN BUSINESS ADMINISTRA-
TION (Filmstrip—Sound). *See* Business

THE MANAGEMENT OF TIME (Motion
Picture—16mm—Sound). Fred A. Niles
Communications Centers/Best Films, 1973.
16mm color, sound film, 8 min.; videocassette also available. #BF04 ($195). Gr. 9–A.
BKL (1/19/78), IDF
Carefully condenses, through basic animation, the three questions every manager
should ask him/herself: "Am I organized?"
"Am I doing my job?" "Do I use my day
efficiently?" Capsulizes the four key steps to
better time management.

MANNERS AND CUSTOMS

THE GREAT COVER UP (Motion Picture—
16mm—Sound). *See* Clothing and dress

MAP DRAWING

EXPLORING THE WORLD OF MAPS
(Filmstrip—Sound). *See* Maps

MAPS

EXPLORING THE WORLD OF MAPS
(Filmstrip—Sound). National Geographic,
1973. 5 color, sound filmstrips; 5 discs or
cassettes, 11–14 min. #03734 Discs ($74.50),
#03735 Cassettes ($74.50). Gr. 5–12. PRV
(11/74), PVB (5/75). 912

Contents: 1. The Messages of Maps. 2. Using Maps. 3. The Round Earth on Flat Paper.
4. Surveying the Earth. 5. The Making of
Maps.

The graphics of mapmaking are shown from
the beginning sketches to the final product.
Students discover what can be learned from
maps and how to read different kinds of
maps.

GHOST OF CAPTAIN PEALE (Motion Picture—16mm—Sound). *See* Metric system

HOW TO USE MAPS AND GLOBES (Filmstrip—Sound). Troll Associates, 1974. 6 col-
or, sound filmstrips, av. 43 fr. ea.; 3
cassettes, 16 min ea.; Teacher's Guide.
($78). Gr. 4–8. INS (2/75). 912

Contents: 1. What Is a Map? 2. What Is a
Globe? 3. How to Read a Map. 4. Latitude
and Longitude. 5. Latitude and Climate. 6.
Longitude and Time Zone.

Stimulates development of mapreading
skills, and helps students understand scale,
symbols, direction, and distance.

MARCEAU, MARCEL

BIP AS A SKATER (Motion Picture—
16mm—Sound). *See* Pantomimes

THE DREAM (Motion Picture—16mm—
Sound). *See* Pantomimes

THE HANDS (Motion Picture—16mm—
Sound). *See* Pantomimes

THE MASKMAKER (Motion Picture—
16mm—Sound). *See* Pantomimes

THE PAINTER (Motion Picture—16mm—
Sound). *See* Pantomimes

PANTOMIME: THE LANGUAGE OF THE
HEART (Motion Picture—16mm—Sound).
See Pantomimes

THE SIDE SHOW (Motion Picture—16mm—
Sound). *See* Pantomimes

YOUTH, MATURITY, OLD AGE, AND
DEATH (Filmstrip—Sound). *See* Pantomimes

MARCHES (MUSIC)

MARCHES (Phonodiscs). Bowmar Publishing. 1 disc; Lesson Guide; Theme Charts.
#607 ($10). Gr. 2–8. AUR (1978). 785.1

Some of the marches on this disc include
"Entrance of the Little Fawns" by Pierne,
"March Militaire" by Schubert, and "The
March of the Siamese Children" by
Rodgers.

MARIJUANA

WEED (MARIJUANA) (Motion Picture—
16mm—Sound). Concept Films/EBEC,
1971. 16mm color, sound film, 24 min. #3009
($325). Gr. 7–C. BKL (3/15/73), FLN (1972),
M&M (1973). 613.8

The legal consequences of smoking pot
come home sharply as viewers identify with
jailed 17-year-old Charlie. Other facts are
examined and separated from fantasies.
Tales of the weed as a cause of insanity and
violence and addiction and death are de-
bunked. So are the claims of great benefits
from marijuana. Tests show marijuana's effect on memory, judgment, coordination and
driving. The fact that no one knows whether
the weed does lasting damage stands out
clearly.

MARINE ANIMALS

OCTOPUS (Film Loop—8mm—Silent). *See*
Octopus

THE UNDERWATER ENVIRONMENT,
GROUP ONE (Filmstrip—Sound). Imperial
Educational Resources, 1971. 4 color, sound
filmstrips; 4 discs or cassettes; Teacher's
Notes. #3RG44700 Discs ($56),
#3KG447700 Cassettes ($62). Gr. 5–12.
BKL (5/15/71), TEA (9/71). 574.92

Contents: 1. Living Corals. 2. Spiny-
Skinned Animals. 3. Fish in Their Environment. 4. Shellfish and Tubeworms.

Four major classes of underwater life are
shown in this set of filmstrips through vivid
underwater photography. Many different
members of each class are identified and described and their relationship with the underwater environment examined.

MARINE BIOLOGY

LIFE IN THE SEA (Filmstrip—Sound). Jean-
Michel Cousteau/BFA Educational Media,
1977. 4 color, sound filmstrips, av. 55 fr.; 4
cassettes or discs, av. 8 min. #VJP00 Cassettes ($70), #VJN00 Discs ($70). Gr. 5–8.
BKL (11/15/77), PRV (2/78), PVB (4/78).
591.92

Contents: 1. Plants and Simple Sea Animals.
2. Sea Animals without Backbones. 3. Ar-
mored and Spiny-skinned Sea Animals. 4.
Sea Animals with Backbones.

MARINE BIOLOGY (cont.)

Taking the most important groups of marine plants and animals, various forms of life are introduced with bold lettering and lots of examples. Ascending the evolutionary ladder, the set examines the most primitive plants and animals and then progresses to the most complex through detailed close-up, captioned frames and verbal descriptions.

NIGHTLIFE (Motion Picture—16mm—Sound). *See* Ocean

SEA HORSE (Film Loop—8mm—Silent). *See* Sea horses

THE UNDERWATER ENVIRONMENT, GROUP ONE (Filmstrip—Sound). *See* Marine animals

THE UNDERWATER ENVIRONMENT, GROUP TWO (Filmstrip—Sound). *See* Marine plants

MARINE PLANTS

THE UNDERWATER ENVIRONMENT, GROUP TWO (Filmstrip—Sound). Imperial Educational Resources, 1972. 2 color, sound filmstrips; 2 cassettes or discs; Teacher's Notes. #3RG 44800 Discs ($30), #3KG 44800 Cassettes ($33). Gr. 5–12. BKL (10/15/72), SCT. 574.92

Contents: 1. Part I—Marine Vegetation. 2. Part II—Marine Vegetation.

The larger forms of marine vegetation are studied in this set of filmstrips. Both underwater and shoreline photography show a variety of specimens in natural environment that range from cold northern waters to the subtropics.

MARINE RESOURCES

THE OCEANS: A KEY TO OUR FUTURE (Filmstrip—Sound). *See* Oceanography

MARITIME LAW

WHO OWNS THE OCEANS? (Filmstrip—Sound). Current Affairs Films, 1978. 1 color, sound filmstrip; 1 cassette; Discussion Guide. #603 ($24). Gr. 7–12. PRV (10/78). 341.45

For centuries the high seas have been open to every nation on an unrestricted basis. But as the oceans grow in importance as a source of food and mineral wealth, each nation with access to the sea is seeking to maximize its advantages, and areas of national control are being extended farther and farther out to sea. This program shows why the question of international regulation of the oceans is so important and outlines the prospects for finding an answer.

MARRIAGE

THE FAMILY IN TRANSITION (Kit). *See* Family life

MARRIAGE AND PARENTHOOD (Kit). *See* Family life

MATE SELECTION: MAKING THE BEST CHOICE (Filmstrip—Sound). *See* Dating (social customs)

PARTNERS (Kit). *See* Love

WHAT ABOUT MARRIAGE? (Filmstrip—Sound). Sunburst Communications, 1973. 3 color, sound filmstrips, 75–78 fr.; 3 cassettes or discs, 14–15 min.; Teacher's Guide. #112 ($85). Gr. 9–C. BKL (9/1/73), MMB (2/74), PVB (5/75). 301.42

Contents: 1. 'Til Death Do Us Part. 2. Romantic Love and Dirty Dishes. 3. Two Case Studies.

Examines the institution of marriage in our past and present society. Helps students come to an understanding of what to expect—and what not to expect—from marriage.

MARS (PLANET)—EXPLORATION

MARS MINUS MYTH (REV.) (Motion Picture—16mm—Sound). Churchill Films, 1977. 16mm color, sound film, 22 min.; Guide. ($330). Gr. 7–C. BKL (3/1/78). 523.43

A scientist explains the major findings and stunning photographs of the *Mariner 9* and *Viking* missions to Mars. Discussed are origins of landforms, discovery of water ice in the polar caps, liquid water in the past, the improbability of life, and other matters.

MARSHES

THE SALT MARSH: A QUESTION OF VALUES (Motion Picture—16mm—Sound). EBEC, 1975 (EBE Biology Program). 16mm color, sound film, 22 min. #3394 ($290). Gr. 10–A. AAS (1977), BKL (1975), CFF (1975). 574.9

For our own survival, we have to recognize the limits of our exploitation of the natural system. This film documents the value of the biological communities that exist between sea and land. The importance of energy flow and food supply in salt marshes and estuaries is emphasized.

UNDERSTANDING NATURAL ENVIRONMENTS: SWAMPS AND DESERTS (Slides). *See* Ecology

MARSUPIALS

KOALA BEAR (Film Loop—8mm—Silent). *See* Koala bears

POUCHED ANIMALS AND THEIR
YOUNG (Film Loop—8mm—Silent). *See*
Animals

MASKS (FACIAL)

THE MASK (Filmstrip—Sound). Educational
Dimensions Group, 1972. 1 color, sound
filmstrip, 85 fr.; 1 cassette, 15 min.; Teacher's Guide. #699 ($24.50). Gr. 5–A. BKL (2/
1/73), PRV (3/73), PVB (5/73). 391

A comprehensive history of the mask from
the powerful and delightful designs created
by primitive cultures to today's protective
masks for sports and industry.

THE MASKMAKER (Motion Picture—
16mm—Sound). *See* Pantomimes

NATIVE ARTS (Filmstrip—Sound). *See* Folk
art

MASS MEDIA

HOW TO WATCH TV (Filmstrip—Sound).
Educational Direction/Xerox Educational
Publications, 1977. 4 color, sound filmstrips,
47–55 fr.; 4 cassettes, 7–8 min.; Teacher's
Guide. #SC02900 ($69.99). Gr. 5–9. PRV (3/
78), PVB (4/78). 301.16

Contents: 1. News and Documentaries. 2.
Drama and Comedy. 3. Advertising. 4.
Learning from Television.

An introduction to serious television viewing, it motivates students to watch critically.

MEDIA AND MEANING: HUMAN EX-
PRESSION AND TECHNOLOGY (Slides/
Cassettes). Center for Humanities, 1974. 160
slides in 2 Carousel cartridges: 2 cassettes; 2
discs also available; Teacher's Guide. #0243
($139.50). Gr. 9–C. BKL (1975), MMB
(1975), NEA (1975). 301.16

The program explores various media in
terms of the impact technology has had on
human expression and our ability to communicate.

THE POWER OF MEDIA (Filmstrip—
Sound). Coronet Instructional Media, 1977.
4 color, sound filmstrips, 76–93 fr.; 2 discs or
4 cassettes, 10–16 min.; Teacher's Guide.
#M347 Cassettes ($68), #S347 Discs ($68).
Gr. 7–12. PRV (12/78). 301.16

Contents: 1. The Mass Media and Government Institutions. 2. The Mass Media and
Violence. 3. The Mass Media and Everyday
Life. 4. Changing Technology and the Future of Mass Media.

An exploration into the influence of mass
media and their effect on our lives. The set
also touches on the future as technology expands the possibilities for using mass media
in far-reaching ways.

THIS BUSINESS CALLED MEDIA (Filmstrip—Sound). EMC, 1975. 5 color, sound

filmstrips, 77–97 fr.; 5 cassettes, 11–15 min.;
Teacher's Guide. #ELC 229000 ($99). Gr. 8–
12. BKL (9/76), LGB (12/76), PVB (4/77).
301.16

Contents: 1. Television: Servant or Master?
2. Radio: Sound Track for Living. 3. Newspapers: The Living Record. 4. Magazines:
Something for Everybody. 5. Advertising:
Target, You!

An open-ended study of TV, radio, newspapers, magazines, and the advertising industry that supports them. Each unit raises
questions regarding the values expressed
and reflected in the particular medium, with
a realistic examination of how the underlying
profit motive influences what we read, see,
and hear.

WHAT'S GOING ON HERE? (Filmstrip—
Sound). *See* Propaganda

WORDS, MEDIA AND YOU (Filmstrip—
Sound). Globe Filmstrips/Coronet Instructional Media, 1974. 6 color, sound filmstrips,
av. 73 fr.; 6 discs or cassettes, av. 12 min.;
Teacher's Guide. #S702 Discs ($99), #M702
Cassettes ($99). Gr. 5–12. BKL (1975), PVB
(4/76). 301.16

Contents: 1. Mass Media—Servant or Master? 2. Words in Advertising. 3. Words in
News. 4. Words in Politics. 5. Words in En-
tertainment. 6. Words in Literature.

Stresses the way the media affect our culture. Specific individuals in various occupations explain how they choose words
carefully to control emotions.

MASTERS, EDGAR LEE

SPOON RIVER ANTHOLOGY (Filmstrip—
Sound). *See* American poetry

MATHEMATICS

DONALD IN MATHMAGIC LAND (Filmstrip—Sound). Walt Disney Educational
Media, 1976. 4 color, sound filmstrips; 4
discs or cassettes, av. 7 min. ea.; Teacher's
Guide. #63-8034L ($81). Gr. 4–8. MDU (4/
77), MTE (9/77). 511

Contents: 1. Math in Music. 2. The Secret of
the Pentagram. 3. Who's Keeping the Score?
4. The Shape of Things.

Based on the award-winning animated classic film. Youngsters learn how mathematical
principles influence architecture, science,
art, music, and even sports.

EXPLORING MATH SERIES (Filmstrip—
Sound). Guidance Associates, 1977. 3 color,
sound filmstrips, av. 87–95 fr. ea.; 3 cassettes or discs, 15–19 min. ea.; 3 Guides.
#9A-305-050 Cassettes ($74.50), #9A-305-
043 Discs ($74.50). Gr. 4–8. BKL (3/15/78).
511

MATHEMATICS (cont.)

Contents: 1. A Finite Number System. 2. Games and Diversions. 3. Patterns and Functions.

Photographs, graphs, diagrams, and matrices make up the visuals in these filmstrips designed to develop youngsters' proficiency with numbers. They explain and direct exploration of concepts such as finite and infinite numbers, digital root, Fibonacci sequence, and algebraic patterns. Two mathematical tricks are dramatized.

INTRODUCING THE ELECTRONIC CALCULATOR (Filmstrip—Sound). *See* Calculating machines

MATHEMATICAL PEEP SHOW (Motion Picture—16mm—Sound). *See* Topology

PROBING WORD PROBLEMS IN MATHEMATICS (Kit). Coronet Instructional Media, 1973. 8 cassettes; 30 Student Books, Teacher's Guide. #K112 ($85). Gr. 7–10. ATE (1974). 511

Contents: 1. Welcome to Mathematics (Introducing the Problem-Solving Strategy). 2. What's the Question? (Analyzing the Problem). 3. A Picture, Please (Diagramming Problems). 4. Meet Sir X (Deriving Equations). 5. Professor Pavlove's Problems (Perimeter, Area and Volume Problems). 6. Mathematical Field Day (Rate, Distance and Time Problems). 7. Mathematical Magic (Numeral Relationship Problems). 8. Backtracking (Checking the Solution).

Students go on a visit to Mathematica, a mythical land of robots, kooky professors, and equation-spouting knights. A math student learns to solve math work problems. He breaks down problem solving into steps—analyzing, diagramming, deriving equations, checking the solutions.

See also Arithmetic

MATTER

SHAPES AND STRUCTURES IN NATURE (Filmstrip—Captioned). BFA Educational Media, 1974. 4 captioned filmstrips. #VV1000 ($32). Gr. 6–9. BKL (5/1/75), PRV. 530.4

Contents: 1. The Shape of Things. 2. What Are Things Made Of? 3. Forces, Shapes and Changes. 4. Shapes and Structures around Us.

Introduces the structure of matter. Develops concepts of atomic and molecular structure and relates molecular arrangements to properties of natural and manufactured shapes and structures.

THE SPECIAL THEORY OF RELATIVITY (Filmstrip—Sound). *See* Relativity (physics)

MAUGHAM, W. SOMERSET

THE RAZOR'S EDGE (Cassette). *See* Self-realization—Fiction

MAY, JULIAN

SPORTS CLOSE-UPS 1 (Kit). *See* Athletes

SPORTS CLOSE-UPS 3 (Kit). *See* Athletes

SPORTS CLOSE-UPS 4 (Kit). *See* Athletes

MAYAS

ONE IMIX, EIGHT POP (Cassette). *See* Mexico—History

MAYFLIES

THE MAYFLY: ECOLOGY OF AN AQUATIC INSECT (Motion Picture—16mm—Sound). EBEC, 1973. 16mm color, sound film, 15 min. #3198 ($185). Gr. 7–C. AAS (1976), BKL (1973), CGE (1975). 595.734

Because the mayfly's life cycle includes both aquatic and aerial states, the mayfly is an excellent indicator of water quality in rivers and lakes. Through photography, this film reveals the mayfly Hexagenia's curious life history, its importance as part of freshwater food chains, and its dependence upon unpolluted water for survival.

MEASUREMENT

GHOST OF CAPTAIN PEALE (Motion Picture—16mm—Sound). *See* Metric system

THE MEASURE OF MAN (Motion Picture—16mm—Sound). *See* Metric system

MEASUREMENT: FROM CUBITS TO CENTIMETERS (Filmstrip—Sound). Guidance Associates, 1975 (Math Matters). 2 color, sound filmstrips, av. 65 fr.; 2 discs or cassettes, av. 11 min.; Teacher's Guide. #1/8B-301-489 Cassettes ($52.50), #1B-301 497 Discs ($52.50). Gr. 4–8. ATE. 389.152

Traces the evolution of measurements from ancient times through the nineteenth century. Explains how the metric system spread through the world.

THE METRIC MOVE (Motion Picture—16mm—Sound). *See* Metric system

PRESENTANDO MEDIDAS (INTRODUCING MEASURING) (Filmstrip—Sound). BFA Educational Media, 1976. 6 color, sound filmstrips, av. 31–42 fr.; 6 cassettes or discs, av. 6–9 min.; Teacher's Guide. #VGY000 Cassettes ($92), #VGX000 Discs ($92). Gr. 4–9. PRV (11/77). 389

Contents: 1. Comparando. 2. Unidades Fijas. 3. Perimetro y Area. 4. Volumen. 5.

Presando Acerca de Esalas. 6. Conceptos Acerca de Medidas: Repaso.

This set shows what measuring is and how it helps find length, perimeter, area, and volume. Why numbers are important and how to use standard units of measure to solve simple problems are presented.

MEDICAL CARE

BUYING HEALTH CARE (Kit). *See* Insurance, Health

CAREERS IN HEALTH SERVICES (Filmstrip—Sound). Associated Press/Pathescope Educational Media, 1973 (Careers In). 2 color, sound filmstrips, 81–84 fr.; 2 cassettes, 12–16 min.; Teacher's Manual. #702 ($50). Gr. 9–12. IFT (1973), PRV (4/19/75). 610.69

Contents: Part 1. Part 2.

Examines the different types of job opportunities in the health services field. Health care facilities are shown; people from allied health professions are interviewed to determine necessary qualifications, job satisfactions, and opportunities available.

GOING BACK UNIT THREE (Motion Picture—16mm—Sound). *See* Occupations

MEDICINE, PSYCHOSOMATIC

PSYCHOSOMATIC DISORDERS (Filmstrip—Sound). Human Relations Media, 1978. 3 color, sound filmstrips, 67–71 fr.; 3 cassettes, 12–13 min.; Guide. #633-A ($90). Gr. 9–A. BKL (2/1/79). 616.08

Teaches about psychosomatic illnesses, physical disorders whose origins are mental and emotional. Explores such common problems as migraine headaches, asthma, insomnia, ulcers, and eczema. Describes personality factors, family conflicts, and stress situations that may trigger psychosomatic reactions.

MEDICINE—STUDY AND TEACHING

INTERN: A LONG YEAR (Motion Picture—16mm—Sound). *See* Mack, Karen

MEDITATION. *See* Transcendental meditation

MEDITERRANEAN SEA

SOUTHERN EUROPE: MEDITERRANEAN LANDS (Filmstrip—Sound). *See* Europe

MELVILLE, HERMAN

BARTLEBY (Motion Picture—16mm—Sound). *See* American fiction

A DISCUSSION OF BARTLEBY (Motion Picture—16mm—Sound). *See* Motion pictures—History and criticism

EPISODES FROM CLASSIC NOVELS (Filmstrip—Sound). *See* Literature—Collections

THE ROMANTIC AGE (Filmstrip—Sound). *See* American literature—History and criticism

MEN—U.S.

AMERICAN MAN: TWO HUNDRED YEARS OF AUTHENTIC FASHION (Kit). *See* Clothing and dress

MALE AND FEMALE ROLES (Filmstrip—Sound). *See* Sex role

MENOTTI, GIAN CARLO

STORIES IN BALLET AND OPERA (Phonodiscs). *See* Music—Analysis, Appreciation

MENTAL ILLNESS

ORIGINS OF MENTAL ILLNESS (Filmstrip—Sound). Human Relations Media, 1978. 2 color, sound filmstrips, 79 fr.; 2 cassettes, 13–14 min.; Guide. #634–99 ($60). Gr. 9–C. BKL (1/1/79). 157

Contents: 1. Personality Development. 2. Stress of Modern Living.

This set delves into the major causes of abnormal human behavior: genetic defects, faulty personality development, environmental stress, and sociological and cultural influences.

MENTALLY HANDICAPPED

PAIGE (Motion Picture—16mm—Sound). WGBH Educational Foundation/EBEC, 1978 (People You'd Like to Know). 16mm color, sound film, 20 min.; Teacher's Guide. #3590 ($185). Gr. 4–8. BKL (1979), LFR (1979). 362.5

Paige, who has Down's Syndrome, is shown interacting in a mainstreamed fourth-grade classroom and in a resource room. Her younger sister narrates Paige's story and explains her own feelings about Paige's condition. "She is different, but she has a lot to offer."

READIN' AND WRITIN' AIN'T EVERYTHING (Motion Picture—16mm—Sound). Kent County Community Mental Health Services Board/Stanfield House, 1976. 16mm color, sound film ($300). Gr. 9–C. CGE (1976), FLN (6/78). 362.5

Covers deinstitutionalization of severely and moderately retarded, community integration, and parental adjustment.

MENTALLY HANDICAPPED—FICTION

FLOWERS FOR ALGERNON BY DANIEL KEYES (Filmstrip—Sound). Current Af-

MENTALLY HANDICAPPED—FICTION (cont.)

fairs Films, 1978. 1 color, sound filmstrip; 1 cassette; Teacher's Edition of the Book; Testing Materials; Discussion Guide. #Lp–626 ($30). Gr. 7–A. BKL (12/78). Fic

The novel is a series of progress reports of an experiment to increase the 68 IQ of Charlie Gordon, age 32, to genius level—an experiment seemingly successful at first, but which brings unpredictable consequences.

MERRILL, JEAN

THE SUPERLATIVE HORSE (Motion Picture—16mm—Sound). *See* Folklore—China

MERRITT, A.

SCIENCE FICTION (Filmstrip—Sound/Captioned). *See* Science fiction

MESA VERDE NATIONAL PARK

INDIAN CLIFF DWELLINGS AT MESA VERDE—SLIDE SET (Slides). *See* Cliff dwellers and cliff dwellings

METAMORPHOSIS

BUTTERFLY: THE MONARCH'S LIFE CYCLE (Motion Picture—16mm—Sound). *See* Butterflies

DON'T (Motion Picture—16mm—Sound). *See* Butterflies

INSECT LIFE CYCLE (PERIODICAL CICADA) (Motion Picture—16mm—Sound). *See* Cicadas

LIFE CYCLE OF COMMON ANIMALS, GROUP 3 (Filmstrip—Sound). *See* Butterflies

METAMORPHOSIS (Motion Picture—16mm—Sound). *See* Butterflies

MONARCH: STORY OF A BUTTERFLY (Film Loop—8mm—Silent). *See* Butterflies

MOTHS—SLIDE SET (Slides). *See* Moths

METEOROLOGY

STORMS: THE RESTLESS ATMOSPHERE (Motion Picture—16mm—Sound). *See* Storms

WEATHER (Kit). Sue Beauregard/Cypress Publishing, 1977 (Reading in the Content Area). 2 color, sound filmstrips; 2 cassettes, 8–12 min. (with narration for filmstrip on Side One and for book on Side Two); 16 Paperbacks; 2 Teacher's Guides. #049–50 ($78). Gr. 5–8. MDI (9/78). 551.6

Contents: 1. Something's in the Air. 2. Stormy Weather.

Remedial reading in the content area, vocabulary and conceptual development, learning activities to teach introductory level of weather are presented. Introduces the concept of the atmosphere as well as showing the evaporation, condensation, and precipitation cycle.

WEATHER FORECASTING (Motion Picture—16mm—Sound). *See* Weather forecasting

WEATHER, SEASONS AND CLIMATE (Filmstrip—Sound). BFA Educational Media, 1977. 4 color, sound filmstrips, 49–52 fr.; 4 discs or cassettes, approx. 6 min.; Teacher's Guide. #VJS000 Cassettes ($68), #VJR000 Discs ($56). Gr. 4–8. PRV (4/78). 551.6

Contents: 1. What Causes Weather? 2. Our Sun and Time and Seasons. 3. The Earth's Climate. 4. Storms and Clouds.

Introduces weather, storms, seasons, and times, their causes and variations, the need for time zones, the reasons for day and night, and seasons and climates.

METRIC SYSTEM

GHOST OF CAPTAIN PEALE (Motion Picture—16mm—Sound). EBEC, 1975 (Math that Counts). 16mm color, sound film, 11 min. #3471 ($150). Gr. 4–8. ATE (4/77), LFR (9/11/76), TEA (3/76). 389.152

Two important concepts—metric measurement and interpreting map scales—are introduced in this whimsical tale of a treasure map, an incompetent ghost, and a clever little girl.

THE MEASURE OF MAN (Motion Picture—16mm—Sound). Halas & Batchelor Animation/Best Films, 1973. 16mm color, animation, sound film, 9 min.; videocassette also available. #BF–15 ($155). Gr. 3–8. AAS (9/19/78), MTE (4/19/78). 389.152

This presentation is designed to introduce the metric system, and to establish further that measurement is arbitrary and counting is not. Summarizes history of measurement from caveman forward. Concludes with simplified explanation of basic metric system.

MEASUREMENT: FROM CUBITS TO CENTIMETERS (Filmstrip—Sound). *See* Measurement

METRIC MEETS THE INCHWORM (Motion Picture—16mm—Sound). Bosustow Productions, 1974. 16mm color, sound, animated film, 10 min.; Leader's Guide. ($175). Gr. 2–A. ATE (4/76), CIF (1975), MTE (4/76). 389.152

Fred Inchworm is understandably depressed with the new metric system. He decides he will go to work somewhere where they've never heard of metric. Poor Fred, suffering from "Metric-phobia" finds that will not be too easy. Almost all other countries of the world use the metric system. Fred's only hope is outer space, but when he joins a NASA space mission, he finds they've been using the metric system for years. Forced to accept the change, he finds the metric system is not so mysterious and terrifying as he thought.

THE METRIC MOVE (Motion Picture—16mm—Sound). Graphic Films/Best Films, 1975. 16mm color, animation, sound film, 15 min.; videocassette also available. #BF–6 ($245). Gr. 5–A. AAS (12/19/77), BKL (10/19/77), MTE (1/19/78). 389.152

Uses contemporary animation, live-action, split-screen, other special effects. Begins by presenting a brief history of measurement. The balance of the film introduces, illustrates, compares, clarifies the fundamental concepts of the SI Metric System: base and derived units, multiple and submultiple prefixes.

METRIC SYSTEM (Filmstrip—Sound). Educational Dimensions Group, 1976. 2 color, sound filmstrips, 60 fr.; 2 cassettes, 15–17 min.; Teacher's Guide. #527 ($37). Gr. 4–8. PRV (12/76), PVB (4/77). 389.152

Contents: 1. Part I. 2. Part II.

These filmstrips show first a comparison of the two systems (the metric and the English) and then go on to discuss the use of the metric system, generally as easy as counting by tens.

METRIC SYSTEM OF MEASUREMENT (Filmstrip—Sound). Library Filmstrip Center, 1975. 1 color, sound filmstrip, 56 fr.; 1 cassette or disc, 14 min. Cassettes ($26), Discs ($24). Gr. 4–12. PRV (11/76). 389.152

Gives a brief history of our use of measurement. It explains the base unit for length and mass.

THE METRIC SYSTEM OF MEASURE-MENT (Filmstrip—Sound). Educational Development/Imperial Educational Resources, 1975. 4 color, sound filmstrips, 49–55 fr.; 4 cassettes or discs, 10–12 min. #3RG 10200 Discs ($56), #3KG 10200 Cassettes ($62). Gr. 2–9. PRV (12/75), PVB (4/76). 389.152

Contents: 1. History of Measurement. 2. Measuring Length. 3. Measuring Weight. 4. Measuring Volume.

An introduction to the actual use of metric units, not to complicated conversion formulas. Concepts of measurements are treated from an historical point, showing that many standards have been devised and improved throughout history. Fundamental methods of measuring length, weight, and volume and applying these methods using metric units are also presented.

METRICS FOR CAREER EDUCATION (Filmstrip—Sound). EBEC, 1976. 6 color, sound filmstrips, av. 71 fr. ea.; 6 cassettes or discs, av. 12 min. ea. #6953K Cassettes ($86.95), #6953 Discs ($86.95). Gr. 7–12. BKL (11/77), LFR (6/77). 389.152

Contents: 1. Metrics for Woodworking. 2. Metrics for Home Economics. 3. Metrics for General Metals. 4. Metrics for Machine Shops. 5. Metrics for Drafting. 6. Metrics for Energy and Power.

Six presentations show how metric measurements may soon be used in many areas of work. Each filmstrip defines new terms, illustrates uses of metric tools, and discusses problems in adapting current machinery to the new system.

METROPOLITAN AREAS

THE AMERICAN URBANIZATION (Filmstrip—Sound). *See* Urban renewal

URBAN WORLD—VALUES IN CONFLICT (Filmstrip—Sound). Globe Filmstrips/Coronet Instructional Media, 1974. 8 color, sound filmstrips, av. 68 fr.; 8 cassettes or discs, 11½ min. #M704 Cassettes ($129), #S704 Discs ($129). Gr. 6–11. M&M (1975). 301.34

Interviews and documentary photos that present causes, effects, and possible solutions to some urban problems.

MEXICANS IN THE U.S.

THE MEXICAN-AMERICAN SPEAKS: HERITAGE IN BRONZE (Motion Picture—16mm—Sound). EBEC, 1972. 16mm color, sound film, 20 min. #3153 ($255). Gr. 7–C. EFL (1973), MER (1973), PRV (1973). 301.45

Focusing upon Mexican-Americans, the film traces the conquest of Indians by Spanish conquistadors and the spread of Spanish dominion in the New World. The rule of Spain disappeared, but the heritage of language and of mixed blood and culture remained.

MEXICAN FOOD (Kit). *See* Cookery, Mexican

MEXICANS IN THE U.S.—FICTION

CHILD OF FIRE (Phonodisc or Cassette). Miller-Brody Productions 1976 (Young Adult Recordings and Sound Filmstrips). 1 cassette or disc, 45 min.; Teacher's Notes. #YA-402C Cassette (7.95), #YAR-402 Disc ($6.95). Gr. 7–A. BKL (5/15/77), CRC (10/19/76). Fic

A dramatized abridgement of Scott O'Dell's book by the same name. It highlights episodes in the short life of an idealistic Chicano youth. Manuel Castillo can't find his "real place" in today's world. His story is one of violence, bred of discrimination, as well as pride, bravery, and commitment to principles.

MEXICO

MEXICO: IMAGES AND EMPIRES (Filmstrip—Sound). Lyceum/Mook & Blanchard, 1974. 1 color, sound filmstrip, 52 fr.; 1 cassette or disc, 8¹/₂ min. #LY35473C Cassette ($25), #LY35473R Disc ($19.50). Gr. 5–A. FLN (4/5/74), PVB (5/75). 917.2

Authors/photographers Ron and Marcia Atwood's perceptive portrait of Mexico promotes an understanding of Mexican history and culture.

MEXICO IN THE TWENTIETH CENTURY (Filmstrip—Sound). BFA Educational Media, 1970. 6 color, sound filmstrips; 6 cassettes or discs. #VS3000 Cassettes ($92), #VE2000 Discs ($92). Gr. 4–8 BKL (1/15/71). 917.2

Contents: 1. Mexico's Physical Heritage. 2. Mexico's History. 3. Mexicans at Work. 4. Mexicans at Play. 5. Mexican Art, Architecture, and Education. 6. Mexicans on the Move.

Studies similarities and differences between the students' own culture and environment and those of contemporary Mexico.

MEXICO—ANTIQUITIES

SENTINELS OF SILENCE (Motion Picture—16mm—Sound). EBEC, 1973. 16mm color, sound film, 19 min. #3176 ($25). Gr. 7–A. AAS (1973), AAW (1974), BKL (11/15/73). 913.372

This film, winner of two Academy Awards, was shot entirely from a helicopter as it hovered over seven archeological sites in Mexico. As the camera discovers massive pyramid mounds, elaborate relief carvings, and steps leading (where?), a spiritual and esthetic impression of those ancient civilizations emerges—a haunting reminder of the "original Americans." Who were they? What were their dreams? Narrated by Orson Welles.

MEXICO—DESCRIPTION AND TRAVEL

MI HERMANO SE CASO EN MEXICO (MY BROTHER WAS MARRIED IN MEXICO) (Kit). Kevin Donovan Films, 1976. 4 filmstrips; 4 cassettes; 4 Reproducible Spanish Scripts; 4 English Scripts; 4 Teacher's Notes. ($95). Gr. 7–C. FLN (9/78), HIS (9/77), PRV (5/77). 917.2

Contents: 1. Principios (Beginnings). 2. Bienvenidos (Welcome). 3. Misterios (Mysteries). 4. Casados (Married).

True family experience told by teenage sister (who narrates the English side) comparing and contrasting life and customs as she travels Mexico with the family and meets Queta and her family. The section on the Indian ruins gives insight to the ancient cultures in Mexico's past. Original music. Valuable for Spanish, social studies, and world history classes.

PASSPORT TO MEXICO (Filmstrip—Sound). EMC, 1976. 5 color, sound filmstrips, 104–113 fr.; 5 cassettes, 21–30 min.; Teacher's Guide. #SP-115000 ($96). Gr. 4–A. PRV (4/78). 917.2

Contents: 1. Preparations and Trip. 2. Renting a Car, at the Bank, at the Post Office, Etc. 3. In the Restaurant, at the Grocery Store, at the Market. 4. Taking a Taxi, at the Bakery, at the Department Store, Etc. 5. Hair Care, Overcoming the Measurement Problem.

This program will acquaint students with the day-to-day experiences of traveling in Mexico, reflecting cultural subtleties of Mexican society.

MEXICO—FICTION

GOLDEN LIZARD: A FOLK TALE FROM MEXICO (Motion Picture—16mm—Sound). *See* Folklore—Mexico

THE PEARL BY JOHN STEINBECK (Filmstrip—Sound). Current Affairs Film, 1978. 1 color, sound filmstrip; 1 cassette; Teacher's Edition of the Book, Testing Materials, Discussion Guide. #LP-638 ($30). Gr. 7–C. BKL (11/1/78). Fic

This is the story of a poor Mexican pearl diver who finds a great pearl. The story concerns the way his find affects his life and his relationships with the people around him.

MEXICO—HISTORY

THE END OF THE LINE (Motion Picture—16mm—Sound). Los Angeles Community College District/International Television, 1977 (The History of Mexico). 16mm, color, sound film, 30 min.; videocassette also available. #13 ($350). Gr. 11–C. BKL (2/10/79). 972

This program explores the ethnicity in colonial Mexico. Scenes and data reveal an extraordinary race-conscious society. It shows that much of Mexico's twentieth-century struggle has dealt with an attempt to bring these diverse societies and cultures together in a unified nation.

ONE IMIX, EIGHT POP (Cassette). Los Angeles Community College District/Instructional Television, 1977 (The History of Mexico). 16mm color, sound film, 30 min.; ³/₄ in. videocassette also available. #5 ($350). Gr. 10–A. BKL (2/1/79). 972

Presents background on the ruins and artifacts of Mexico and provides information on the religion, society, and economy of the Mayan civilization.

MEXICO—SOCIAL LIFE AND CUSTOMS

THE END OF THE LINE (Motion Picture—16mm—Sound). *See* Mexico—History

MICE—FICTION

THE ISLAND OF THE SKOG (Filmstrip—Sound). Weston Woods Studios, 1976. 1 color, sound filmstrip, 49 fr.; 1 cassette, 11 min.; Guide. ($12.75). Gr. 4–A. BKL (1/15/77). Fic

Based on Steven Kellogg's book, the mice have an exciting adventure while learning the importance of diplomacy, comradeship, and the necessity of communicating.

MICROORGANISMS

THE PROTISTS (Filmstrip—Sound). EBEC, 1976. 4 color, sound filmstrips, av. 85 fr. ea.; 4 cassettes or discs, av. 15 min. ea.; Teacher's Guide. #6932K Cassettes ($57.95), #6932 Discs ($57.95). Gr. 9–12. ABT, LFR (2/78).

Contents: 1. How Living Things Are Classified. 2. Bacteria, Blue-Green Algae, and Viruses. 3. Algae. 4. The Decomposers: Fungi, Slime, Molds, and Protozoa.

This set explores the problems biologists face in classifying living things and probes the problems and peculiarities of protists, the oldest and simplest organisms. It explores the incredible variety and complexity of algae; and explains how fungi, slime molds, and protozoa recycle dead organisms.

MIDDLE AGES

THE ART OF THE MIDDLE AGES (Filmstrip—Sound). *See* Art, Medieval

MEDIEVAL EUROPE (Filmstrip—Sound). Society for Visual Education, 1977. 4 color, sound filmstrips, av. 40–48 fr.; 4 cassettes or discs, av. 8–10 min. #A380-SCTC Cassettes ($70), #A380-SCR Discs ($70). Gr. 4–8. BKL (10/15/77). 940.1

Contents: 1. People and Migration. 2. Society. 3. Religion. 4. Towns and Cities.

The migration of various people and tribes helped determine the character of medieval Europe. The nature of feudal society, the participation in the Crusades, the role of religion in preserving cultural achievements, and the appearance of cities and towns are some of the facets of medieval life that are presented.

MEDIEVAL THEATER: THE PLAY OF ABRAHAM AND ISAAC (Motion Picture—16mm—Sound). *See* Theater—History

MIDDLE AGES—FICTION

THE PRINCE AND THE PAUPER (Phonodisc or Cassette). *See* Fantasy—Fiction

MIDDLE EAST

ISLAM: THE PROPHET AND THE PEOPLE (Motion Picture—16mm—Sound). *See* Islam

THE MIDDLE EAST: FACING A NEW WORLD ROLE (Filmstrip—Sound). Society for Visual Education, 1976. 6 color, sound filmstrips, av. 68 fr.; 6 discs or cassettes, av. 14 min.; Guide. #256-SATC Cassettes ($105), #256-SAR Discs ($105). Gr. 4–9. BKL (6/77), PRV (4/76). 915.6

Contents: 1. Egypt: Balancing Its Past and Future. 2. Cairo: Restless Center in the Arab World. 3. Jordan: A New Nation in an Ancient Land. 4. Jordan River Valley Project: Planning the Future. 5. Saudi Arabia: Custodian of Tradition and Oil. 6. Saudi Arabia: Petro Power.

This set examines the daily lives of a cross section of citizens of Jordan, Egypt, and Saudi Arabia. The photographs, narration, and authentic music introduce the changing world of the Arab Middle East. Viewers study attempts by Jordan to establish a new settlement in the Jordan River Valley, visit Cairo and the Suez Canal, see how oil is changing the lives of the Saudis, and review the Arab-Israeli conflict.

THE MIDDLE EAST: LANDS IN TRANSITION (Kit). Education Enrichment Materials, 1975. 6 color, sound filmstrips, 60–75 fr.; 6 cassettes or discs, 12–15 min.; 5 Wall Charts; 6 Spirit Masters; Teacher's Guide; Paperback: *The Middle East; History, Culture, People.* #41089 C or R ($126). Gr. 6–12. PRV (12/76). 915.6

Contents: 1 History and Heritage. 2. Islam: The Religion of Mohammed. 3. Natural Resources. 4. Agriculture and Change. 5. Changing Social Patterns. 6. A Political Perspective.

A broad overview to provide students with the background necessary to understand the social, economic, and cultural patterns in the Middle East and their implications on present-day turmoil.

THE MIDDLE EAST: A UNIT OF STUDY (Kit). United Learning, 1977. 8 color, sound filmstrips, 64–87 fr.; 8 cassettes, 10–21 min.; 2 Cassette Interview Tapes; Student Activities; Duplicating Materials; Teacher's Guide. #2210 ($135). Gr. 7–12. BKL (9/15/78). 915.6

Contents: 1. Geography of the Middle East. 2. Religions. 3. History. 4. The Arab-Israeli Conflict. 5. Nomadic and Village Life. 6. City Life in the Middle East. 7. Economy of the Middle East Region. 8. Middle East Politics and International Relations.

Utilizing a unique regional approach, this filmstrip unit presents the history, geography, religions, economy, and politics that have been both the unifying strengths and destructive weaknesses of the Middle East. The great importance of the region to West-

MIDDLE EAST (cont.)

ern democracies is obvious. Two interview cassettes contain interviews with teenagers from the region who talk about their life-styles, dating, school, work, and expectations.

THE PALESTINIANS: PROBLEM PEOPLE OF THE MIDDLE EAST (Filmstrip—Sound). Current Affairs Film,s 1976. 1 color, sound filmstrip, 74 fr.; 1 cassette, 15 min.; Teacher's Guide. #524 ($24). Gr. 7–A. PRV (4/77). 915.6

The Palestinians, displaced from their homeland during the 1948 war that followed the creation of Israel, and scattered in refugee camps throughout several Arab countries, constitute an open, long-festering wound that must be healed before there is a hope for any lasting peace in the Middle East. The filmstrip examines the Palestinians both as a people and as a political force.

MIDDLE EAST—SOCIAL LIFE AND CUSTOMS

FAMILIES OF THE DRY MUSLIM WORLD (Filmstrip—Sound). EBEC, 1973. 5 color, sound filmstrips, av. 70 fr. ea.; 5 cassettes or phonodiscs, av. 6 min. ea.; Teacher's Guide. #6488K Cassettes ($72.50), #6488 Discs ($72.50), each filmstrip ($17). Gr. 5–9. BKL (12/15/74), PRV (3/74), PVB (5/75). 915.6

Contents: 1. Village Life in Pakistan. 2. Oil Worker of Kuwait. 3. Cooperative Farming in Iran. 4. A Berber Village in Morocco. 5. Nomads of Morocco.

The differing life-styles of peoples of the Muslim world are depicted in this series. Through unposed photographs and narration, you become well acquainted with five families. Observing the differences, you also discover similarities among these areas.

MIDDLE WEST

FOCUS ON AMERICA, THE MIDWEST (Filmstrip—Sound). See U.S.—Economic conditions

MIGRANT LABOR—FICTION

CHILD OF FIRE (Phonodisc or Cassette). See Mexicans in the U.S.—Fiction

MILLAY, EDNA ST. VINCENT

CHILDHOOD OF FAMOUS WOMEN, VOLUME THREE (Cassette). See Women—Biography

MILLER, ALFRED JACOB

A. J. MILLER'S WEST: THE PLAINS INDIAN—1837—SLIDE SET (Slides). See Indians of North America—Paintings.

MILNE, A. A.

MR. SHEPARD AND MR. MILNE (Motion Picture—16mm—Sound). See Authors, English

MIND AND BODY

PSYCHOSOMATIC DISORDERS (Filmstrip—Sound). See Medicine, Psychosomatic

MINERALOGY

ROCKS AND MINERALS (Filmstrip—Sound). See Geology

MINORITIES

AMERICA'S ETHNIC HERITAGE—GROWTH AND EXPANSION (Filmstrip—Sound). BFA Educational Media, 1976. 4 color, sound filmstrips; 4 cassettes or discs. #VJG000 Cassettes ($80), #VJH000 Discs ($68). Gr. 4–8. PRV (4/78). 301.45

Contents: 1. Growth and Expansion. 2. Scandinavian. 3. Irish. 4. Chinese and Japanese.

This series chronicles America's growth and development during 1800–1880 and the contributions of America's ethnic groups to this development.

CHINESE FOOD (Kit). See Cookery, Chinese

GERMAN FOOD (Kit). See Cookery, German

ITALIAN FOOD (Kit). See Cookery, Italian

LIVING TOGETHER AS AMERICANS (Filmstrip—Sound). See Human relations

THE MANY AMERICANS, UNIT ONE (Filmstrip—Sound). See Biculturalism

THE MANY AMERICANS, UNIT TWO (Filmstrip—Sound). See Biculturalism

THE MEXICAN-AMERICAN SPEAKS: HERITAGE IN BRONZE (Motion Picture—16mm—Sound). See Mexicans in the U.S.

MEXICAN FOOD (Kit). See Cookery, Mexican

MINORITIES—USA (Filmstrip—Sound). Globe Filmstrips/Coronet Instructional Media, 1975. 8 color, sound filmstrips, 60–78 fr.; 8 discs or cassettes, 9–15 min.; Teacher's Guide. #M703 Cassettes ($118), #S703 Discs ($118). Gr. 6–12. BKL (12/15/75), PRV (10/76), PVB (4/77). 301.45

Contents: 1. The American Dilemma. 2. Who Am I? (Native Americans). 3. A Piece of the Pie (Black Americans). 4. La Causa (Mexican Americans). 5. Executive Order 9066 (Asian Americans). 6. Two Different Worlds (Puerto Rican Americans). 7. You

Breathe Free (Religious Minorities). 8.
Bringing about Change.

Candid case studies, documentary interviews, poetry, and photography develop the key concepts of prejudice, discrimination, and scapegoating. Stimulates thinking about the conflict between the American creed and how minorities are treated.

THE OTHER AMERICAN MINORITIES, PART II (Filmstrip—Sound). Educational Enrichment Materials, 1975. 4 color, sound filmstrips; 4 cassettes or discs; Teacher's Guide. #41082 C or R ($76). Gr. 7–12. BKL (9/15/76), LNG, M & M. 301.45

Contents: 1. The Irish Americans. 2. The Jewish-Americans. 3. The Italian-Americans. 4. The German-Americans.

Describes the immeasurable contribution of each group to the American social fabric. Discusses the special problems and aspirations, with emphasis on movements that seek to attain minority rights that are equal to those accorded to society's majority.

PREJUDICE IN AMERICA—THE JAPANESE AMERICANS (Filmstrip—Sound). *See* Japanese in the U.S.

SOUL FOOD (Kit). *See* Cookery, Black

THEY CHOSE AMERICA: VOLUME ONE (Cassette). *See* Immigration and emigration

THEY CHOSE AMERICA: VOLUME TWO (Cassette). *See* Immigration and emigration

MOBILES (SCULPTURE)

MOBILES: ARTISTRY IN MOTION (Filmstrip—Sound). Warner Educational Productions, 1975. 2 color, sound filmstrips, 80–90 fr.; 1 cassette 10 to 12 min.; Teaching Guide. #710 ($47.50). Gr. 6–A. BKL (2/1/76), PRV (1/77). 731.55

Combines information about the creation of mobiles by Alexander Calder in 1932 with a brief history and analysis of the art form's technique. The principles of motion—balance, motion, form, and color—are explained. Step-by-step designing, planning, and constructing a mobile cover materials and tools needed. Creative ideas and uses of mobiles are shown.

MOBY DICK

EPISODES FROM CLASSIC NOVELS (Filmstrip—Sound). *See* Literature—Collections

MOHAMMED

ISLAM: THE PROPHET AND THE PEOPLE (Motion Picture—16mm—Sound). *See* Islam

MOHAVE (INDIAN TRIBE)

PETROGLYPHS: ANCIENT ART OF THE MOJAVE (Filmstrip—Sound). *See* Indians of North America—Art

MONEY

MONEY: FROM BARTER TO BANKING (Filmstrip—Sound). Guidance Associates, 1975 (Math Matters). 2 color, sound filmstrips, av. 69 fr.; 2 discs or cassettes, av. 9 min.; Teacher's Guide. #1B-301-406 Discs ($52.50), #1B-301-398 Cassettes ($52.50). Gr. 4–8. ATE. 332.4

Traces the use of barter, commodity, object, and metal money in early civilizations. Discusses development of coinage, paper money, and personal credit.

MONROE, JAMES

PRESIDENTS AND PRECEDENTS (Kit). *See* Presidents—U.S.

MONSTERS—FICTION

THE FOGHORN BY RAY BRADBURY (Filmstrip—Sound). *See* Science fiction

MONTEREY, CALIFORNIA—HISTORY

CANNERY ROW, LIFE AND DEATH OF AN INDUSTRY (Filmstrip—Sound). Kenneth E. Clouse, 1976. 1 color, sound filmstrip, 59 fr.; 1 cassette, 15 min.; Two-Part Study Guide (science and literature emphases). ($19.50). Gr. 8–12. PRV (3/77), SCT (3/77). 979.476

Historical photographs show the development and decline of the sardine fishing and canning industry at Cannery Row in Monterey, California. This setting subsequently became the inspiration for some of John Steinbeck's most poignant stories.

MONTREAL

TOMORROW'S CITIES TODAY (Filmstrip—Sound). *See* Cities and towns

MOON

THE MOON: A GIANT STEP IN GEOLOGY (Motion Picture—16mm—Sound). *See* Lunar geology

MOORE, MARIANNE

POETS OF THE TWENTIETH CENTURY (Filmstrip—Sound). *See* American literature—History and criticism

MORALITY. *See* Ethics; Human behavior

MOSAICS

THE ART OF MOSAICS (Filmstrip—Sound). Educational Audio Visual, 1977 (Art History and Techniques). 2 color, sound filmstrips, 89–101 fr.; 2 discs or cassettes, 12–16 min. Teacher's Guide. #7RF0044 Discs ($40), #7KF0044 Cassettes ($44). Gr. 5–12. PRV (1/78), PVB (4/78). 729

Contents: 1. Techniques. 2. History.

Designed to develop interest in the making of mosaics in the past and present. Also presents a step-by-step procedure for mosaic making and shows examples of completed mosaics. It traces the materials used and also covers the styles as well as the subject of mosaics.

MOSCOW

THE IDEOLOGICAL CAPITALS OF THE WORLD (Filmstrip—Sound). *See* Cities and towns

MOTHERS

WHY MOTHERS WORK (Motion Picture—16mm—Sound). *See* Women—Employment

MOTHS

MOTHS—SLIDE SET (Slides). National Film Board of Canada/Donars Productions, 1972. 10 cardboard mounted, color slides in plastic storage page. Script in English and French. ($10). Gr. 4–A. BKL (2/1/73). 595.781

Presents the life cycle of the moth through its four stages; beginning with the egg, then the larva, into the pupa, until it finally reaches the last stage, the adult moth.

MOTION PICTURE PHOTOGRAPHY

PRACTICAL FILM MAKING (Motion Picture—16mm—Sound). Julian Films/EBEC, 1972. 16mm color, sound film, 19 min. #3154 ($255). Gr. 7–C. EFL (1973), EGT (1975), STE (1973). 778.s

Explained briefly are the functions of the director, the cameraman, sound man, script supervisor, and a number of others responsible for making a film. This film shows how hard work, teamwork, and organization transform into a successful, finished production.

MOTION PICTURES

ACTING FOR FILM (LONG CHRISTMAS DINNER) (Motion Picture—16mm—Sound). *See* Acting

EDITING A FILM (FROM WELL OF THE SAINTS) (Motion Picture—16mm—Sound).

EBEC, 1976. 16mm color, sound film, 12 min. #47821 ($220). Gr. 9–C. EGT (1976), LFR (1977), PRV (1977). 791.43

The job of the film editor is illustrated by examples of editing of the filmed dramatization, with commentary explaining how the final cuts were made in the completed film and, perhaps more importantly, the reasons behind the decision to cut.

THE FIRST MOVING PICTURE SHOW (Motion Picture—16mm—Sound). Phoenix Films, 1973. 16mm color, sound film, 7 min. #73-702740 ($125). Gr. K–8. PRV (11/74). 791.43

Uses an animated clay caveman as a technique to present the story of the development of one of our most fascinating art forms—the motion picture.

MOTION PICTURES, ANIMATED

ANIMATED WOMEN (Motion Picture—14mm—Sound). *See* Women—U.S.

CLOSED MONDAYS (Motion Picture—16mm—Sound). *See* Art—Exhibitions—Fiction

DE FACTO (Motion Picture—16mm—Sound). *See* Decision making

FIRE! (Motion Picture—16mm—Sound). *See* Forest Fires

FRANK FILM (Motion Picture—16mm—Sound). *See* Mouris, Frank

GENE DEITCH: THE PICTURE BOOK ANIMATED (Motion Picture—16mm—Sound). *See* Deitch, Gene

THE GIVING TREE (Motion Picture—16mm—Sound). *See* Friendship—Fiction

HOW BEAVER STOLE FIRE (Motion Picture—16mm—Sound). *See* Indians of North America—Legends

MINDSCAPE (Motion Picture—16mm—Sound). *See* Painters—Fiction

MOVIN' ON (Motion Picture—16mm—Sound). *See* Transportation—History

SISYPHUS (Motion Picture—16mm—Sound). *See* Mythology, Classical

THE STORY OF CHRISTMAS (Motion Picture—16mm—Sound). *See* Christmas

UP IS DOWN (Motion Picture—16mm—Sound). *See* Individuality—Fiction

MOTION PICTURES—BIOGRAPHY

FRANK FILM (Motion Picture—16mm—Sound). *See* Mouris, Frank

WOODY ALLEN (Cassette). *See* Allen, Woody

MOTION PICTURES—ECONOMIC ASPECTS

THE BUSINESS OF MOTION PICTURES (Cassettes). Cinetel/Jeffrey Morton Publishers, 1977. 8 cassettes with booklet. #11020 ($125). Gr. 10–A. BKL (9/15/78). 791.43
Contents: 1. Hollywood in the 30's. 2. Hollywood and the Black Actor. 3. Movies, Movies, Movies. 4. Women in Film. 5. Humphrey Bogart. 6. The Silent Clowns. 7. Making Mystery Movies. 8. The Silent Outcry.
"The Business of Motion Pictures: An Insider's Look at the Industry" originated in a lecture series. These tapes reveal numerous aspects of the creation and distribution of feature films. Insights and anecdotes on financing and finding scripts are included. Distribution deals, percentages, advertising, ratings, profits from concessions, and why the local movie house doesn't get the biggie are covered in knowing detail. How does art emanate from this business hodgepodge?

MOTION PICTURES—HISTORY AND CRITICISM

CHAPLIN—A CHARACTER IS BORN (Motion Picture—16mm—Sound). *See* Chaplin, Charles Spencer

DIRECTING A FILM (IONESCO'S: THE NEW TENANT) (Motion Picture—16mm—Sound). EBEC, 1975. 16mm color, sound film, 17 min. #47817 ($275). Gr. 9–A. EFL (1977), EGT (1976), PRV (1977). 791.43
The producer and director, Larry Yust, takes viewers behind the scenes of this production of Ionesco's "The New Tenant" to illustrate the decisions a director must make.

A DISCUSSION OF BARTLEBY (Motion Picture—16mm—Sound). EBEC, 1969 (Short Story Showcase). 16mm color, sound film, 10 min. #47753 ($359). Gr. 7–C. M&M (1974). 791.43
Commentary written and presented by Dr. Charles Van Doren, author, associate director of the Institute for Philosophical Research, and executive editor of the *Annals of America*.

A DISCUSSION OF DR. HEIDEGGER'S EXPERIMENT (Motion Picture—16mm—Sound). EBEC, 1969 (Short Story Showcase). 16mm color, sound film, 11 min. #47752 ($135). Gr. 7–C. LFR (1970). 791.43
Commentary written and presented by Clifton Fadiman, author and general editor of the EBEC Humanities Program.

A DISCUSSION OF MY OLD MAN (Motion Picture—16mm—Sound). EBEC, 1970 (Short Story Showcase). 16mm color, sound film, 11 min. #47768 ($135). Gr. 7–C. M&M (1974). 791.43

Commentary written by Blake Nevius, professor of English, University of California at Los Angeles.

A DISCUSSION OF THE CROCODILE (Motion Picture—16mm—Sound). EBEC, 1973 (Short Story Showcase). 16mm color, sound film, 11 min. #47788 ($135). Gr. 7–C. BKL (1973), SLJ (1973), STE (1973). 791.43
Commentary written and presented by Clifton Fadiman, author and general editor of the EBEC Humanities Program. He shares his insights and ideas about the story

A DISCUSSION OF THE FALL OF THE HOUSE OF USHER (Motion Picture—16mm—Sound). Avatar Productions, 1975. 16mm color, sound film, 12 min. #47823 ($205). Gr. 7–C. BKL (1/1/77), LFR (9/19/76). 701.43
Science fiction writer Ray Bradbury comments on the story, compares this screenplay to the written work, and discusses the gothic tradition and Poe's influence on contemporary science fiction.

A DISCUSSION OF THE HUNT (Motion Picture—16mm—Sound). EBEC, 1975 (Short Story Showcase). 16mm color, sound film, 16 min. #47824 ($255). Gr. 7–C. LFR (1972). 791.43
Producer/director David Deverell meets with the three actors from *The Hunt* to discuss characterizations, acting techniques, and themes. Film flashbacks illustrate.

A DISCUSSION OF THE LADY OR THE TIGER? (Motion Picture—16mm—Sound). EBEC, 1969 (Short Story Showcase). 16mm color, sound film, 11 min. #47756 ($135). Gr. 7–C. LFR (1972), M&M (11/74). 791.43
Commentary written and presented by Clifton Fadiman.

A DISCUSSION OF THE LOTTERY (Motion Picture—16mm—Sound). EBEC, 1969 (Short Story Showcase). 16 mm color, sound film, 10 min. #47758 ($135). Gr. 7–C. M&M (11/74), STE (1974). 791.43
Commentary written and presented by Dr. James Durbin, associate professor of English, University of Southern California.

A DISCUSSION OF THE SECRET SHARER (Motion Picture—16mm—Sound). EBEC, 1973 (Short Story Showcase). 16mm color, sound film, 11 min. #47786 ($135). Gr. 7–C. M&M (11/74). 791.43
Commentary written and presented by Dr. Charles Van Doren, vice president, editorial, Encyclopaedia Britannica, Inc., and executive editor of the *Annals of America*.

STORY INTO FILM: CLARK'S THE PORTABLE PHONOGRAPH (Motion Picture—16mm—Sound). EBEC, 1977 (Short Story Showcase). 16mm color, sound film, 11 min. #47828 ($200). Gr. 7–A. BKL (1978), EGT (1978), PRV (1979). 791.43

MOTION PICTURES—ECONOMIC ASPECTS (cont.)

Producer/director John Barnes discusses the process and problems of translating the short story to a filmed dramatization. What are the differences between film and print? What does it mean to be faithful to the original in making a film of a short story?

MOTIVATION (PSYCHOLOGY)

THE ART OF MOTIVATION (Motion Picture—16mm—Sound). Fred A. Niles Communications Centers/Best Films, 1972. 16mm color, sound film, 10 min.; or videocassette BF-2. ($195). Gr. 9–A. BKL (1/78). 153.4

Using live-action animation, the film summarizes both MacGregor's Theory X and Y, as well as Herzerg's theory on motivation. Discusses the relationship of individuals, groups, environment, and feedback

WHY WE DO WHAT WE DO! HUMAN MOTIVATION (Filmstrip—Sound). Human Relations Media Center, 1977. 3 color, sound filmstrips, 61–70 fr.; 3 cassettes or discs, 11–12 min. ea.; Teacher's Guide. #623 ($90). Gr. 9–A. BKL (5/1/78), M&M (4/78). 153.4

Contents: 1. Physiological Drives. 2. Psychological Drives. 3. Learned Drives.

Specifies the characteristics and functions of basic human and animal needs—hunger, thirst, sex, avoidance of pain, sleep, curiosity, and maternity. Then concentrates on the distinctly human drives of approval, money, prestige, achievement, aggression, and fear of success. Gives students a useful perspective on feelings and motivations to sharpen their insight into their own behavior.

MOTORCYCLE RACING

MOTOCROSS RACING (Kit). *See* Spanish language—Reading materials

RACING NUMBERS (Kit). *See* Automobile racing

MOTORCYCLES

MOTORCYCLE SAFETY (Kit). Cal Industries/Pathescope Educational Media, 1976. 3 color, sound filmstrips, 57–91 fr.; 3 cassettes, 7–11 min.; 10 Spirit Masters; Teacher's Guide. #204 ($75). Gr. 7–12 BKL (3/1/77), PRV (12/77). 614.8

Contents: 1. Rider Protection and Motorcycle Maintenance. 2. Hazards of the Road. 3. The Motorcyclist in Traffic.

A new rider, an experienced rider, and a mechanic discuss proper protection for the rider, maintenance of the motorcycle, the hazards of the road, traffic regulations, and safe riding in traffic.

NOT SO EASY (MOTORCYCLE SAFETY) (Motion Picture—16mm—Sound). FilmFair Communications, 1973. 16mm color, sound film, 17-1/4 min. ($225). Gr. 4–A. CFF (1973), JEH (5/6/75), LFR (11/73). 796.75

Narrated by Peter Fonda and featuring Evel Knievel, this film demonstrates the essential safety rules of motorcycle riding. Commonsense reasons are given for wearing leather clothing, boots, a helmet, and face covering. Checking out a bike new to the rider and what to check are shown, as well as safety tips for street and highway riding.

MOUNTAINEERING

CLIMB (Motion Picture—16mm—Sound). Churchill Films, 1974. 16mm color, sound film, 22 min. ($320). Gr. 12–A. BKL (2/15/75). 796.522

The thoughts of two men transform an exquisitely suspenseful rock climbing film into a parable of living, of struggle, of independence and interdependence.

MOUNTAINEERING (Filmstrip—Sound). Great American Film Factory, 1975 (Outdoor Education). 1 color, sound filmstrip; 1 cassette, 18 min. #FS-2 ($35). Gr. 7–A. MMB (1/76). 796.522

This filmstrip introduces the student to the ascent of a mountain, highlighted with the words of Ralph Waldo Emerson. Stresses "living with nature," not its conquest.

SOLO (Motion Picture—16mm—Sound). Pyramid Films, 1972. 16mm color, sound film, 15 min.; 3/4 in. videocassette also available. ($2.50). Gr. 5–A. EGT (1973), M&M (1973), NEF (1972). 796.522

A visual essay on mountain climbing and the efforts and exhilarations of individual effort. Photographed during several ascents, this film shows a lone climber achieving the rewards and personal satisfactions of his spectacular climbs.

MOURIS, FRANK

FRANK FILM (Motion Picture—16mm—Sound). Pyramid Films, 1973. 16mm color, collage animation, sound film, 9 min.; 3/4 in. videocassette also available. ($190). Gr. 7–A. AFF (1974), EGT (1974), LFR (12/19/74). 921

The filmmaker tells the story of his life and times through an incredible collection of color cutouts and a dual soundtrack—one track giving biographical data and the other dispensing words in free association with the visual images. This multiaward-winning film has applications in all curriculum areas.

MOZART, WOLFGANG

MOZART, THE MARRIAGE OF FIGARO (Filmstrip—Sound). *See* Operas

MUIR, JOHN

JOHN MUIR: CONSERVATION (Film-strip—Sound). The Great American Film Factory, 1976 (Outdoor Education). 1 color, b/w, and sepia, sound filmstrip, 52 fr.; 1 cassette, 15 min. #FS-6 ($35). Gr. 5–A. BKL (2/19/76), PRV (1977). 921

Words like "ecology" and "conservation" are taken for granted today. But it was John Muir and his efforts to promote the preservation of national land for recreational use that led to the development of the National Park Service. This filmstrip provides information on John Muir and conservation.

JOHN MUIR'S HIGH SIERRA (Motion Picture—16mm—Sound). Pyramid Films, 1974. 16mm color, sound film, 27 min.; ³/₄ in. videocassette also available. ($375). Gr. 7–12. BKL (1975), LFR (1975), LGB (1975). 921

Combines readings from Muir's poetic prose journals with scenes of the landscape he wrote about—the Sierra Nevada. Includes biographical information about John Muir and his role as an environmentalist.

MURLOCK, DINAH MARIE

THE LITTLE LAME PRINCE (Phonodisc or Cassette). *See* Fairy tales

MUSEUMS

MUSEUMS AND MAN (Filmstrip—Sound). EBEC, 1974. 5 color, sound filmstrips, av. 89 fr. ea.; 5 discs or cassettes, av. 13 min. ea.; Teacher Guide. #6905 Discs ($82.95), #6905K Cassettes ($82.95). Gr. 5–A. CPR (10/76), LFR (12/75), PRV (4/76). 069

Contents: 1. What Is a Museum? 2. An Exhibit: Behind the Scenes. 3. Museum Conservation: Preserving Our Heritage. 4. The Zoo: A Living Collection. 5. Museums: New Directions.

Using the Smithsonian museums as the focus, this series depicts the modern museum as a place to "do" as well as look. Students glimpse the museum's role as a restorer of precious and sometimes fragile objects.

SMITHSONIAN INSTITUTION (Motion Picture—16mm—Sound). EBEC, 1973. 16mm b/w, sound film, 21 min. #3188 ($130). Gr. 7–A. AAS (1976), BKL (1974), PRV (1975). 069

This inside look at the "nation's attic" gives the viewer an idea of the scale of the Institution, a sense of its present-day goals, and a knowledge of its early history. This film provides a close-up view of some of the museum's fascinating collections, such as Abraham Lincoln's silk hat and sixteen million insect specimens.

MUSIC, AMERICAN

DECLARATION OF INDEPENDENCE AND PROGRAM OF PATRIOTIC MUSIC (Phonodisc or Cassette). *See* U.S. Declaration of Independence

MUSICAL VISIONS OF AMERICA (Film-strip—Sound). Chevron School Broadcast/Cypress Publishing, 1977. 4 color, sound filmstrips, 93–119 fr.; 4 cassettes, 13–19 min. #086 ($96). Gr. 6–C. BKL (7/1/78). 780.9

Contents: 1. The Mix. 2. The Mix at Work. 3. The Mix Heats Up. 4. The Mix Moves On.

The history of American music. Reveals the contribution of all ethnic groups. Recreates historical times and places; shows styles and transitions from one musical form to another. Includes ballads, psalms, field hollers, fiddle tunes, marches, ragtime, minstrel, folk, work songs, spirituals, dixie, lullabies, chanties, hymns, gospel, blues, musicals, Charleston, jazz, country western, soul, swing, improvisation, rock, and more.

MUSIC—ANALYSIS, APPRECIATION

THE CONDUCTOR (Filmstrip—Sound). *See* Conductors (music)

DESIGN IN MUSIC (Phonodiscs). Bowmar Publishing. 1 disc and overhead transparencies. #495 ($9.45). Gr. 4–8. AUR (1978). 780.1

A collection of compositions that presents the ABA form, theme and variations, rondo, and the symphony. Teaching notes help to explain musical form.

MUSIC APPRECIATION: TRAVELING SOUND TO SOUND (Filmstrip—Sound). Society for Visual Education, 1977. 5 color, sound filmstrips, 30–71 fr.; 5 cassettes or discs, 5–10 min.; Teacher's Guide. #A677-SATC Cassettes ($80), #A677 SATR Discs ($80). Gr. 4–12. BKL (6/1/78), PRV (1/79). 780.1

Contents: 1. Sound Waves. 2. Sounds of the Sea. 3. Sonar. 4. Bell Buoys. 5. Sounds of Silence.

An interdisciplinary approach, using historic prints, cartoon graphics, photographs, and paintings, as well as instrumental and vocal compositions specifically composed or selected for the musical insights to other subject areas. Word definitions, animal communications, physical properties of sound, types of sonor, origin of the bell buoy, and silent thought and communications are all described.

THE MUSIC MAKERS (Phonodiscs). *See* Musicians

MUSICAL VISIONS OF AMERICA (Film-strip—Sound). *See* Music, American

STORIES IN BALLET AND OPERA (Pho-nodiscs). Bowmar Publishing. 1 disc; Lesson

MUSIC—ANALYSIS, APPRECIATION (cont.)

Guides; Theme Chart. #071 ($10). Gr. 2–8. AUR (1978). 780.1

This disc presents samples from ballet and opera such as the Suite from *Amahl and the Night Visitors* by Menotti, *Hansel and Gretel* overture by Humperdinck, the *Nutcracker Suite* by Tchaikovsky, and others.

MUSIC—HISTORY AND CRITICISM

MUSICAL VISIONS OF AMERICA (Filmstrip—Sound). *See* Music, American

MUSIC, POPULAR

JAZZ, ROCK, FOLK, COUNTRY (Filmstrip—Sound). Troll Associates, 1976 (Troll Jam Sessions). Each of four modules contains 1 cassette tape, av. 12 min.; 1 color sound filmstrip, av. 34 frames; 10 Illus. Soft-Cover Books, 4 Duplicating Masters; Teacher's Guide. ($42.95). Gr. 5–8. BKL (1/15/77), PRV (10/77), TEA (12/77). 785.4

Contents: 1. Jazz. 2. Rock. 3. Folk. 4. Country.

Multimedia supplementary reading program based on the topics of jazz, rock, folk, and country music, their beginnings, development, and stars. Tape follows text in read-along book.

MUSIC IN EDUCATION

THE MUSIC CHILD (Motion Picture—16mm—Sound). *See* Handicapped children—Education

MUSICAL INSTRUMENTS

HARPSICHORD BUILDER (Motion Picture—16mm—Sound). *See* Harpsichord

THE MUSIC MAKERS (Phonodiscs). *See* Musicians

MUSICIANS

MEN BEHIND THE BRIGHT LIGHTS (Kit). EMC, 1976. 4 cassettes, 22–27 min.; 4 Paperbacks; Teacher's Guide. #ELC-231000 ($54). Gr. 4–8. BKL (2/15/77). 920

Contents: 1. Jim Croce. 2. Stevie Wonder. 3. John Denver. 4. Elton John.

The cassettes contain the text from the paperbacks that focus on incidents from the lives of four successful and familiar musicians.

THE MUSIC MAKERS (Phonodiscs). Chevron School Broadcast/Cypress Publishing, 1977. 10 discs 12 in. in six albums; Teacher's Guide. #078 ($90). Gr. 6–C. TEA (2/79). 780.92

Contents: 1. Strings (2 discs). 2. Keyboard (2 discs). 3. Woodwinds and Reeds (1 disc). 4. Percussion (2 discs). 5. Guitar (2 discs). 6. Brass (1 disc).

Contemporary artist-musicians talk about their work, their music, and their instruments, and give advice about learning. The musicians teach about musical styles as well, including classical, rock, pop, blues, jazz, bluegrass, and country. Commentary by Dizzie Gillespie, Herb Ellis, Joe Pass, Louis Bellson, Stan Getz, George Shearing, Andre Watts, and many others.

See also Names of individual musicians, e.g., Mozart, Wolfgang, etc.

MUSICIANS—FICTION

THE CONCERT (Motion Picture—16mm—Sound). *See* Pantomimes

MUSLIMS

ISLAM: THE PROPHET AND THE PEOPLE (Motion Picture—16mm—Sound). *See* Islam

MY OLD MAN

A DISCUSSION OF MY OLD MAN (Motion Picture—16mm—Sound). *See* Motion pictures—History and criticism

MYSTERY AND DETECTIVE STORIES

THE ADVENTURE OF THE SPECKLED BAND BY DOYLE (Cassette). Jimcim Recordings, 1978. 1 cassette, 46 min.; Script. ($6.95). Gr. 7–A. BKL (1/15/79). Fic

A Sherlock Holmes mystery is narrated in a precise, English manner. No music is used, and only little sound effects.

THE BEST OF ENCYCLOPEDIA BROWN (Filmstrip—Sound). Miller-Brody Productions, 1978. 4 color, sound filmstrips, 41–54 fr. ea.; 4 cassettes, 7-min. ea. #EBC1-4 ($64). Gr. 5–8. BKL (3/1/78), MNL (5/8/78), TEA (9/78). Fic

Contents: 1. The Case of Natty Nut. 2. The Case of the Scattered Cards. 3. The Case of the Hungry Hitchhiker. 4. The Case of the Whistling Ghost.

Adaptations of the books by Donald Sobol. Colorful drawings, animated reading, and dramatizations and lighthearted music provide the background for unraveling the mysteries of this favorite character.

TEN TALES OF MYSTERY AND TERROR (Cassette). *See* Literature—Collections

MYTHOLOGY

HEROIC ADVENTURES (Filmstrip—Sound). *See* Legends

MYTHOLOGY, CLASSICAL

GREAT MYTHS OF GREECE (Filmstrip—Sound). EBEC, 1972. 4 color sound filmstrips, av. 60 fr. ea.; 4 discs or cassettes, 8 min. ea.; Teacher's Guide. #6462 Discs ($57.95), #6462K Cassettes ($57.95). Gr. 4–8. BKL (5/1/73), PRV (2/74). 292

Contents: 1. An Introduction to Greek Mythology and the Myth of Narcissus. 2. Demeter and Persephone. 3. Phaeton. 4. Orpheus and Eurydice.

Themes as new as today shine through the enchanting stories of these famous myths. These filmstrips will be enjoyed and of benefit on many levels.

HEROES OF THE ILIAD (Cassettes). Children's Classics on Tape, 1973. 2 cassettes (4 parts); Detailed Study Guide. #CCT-143 ($21.95). Gr. 4–8. BKL (4/1/75), PRV (9/75). 292

Contents: Part 1—How the Greek and Trojan war began. Quarrel between Achilles and Agamemnon. Part 2—Hector's men storm Greek camp. Achilles refuses to help Agamemnon. Part 3—Death of Patroclus. Hector and Achilles meet in battle. Part 4—Hector's funeral.

Presents the story of the Trojan horse and fall of Troy. Heinrich Schliemann's discovery is also discussed.

HOMER'S MYTHOLOGY: TRACING A TRADITION (Filmstrip—Sound). *See* Greek poetry

ROMAN MYTHOLOGY (Filmstrip—Sound). Coronet Instructional Media, 1974, 4 color, sound filmstrips, av. 55 fr.; 2 discs or 4 cassettes, av. 15 min. #M245 Cassettes ($65), #S245 Discs ($65). Gr. 7–12. PRV (1975). 292

Contents: 1. Origins and Development. 2. The Gods and Their Powers. 3. Metamorphosis 4. The State and Its Heroes.

Art and photographs from Egypt, Persia, and Greece disclose major differences in Greek and Roman mythology. Historical and mythical influences in art and literature and religion within the Roman Empire are explained.

SISYPHUS (Motion Picture—16mm—Sound). Pannonia Studios, Budapest/Pyramid Films, 1975. 16mm b/w, animated, sound film, 3 min.; 3/4 in. videocassette also available. ($60). Gr. 7–A. BKL (1977), EGT (1977), LNG (1977). 292

An animated interpretation of the classic Greek myth about a man who pushes an ever-growing boulder up a mountain. This version has an unconventional ending, illustrating the nature of human endeavor. The bold brush strokes suggest Oriental artwork.

SPLENDOR FROM OLYMPUS (Kit). EMC, 1972. 2 color, sound filmstrips, 67–82 fr.; 2 cassettes, 13–14 min.; 8 Cassette Dramatizations, 1 Wall Map; Spirit Masters; Teacher's Guide. #EL208000 ($110). Gr. 6–10. BKL (5/72), EGT (5/72), IDF (1972). 292

Contents: Sound filmstrips: 1. Of Gods and Heroes. 2. Mythology Now. Cassettes: 1. Perseus Slays the Gorgon. 2. Theseus Slays the Minotaur. 3. Pyrrha and Deucalion. 4. The Story of Midas. 5. Philemon and Baucis. 6. The Story of Phaethon. 7. Jason and the Golden Fleece, Parts 1 and 2.

A sight and sound approach to classical mythology and its relevance to present-day living. The gods and heroes of ancient Greece and Rome are introduced plus the countless classical allusions in our everyday world.

NAPOLEON I

NAPOLEON BONAPARTE (Cassette). Ivan Berg Associates/Jeffrey Norton Publishers, 1976 (History Makers). 1 cassette, approx. 75 min. #41003 (11.95). Gr. 10–A. BKL (9/15/77). 921

Napoleon Bonaparte's fascinating career and his rise to unparalleled heights, his triumphs, the disaster of the Russian steppes, his final defeat at Waterloo, and a lonely death on the remote island of St. Helena are presented in this recording.

NATIONAL CHARACTERISTICS, AMERICAN

THE AMERICAN SPIRIT (Motion Picture—16mm—Sound). *See* U.S.—History

AMERICANS ON AMERICA: OUR IDENTITY AND SELF IMAGE (Slides/Cassettes). *See* American literature—Study and teaching.

AUNT ARIE (Motion Picture—16mm—Sound). *See* Carpenter, Aunt Arie

THE ROOTS OF THE AMERICAN CHARACTER (Filmstrip—Sound.) *See* U.S.—History

NATIONAL PARKS AND RESERVES—U.S.

JOHN MUIR: CONSERVATION (Filmstrip—Sound). *See* Muir, John

JOHN WESLEY POWELL: DISCOVERY (Filmstrip—Sound). *See* Powell, John Wesley

NATIONAL SOCIALISM

ADOLF HITLER (Cassette). *See* Hitler, Adolf

NATIONAL SONGS, AMERICAN

DECLARATION OF INDEPENDENCE AND PROGRAM OF PATRIOTIC MUSIC (Phonodisc or Cassette). *See* U.S. Declaration of Independence

NATURAL HISTORY

ALONG NATURE TRAILS (Film Loop—8mm—Silent). Troll Associates. 6, 8mm color, silent film loops, approx. 4 min. ea.; available separately. ($149.70). Gr. K–8. AUR (1978). 574.5

Contents: 1. A Nature Walk. 2. Pattern and Design in Nature. 3. Life in a Pond. 4. Fireflies and Other Night Insects. 5. Importance of Weeds. 6. Nature Homebuilders.

These loops allow the observer to witness fascinating activities in nature's hidden world. Close-ups reveal the lives and habits of insects and other creatures.

BEAR COUNTRY AND BEAVER VALLEY (Filmstrip—Sound). Walt Disney Educational Media, 1975. 2 color, sound filmstrips; 2 cassettes. #63-00212L ($18). Gr. 5–9. PRV (11/76). 574.5

Contents: 1. Beaver Valley. 2. Bear Country.

Adapted from films of the same name, this set takes students on a field trip to see the black bears and beavers. It shows the interrelationship of a variety of animals from the bear's and beaver's environment.

POWERS OF NATURE (Filmstrip—Sound). *See* Geology

SEA, SAND AND SHORE (Filmstrip—Sound). *See* Ocean

SPACESHIP EARTH (Filmstrip—Sound). Hawkhill Associates, 1973. 6 color, sound filmstrips, 78–92 fr.; 6 cassettes or discs, 13–20 min.; Teacher's Guide. ($135). Gr. 5–9. BKL (6/74), PVB (5.74). 574.5

Contents: 1. The Universe. 2. The Biosphere. 3. Living Thing. 4. Cells. 5. Atoms and Molecules. 6. A Little While Aware.

Stresses the unity and harmony of nature by combining scientific facts and philosophy with the arts of photography, poetry, and music. Introduces students to current scientific knowledge in astronomy, ecology, biology, chemistry, and physics and their interdependence.

WHAT IS ECOLOGY? (SECOND EDITION) Motion Picture—16mm—Sound). *See* Ecology

NATURAL RESOURCES

THE EARTH'S RESOURCES (Filmstrip—Captioned). Multi-Media Productions/McGraw-Hill Films, 1970. 6 color, captioned filmstrips, av. 40 fr. ea.; Teacher's Guide. #619020 ($49). Gr. 4–8. PVB (5/73), PRV (1/73). 551

Contents: 1. Ocean Resources. 2. Fresh Water Resources. 3. Rock Resources. 4. Fuel Resources. 5. Soil Resources. 6. Atmospheric Resources.

These filmstrips present earth science concepts associated with some of the earth's natural resources, and emphasize the methods that allow scientists to formulate these concepts, and the importance of these resources to the environment.

NATURALISTS

JOHN MUIR: CONSERVATION (Filmstrip—Sound). *See* Muir, John

JOHN MUIR'S HIGH SIERRA (Motion Picture—16mm—Sound). *See* Muir, John

NATURE, EFFECT OF MAN ON. *See* Man—Influence on nature

NATURE IN ORNAMENT. *See* Design, Decorative

NATURE IN POETRY

THE CRYSTAL CAVERN (Filmstrip—Sound). *See* Caves

DAYBREAK (Motion Picture—16mm—Sound). EBEC, 1976. 16mm color, sound film, 9 min.; Guide. #3392 ($135). Gr. 5–A. BKL (11/1/77), EGT (1978), LNG (1977). 808.81

''Daybreak comes first/in thin splinters shimmering . . .''—Carl Sandburg's poem sets the stage for this view of the sunrise over the southwestern desert of Monument Valley. A haunting orchestral accompaniment and Indian poetry underscore the visual expression of daybreak as a natural phenomenon.

NAVAJO

BLESSINGWAY: TALES OF A NAVAJO FAMILY (Kit). EMC, 1976. 4 cassettes, av. 21 min.; 4 Paperbacks; Teacher's Guide. #ELC-129000 ($52). Gr. 3–8. BKL (4/15/77). 970.3

Contents: 1. The Spiderweb Stone. 2. Tall Singer. 3. The Secret of the Mask. 4. The Magic Bear.

Stories of a contemporary Indian family showing some of the problems as Indians move to Anglo communities. The beliefs of the Navajo are treated respectfully and told from the standpoint of the Indian.

NAVAJO SAND PAINTING CEREMONY (Film Loop—8mm—Silent.) *See* Indians of North America—Art

NAVAJO—FICTION

SING DOWN THE MOON (Filmstrip—Sound). Miller-Brody Productions, 1975 (Newbery Filmstrips). 2 color, sound filmstrips, av. 98–103 fr. ea.; 1 cassette or disc, av. 18 min. Phonodisc ($24), Cassette ($28). Gr. 4–8. BKL (9/15/75), PVB (4/78). Fic

With ruggedly bold illustrations whose strokes are as strong as the spirit of the Navajo people they portray, this set capsulizes and dramatically presents Scott O'Dell's award-winning historical novel.

NAZI MOVEMENT. *See* National socialism

NEAGLE, JOHN

AMERICAN CIVILIZATION: 1783–1840 (Filmstrip—Sound). *See* Art, American

NEAR EAST. *See* Middle East

NEGROES. *See* Blacks

NEMATODES

NEMATODE (Motion Picture—16mm—Sound). EBEC, 1973. 16mm color, non-narrated, sound film, 11 min. #3192 ($150). Gr. 7–A. BKL (12/15), PRV (1974), SCT (1974). 595.18

This nonnarrated film is an unconventional treatment of a species of roundworms, or nematodes, commonly found in plant roots. Using microscopes and extreme close-up lenses, the film depicts locomotion, feeding, reproduction, elimination, and other behavior of these tiny, plentiful worms. Use of captions and photography of toys helps to emphasize major concepts.

NEMEROV, HOWARD

HOWARD NEMEROV (Cassette). Tapes for Readers, 1978. 1 cassette, 23 min. #LIT-052A ($10.95). Gr. 11–A. BKL (1/15/79). 921

An interview with the prize-winning poet Nemerov. This tape also includes readings by the poet of some of his works.

NEW ZEALAND

AUSTRALIA AND NEW ZEALAND (Filmstrip—Sound). *See* Australia

NEWBERY MEDAL BOOKS

ELIZABETH YATES (Filmstrip—Sound). *See* Yates, Elizabeth

JULIE OF THE WOLVES (Phonodisc or Cassette). *See* Eskimos—Fiction

NEWBERY MEDAL BOOKS—BIOGRAPHY

ELEANOR ESTES (Filmstrip—Sound). *See* Authors

JEAN CRAIGHEAD GEORGE (Filmstrip—Sound). *See* Authors

LLOYD ALEXANDER (Filmstrip—Sound). *See* Authors

VIRGINIA HAMILTON (Filmstrip—Sound). *See* Hamilton, Virginia

WILLIAM H. ARMSTRONG (Filmstrip—Sound). *See* Armstrong, William

NEWSPAPERS

NEWSPAPER STORY (SECOND EDITION) (Motion Picture—16mm—Sound). EBEC, 1973 (World of Work). 16mm color, sound film, 27 min. #3200 ($360). Gr. 7–C. CFF (1975). EGT (1974), M&M (1974). 070

This film traces a 24-hour period in the life of the *Los Angeles Times*. As the camera shows how news is gathered, written, and edited and how newspapers are printed, the people who are being photographed describe what they do.

WHAT IS JOURNALISM? (Filmstrip—Sound). *See* Journalism

YOUR NEWSPAPER (Filmstrip—Sound). Globe Filmstrips/Coronet Instructional Media, 1976. 6 color, sound filmstrips; 6 cassettes or discs. #M721 ($99), #S721 ($99). Gr. 9–12. PRV (2/77), PVB (7/77). 070

Contents: 1. What Is a Newspaper? 2. Preparation and Production. 3. The News Story. 4. Features and Columns. 5. The Editorial Page. 6. How Free Is Today's Press?

This set shows the operation and function of a newspaper. Each strip analyzes aspects of the free press. Technical information is illustrated and the importance of following the traditional form and style of news writing is illustrated.

NEWTON, ISAAC

ISAAC NEWTON (Cassette). Ivan Berg Associates/Jeffrey Norton Publishers, 1976 (History Makers). 1 cassette, approx. 57 min. #41002 ($11.95). Gr. 10–A. BKL (5/15/78). 921

This program traces the life of a modest man who is regarded as the greatest scientist the world has ever known from his schooldays and his earliest experiments, through the terrible plague years when he began the work that was to make his name famous forever, to the publication of his book *The Principia*, which describes the workings of the universe.

NICHOLSON, NIGEL

THE LETTERS OF VIRGINIA WOOLF (Cassette). *See* Woolf, Virginia

NIGERIA

NIGERIA AND THE IVORY COAST: ENTERING THE 21ST CENTURY (Filmstrip—Sound). *See* Africa

NIGHTINGALE, FLORENCE

FLORENCE NIGHTINGALE (Cassette).
Ivan Berg Associates/Jeffrey Norton Publishers, 1976 (History Makers). 1 cassette, approx. 73. #41011 ($11.95). Gr. 7–A. BKL (2/15/78), PRV (5/19/78). 921

The accepted picture of Florence Nightingale is of a slim woman in a crinoline dress walking quietly through the crowded wards of a hospital in the Crimea, with a simple lamp held aloft, bringing comfort to the hundreds of wretched soldiers who stared gratefully at her as she passed. She was their "Angel of Mercy," their "Lady with the Lamp." Florence Nightingale's life from her childhood to her early interest in nursing and to the fever-haunted and filthy wards of Scutari and Balaclava and on through the years of inspired endeavor that followed are presented. This is the absorbing story of the "Lady with the Lamp."

NOBEL, ALFRED

ALFRED NOBEL—THE MERCHANT OF DEATH? (Motion Picture—16mm—Sound). Chatsworth Film Distributors/Centron Films, 1978 (The Prizewinners Series). 16mm color, sound film, 26-1/2 min. ($445). Gr. 7–A. BKL (9/15/78). 921

The film recounts the life of Alfred Nobel: his brilliant inventions; his early disasters with explosives; his attainment of vast wealth and industrial power; his tender love for Countess Bertha Kinsky; his romantic entanglements with Sophie Hess; and his consummate achievement—endowment of the fabulous Nobel prizes.

NOBEL PRIZES

ALFRED NOBEL—THE MERCHANT OF DEATH? (Motion Picture—16mm—Sound). See Nobel, Alfred

LINUS PAULING: SCIENTISTS AND RESPONSIBILITY (Cassette). See Pauling, Linus

NOBLE, ELAINE

A WOMAN'S PLACE IS IN THE HOUSE (Motion Picture—16mm—Sound). Texture Films, 1977. 16mm color, sound film, 30 min. ($350). Gr. 11–A. AFF (1977), ALA (1978), BKL (1977). 921

Elaine Noble is a young, intelligent, idealistic American woman in politics. Elected to the Massachusetts House of Representatives, her constituency combines the affluent Boston Back Bay area with the multiracial, working-class Fenway district. In addition, Elaine Noble is gay. Because of existing prejudice and fear of homosexuality, she feels it important to be open about her gayness and to confront the issue with absolute candor. This is a portrait of a contemporary woman dealing with some of the most complex and crucial issues of our day.

NOISE POLLUTION

NOISE AND ITS EFFECTS ON HEALTH (Motion Picture—16mm—Sound). FilmFair Communications, 1973. 16mm color, sound film, 20 min. ($275). Gr. 7–A. CFF (1973), FHE (10/73), LFR (10/73). 614.7

Defines the physical and psychological effects of noise on health and suggests what can and should be done to combat it.

WHO STOLE THE QUIET DAY? (Motion Picture—16mm—Sound). Alfred Higgins Productions, 1975. 16mm color, sound film, 15-1/2 min. ($225). Gr. 1–A. AAS (5/75), CFF (1975), PRV (3/75). 614.7

This film shows how loud noises and high frequencies destroy nerve cells within the ear—at any age. It points out that hearing loss is often not the result of aging, but a process that begins through thoughtless exposure to loud noises. Suggestions are made on how to protect our hearing.

NORTH AMERICA

HOW OUR CONTINENT WAS MADE (Filmstrip—Sound). Educational Dimensions Group, 1976. 2 color, sound filmstrips, av. 60 fr.; 2 cassettes, 18 min. ea.; Guide. #531 ($39). Gr. 6–10. BKL (6/1/77). 551.4

Contents: 1. Part 1: The East. 2. Part 2: The West.

A geological history of our continent and an analysis of those forces that build up and wear down—change and re-form—the continent. Aerial photographs, satellite photographs, stylized maps, split-frame comparisons, and diagrams are used in the visuals.

NUCLEAR PHYSICS

FERMILAB (Filmstrip—Sound). See Atom smashers

MATTER (Filmstrip—Sound). Scholastic Book Services, 1977 (Adventures in Science). 4 color, sound filmstrips, av. 63–76 fr.; 4 discs or cassettes, 11–15 min. #8627 Discs ($69.50), #8628 Cassettes ($69.50). Gr. 4–8. BKL (12/15/77). 539.7

Contents: 1. Atoms: Their Origin. 2. Atoms: Their Order. 3. Atoms: Reacting. 4. Atoms: The Nucleus.

Content of this filmstrip ranges from the hypothetical answers to queries on the origin of the stars to an explanation of Einstein's theory of the interchangeability of mass and energy; and clarifies the subject of nuclear fission.

NUMBERS. See Arithmetic; Counting

NURSES

FLORENCE NIGHTINGALE (Cassette). *See* Nightingale, Florence

NURSING

CAREERS IN NURSING (Filmstrip—Sound). Associated Press/Pathescope Educational Media, 1973 (Careers In). 2 color, sound filmstrips, 80 fr. ea.; 2 cassettes, 12–20 min.; Teacher's Manual. #701 ($50). Gr. 9–A. BKL (12/1/73), PRV (1/75). 610.73

Interviews with people on the job prepare the future nurse for the responsibilities, studies, and personal commitment required. A vast range of opportunities for both men and women in a variety of duties is shown.

NUTRITION

ANIMAL AND PLANT PROTEIN (Kit). *See* Cookery

BEFORE YOU TAKE THAT BITE (Motion Picture—16mm—Sound). FilmFair Communications, 1974. 16mm color, sound film, 14 min. #C249 ($185). Gr. 4–12. LFR (3/75), PRV (10/75). 613.2

An overweight teenager begins her day by missing breakfast and later has quick, sweet junk foods. At lunch she chooses desserts instead of foods with nutritional value and later at a basketball game again snacks on empty foods. Besides weight problems, other consequences of an unbalanced diet are illustrated such as skin problems, tooth decay, etc. The film provides information on the relative nutritional values of certain foods, and encourages a balanced diet as a primary step to good physical and emotional health. Finally, the film advises viewers to "read the label" and thus use their nutrition knowledge.

BREADS AND CEREALS (Kit). *See* Cookery

EAT, DRINK, AND BE WARY (Motion Picture—16mm—Sound). Churchill Films, 1975. 16mm color, sound film, 21 min; Guide. ($315). Gr. 7–A. BKL (11/1/75), EFL (1975). 613.2

A critical examination of our eating habits, of nutritional losses in food processing, of food additives, and of the role of food manufacturers in changing our diets.

THE EATING ON THE RUN FILM (Motion Picture—16mm—Sound). Alfred Higgins Productions, 1975. 16mm color, sound film, 15-¹/₂ min. ($225). Gr. 7–A. CFF (1975), LFR (5/75), PRV (10/75). 613.2

This humorous film shows how it is possible to have a well-balanced diet even while "eating on the run."

FOOD AND NUTRITION (Filmstrip—Sound). *See* Child development

FOOD AND NUTRITION: DOLLARS AND SENSE (Kit). Butterick Publishing, 1975. 4 color, sound filmstrips, av. 75 fr. ea.; 4 cassettes or discs, av. 8 min. ea.; Spirit Masters; Teacher's Guide. #C-503-2 Cassettes ($85), #R-50204 Discs ($85). Gr. 7–12. BKL (6/15/76). 613.2

Important background information about nutrition and metabolism is provided. Kitchen planning, food management and preparation, proper serving techniques, and good consumer practices are included.

FOOD: HEALTH AND DIET (Filmstrip—Sound). Sunburst Communications, 1976. 2 color, sound filmstrips, 83–93 fr.; 2 cassettes or discs, 15 min.; Teacher's Guide. #231 ($59). Gr. 9–12. BKL (3/15/77), CPR (1/77), PRV (10/77). 613.2

Contents: 1. Nutritional Needs of Your Body. 2. How to Diet Sensibly.

Nutrition is examined from the teenage point of view, stressing the importance of balanced meals to maintain good health. Specific nutrients are identified and their sources located. How the body utilizes its daily requirements and adolescent weight problems are explored. So-called energy diets for athletes, vitamin supplements, and diets for vegetarians are also investigated.

FRUITS AND VEGETABLES (Kit). *See* Cookery

KITCHEN WITH A MISSION: NUTRITION (Filmstrip–Sound). Encore Visual Education, 1977. 1 color, sound filmstrip, 74 fr.; 1 cassette, 21 min.; Teacher's Guide. #94 ($21). Gr. 7–A. PRV (1/79). 641.1

Betty Kamen, a natural foods nutritionist, presents nutritious eating as she shows how to prepare a healthful meal. The narration also includes interesting food facts.

MILK AND DAIRY PRODUCTS (Kit). *See* Cookery

NUTRITION AND GOOD HEALTH (Kit). Society for Visual Education, 1978. 4 color, sound filmstrips, 57–68 fr. ea.; 4 cassettes, 10–12 min.; 30 Skill Extenders, Teacher's Guide. #A576-SATC ($99). Gr. 9–12. BKL (1/15/79). 613.2

This kit encourages students to understand that good eating habits are necessary and gives them incentives to evaluate and improve their eating habits.

NUTRITION BASICS (Kit). *See* Cookery

NUTRITION: FOOD VS. HEALTH (Filmstrip—Sound). Sunburst Communications, 1975. 2 color, sound filmstrips, 80–102 fr.; 2 cassettes or discs, 9–14 min.; Teacher's Guide. #216 ($59). Gr. 8–12. BKL (11/1/75), LGB (12/75), M&M (5/76). 613.2

Contents: 1. The Food Game. 2. A Consumer Viewpoint.

In a quiz-show format, this unit examines the eating habits of Chuck America and

NUTRITION (cont.)

Susie Tomorrow as they vie for years of good health on "The Food Game." Explores why Americans have little awareness of their nutritional needs. Examines the role of manufacturers, advertisers, and governmental agencies in shaping American eating habits.

NUTRITION FOR YOUNG PEOPLE (Filmstrip—Sound). Guidance Associates, 1976. 4 color, sound filmstrips, 37–52 fr. ea.; 4 cassettes, 6–8 min. ea.; Discussion Guide. #6L-303-907 Discs ($94.50), #6L-303-915 Cassettes ($94.50). Gr. 5–8. BKL (12/1/76). 641.1

Contents: 1. Why People Eat What They Do. 2. How Food Becomes Part of You. 3. What Foods People Need. 4. What Is Food?

An overview of nutritional concepts, it treats seriously and thoroughly the various aspects of nutrition for middle-school students. Each strip approaches its subject a little differently, all combine five action photos, paper cutout graphics, and voice-over commentary to explore the importance of human nutrition.

TOO MUCH OF A GOOD THING (Filmstrip—Sound). Marshfilm. 1 color, sound filmstrip, 50 fr.; 1 disc or cassette, 15 min.; Teacher's Guide. #1118 ($21). Gr. 4–9. AIF (1975), PRV (3/75), PVB. 641.1

Indirectly attacking the problem of obesity in the young, parallels are noted between the national energy crisis and a human energy crisis, comparing the body to a power plant.

O PIONEERS!

EPISODES FROM CLASSIC NOVELS (Filmstrip—Sound). *See* Literature—Collections

OCCUPATIONS

ADVENTURES IN THE WORLD OF WORK, SET ONE (Kit). *See* Vocational guidance

ADVENTURES IN THE WORLD OF WORK, SET TWO (Kit). *See* Vocational guidance

AUTO RACING: SOMETHING FOR EVERYONE (Kit). *See* Automobile racing

BECAUSE IT'S JUST ME UNIT TWO (Motion Picture—16mm—Sound). WNVT (Northern Virginia Educational TV)/EBEC, 1973 (Watcha Gonna Do?). 16mm color, sound film, 15 min.; Teacher's Manual. #3352 ($200). Gr. 5–9. BKL (5/15/75). CPR (11/74), LFR (4/12/74). 331.702

Students are encouraged to take a self-inventory and to think about how their present abilities, interests, and hobbies might extend into career roles or to future leisure-time activities. Three individuals discuss artistic

avocations as an actor, puppeteer, or a musician.

BEGINNING CONCEPTS: PEOPLE WHO WORK, UNIT ONE (Kit). *See* Vocational guidance

A CAREER IN COMPUTERS (Filmstrip—Sound). *See* Computers

A CAREER IN SALES (Filmstrip—Sound). *See* Sales Personnel

CAREER TRAINING THROUGH THE ARMED FORCES (Filmstrip—Sound). *See* U.S.—Armed forces

CAREERS IN BANKING AND INSURANCE (Filmstrip—Sound). *See* Banks and banking

CAREERS IN BUSINESS ADMINISTRATION (Filmstrip—Sound). *See* Business

CAREERS IN ENGINEERING (Filmstrip—Sound). *See* Engineering

CAREERS IN FOOD SERVICE (Filmstrip—Sound). *See* Cookery, Quantity

CAREERS IN GRAPHIC ARTS (Filmstrip—Sound). *See* Graphic arts

CAREERS IN HEALTH SERVICES (Filmstrip—Sound). *See* Medical care

CAREERS IN HOME ECONOMICS (Filmstrip—Sound). *See* Home economics

CAREERS IN LAW ENFORCEMENT (Filmstrip—Sound). *See* Law enforcement

CAREERS IN NURSING (Filmstrip—Sound). *See* Nursing

CAREERS IN SOCIAL WORK (Filmstrip—Sound). *See* Social work

GETTING READY UNIT ONE (Motion Picture—16mm—Sound). WNVT (Northern Virginia Educational TV)/EBEC, 1973 (Watcha Gonna Do?). 16mm color, sound film, 15 min.; Teacher's Manual. #3358 ($200). Gr. 5–9. BKL (5/15/75), CPR (11/74), LFR (12/74). 331.702

Careers require preparation and planning. This film introduces a gymnast, an archeologist, and a policewoman, and shows how each of them goes about preparing for his or her career.

GOING BACK UNIT THREE (Motion Picture—16mm—Sound). WNVT (Northern Virginia Educational TV)/EBEC, 1973 (Whatcha Gonna Do?). 16mm color, sound film, 15 min. #3355 ($200). Gr. 5–A. BKL (5/15/75), LFR (12/74), PRV (10/76). 331.702

This film focuses on how one's environment (family, peers, culture, and geography) plays a role in career decisions.

JOBS IN AUTOMOTIVE SERVICES (Filmstrip—Sound). *See* Automobiles—Maintenance and Repair

JOBS IN DATA PROCESSING (Filmstrip—Sound). *See* Computers

NIGHT PEOPLE'S DAY (Motion Picture—16mm—Sound). *See* Cities and towns

NON-TRADITIONAL CAREERS FOR WOMEN (Filmstrip—Sound). *See* Women—Employment

PARTNERS, UNIT 3 (Motion Picture—16mm—Sound). WNVT (Northern Virginia Educational TV)/EBEC, 1973 (Watcha Gonna Do?). 16mm color, sound film, 15 min.; Teacher's Manual. #3356 ($200). Gr. 5–A. BKL (5/15/75), CGE (1974), LFR (12/74). 331.702

Some people want to be their own bosses and look for ways to create their own jobs. Many small businesses have been created through the imagination of one or two people. This film introduces students to how a business is run; it also shows them how, with imagination and some risk, one can carve out a job for oneself that doesn't already exist.

THE ROAD NEVER ENDS (Motion Picture—16mm—Sound). *See* Truck drivers

A TIME OF CHANGES UNIT THREE (Motion Picture—16mm—Sound). WNVT (Northern Virginia Educational TV)/EBEC, 1973 (Watcha Gonna Do?). 16mm color, sound film, 15 min.; Teacher's Guide. #3359 ($200). Gr. 5–A. AIF (1974), CPR (11/74), LFR (12/74). 331.702

How technological changes can affect career decisions is clearly emphasized in this film. Technology has made some working roles obsolete but has created others. The student's imagination will be spurred to consider jobs that might exist in the future although they do not exist today.

VOCATIONAL SKILLS FOR TOMORROW (Filmstrip—Sound). *See* Vocational guidance

WEATHER FORECASTING (Motion Picture—16mm—Sound). *See* Weather forecasting

WHAT DO YOU THINK? UNIT ONE (Motion Picture—16mm—Sound). WNVT (Northern Virginia Educational TV)/EBEC, 1973 (Watcha Gonna Do?). 16mm color, sound film, 15 min.; Teacher's Guide. #3351 ($200). Gr. 5–A, CPR (11/74), LFR (1974), PRV (12/74). 331.702

Career choice is presented as a developmental process, a continuous series of decisions about one's self and what one wants to be. An important underlying theme in the film is that each person is presented with choices that give him or her some control in shaping the future.

WHAT'S THE LIMIT? UNIT TWO (Motion Picture—16mm—Sound). WNVT (Northern Virginia Educational TV)/EBEC, 1973 (Watcha Gonna Do?). 16mm color, sound film, 15 min.; Teacher's Manual. #3357 ($200). Gr. 5–A. BKL (5/15/75), CPR (11/74), LFR (12/74). 331.702

All of us have limitations, and part of growing up is accepting and learning to deal with them. In this film, the student will observe two individuals who did not become what they had wanted to be. However, they have overcome disappointment and are happy in other occupations better suited to their abilities.

WHO WORKS FOR YOU? (Kit). *See* Vocational guidance

WHY MOTHERS WORK (Motion Picture—16mm—Sound). *See* Women—Employment

WOMEN AT WORK: CHANGE, CHOICE, CHALLENGE (Motion Picture—16mm—Sound). *See* Women—Employment

WORKING IN U.S. COMMUNITIES, GROUP ONE (Filmstrip—Sound). Society for Visual Education, 1970. 4 color, sound filmstrips, 43–54 fr.; 2 cassettes or discs, 9–13 min.; 4 Guides. #201-SAR Discs ($54), #201-SATC Cassettes ($54). Gr. 3–8. BKL (12/72). 331.7

Contents: 1. Old Sturbridge and Mystic Seaport: Historic Communities. 2. Douglas, Wyoming: Ranch Community. 3. Rockland, Maine: Coastal Community. 4. Flagstaff, Arizona: Service Community.

These filmstrips emphasize elementary economics. On-location photography shows students the diversity of business and industry, both large and small, and how they meet people's needs.

OCEAN

EARTH AND UNIVERSE SERIES, SET 2 (Filmstrip—Sound). *See* Geology

NIGHTLIFE (Motion Picture—16mm—Sound). Phoenix Films, 1976. 16mm color, sound film, 12 min. ($250). Gr. 3–A. AFF (1977). 574.5

Exploration of the wonders that exist in the depths of the silent, mysterious ocean. See underwater life at close range, and experience the quiet, shadowy world these intriguing sea creatures inhabit.

SEA, SAND AND SHORE (Filmstrip—Sound). Lyceum/Mook & Blanchard, 1969. 3 color, sound filmstrips, 44–48 fr.; 3 cassettes or discs, 6–11 min. #LY35469SC Cassettes ($71), #LY35469SR Discs ($56). Gr. 3–12. BKL (9/1/69), FLN (10/69), INS (11/6). 551.46

Contents: 1. Sea, Sand and Shore. 2. Signatures in the Sand. 3. The Art of the Sea.

This set tells of the working of the sea on the shore, recounting endless pounding, carving, and smoothing of the land's edge. Written and photographed by Ann Atwood.

OCEAN (cont.)

WHO OWNS THE OCEANS? (Filmstrip—Sound). *See* Maritime law

OCEANOGRAPHY

THE NEW OCEANS (Filmstrip—Sound). Educational Dimensions Group, 1973. 4 color, sound filmstrips, av. 60 fr. ea.; 5 cassettes, av. 16 min. ea.; Guide. #526 ($74). Gr. 6–8. BKL (1/15/77), PVB (1976/77). 551.46
This set explores the role of the ocean in the earth's energy cycle, the ocean as a physical environment, and humanity's exploration of the ocean today for tomorrow.

OCEANOGRAPHY (Filmstrip—Sound). Coronet Instructional Media, 1971. 6 color, sound filmstrips, av. 52 fr.; 3 discs or 6 cassettes, av. 13¹/₂ min. #M182 Cassettes (95). Gr. 7–12. BKL (1972), PRV (1972). 551.46
Contents: 1. How We Study Waves. 2. How We Study Tides. 3. How We Study Ocean Currents. 4. How We Study the Ocean Floor. 5. How We Study the Marine Sediments. 6. How We Study Marine Life.
This set is a photographic trip into the ocean to experience methods of measuring and charting effects of waves and tides on coastlines, the characteristics of currents, and ocean ecology.

THE OCEANS: A KEY TO OUR FUTURE (Filmstrip—Sound). United Learning, 1977. 5 color, sound filmstrips, av. 38–52 fr.; 5 cassettes, av. 9–10 min.; Teacher's Guide. #4510 ($82). Gr. 5–8. BKL (1/1/78), PRV (10/77). 551.46
Contents: 1. The Oceans: Past, Present and Future. 2. The Oceans: A Storehouse of Food. 3. The Oceans: A Storehouse of Raw Materials. 4. The Oceans: A Storehouse of Power. 5. Changing the Oceans.
Designed to relate basic earth and life science concepts to the world's oceans, this program portrays the oceans as a major resource for the future development of human society and economics. The human interrelationship with the sea is examined in historical and present-day perspectives with emphasis placed upon sensible and ecologically sound methods of utilizing the potential of the sea.

SCRUB (Filmstrip—Sound). *See* Scuba diving

THE WORLD OF INNERSPACE (Filmstrip—Sound). Lyceum/Mook & Blanchard, 1973. 2 color, sound filmstrips, av. 53–58 fr.; 2 cassettes or discs. #LY35773SC Cassettes ($46). #LY35773SR Discs ($37). Gr. 5–A. BKL (7/15/74), PRV (10/74), PVB (5/75). 551.46
Contents: 1. From Aquapedes to Aquanauts. 2. Windows in the Sea.
Presents an illuminating voyage in the ocean depths aboard the newest undersea research vessel. Discusses the possibilities of developing the riches of the seas.

OCTOPUS

OCTOPUS (Film—8mm—Silent). Walt Disney Educational Media, 1966. 8mm color, silent film loop, approx. 4 min. #62-5455L ($30). Gr. K–12. MDU (1974). 591.92
Shows a female octopus caring for her nest of eggs and the hatching of the eggs, as well as the octopus hunting.

O'DELL, SCOTT

CHILD OF FIRE (Phonodisc or Cassette). *See* Mexicans in the U.S.—Fiction

THE KING'S FIFTH (Filmstrip—Sound). *See* Discoveries (in geography)—Fiction

SING DOWN THE MOON (Filmstrip—Sound). *See* Navajo—Fiction

OFFICE MANAGEMENT

YOU AND OFFICE SAFETY (Motion Picture—16mm—Sound). Academy-McLarty Productions/Xerox Educational Publications, 1969. 16mm, color, sound film or videocassette, 8¹/₂ min. #2101 Film ($185), #2101 Videocassette ($125). Gr. 10–A. CGE (1970), CRA (1970). 651
Using visual hyperbole and laughter, this film illustrates how easy it is to avoid many common office accidents. The viewer quickly sees that safety in the office can be assured by using a little common sense, by thinking ahead, and by considering the consequences of one's actions. (Also available in Spanish.)

OLD AGE

AUNT ARIE (Motion Picture—16mm—Sound). *See* Carpenter, Aunt Arie

GRAMP: A MAN AGES AND DIES (Filmstrip—Sound). Sunburst Communications, 1976. 1 color, sound filmstrip, 86 fr.; 1 cassette or disc, 16 min.; Teacher's Guide. #230 ($29). Gr. 9–A. BKL (3/77) M&M.
Gramp's family decided to care for him at home even though he was senile and had decided to hasten death by not eating or drinking. This filmstrip is an adaptation of the book *Gramp* by Mark and Dan Jury. The warm family relationship is developed through black-and-white photographs that document the man's life. Special emphasis is placed on the last three years during which the family coped with his erratic behavior and deteriorating physical capabilities.

GROWING OLD WITH GRACE (Cassette). Cinema Sound/Jeffrey Norton Publishers, 1975 (Avid Reader). 1 cassette, approx. 55

min. #40205 ($11.95). Gr. 10–A. BKL (9/15/78). 301.43

Dr. Leopold Bellak, author of *The Best Years of Your Life,* and Dr. Robert Butler, author of *Why Survive?,* discuss how to grow old and face it with equanimity, both for the old person and for the family. They describe some of the services available, and some of the techniques of postponing the inevitable as long as possible.

LEE BALTIMORE: NINETY NINE YEARS (Motion Picture—16mm—Sound). *See* Baltimore, Lee

RUTH STOUT'S GARDEN (Motion Picture —16 mm—Sound). *See* Stout, Ruth

OLMSTED, FREDERICK LAW

AMERICANS WHO CHANGED THINGS (Filmstrip—Sound). *See* U.S.—Civilization—Biography

O'NEILL, EUGENE

INTERVIEWS WITH PLAYWRIGHTS: SOPHOCLES, SHAKESPEARE, O'NEILL (Cassette). *See* Dramatists

OPERAS

MOZART, THE MARRIAGE OF FIGARO (Filmstrip—Sound). Educational Audio Visual, 1977. 3 color, sound filmstrips, 65–82 fr.; 3 discs, 48 min.; Teacher's Guide. ($80). Gr. 8–12. PRV (1/79). 782.1

An abridged version of the Italian comic opera. The music and photography are taken from an actual performance by the New York City Opera Company. The interspliced narration on the recording correlates with the captions when needed for clarification on the screen. The teacher's manual includes a brief summary of the opera, background information, and suggestions for utilization.

OPTICAL ILLUSIONS

SENSES AND PERCEPTION: LINKS TO THE OUTSIDE WORLD (Motion Picture—16mm—Sound). *See* Senses and sensation

ORCHESTRAL MUSIC

AMERICAN SCENES (Phonodiscs). Bowmar Publishing, 1965. 1 disc; 6 in. × 36 in. Theme Charts. #074 ($10). Gr. 3–8. AUR (1978). 785.8

Contents: 1. Grand Canyon Suite, by Grofe. 2. Mississippi Suite, by Grofe.

Scenes of the Grand Canyon and the Mississippi are presented in dramatic music. Teaching suggestions are designed to aid students in understanding musical notation and concepts of melody, rhythm and form.

CLASSICAL MUSIC FOR PEOPLE WHO HATE CLASSICAL MUSIC (Phonodiscs). Listening Library. 1 disc, 12 in. 33¹/₃ rpm. #AM 20 R ($6.95). Gr. 2–A. MDU (1978). 785

Arthur Fiedler conducts the Boston Pops Orchestra in excerpts from such pieces as Schubert's *Unfinished Symphony* and Chopin's *Les Sylphides.*

STRAVINSKI: GREETING PRELUDE, ETC. (Phonodiscs). Columbia Records (Special Products). 1 disc, 33¹/₃ rpm. #M-31729 ($6.98). Gr. 4–A. PRV (9/75). 785

Contents: 1. Greeting Prelude. 2. Dumbarton Oaks Concerto. 3. Circus Polka. 4. Eight Instrumental Miniatures for 15 Players. 5. Four Etudes for Orchestra. 6. Two Suites for Small Orchestra.

Columbia Symphony Orchestra and the Columbia Chamber Orchestra, with Igor Stravinski conducting, present this enjoyable disc.

OROZCO, RAFAEL

PIANO CONCERTO NO. 3 IN D MINOR (Phonodiscs). *See* Concertos

ORWELL, GEORGE

ANIMAL FARM BY GEORGE ORWELL (Filmstrip—Sound). *See* Fantasy—Fiction

OSMOSIS

DIFFUSION AND OSMOSIS (2ND EDITION) (Motion Picture—16mm—Sound). EBEC, 1973 (Biology Program). 16mm color, sound film, 14 min. #3207 ($185). Gr. 7–12. BKL (12/15/73), PRV (11/74). 574.875

This film is a study of diffusion, the random movement of molecules by which matter is transported as it takes place in living cells. Photomicrography, time-lapse photography, simple lab demonstrations, and a dynamic model help to explain a diffusion, clarify how a selectively permeable membrane functions, and demonstrates the effects of osmosis on plant cells.

OUTDOOR LIFE

FOUR WHEELS (Filmstrip—Sound). *See* Camping

LIFE IN RURAL AMERICA (Filmstrip—Sound). *See* U.S.—Social life and customs

MOUNTAINEERING (Filmstrip—Sound). *See* Mountaineering

WEATHER IN THE WILDERNESS (Filmstrip—Sound). *See* Survival

OWLS—FICTION

LA LECHUZA—CUENTOS DE MI BARRIO (Cassette). Educational Activities. 1

OWL—FICTION (cont.)

cassette; Text. #BC-148 ($12.90). Gr. 5–8. PRV (11/76). 468.6

Four short stories in Spanish of mystery and suspense of the owl. No English translation included.

PACIFIC STATES

FOCUS ON AMERICA, ALASKA AND HAWAII (Filmstrip—Sound). *See* U.S.—Economic conditions

PAINE, THOMAS

THOMAS PAINE (Motion Picture—16mm— Sound). EBEC, 1975. 16mm color, sound film, 13 min. #3427 ($185). Gr. 6–12. EFL (1977), LFR (1/76), PVR (3/76). 921

The life of America's revolutionary journalist is portrayed in animation featuring original watercolor by the artist A. N. Wyeth. Surprising facts are included: Paine's active role in the French Revolution, a cause he embraced fervently after seeing our own struggle for independence safely through, and his scientific interests, for example.

PAINTERS, AMERICAN

JACK LEVINE (Motion Picture—16mm— Sound). *See* Levine, Jack

PAINTERS—FICTION

MINDSCAPE (Motion Picture—16mm— Sound). National Film Board of Canada/ Pyramid Films, 1976. 16mm b/w, sound film, pinscreen animation, 8 min.; $3/4$ in. videocassette also available. ($150). Gr. 10–A. BKL (1976), FLQ (1976), MDM (1976). Fic

A painter steps into the scene of the landscape he is painting and travels the regions of the mind. The images of this dreamlike, symbolic journey were created by manipulating 240,000 pins on a perforated screen.

PAINTERS, FRENCH

DEGAS (Motion Picture—16mm—Sound). *See* Degas, Edgar

PAINTERS, ITALIAN

FUTURISM (Motion Picture—16mm— Sound). Texture Films, 1972. 16mm color, sound film, 20 min. ($260). Gr. 10–A. AFF (73). 759.5

The Italian Futurists, in 1910, were the first artists to celebrate modern technology: figures in violent motion, complex interplay between objects and space. This film captures the restless, experimental energy of Futurism and portrays the work of its major art-

ists: Marinetti, Balla, Boccione, Carra, Russolo, Severini.

PAINTINGS, AMERICAN

A. J. MILLER'S WEST: THE PLAINS INDIAN—1837—SLIDE SET (Slides). *See* Indians of North America—Paintings

AUDUBON'S SHORE BIRDS (Motion Picture—16mm—Sound). *See* Water birds

JOHN SINGLETON COPLEY (Motion Picture—16mm—Sound). *See* Copley, John Singleton

PAINTINGS—COLLECTIONS

THE IMPRESSIONIST EPOCH (Filmstrip— Sound). *See* Impressionism (art)

PAINTINGS, DUTCH

REMBRANDT (Motion Picture—16mm— Sound). *See* Rembrandt

PAINTINGS, ENGLISH

JOSEPH MALLORD WILLIAM TURNER, R. A. (Motion Picture—16mm—Sound). *See* Turner, Joseph M. W.

PAINTINGS, FRENCH

FRAGONARD (Motion Picture—16mm— Sound). *See* Fragonard, Jean Honore

GEORGES ROUAULT (Motion Picture— 16mm—Sound). *See* Rouault, Georges

RENOIR (Motion Picture—16mm—Sound). *See* Renoir, Pierre Auguste

PAINTINGS, SPANISH

GOYA (Motion Picture—16mm—Sound). *See* Goya, Francisco

PAKISTAN

SOUTH ASIA: REGION IN TRANSITION (Filmstrip—Sound). *See* Asia

PALEONTOLOGY. *See* Fossils

PANTOMIMES

BIP AS A SKATER (Motion Picture— 16mm—Sound). EBEC, 1975 (The Art of Silence: Pantomimes with Marcel Marceau). 16mm color, sound film, 8 min.; Teacher's Guide. #47803 ($135). Gr. 1–A. BKL (11/ 75), LFR (3/4/76), PRV (1/76). 792.3

Marceau created Bip as his "silent alter ego." In this Bip pantomime, we see Bip stumble, fight for balance, and joyously strive to become a great skater—on illusionary ice.

THE CONCERT (Motion Picture—16mm—Sound). Pyramid Films, 1975. 16mm color, sound film, 12 min.; 3/4 in. videocassette also available. ($200). Gr. K-A. LGB (1975), NEF (1975), WLB (4/75). 792.3

Utilizing his genius for pantomime, Julian Chagrin creates a fantasy of a London musician who performs an extraordinary concert on the black and white stripes of a pedestrian crosswalk behind London's Royal Albert Hall.

THE DREAM (Motion Picture—16mm—Sound). EBEC, 1975. 16mm color, sound film, 9 min. #47807 ($140). Gr. 7-C. BKL (1975), LAM (1976), PRV (1976). 792.3

"How am I to express my dreams?" Marceau answers this self-posed question in pantomime. He interprets the idea of dreaming through dancelike symbolic movement, which is perhaps closer to the sensation of dreaming than words could be.

THE HANDS (Motion Picture—16mm—Sound). EBEC, 1975 (The Art of Silence: Pantomimes with Marcel Marceau). 16mm color, sound film, 7 min. #47809 ($135). Gr. 7-A. CFF (1975), LGB (1975), MNL (1977). 792.3

This pantomime is an allegory of good and evil. The hands are the struggle of good and evil—revealed by the force of the hands themselves through symbolic movements. Marceau's selection of dramatic music complements his motion and helps to create the changing moods.

THE MASKMAKER (Motion Picture—16mm—Sound). EBEC, 1975. 16mm color, sound film, 9 min. #47805 ($140). Gr. 7-C. BKL (1975), FLN (1977), LAM (1976). 792.3

For Marcel Marceau, the maskmaker is one who represents humanity, with all the faces humanity can possess. Among the masks that Marceau impersonates are included the classic symbols for comedy and tragedy.

PANTOMIME: THE LANGUAGE OF THE HEART (Motion Picture—16mm—Sound). EBEC, 1975 (The Art of Silence: Pantomimes with Marcel Marceau). 16mm color, sound film, 10 min.; Teacher's Guide. #47813 ($150). Gr. 1-A. BKL (11/75), FLN (3/4/77), PRV (3/76). 792.3

In this film, Marceau talks about mime—how the body movement and gestures communicate attitudes and emotions. He calls pantomime the language of the heart. Brief clips from many of the pantomimes in the series vividly illustrate his words.

THE SIDE SHOW (Motion Picture—16mm—Sound). EBEC, 1975 (The Art of Silence: Pantomimes with Marcel Marceau). 16mm color, sound film, 9 min. #47810 ($140). Gr. 1-A. BKL (11/1/77), FLN (3/4/77), PRV (1/76). 792.3

In this pantomime, Marceau shows circus performers demonstrating their skills: a juggler, an acrobat, and a clown pulling ropes without ropes. The major performer is a tightrope walker. Why is it more frightening to see Marceau walk a tightrope on the floor than a real tightrope many feet in the air? The answer is found in Marceau's introductory remarks ". . . Illusion is stronger than reality sometimes."

YOUTH, MATURITY, OLD AGE, AND DEATH (Filmstrip—Sound). EBEC, 1975. 16mm color, sound film, 8 min. #47802 ($135). Gr. 7-C. BKL (1975), FLN (1977), LAM (1976). 792.3

In four minutes, and not moving from one spot, Marcel Marceau expresses the seven ages of humankind—beginning with the life emergence from a mother's womb, through the cycles of life, to the grave. He believes "this pantomime reveals what mime can do in condensing time."

PAPER CRAFTS

REDISCOVERY: ART MEDIA SERIES, SET 2 (Filmstrip—Sound). *See* Handicraft

PARENT AND CHILD

ADOLESCENT CONFLICT: PARENTS VS. TEENS (Filmstrip—Sound). *See* Adolescence

BECOMING A PARENT: THE EMOTIONAL IMPACT (Filmstrip—Sound). *See* Youth as parents

BUILDING A FUTURE (Filmstrip—Sound). *See* Youth as parents

COPING WITH PARENTS (Motion Picture—16mm—Sound). *See* Conflict of generations

DEALING WITH PRACTICAL PROBLEMS OF PARENTHOOD (Filmstrip—Sound). *See* Youth as parents

THE FAMILY (Kit). *See* Family life

INFANT CARE AND UNDERSTANDING (Kit). *See* Infants—Care and hygiene

LEARNING AWAY FROM HOME (Kit). *See* Family life

LEARNING IN THE HOME (Kit). *See* Family life

LEARNING THROUGH PLAY (Kit). *See* Family life

LOOKING TOWARDS ADULTHOOD (Filmstrip—Sound). *See* Adolescence

MY PARENTS ARE GETTING A DIVORCE (Filmstrip—Sound). *See* Divorce

THE PARENT AS A TEACHER (Kit). *See* Family life

PORTRAIT OF PARENTS (Filmstrip—Sound). *See* Adolescence

PARENT AND CHILD (cont.)

PORTRAIT OF TEENAGERS (Filmstrip—
Sound). *See* Adolescence

PREPARATION FOR PARENTHOOD (Filmstrip—Sound). Sunburst Communications,
1975. 3 color, sound filmstrips, 72–79 fr.; 3
cassettes or discs, 9–11 min.; Teacher's
Guide. #212 ($85). Gr. 9–C. BKL (11/15/75),
PRV (5/76), TSS (1/78). 301.42
Contents: 1. The Decision. 2. The Alternatives. 3. Memories.
Investigates qualities that lead to effective
parenting and stresses that modern methods
for family planning and greater individual
freedoms have made parenthood a conscious, personal choice. Traditional attitudes toward parenthood and how changes
in contemporary life have affected the job of
today's parents are examined.

RIGHTS AND OPPORTUNITIES (Filmstrip—Sound). *See* Youth as parents

THE STRUGGLE FOR INDEPENDENCE
(Filmstrip—Sound). *See* Adolescence

PARENT AND CHILD—FICTION

THE GRIZZLY (Kit). *See* Bears—Fiction

PARENTS AND TEACHERS. *See* Home and
school

PARKS, ROSA

TALKING WITH THOREAU (Motion Picture—16mm—Sound). *See* Philosophy,
American

PASSION PLAYS

AGUA SALADA (Motion Picture—16mm—
Sound). Caltari Films/Texture Films, 1975.
16mm b/w, sound film, 12 min. ($135). Gr. 8–
A. CIF (1976), LFR (1976). 792.1
This modern-day passion play by a young
Latin American director tells the tragic tale
of two fishermen, father and son. Shot on location along the turbulent Peruvian seacoast, the film is dramatically constructed to
the powerful music of J. S. Bach's "Passion
According to St. John."

PATRIOTIC SONGS. *See* National Songs—
American

PATRIOTISM

THE AMERICAN SPIRIT (Motion Picture—
16mm—Sound). *See* U.S.—History

PAULING, LINUS

LINUS PAULING: SCIENTISTS AND RE-
SPONSIBILITY (Cassette). Jeffrey Norton
Publishers, 1976. 1 cassette, approx. 26 min.

#33032 ($13.95). Gr. 10–A. BKL (12/15/77),
PRV (1/19/78). 921
Awarded the 1954 Nobel Prize for Chemistry
and the 1962 Nobel Peace Prize, Linus Pauling speaks about the explosion of atomic
bombs in Japan in 1945 and his feelings regarding the elimination of war. Reporter Pat
Williams elicits information on the life and
career of Pauling.

PEARLE, CHARLES WILLSON

AMERICAN CIVILIZATION: 1783–1840
(Filmstrip—Sound). *See* Art, American

PEARL DIVERS AND DIVING—FICTION

THE PEARL BY JOHN STEINBECK (Filmstrip—Sound). *See* Mexico—Fiction

PELICANS

PELICANS (Film Loop—8mm—Silent). Walt
Disney Educational Media, 1966. 8mm col-
or, silent film loop, approx. 4 min. #62–
5407L ($30). Gr. K–12. MDU (1974). 598.4
Both species of American pelicans are examined flying, diving, brooding, and feeding
their young. The nest is shown and young
are observed.

PENGUINS

ANTARCTIC PENGUINS (film Loop—
8mm—Silent). Walt Disney Educational Media, 1966. 8mm color, silent film loop, approx. 4 min. #62–5370L ($30). Gr. K–12.
MDU (1974). 598.4
Penguins are viewed on land and as they
swim and play in the icy water. Close-ups of
penguins and their young are shown.

PEOPLE'S REPUBLIC OF CHINA. *See* China

PERCEPTION

LISTENING BETWEEN THE LINES (Motion Picture—16mm—Sound). *See* Listening

OUR VISUAL WORLD (Filmstrip—Sound).
See Art appreciation

SENSES AND PERCEPTION: LINKS TO
THE OUTSIDE WORLD (Motion Picture—
16mm—Sound). *See* Senses and senation

PERFORMING ARTS

BECAUSE IT'S JUST ME UNIT TWO (Motion Picture—16mm—Sound). *See* Occupations

PERSIA. *See* Iran

PERSONALITY

MASKS: HOW WE HIDE OUR EMOTIONS
(Filmstrip—Sound). *See* Emotions

PERSONALITY: ROLES YOU PLAY (Film-strip—Sound). *See* Self

PERSONNEL MANAGEMENT

THE ART OF MOTIVATION (Motion Pic-ture—16mm—Sound). *See* Motivation (psychology)

PETER, PAUL, AND MARY

FOLK SONGS IN AMERICAN HISTORY, SET ONE: 1700-1864 (Filmstrip—Sound). *See* Folk songs—U.S.

FOLK SONGS IN AMERICAN HISTORY, SET TWO: 1865-1967 (Filmstrip—Sound). *See* Folk songs—U.S.

PETROGLYPHS. *See* Indians of North America—Art

PETS

PLAYING IT SAFE WITH ANIMALS (Film-strip—Sound). *See* Animals—Habits and behavior

THE SNAKE: VILLAIN OR VICTIM? (Motion Picture—16mm—Sound). *See* Snakes

PHILIPPINE ISLANDS

THE TASADAY: STONE AGE PEOPLE IN A SPACE AGE WORLD (Filmstrip—Sound). *See* Man, Nonliterate

PHILOSOPHY, AMERICAN

TALKING WITH THOREAU (Motion Pic-ture—16mm—Sound). Signet Productions/EBEC, 1975. 16mm color, sound film, 29 min. ($395). Gr. 8-A. EFL (1977), EGT (1976), M&M (1975). 191
Borrowing a "time travel" technique from science fiction, four contemporary Americans drop in at Walden Pond for a chat with Henry David Thoreau. The visitors are David Brower, conservationist; B. F. Skinner, controversial Harvard psychologist; Rosa Parks, the black woman who, in 1955, sparked the modern civil rights movement; and Elliot Richardson, former U.S. attorney general. Searching questions are raised as Thoreau's ideas are sometimes attacked, sometimes defended, but always found to be relevant.

PHONETICS

PHONETIC RULES IN READING (Kit). *See* Reading—Study and teaching

PHONICS. *See* Reading—Study and teaching

PHOTOGRAPHY

BASIC PHOTOGRAPHY (Filmstrip—Sound). Coronet Instructional Media, 1975. 6 color, sound filmstrips, 58–66 fr.; 6 cassettes or 3 discs, 10–14 min. #M304 Cassettes ($95), #S304 Discs ($95). Gr. 4-A. BKL (10/15/75), PRV. 770.28
Contents: 1. Which Camera for You? 2. You're in Control. 3. Making Film Work. 4. Adding Light. 5. Darkroom Skills. 6. Taking Better Pictures.
Detailed photographs introduce students to many types of cameras, use of their controls, and the reasons for choosing different kinds of cameras and darkroom equipment.

PRACTICAL FILM MAKING (Motion Pic-ture—16mm—Sound). *See* Motion picture photography

SEEING . . . THROUGH A LENS (Film-strip—Sound). ACI Media/Paramount Communications, 1975. 4 color, sound filmstrips; 4 cassettes av. 8 min. ea.; Teacher's Guide. #9124 ($78). Gr. 4-8. BKL (4/75), PVB (1977), TEA (5/6/76). 770
Contents: 1. Click—The Fun of Photography. 2. The Process—How It Works. 3. The Art—Light and Composition. 4. Variations and Applications.
This set explains the process and the art of photography and at the same time develops the basic ability to see creatively and with discrimination.

PHOTOGRAPHY—HISTORY

MATHEW BRADY: PHOTOGRAPHER OF AN ERA (Motion Picture—16mm—Sound). Texture Films, 1976. 16mm b/w, sound film, 12 min. ($135). Gr. 8-A. BKL (1977), CGE (1977). 770.9
Mathew Brady was the most fashionable daguerreotype portrait photographer of his time. In 1860, when the nation was divided by civil war, Brady left his galleries for the battlefront. He abandoned the daguerreotype for wet-plate photography, with its multiple prints. He trained the world's first combat crew, equipped his men with traveling darkrooms, and used the camera as it had never been used before: to record the history of a nation; to create a passionate social document.

PHOTOTROPISM

ISOLATION AND FUNCTION OF AUXIN (Film Loop—8mm—Silent). BFA Educational Media, 1973 (Plant Behavior). 8 mm color, silent film loop, approx. 4 min. #481414 ($30). Gr. 4-9. BKL (5/1/76). 581.18
This demonstrates experiments, phenomena, and behavior in plants.

MECHANICS OF PHOTOTROPISM (Film Loop—8mm—Captioned). BFA Education-

PHOTOTROPISM (cont.)

al Media, 1973 (Plant Behavior). 8mm color, captioned film loop, approx. 4 min. #481423 ($30). Gr. 4–9. BKL (5/1/76). 581.18

This demonstrates experiments, phenomena, and behavior in plants.

THE NATURE OF PHOTOTROPISM (Film Loop—8mm—Captioned). BFA Educational Media, 1973 (Plant Behavior). 8mm color, captioned film loop, approx. 4 min. #481422 ($30). Gr. 4–9. BKL (5/1/76). 581.18

This demonstrates experiments, phenomena, and behavior in plants.

PHYSICAL EDUCATION AND TRAINING

PHYSICAL FITNESS: IT CAN SAVE YOUR LIFE (motion Picture—16mm—Sound). EBEC, 1977. 16mm color, sound film, 23 min. #3575 ($320). Gr. 5–12. LFR (1979), PRV (1978). 613.7

Add overeating to inactivity and you have poor physical health. The film suggests a long-range alteration of eating habits, and daily exercise. Although San Francisco 49er Gene Washington appears in the film, the focus is not on becoming a football star, but picking activities that fit into one's own life. Fad diets are definitely out. Stressed is the importance of long-term diet habits.

PHYSICALLY HANDICAPPED

CHANGES (Motion Picture—16mm—Sound). Stanfield House, 1977. 16mm color, sound film ($335). Gr. 9–C. CGE (1978), IFT (1977). 362.4

A documentary about a group of physically handicapped university students who speak frankly about their lives, their capabilities, and their needs, in an effort to break down architectural and attitudinal barriers.

GRAVITY IS MY ENEMY (Motion Picture—16mm—Sound). See Hicks, Mark

MY SON, KEVIN (Motion Picture—16mm—Sound). See Handicapped children

PORTRAIT OF CHRISTINE (Motion Picture—16mm—Sound). Film Arts Production/Wombat Productions, 1977. 16mm color, sound film, 25 min. ($375). Gr. 9–A. BKL (10/1/78), LFR (9/10/78). 362.4

Orthopedically handicapped since infancy, twenty-eight year old Christine Karcza is compelling proof that given the "opportunity of risk," the disabled can become fully involved members of society.

PHYSICALLY HANDICAPPED— BIOGRAPHY

LEO BEUERMAN (Motion Picture—16mm—Sound). See Beuerman, Leo

PHYSICALLY HANDICAPPED—FICTION

THE LITTLE LAME PRINCE (Phonodisc or Cassette). See Fairy tales

PHYSICIANS

INTERN: A LONG YEAR (Motion Picture—16mm—Sound). See Mack, Karen

PHYSICS

STANDING WAVES AND THE PRINCIPLES OF SUPERPOSITION (Motion Picture—16mm—Sound). See Waves

PHYSIOLOGY

THE BODY (Filmstrip—Sound). Scholastic Book Services, 1977 (Adventures in Science). 4 color, sound filmstrips, av. 54–70 fr.; 4 cassettes or discs, av. 10–12 min.; Teacher's Guide. Discs #8623 ($69.50), Cassettes #8624 ($69.50). Gr. 4–8. BKL (12/15/77). 612

Contents: 1. Circulation and Respiration. 2. Genetics and Blood. 3. Digestion and Hunger. 4. Muscles and Nerves.

Respiratory problems suffered by some athletes in the 1968 Mexico City Olympics lead into the study of the circulation and respiration of the body. The discovery of sickle-cell anemia and the resulting research in genetics and blood are relived. Experiments uncovered the origins of hunger sensation while curiosity over the paralyzing effects of curare led to the discovery of medicinal uses for that lethal substance.

THE HUMAN BODY (Filmstrip—Sound). See Anatomy

THE HUMAN BODY, SET ONE (Filmstrip—Sound). See Anatomy

THE HUMAN BODY, SET TWO (Filmstrip—Sound). See Anatomy

HUMAN BODY AND HOW IT WORKS (filmstrip—Sound). See Anatomy

THE HUMAN MACHINE (Filmstrip—Sound). See Anatomy

THE MENTAL/SOCIAL ME (Kit). See Anatomy

THE PHYSIOLOGY OF EXERCISE (Filmstrip—Sound). See Exercise

PSYCHOACTIVE (Motion Picture—16mm—Sound). See Drugs—Physiological effect

REGULATING BODY TEMPERATURE (SECOND EDITION) (Motion Picture—16mm—Sound). See Body temperature

WORK OF THE KIDNEYS (SECOND EDITION) (Motion Picture—16mm—Sound). See Kidneys

THE WORLD OF ME (Kit). *See* Anatomy

PICTURE WRITING

PETROGLYPHS: ANCIENT ART OF THE MOJAVE (Filmstrip—Sound). *See* Indians of North America—Art

PIERNE (HENRI CONSTANT) GABRIEL

MARCHES (Phonodiscs). *See* Marches (music)

PINES, MARK

THE HUNGRY PLANT (Filmstrip—Sound). *See* Plants—Nutrition

PIONEER LIFE. *See* Frontier and pioneer life

PIPPA THE CHEETAH

PIPPA THE CHEETAH (Filmstrip—Sound). *See* Cheetahs

PIRATES—FICTION

TREASURE ISLAND (Cassette). Jabberwocky Cassette Classics, 1972. 2 cassettes, 120 min. #1011-1021 ($15.96). Gr. 6-9. BKL (3/15/74), JOR (11/74), PRV (11/73). Fic
Contents: 1. Part One. 2. Part Two.
Stevenson's favorite tale of the search for hidden treasure is narrated by a boy who portrays the character of Jim Hawkins. He relates the happenings on the voyage with Long John Silver as the few sound effects add to the telling of this tale.

TREASURE ISLAND (Filmstrip—Sound). Listening Library, 1977 (Visu-Literature). 2 color, sound filmstrips, av. 85 fr.; 2 cassettes, av. 23 min. ea.; Teacher's Guide. ($29.95). Gr. 6-8. LGB (1977). Fic
Young Jim Hawkins and his friends find themselves with blood-thirsty pirates in a race for buried treasure. Narration adapted from the novel.

PITCHER, MOLLY

FAMOUS PATRIOTS OF THE AMERICAN REVOLUTION (Filmstrip—Sound). *See* U.S.—History—Biography

PLANT PHYSIOLOGY

ANGIOSPERMS: THE LIFE CYCLE OF A STRAWBERRY (Film Loop—8mm—Captioned). *See* Growth (plants)

ISOLATION AND FUNCTION OF AUXIN (Film Loop—8mm—Silent). *See* Phototropism

MECHANICS OF PHOTOTROPISM (Film Loop—8mm—Captioned). *See* Phototropism

THE NATURE OF PHOTOTROPISM (Film Loop—8mm—Captioned). *See* Phototropism

PLANT PROPAGATION

HOW FLOWERS REPRODUCE: THE CALIFORNIA POPPY (Study Print). Kenneth E. Clouse, 1974. 12 heavy stock cardboard prints, unmounted, color 11 in. × 14 in. ($14). Gr. 5-9. PRV (10/75). 581.16
The photographs show reproduction of flowers by seed formation using the California poppy. Text and a labeled drawing of the photograph are adjacent to the colored photograph on the same side of the print so the entire process can be followed in detail.

SEED DISPERSAL (Film Loop—8mm—Silent). *See* Seeds

SEEDS SPROUTING (Film Loop—8mm—Silent). *See* Seeds

SELF PLANTING SEEDS (Film Loop—8mm—Silent). *See* Seeds

PLANTS

THE PLANTS, SERIES TWO (Filmstrip—Sound). *See* Botany

PLANTS—NUTRITION

CARNIVOROUS PLANTS (Motion Picture—16mm—Sound). *See* Insectivorous plants

THE HUNGRY PLANT (Filmstrip—Sound). Lyceum, 1972. 1 filmstrip, 50 fr.; 1 cassette, 12$1/2$ min. LY 35770C ($25), LY 35770R ($19.50). Gr. 4-9. PRV (9/72), STE (10/72). 631.5
This filmstrip explores the reasons why some plants are healthy and others stunted. It investigates how soil chemistry affects plant growth. Students are shown experimenting with growing corn and controlling nutrient, nitrogen, phosphorous, and potassium content of the soil. Soil is tested to determine which nutrients it needs and what proportions.

PLATE TECTONICS

THE MOVING EARTH: NEW THEORIES OF PLATE TECTONICS (Filmstrip—Sound). Educational Dimensions Group, 1977. 2 color, sound filmstrips, 66 fr.; 2 cassettes, 16-18 min.; 2 scripts; Teacher's Guide. #1208 ($40). Gr. 7-12. PRV (5/78). 551.5
This program expands on the most recent development and scientific testing to show Geo-Art and new techniques of mapping seismic and volcanic activity. The discovery that plate divisions mark the leading edges of the continents as they drift away from their starting point in large concentric circles is

PLATE TECTONICS (cont.)

explored as well as the ''hot spots'' of unusual volcanic activity that record the passage of plates over the face of the earth.

PLAY

LEARNING THROUGH PLAY (Kit). *See* Family life

TOPS (Motion Picture—16mm—Sound). *See* Toys

POE, EDGAR ALLAN

A DISCUSSION OF THE FALL OF THE HOUSE OF USHER (Motion Picture—16mm—Sound). *See* Motion pictures—History and criticism

EDGAR ALLAN POE'S THE BLACK CAT (Filmstrip—Sound). *See* Devil—Fiction

THE FALL OF THE HOUSE OF USHER (Motion Picture—16mm—Sound). *See* American literature—Study and teaching

THE RAVEN (Motion Picture—16mm—Sound). *See* American poetry

THE ROMANTIC AGE (Filmstrip—Sound). *See* American literature—History and criticism

POETRY

JAMES DICKEY: POET (LORD LET ME DIE, BUT NOT DIE OUT) (Motion Picture—16mm—Sound). *See* Dickey, James

POETRY—COLLECTIONS

AMERICAN INDIAN POETRY (Filmstrip—Sound). *See* Indians of North America—Poetry

AMERICAN POETRY TO 1900 (Phonodiscs). *See* American Poetry

DAYBREAK (Motion Picture—16mm—Sound). *See* Nature in poetry

MORE SILVER PENNIES (Phonodisc or Cassette). Caedmon Records, 1976. 1 cassette or disc, 32 min. #TC-1511 Disc ($6.98), #CDL-51560 Cassette ($7.95). Gr. K–A. BKL (11/17/77). 808.81
A follow-up anthology of Blanche Jenning's ''Silver Pennies''; its selections feature poets who became famous as well as lesser-known poets. The short poems are alternately read by Claire Bloom and Cyril Ritchard.

POETRY OF THE SEASONS (Filmstrip—Sound). *See* Seasons—Poetry

POETRY—HISTORY AND CRITICISM

HOWARD NEMEROV (Cassette). *See* Nemerov, Howard

POETRY, JAPANESE

HAIKU: THE HIDDEN GLIMMERING (Filmstrip—Sound). *See* Haiku

POETS

JAMES DICKEY: POET (LORD LET ME DIE, BUT NOT DIE OUT) (Motion Picture—16mm—Sound). *See* Dickey, James

POETS, AMERICAN

HOWARD NEMEROV (Cassette). *See* Nemerov, Howard

NINETEENTH CENTURY POETS (Filmstrip—Sound). *See* American literature—History and criticism

POETS OF THE TWENTIETH CENTURY (Filmstrip—Sound). *See* American literature—History and criticism

WALT WHITMAN (Filmstrip—Sound). *See* Whitman, Walt

WALT WHITMAN: POET FOR A NEW AGE (Motion Picture—16mm—Sound). *See* Whitman, Walt

POETS, ENGLISH

COLERIDGE: THE FOUNTAIN AND THE CAVE (Motion Picture—16mm—Sound). *See* English poetry

JOHN KEATS: POET (Motion Picture—16mm—Sound). *See* Keats, John

POLAND

POLAND (Filmstrip—Sound). United Learning, 1977. 4 color, sound filmstrips 64–76 fr.; 4 cassettes, 10–13 min.; Teacher's Guide. #54902 ($68). Gr. 7–10. BKL (3/78), PRV (12/78). 914.38
Contents: 1. Its Land and People. 2. Its History and Culture. 3. Its Agriculture and Industry. 4. Three Cities.
These filmstrips examine the history and development of Poland as well as portraying daily life as it is in the cities and countryside today.

POLAND—FICTION

ZLATEH THE GOAT (Motion Picture—16mm—Sound). *See* Folklore—Poland

POLICE

CAREERS IN LAW ENFORCEMENT (Filmstrip—Sound). *See* Law enforcement

THE POLICE AND THE COMMUNITY (Kit). Pathescope Educational Media, 1976 (Let's Find Out). 1 color, sound filmstrip, 60 fr.; 1 cassette, 6$\frac{1}{2}$ min.; 5 Spirit Masters;

Teacher's Manual. #9503 ($28). Gr. 7–9. PRV (4/78). 363.2

This program covers the role of the police, the courts, and correction agencies and involves students in thinking about the problems related to crime and in contemplating their possible solutions.

POLITICAL ETHICS

CORRUPTION IN AMERICA: WHERE DO YOU STAND? (Filmstrip—Sound). *See* Corruption in politics

POLITICAL PARTIES

POLITICAL PARTIES IN AMERICA: GETTING THE PEOPLE TOGETHER (Motion Picture—16mm—Sound). EBEC, 1976. 16mm color, sound film, 20 min. #3375 ($275). Gr. 9–A. BKL (11/15/76), LFR (1977), PRV (1977). 329.02

Shows how and why political parties function by examining them at work from the grass roots to the national level. Narrated segments are interspersed with candid interviews with ordinary voters and their elected representatives. Each tells how he or she views the party system and the individual's place in it. A brief historical survey shows how the role of the political party changes as the political environment in which it functions also changes.

POLITICAL SCIENCE

GOVERNMENT: HOW MUCH IS ENOUGH? (Filmstrip—Sound). Current Affairs Films, 1978. 4 color, sound filmstrips, av. 80 fr.; 4 cassettes, 15 min. ea.; Teacher's Guide. #591 ($96). Gr. 7–12. BKL (2/1/78), PRV (5/78). 320.2

Contents: 1. Government: The Economy. 2. The Welfare State. 3. The Regulators. 4. Personal Freedom.

The four filmstrips in this series examine some of the reasons why people might be disenchanted with big government. Each strip develops historical background, shows how the system operates today, and closes with a thought-provoking examination of different viewpoints.

GOVERNMENT AND YOU (Filmstrip—Sound). EBEC, 1976. 5 color, sound filmstrips, av. 81 fr. ea.; 5 discs or cassettes, 15 min. ea.; Teacher's Guide. #6921K Cassettes ($72.50), #6921 Discs ($72.50). Gr. 7–12. MDU (1/77), SOC (5/77). 320.4

Contents: 1. The Presidency. 2. The Congress. 3. The Federal Courts. 4. State Governments. 5. Local and Municipal Governments.

Surveys the workings of American government from the presidency down to the township sanitary district. Demonstrates how our state governments both reflect and supplement the federal system and concludes with a close look at the conglomerate of American municipal government organizations and services.

JOURNEY TO DEMOCRACY (Filmstrip—Sound). Macmillan Library Services, 1975. 4 color, sound filmstrips, av. 65 fr.; 4 discs or cassettes, 8–10 min.; Teacher's Guide/Script. #48800 Cassettes ($90), #48804 Discs ($90). Gr. 4–8. BKL (12/1/75), PRV (11/76). 351

Contents: 1. Anarchy. 2. Totalitarianism. 3. Aristocratic Privilege. 4. Democracy.

Introduces alternative governments via two cartoon youngsters traveling to make-believe lands. Exaggerated situations in societies based on the principles of anarchy, totalitarianism, and aristocratic privilege give the viewer a simplified picture of the weaknesses inherent in these forms. The final strip involves the forming of democracy in the United States.

POLITICIANS

A WOMAN'S PLACE IS IN THE HOUSE (Motion Picture—16mm—Sound). *See* Noble, Elaine

POLITICS, PRACTICAL

HOW TO GET ELECTED PRESIDENT (Filmstrip—Sound). *See* Elections

LOBBYING: A CASE HISTORY (2ND EDITION) (Motion Picture—16mm—Sound). *See* Lobbying

POLLUTION

ECOLOGY: BARRY COMMONER'S VIEWPOINT (Motion Picture—16mm—Sound). *See* Ecology

ENERGY AND OUR ENVIRONMENT (Filmstrip—Sound). *See* Power resources

LET NO MAN REGRET (Motion Picture—16mm—Sound). *See* Man—Influence on nature

See also Man—Influence on nature

POLLUTION—LAW AND LEGISLATION

LAW AND THE ENVIRONMENT (Kit). *See* Environment—Law and legislation

POMPEII

THE POMPEIAN EXPERIENCE (Filmstrip—Sound). Industrial Film Center/Wilson Educational Media, 1977. 4 color, sound filmstrips; 4 cassettes or discs. ($85). Gr. 9–A. BKL (6/15/78), PRV (2/79). 913.03

The viewer inspects shops looking for olive oil and mulled wine and then steps into the

POMPEII (cont.)

public baths of Pompeii. The presentation uses Pompeian artifacts to help present the city that was before the eruption of Vesuvius.

PONY EXPRESS

HURRYING HOOF BEATS: THE STORY OF THE PONY EXPRESS (Filmstrip—sound). Perfection Form Company, 1976. 1 color, sound filmstrip, 104 fr.; 1 cassette or disc, 14 min. #KH95260 Disc ($21.45), #KH95261 Cassette ($22.95). Gr. 1–A. BKL (7/1/77). 383

Photos and art work describe the pony express, which operated briefly from 1860–1861. Stirring music and an excellent narration capture the spirit of this brief chapter in American history.

POPULATION

ON POPULATION (Filmstrip—Sound). Walt Disney Educational Media, 1976. 6 color, sound filmstrips, av. 38–53 fr.; 6 cassettes or discs, av. 6–10 min.; Guide. #63–8037 ($105). Gr. 4–8. BKL (12/1/77), M&M (4/77). 301.32

Contents: 1. Exploding Numbers. 2. Increasing Harvests. 3. Dwindling Resources. 4. Struggling Economics. 5. Deteriorating Environment. 6. Achieving Stability.

In this set are gathered key issues that will inform students about the dilemma we face with increasing population and decreasing resources.

POPULATION: THE PEOPLE PROBLEM (Filmstrip—Sound). Multi-Media Productions, 1975. 2 color, sound filmstrips, 61–64 fr.; 1 cassette or disc, 13–18 min. Teacher's Manual. #7141 R or C ($17.95). Gr. 9–12. PVB (9/76). 301.32

Examines the world's people problem—history, current attitudes, possible solutions, and future predictions.

THE POPULATION DEBATE (Filmstrip—Sound). Westport Communications/Sunburst Communications, 1975. 4 color, sound filmstrips, 83–102 fr.; 4 cassettes or discs, 18–22 min.; Teacher's Guide. #41 ($99). Gr. 7–12. IFP (1975), MDM (11/77). 301.32

Contents: 1. Demography. 2. Ecology and Food. 3. Distribution and Economics. 4. People.

Presents a gripping look at the world's astonishing population growth. Considers potential solutions, but shows how explosive religious, ethical, moral, legal, and political factors impose constraints and inhibit easy resolution of the problems.

POSTAGE STAMPS

AMERICAN HISTORY ON STAMPS (Kit). TransMedia International/Donovan Films, 1976. 4 color, sound filmstrips, 54–69 fr.; 4 cassettes, 6–10 min.; 1 Game; 1 Guide. ($95). Gr. 5–9. BKL (11/15/76), FLN (5/76), PVB (4/78). 383

Contents: 1. Colonial America. 2. The New Nation. 3. America Grows Up. 4. The 20th Century.

The dramatic history of America and prominent Americans is presented through the use of stamps. Some of the rare ones were photographed in the vaults of the Smithsonian Institution.

POSTAL SERVICE—U.S.

AMERICAN HISTORY ON STAMPS (Kit). *See* Postage stamps

HURRYING HOOF BEATS: THE STORY OF THE PONY EXPRESS (Filmstrip—sound). *See* Pony Express

POTTERY

POTTERY: ARTISTRY WITH EARTH (Filmstrip—Sound). Warner Educational Productions, 1977. 6 color, sound filmstrips, 66–147 fr.; 3 cassettes, 7–30 min.; Teaching Guide. #235 ($106.50). Gr. 7–A. PRV (12/19/78). 738.3

Pottery through the ages, the pottery of the American Indian, and the modern potter are all discussed. The pinch method, the coil technique, the slab method, using the potter's wheel, and glazing are explained.

TECHNIQUES OF POTTERY (Filmstrip—Sound). Educational Dimensions Group, 1974 (How to Do). 2 color, sound filmstrips, 79–83 fr.; 2 cassettes, 16–19 min.; Teacher's Guide. #654 ($60). Gr. 7–C. BKL (1976), PVB (4/76). 738

Contents: 1. Coil Technique. 2. Slab Technique.

Detailed directions help the beginner and refresh the experienced student in these pottery techniques.

POTTERY, INDIAN

POTTERY DECORATION (Film Loop—8mm—Silent). Thorne Films/Prentice-Hall Media, 1972 (Southwest Indians). 8mm color, silent film loop, approx. 4 min. #HAT 288 ($26). Gr. 4–A. AUR (1978). 738.1

This loop demonstrates the decorations used on the pottery of Southwest Indians.

POTTERY FIRING (Film Loop—8mm—Silent). Thorne Films/Prentice-Hall Media, 1972 (Southwest Indians). 8 mm color, silent film loop, approx. 4 min. #HAT 289 ($26). Gr. 4–A. AUR (1978). 738.1

This loop demonstrates the firing of pottery by Southwest Indians.

POTTERY MAKING (Film Loop—8mm—Silent). Thorne Films/Prentice-Hall Media, 1972 (Southwest Indians). 8mm color, silent film loop, approx. 4 min. #HAT 287 ($26). Gr. 4–A. AUR (1978). 738.1

This film loop demonstrates the making of pottery by the Southwest Indians.

POVERTY

MEET MARGIE (Motion Picture—16mm—Sound). *See* Finance, Personal

THE MOUNTAIN PEOPLE (Motion Picture—16mm—Sound). *See* Appalachian Mountains

POWELL, JOHN WESLEY

JOHN WESLEY POWELL: DISCOVERY (Filmstrip—Sound). The Great American Film Factory, 1976 (Outdoor Education). 1 color and b/w, sound filmstrip, 76 fr.; 1 cassette, 17 min. #FS-5 ($35). Gr. 7–12. PRV (12/77), SCT (12/76). 921

A serious filmstrip that recounts the discovery of the Colorado River and Grand Canyon in Powell's own words. Photographs are all original pictures taken by Hillers and Baman in 1869 (largest published collection).

POWER RESOURCES

ENERGY: A MATTER OF CHOICES (Motion Picture—16mm—Sound). EBEC, 1973 (Environmental Studies Program). 16mm color, sound film, 22 min. #3301 ($290). Gr. 7–A. BKL (9/15/74), CFF (1975), SCT (11/19/74). 333.7

This film aims to help students interpret the energy crisis and its effects on society. Focusing on the shortage of electrical power, the film examines the social attitudes and technological forces that have led to the crisis—then suggests alternatives that could help overcome it.

ENERGY: CRISIS AND RESOLUTION (Filmstrip—sound). United Learning, 1974. 6 color, sound filmstrips, av. 47–63 fr. ea.; 6 cassettes, av. 11–14 min. ea.; Guide. #46 ($85). Gr. 6–10. PRV (2/76). 333.7

Contents: 1. Introduction to Energy. 2. Fossil Fuels—Coal. 3. Fossil Fuels—Oil and Natural Gas. 4. Electrical Energy. 5. Nuclear Energy. 6. Alternatives for the Future.

An historical perspective on the world's energy sources, past and present, enhances the introductory strips. Deals with energy issues and future solutions.

ENERGY AND OUR ENVIRONMENT (Filmstrip—Sound). Coronet Instructional Media, 1974. 4 color, sound filmstrips, av. 51 fr. ea.; 4 cassettes or 2 discs, av. 11½ min. ea.; Guide. #M290 Cassettes ($65), #S290 Discs ($65). Gr. 7–12. BKL (6/15/75), PRV (1975), PVB (1976). 333.7

Contents: 1. Man and the World's Energy. 2. Waste and Pollution. 3. Our Growing Use of Energy. 4. The Future.

This set shows the sources of the world's energy and our use and misuse of them. Waste, pollution, consumption, and aspects of supply and demand are explored, and the need for new energy sources and ways to conserve existing sources is stressed.

ENERGY AND THE EARTH (Filmstrip—Sound). *See* Conservation of natural resources

ENERGY FOR THE FUTURE (Motion Picture—16mm—Sound). EBEC, 1974. 16mm color, sound film, 17 min. #3348 ($220). Gr. 4–A. BKL (5/15/75), LFR (4/75), PRV (10/75). 33.7

Constantly growing energy needs, dwindling reserves of fossil fuels, recent fuel shortages, and increasing fuel costs—such conditions have established a need for new energy sources. This film examines energy alternatives for the future, including processed coal, shale oil, geothermal heat, nuclear fission and fusion, wind and solar heat.

LIVING WITH A LIMIT (Slides/Cassettes). Science and Mankind, 1978. 160 slides in 2 carousel trays; 2 cassettes or discs, 30 min. ea; Guide. #1020 ($160). Gr. 7–A. BKL (9/1/78). 333.7

President Carter narrates the introduction to this presentation designed to present methods of saving energy in schools, homes, automobiles, and businesses. Its focus is on how the individual can help.

PREGNANCY

A LIFE BEGINS . . . LIFE CHANGES . . . THE SCHOOL-AGE PARENT (Filmstrip—Sound). Parents' Magazine Films, 1977. 5 color, sound filmstrips, approx. 60 fr.; 3 cassettes or discs, av. 8 min.; 4 scripts; Teacher's Guide. ($220). Gr. 8–A. BKL (12/1/77), PRV (5/77). 301.42

Contents: 1. Becoming a Parent: The Emotional Impact. 2. Dealing with Practical Problems of Parenthood. 3. Rights and Opportunities. 4. Building a Future.

This set provides school-age parents with information and some practical assistance regarding the emotional, legal, financial, and medical problems facing them. A group of parents of various minority and cultural backgrounds appears in each set. A sketchy overview of a very complex problem, but a source of valuable information.

PREGNANCY (Filmstrip—Sound). Guidance Associates, 1977 (Parenthood Series). 2 color, sound filmstrips, 73–109 fr. ea.; 2 cas-

PREGNANCY (cont.)

settes or 2 discs, 12–21 min. ea.; Guide.
#9A–104 057 Cassettes ($52.50), #9A–104
040 Discs ($52.50). Gr. 10–A. PRV (12/78).
618

Contents: 1. The First Trimester. 2. The
Second Trimester. 3. The Third Trimester.

Explains and illustrates the three basic
stages of normal pregnancy; details changes
in the mother and in the growth of the fetus.
Expectant mothers and fathers relate each of
these stages to a variety of emotional
changes, anxieties, and satisfactions.

PREPARING FOR PARENTHOOD (Filmstrip—Sound). Guidance Associates, 1971
(Parenthood Series). 2 color, sound filmstrips, 73–109 fr. ea.; 2 cassettes or discs,
12–21 min. ea.; Guide. #9A–104 032 Cassettes ($52.50), #9A–104 024 Discs ($52.50).
Gr. 10–A. PRV (12/78). 618

Contents: 1. Having Children. 2. Being a
Parent.

Documentary case histories and interviews
present the feelings, beliefs, attitudes, and
experiences of parents of different ages from
various ethnic and economic groups.

PREHISTORIC ANIMALS. *See* Fossils

PREJUDICES AND ANTIPATHIES

BALABLOK (Motion Picture—16mm—
Sound). National Film Board of Canada/
EBEC, 1972. 16mm color, sound film, 8
min.; Spanish version available. #3338
($115). Gr. 4–A. AVI (1975), M&M (1976),
PRV (1975). 301.6

What starts a fight? Differences between
people with various cultures, religions, races,
attitudes. This nonnarrated film reduces human conflict to simple forms. Animated
blocks squeak friendly greetings as they
walk past each other. A different shape (a
ball) appears; they taunt it, and the ball re-
sponds. Both sides call up reserves, and
physical violence develops. They batter
each other until all are hexagonal and then
friendly.

BILL COSBY ON PREJUDICE (Motion Picture—16mm—Sound). Bill Cosby/Pyramid
Films, 1972. 16mm color, sound film, 24
min.; ¾-in. videocassette also available.
($325). Gr. 7–A. LFR (1974), M&M (1974),
PRV (1975). 301.6

A biting, satirical monologue by Bill Cosby
portraying America's composite bigot. Underlying Cosby's deliberate characterization
is a serious comment that stimulates honest
discussion and self-examination.

EYE OF THE STORM (Motion Picture—
16mm—Sound). American Broadcasting
Companies/Xerox Educational Publications,
1970. 16mm color, sound film, 25 min.; ¾-
in. videocassette also available. #6201 Film

($405), #6201 Videocassette ($270). Gr. 3–
A. AFF (1971), LFR (1971). 152.4

This award-winning documentary records an
Iowa teacher's highly successful attempts to
introduce her third-grade class to the realities of prejudice. She labels her blue-eyed
students "superior" one day and her brown-
eyed students "superior" the next day. Nor-
mally cooperative children become nasty,
vicious, and discriminating.

PERCEPTION AND PREJUDICE: THE
SETTINGS OF INTOLERANCE (Kit).
Newsweek, 1977. 4 color, sound filmstrips,
75–95 fr. ea.; 4 cassettes or discs, 14–16 min.
ea.; Teacher's Guide; Resource Book.
#707C ($95), #707 ($90). Gr. 9–C. BKL (4/
15/78), PRV (5/78). 301.6

Contents: 1. Causes of Prejudice. 2. Nature
of Prejudice. 3. Social Settings of Intolerance. 4. Coping with Differences.

This set helps us to understand the psychological basis for prejudice, recognize prejudice and discrimination in our
environments, and cope with it in ourselves
and others. Further sources of information
are given in suggested readings and bibliog-
raphy. The resource book has duplicating visuals and transparency visuals. Current
photographs and historical news photos give
insight into the violence precipitated by prejudice.

PREJUDICE IN AMERICA—THE JAPA-
NESE AMERICANS (Filmstrip—Sound).
See Japanese in the U.S.

SCAPEGOATING/THE IMPACT OF PREJ-
UDICE (Filmstrip—Sound). Sunburst Communications, 1973. 2 color, sound filmstrips,
64–74 fr.; 2 cassettes or discs, 12–15 min.;
Teacher's Guide. #102 ($59). Gr. 7–12. PRV
(9/74), PVB (5/75), STE (11/74). 152.4

Contents: 1. Scapegoating. 2. The Impact of
Prejudice.

The study of three specific cases—the Salem
witch trials, the Warsaw ghetto, and the rise
of McCarthyism—shows what motivates individuals and societies to employ the mecha-
nism of scapegoating. Interviews with
people of diverse backgrounds increase student awareness of personal, social, and economic effects of prejudice.

STEREOTYPING/THE MASTER RACE
MYTH (Filmstrip—Sound). Sunburst Communications, 1973. 2 color, sound filmstrips;
2 cassettes or discs; Teacher's Guide. #101
($59). Gr. 7–12. PRV (9/74), PVB (1975),
STE (11/74). 152.4

Contents: 1. Stereotyping. 2. The Master
Race Myth.

Probes the workings of prejudice and stereotyping—what they are, how they grow, how
they are mutually destructive to all concerned, and how they can be eradicated.
Stresses the importance of facts and shows
that prejudice has no basis in fact.

UNDERSTAND INSTITUTIONAL RAC-
ISM (Kit). Racism/Sexism Resource Center
for Educators, 1978. 1 color and b/w, sound
filmstrip, 133 fr.; 1 cassette, 17 min.; Discussion Guide; Pamphlets and Booklets.
($32). Gr. 7–A. PRV (1/79). 301.6

A resource to help understand the causes
and effects of racism in our social system.
Distinguishes between institutional racism
and prejudice as it examines and describes
racist policies and practices in the areas of
justice, education, employment, and housing. The presentation is calm and factual.

WALLS AND WALLS (Motion Picture—
16mm—Sound). FilmFair Communications,
1973. 16mm color, sound film, 9¹/₂ min.
($135). Gr. 7–A. BKL (5/74), CFF (1974),
LFR (3/74). 301.6

Man first built walls against physical dangers. This film traces the evolution of wall
building through walled cities, the China
Wall, and modern walled communities to
such symbolic walls as flags, the Nazi
swastika, etc., and their results. It then concentrates on the walls represented by attitudes erected as protection against
appearing foolish or being hurt. The film
points out the use of these devices as a deter-
rent to the flow of ideas that not only walls
others out, but walls ourselves in.

PRESIDENTS—U.S.

THE AMERICAN PRESIDENCY (Kit).
Films/EBEC, 1976 (Ongoing Issues in American Society). 3 color, sound filmstrips, 85–
103 fr.; 3 cassettes, 13–14 min.; 30 Student
Resource Readers; 12 Duplicating Masters;
16 Mini-prints; 2 Cassettes, 27–36 min.;
Guide. #6850K ($133.50). Gr. 8–12. BKL (5/
15/77). 353

Contents: 1. The President as Administrator
in Chief. 2. The President as Legislator in
Chief. 3. Presidential Styles.

How much power should we allow a president? Students will discover that this and
other questions revolving around the American presidency have had different answers at
different times in history. They will find, too,
that every president has in some way rede-
fined and reshaped the office. Uses and
abuses of presidential power and whether
or not the office of the presidency should be
modified are discussed.

GEORGE WASHINGTON (Cassette). *See*
Washington, George

GEORGE WASHINGTON—THE COUR-
AGE THAT MADE A NATION (Motion
picture)—16mm—Sound). *See* Washington,
George

HOW TO GET ELECTED PRESIDENT
(Filmstrip—Sound). *See* Elections

THE PICTORIAL LIFE-STORY OF
GEORGE WASHINGTON (Kit). *See*
Washington, George

THE PICTORIAL LIFE-STORY OF THOM-
AS JEFFERSON (Kit). *See* Jefferson,
Thomas

THE PRESIDENCY (Kit). Pathescope Educational Media, 1977 (Let's Find Out). 1 color,
sound filmstrip, 60 fr.; 1 cassette, approx. 7
min.; 6 Spirit Masters; Teacher's Manual.
#9516 ($28). Gr. 5–9. LGB (1977), PRV (3/
78). 353

This program introduces the duties, responsibilities and powers of the presidency. Photography is current, which helps to keep the
student's attention. The teacher's manual
and the spirit masters provide vocabulary,
observation and listening clues, and a variety of follow-up activities.

PRESIDENTIAL STYLES: SOME GIANTS
AND A PYGMY (Cassette). Cinema Sound/
Jeffrey Norton Publishers, 1976 (Avid Read-
er). 1 cassette, approx. 55 min. #40244
($11.95). Gr. 10–A. BKL (12/15/78). 353

Anecdotes that link presidential personalities to their leadership images and political
actions.

PRESIDENTS AND PRECEDENTS (Kit).
Educational Enrichment Materials, 1976. 10
color & b/w, sound filmstrips, 71–89 fr.; 10
cassettes, or discs; 10 Spirit Masters; 10 Paperbacks: *Our Presidents & Their Times*; 1
Wall Chart; 6 Front Pages of the *New York
Times*. #41098 C or R ($210). Gr. 7–12. PRV
(5/77). 353

Contents: 1. George Washington. 2. Harry
S Truman. 3. Andrew Jackson. 4. Abraham
Lincoln. 5. Thomas Jefferson. 6. Woodrow
Wilson. 7. Theodore Roosevelt. 8. Franklin
D. Roosevelt. 9. James Monroe. 10. John F.
Kennedy.

Analyzes the way our presidents set the pattern and forged the precedents that formu-
lated American political traditions. Explores
how each leader exerted influence and applied solutions to their problems—techniques that would become role models for
their successors.

THOMAS JEFFERSON (Motion Picture—
16mm—Sound). *See* Jefferson, Thomas

TO LEAD A NATION (Motion Picture—
16mm—Sound). EMC, 1974. 4 color, sound
filmstrips. 87–113 fr.; 4 cassettes, 15–21
min.; Resource Book; Teacher's Guide.
#SS216000 ($84). Gr. 9–C. BKL (6/75), HST
(11/75), PVB (4/76). 353

Contents: 1. A Leader for the People. 2. The
Shaping of the Presidency. 3. Bold Lead-
ership; The 20th Century. 4. Four Score and
Seven Years from Now.

This set examines the presidency and its
growth from its constitutional inception to
the present and beyond. A series of minicase
studies shows how the office has grown.

WOODROW WILSON AND THE SEARCH
FOR PEACE (Filmstrip—Sound). *See* Wilson, Woodrow

PRIDE AND PREJUDICE

EPISODES FROM CLASSIC NOVELS
(Filmstrip—Sound). *See* Literature—Collections

PRIMITIVE MAN. *See* Man, Nonliterate

PRINTS

THE PRINT-MAKER (Filmstrip—Sound).
Warner Educational Productions, 1977. 6
color, sound filmstrips, 70–148 fr.; 3 cassettes, 9–19 min.; Teaching Guide. #602
($105.50). Gr. 7–A. PRV (12/19/78). 769
How to make your own paper; how to do relief printing (including woodcuts, linocuts,
carved plaster and wax blocks); how to do
intaglio, drypoint, engraving, and etching;
how to do silk screening; how to do lithography; how to sign an edition, mount and
display it; and how to select materials and
tools are all presented in this set. In addition
great prints and print makers are shown.

PRINTS, AMERICAN

THE AMERICA OF CURRIER AND IVES
(Filmstrip—Sound). Bill Boal Productions/
Coronet Instructional Media, 1974. 4 color,
sound filmstrips, 56–63 fr.; 4 cassettes or
discs, 11½–14 min. #M279 Cassettes ($65),
#S279 Discs ($65). Gr. 7–A. BKL (12/15/74),
PRV (12/15/74), TEA (1974). 769.973
Contents: 1. Printmakers to the American
People. 2. Rural America. 3. From Sail to
Steam. 4. Urban America in the Making.
This set shows more than 200 historic prints,
some famous, some that are little known,
providing a graphic view of 19th-century
America. The commentary helps students
understand the values and interests of the
period being viewed.

PRISONS

PROJECT AWARE (Motion Picture—
16mm—Sound). *See* Juvenile delinquency

PRIVACY, RIGHT OF

PRIVACY: PROTECTING THE SENSE OF
SELF (Filmstrip—Sound). Harper and Row
Publishers/Sunburst Communications, 1974.
2 color, sound filmstrips, 67–69 fr. ea.; 2 cassettes or discs, approx. 13 min. ea.; Teacher's Guide. #84 ($59). Gr. 7–12. PRV (5/76),
PVB (1977). 323.44
Contents: 1. Society's View. 2. Personal
View.
Explores concept of privacy as a basic individual right and examines the obligation of
society to protect that right. Excerpts from
literature and interviews with a broad range
of people—a U.S. senator, an anthropologist, high school students, journalists—provide a provocative cross section of views on
this controversial subject.

PRIVACY UNDER ATTACK (Filmstrip—
Sound). Current Affairs Films, 1977. 1 color,
sound filmstrip, 74 fr.; 1 cassette, 16 min.;
Teacher's Guide. #568 ($24). Gr. 8–12. PRV
(9/77). 323.44
This filmstrip deals with the questions of privacy in relation to freedom of the press,
needs of a complex society, social values, legal rights, and personal freedom. Discusses
techniques used to store personal information about individuals.

PROBLEM SOLVING

A BETTER TRAIN OF THOUGHT (Motion
Picture—16mm—Sound). *See* Creativity—
Fiction

DECISIONS, DECISIONS (Filmstrip—
Sound). *See* Decision making

LET'S LOOK AT LOGIC (Filmstrip—
Sound). *See* Logic

PROBING WORD PROBLEMS IN MATHE-
MATICS (Kit). *See* Mathematics

VIEW FROM THE PEOPLE WALL (Motion
Picture—16mm—Sound). *See* Thought and
thinking

PROBLEM SOLVING—FICTION

GIRL STUFF (Kit). *See* Girls—Fiction

PROFIT

PROFIT: A LURE/A RISK (Motion Picture—
16mm—Sound). Sutherland Learning Associates/EBEC, 1977. 16mm color, sound film,
8 min. #3540 ($125). Gr. 7–12. BKL (1978),
FHE (1978), PRV (1978). 332
Advertising and musical comedy techniques
combine to explain the role of profit and loss
in economic theory. Viewers hear from all
sides involved in a business venture—the entrepreneur, the workers, the banker. Economic terms are seen in the viewer's mind as
more than definitions.

PROHIBITION

THE LONG THIRST: PROHIBITION IN
AMERICA (Cassette). Cinema Sound/Jeffrey Norton Publishers, 1976 (Avid Reader).
1 cassette, approx. 55 min. #40225 ($11.95).
Gr. 11–A. BKL (6/15/77). 973.914
In this radio interview Heywood Hale Broun
and Thomas M. Coffey, authors of *The Long
Thirst: Prohibition in America* (Norton,
1975), and Alan Churchill, author of *The
Theatrical Twenties* (McGraw, 1975),
recount anecdotes on prohibition and American culture in the 1920s.

PROPAGANDA

WHAT'S GOING ON HERE? (Filmstrip—Sound). EMC, 1973. 5 color, sound filmstrips, 46–75 fr.; 5 discs or cassettes, 9–13 min.; Teacher's Guide. #EL210000 ($110). Gr. 7–12. BKL (10/15/73). 301.154

Contents: 1. Media and Its Impact. 2. Fact or Otherwise. 3. Semantics: The Nuts and Bolts of Persuasion. 4. Packaging the Message. 5. The "You" Factor.

Defines, relates, and discusses the broad topic of media and propaganda. Emphasizes the nature and techniques of propaganda as related to mass communications.

PSYCHOLOGY

THE MENTAL/SOCIAL ME (Kit). *See* Anatomy

PSYCHOLOGY, ABNORMAL. *See* Psychology, Pathological

PSYCHOLOGY, APPLIED

PSYCHOLOGICAL DEFENSES: SERIES A (Filmstrip—Sound). *See* Human behavior

PSYCHOLOGICAL DEFENSES: SERIES B (Filmstrip—Sound). *See* Human behavior

PSYCHOLOGY—HISTORY

LANDMARKS IN PSYCHOLOGY (Filmstrip—Sound). Human Relations Media Center, 1976. 3 color, sound filmstrips, 69–76 fr.; 3 cassettes or discs, 16–18 min.; Teacher's Guide. #221 ($90). Gr. 10–C. BKL (11/1/76), MMB (4/77), PRV (3/77). 150.9

Contents: 1. Origins. 2. Views of Man. 3. Views of Man Continued.

Traces development of psychology by focusing on the key contributions of Freud, Jung, Adler, Pavlov, Sullivan, Maslow, Watson, and Skinner. Examines their individual theories and methods and compares the psychoanalytic, humanistic, interpersonal, and behavioral schools they represent.

PSYCHOLOGY, PATHOLOGICAL

ORIGINS OF MENTAL ILLNESS (Filmstrip—Sound). *See* Mental illness

PSYCHOSOMATIC DISORDERS (Filmstrip—Sound). *See* Medicine, Psychosomatic

WHO'S OK, WHO'S NOT OK: AN INTRODUCTION TO ABNORMAL PSYCHOLOGY (Filmstrip—Sound). Human Relations Media Center, 1976. 3 color, sound filmstrips, 67–77 fr.; 3 cassettes or 2 discs, approx. 13 min. ea.; Teacher's Guide with Script. #600 ($90). Gr. 9–C. BKL (7/77), M&M (4/77), PRV (1/78). 157

Contents: 1. Abnormal vs. Normal. 2. Neurotic Behavior. 3. Psychotic Behavior.

Distinguishes normal behavior from neurotic and psychotic behavior, developing criteria for normal and abnormal behavior. Identifies and presents concepts such as phobias, depressions, obsessive-compulsive personalities, schizophrenia, paranoia, manic-depression, hypochondria, and melancholia.

PUBLISHERS AND PUBLISHING

FIVE THOUSAND BRAINS (Motion Picture—16mm—Sound). *See* Encyclopedias

HOW A PICTURE BOOK IS MADE (Filmstrip—Sound). *See* Books

PULITZER PRIZE BOOKS

KARL SHAPIRO'S AMERICA (Motion Picture—16mm—Sound). *See* American poetry

TO KILL A MOCKINGBIRD (Kit). *See* Southern states—Fiction

PULSARS

QUASARS, PULSARS AND BLACK HOLES (Filmstrip—Sound). *See* Radio astronomy

PUNCTUATION

ON YOUR MARKS: PART I (Motion Picture—16mm—Sound). Communications Group West, 1971. 16mm color, sound, animated film, 7 min.; also available in videotape format. Gr. 4–A. LFR (12/71). 421

Animation is used to show the proper usage of punctuation, as verse narration explains what each mark does.

ON YOUR MARKS: PART II (Motion Picture—16mm—Sound). Communications Group West, 1971. 16mm color, sound, animated film, 7 min.; also available in videotape format. Gr. 4–A. LFR (12/71). 421

Animation is used to demonstrate how to use punctuation marks and how they affect the meaning of sentences.

PUNCTUATION PROBLEMS (Filmstrip—Sound). United Learning, 1974. 4 color, sound filmstrips; 2 cassettes; 1 Script; Teacher's Guide. #77602/01 ($60). Gr. 4–8. PRV (1975). 421

Contents: 1. End Punctuation. 2. Comma. 3. Semicolon, Parenthesis, Dash, and Colon. 4. Quotation Marks.

Presents the most common punctuation problems, and helps students, step-by-step, identify, select alternative forms of ex-

PUNCTUATION (cont.)

pression, and verify the correctness of their replies.

THE STRUCTURE OF LANGUAGE (Filmstrip—Sound). *See* English language—Grammar

PUPPETS AND PUPPET PLAYS

PUPPETRY: MINIATURES FOR THEATRE (Filmstrip—Sound). Warner Educational Productions, 1978. 6 color, sound filmstrips, 80–126 fr.; 3 cassettes, 12–18 min.; Teaching Guide. #135 ($105.50). Gr. 4–A. BKL (3/15/79). 791.53

This set gives the history and origin of puppetry and delves into multicultural puppetry as well as looking at the art today. Directions are given for making puppets and marionettes and their costumes and for staging and manipulating them.

PYLE, HOWARD

LIBRARY 3 (Cassette). *See* Literature—Collections

QUASARS

QUASARS, PULSARS AND BLACK HOLES (Filmstrip—Sound). *See* Radio astronomy

QUEENS

CATHERINE DE'MEDICI (Cassette). *See* Catherine de'Medici

RACE PROBLEMS. *See* Race relations

RACE RELATIONS

WHITE ROOTS IN BLACK AFRICA (Filmstrip—Sound). *See* South Africa—Race relations

RACE RELATIONS—FICTION

TO KILL A MOCKINGBIRD (Kit). *See* Southern states—Fiction

RACHMANINOFF, SERGEI WASSILIEVITCH

PIANO CONCERTO NO. 3 IN D MINOR (Phonodiscs). *See* Concertos

RACING. *See* Names of types, e.g., Automobile, Motorcycle, etc.

RADIO ASTRONOMY

QUASARS, PULSARS AND BLACK HOLES (Filmstrip—Sound). Educational

Dimensions Group, 1978. 2 color, sound filmstrips, 60 fr.; 2 cassettes; Teacher's Guide. #1224 ($37). Gr. 7–12. BKL (9/1/78), PRV (2/79). 523

Contents: 1. Red Giants and White Dwarfs. 2. Neutron Stars and Black Holes.

The properties of a star are examined and then its death. Actual photos from the space exploration program are explained with diagrams. New information on neutron stars and black holes is included.

RADIO PLAYS

MEDIA CLASSIC ADAPTATIONS (Filmstrip—Sound). *See* Literature—Collections

RADLAUER, ED

BOBSLEDDING: DOWN THE CHUTE (Kit). *See* Winter sports

RAILROADS—HISTORY

METROLINER (Motion Picture—16mm—Sound). Victoria Hochberg Productions, 1976. 16mm color, sound film, 35 min. ($400). Gr. 9–A. BKL (1/15/77). 385

This film combines an ordinary topic with some unique perceptions. Actual footage of a highspeed train is intercut with the preparation of the train. As the train moves along, glimpses of the countryside, towns, and cities and archival stills of railroad history scenes are shown.

RAILROADS WEST (Filmstrip—Sound). *See* U.S.—History—1865–1898

RAIN AND RAINFALL

WHAT MAKES RAIN? (Motion Picture—16mm—Sound). EBEC, 1975. 16mm color, sound film, 22 min.; Teacher Guide. #3474 ($290). Gr. 4–12. BKL (1/76), INS (2/77), PRV (9/76). 551.5

The phenomenon of rain still holds mysteries—even for scientists. Through a blend of animation and live action, students learn to view clouds as physicists do: massed drops of moisture clinging to minute solid particles. From there, simple logical steps reveal what makes the moisture fall as rain, as well as other related forms of precipitation, and how industrial pollution is increasingly affecting our weather.

RAIN FORESTS

AMAZON JUNGLE (Film Loop—8mm—Silent). *See* Forests and forestry

THE ECOLOGY OF A TEMPERATE RAIN FOREST (Filmstrip—Sound). Imperial Educational Resources, 1973. 2 color, sound filmstrips, 36 fr.; 2 discs or cassettes, 11–12 min.; Teacher's Guide. #3RG44200 Disc ($56), #3KG44200 Cassette ($62). Gr. 4–12. BKL (3/1/74), PRV (1/75), PVB (5/75). 634.9

Contents: 1. Olympic Rain Forest—Part One. 2. Olympic Rain Forest—Part Two.

A perceptive study of rain forest ecology that demonstrates how all life forms within this environmental system work together to sustain the cycle of forest life.

RAND, AYN

AYN RAND—INTERVIEW (Videocassette). Videotape Network/Jeffrey Norton Publishers, 1976 (People on Tape). Videocassette, ³/₄-in. U-Matic, ¹/₂-in. Betamax, or ¹/₂-in. EIAJ (b/w); 25 min. #71049 ($260). Gr. 11–A. BKL (3/1/78). 921

Ayn Rand's novels *Atlas Shrugged* and *The Fountainhead* and her unique philosophy of objectivism have gained her a worldwide audience. In this interview program, Rand describes her point of view and the background and influences that helped form her philosophy.

RARE AMPHIBIANS

ENDANGERED SPECIES: REPTILES AND AMPHIBIANS (Study Print). *See* Rare reptiles

RARE ANIMALS

ECOLOGY AND THE ROLE OF MAN (Filmstrip—Sound). *See* Wildlife conservation

ENDANGERED ANIMALS: WILL THEY SURVIVE? (Motion Picture—16mm—Sound). Avatar Learning/EBEC, 1976 (Wide World of Adventure). 16mm color, sound film, 24 min. #3510 ($320). Gr. 3–12. AVI (1977), LFR (1977), PRV (1978). 591.42

In the last three and one-half centuries more than 270 species have become extinct. This film shows the steps being taken today to save our vanishing wildlife.

ENDANGERED SPECIES: MAMMALS (Study Print). EBEC, 1976. 8 study prints, 13 in. × 18 in. unmounted, in heavy vinyl pouch; Study Guide. #18040 ($19.50). Gr. 4–12. MDU (4/77). 591

Contents: 1. Timber Wolf. 2. Rocky Mountain Bighorn Sheep. 3. Siberian Tiger. 4. Cheetah. 5. Square-lipped Rhinoceros. 6. Grizzly Bear. 7. Orangutan. 8. Mountain Gorilla.

The series explores the reasons these animals have become endangered species by discussing ecological niches, importance of nature's food chain, factors relating to the perpetuation of species, and the possibilities of preservation.

WHALES: CAN THEY BE SAVED? (Motion Picture—16mm—Sound). *See* Whales

RARE BIRDS

ECOLOGY AND THE ROLE OF MAN (Filmstrip—Sound). *See* Wildlife conservation

ENDANGERED SPECIES: BIRDS (Study Print). EBEC, 1976. 8 study prints, 13 in. × 18 in. unmounted, in heavy vinyl pouch; Study Guide. #6890 ($19.50). Gr. 4–A. MDU (4/77). 598.2

Contents: 1. American Peregrine Falcon. 2. Bald Eagle. 3. Brown Pelican. 4. California Condor. 5. Hawaiian Goose (Nene). 6. Kirtland's Warbler. 7. Mississippi Sandhill Crane. 8. Whooping Crane.

Basic facts about the life cycle, physical characteristics, behavior, and status of 8 American birds are presented. It is designed to create awareness of progressive destruction of the animal's habitat and the use of chemical pollutants by humans.

RARE PLANTS

ECOLOGY AND THE ROLE OF MAN (Filmstrip—Sound). *See* Wildlife conservation

RARE REPTILES

ENDANGERED SPECIES: REPTILES AND AMPHIBIANS (Study Print), EBEC, 1976. 8 study prints, 13 in. × 18 in. unmounted, in heavy vinyl pouch; Study Guide. #18120 ($19.50). Gr. 4–12. MDU (1977). 598.1

Contents: 1. American Crocodile. 2. San Francisco Garter Snake. 3. Giant Galapagos Tortoise. 4. Hawksbill Turtle. 5. Gila Monster. 6. Tuatara. 7. Houston Toad. 8. Santa Cruz Long-toed Salamander.

These prints are backed by text, range maps, and illustrations presenting 8 reptiles and amphibians. The study prints explain how the animals' decline has resulted from their inability to adapt to fundamental changes in their living conditions. Also described are the animals' feeding habits, physical appearance, reproductive processes, and special habitats.

READING—REMEDIAL TEACHING

MARK (Motion Picture—16mm—Sound). *See* Handicapped children—Education

READING MATERIALS

AUTO RACING: SOMETHING FOR EVERYONE (Kit). *See* Automobile racing

BOBSLEDDING: DOWN THE CHUTE (Kit). *See* Winter sports

READING MATERIALS (cont.)

DRAGSTRIP CHALLENGE (Kit). *See* Automobile racing

I COULDN'T PUT IT DOWN: HOOKED ON READING (Kit). *See* Literature—Collections

MODEL AIRPLANES (Kit). *See* Airplanes—Models

SPORTS CLOSE-UPS 1 (Kit). *See* Athletes

SPORTS CLOSE-UPS 3 (Kit). *See* Athletes

SPORTS CLOSE-UPS 4 (Kit). *See* Athletes

STORIES OF ADVENTURE AND HERO-ISM (Filmstrip—Sound). Educational Direction/Learning Tree Filmstrips, 1976 (Let's Read). 4 color, sound filmstrips, av. 58 fr. ea.; 2 cassettes, av. 9 min. ea.; Teacher's Guide. ($58). Gr. 3-8 BKL (11/1/76), PRV (2/77), RTE (5/77). 372.4
Contents: 1. The Day I was Invisible. 2. The Girl Who Could Fly. 3. The Runaway. 4. The Spinning Wheel.
Four stimulating stories are presented to deal with the major components involved in learning to read: sight vocabulary, silent reading practice, listening practice, comprehension, and oral reading.

SURFING, THE BIG WAVE (Kit). *See* Surfing

TWO FOR ADVENTURE (Kit). *See* Animals—Fiction

WEATHER (Kit). *See* Meteorology

WOMEN BEHIND THE BRIGHT LIGHTS (Kit). *See* Entertainers

WOMEN WHO WIN, SET 1 (Kit). *See* Athletes

WOMEN WHO WIN, SET 2 (Kit). *See* Athletes

WOMEN WHO WIN, SET 3 (Kit). *See* Athletes

WOMEN WHO WIN, SET 4 (Kit). *See* Athletes

READING—STUDY AND TEACHING

BOXING (Kit). *See* Boxing

I COULDN'T PUT IT DOWN: HOOKED ON READING, SET TWO (Kit). *See* Books and reading—Best books

I COULDN'T PUT IT DOWN: HOOKED ON READING, SET THREE (Kit). *See* Books and reading—Best books

JAZZ, ROCK, FOLK, COUNTRY (Filmstrip—Sound). *See* Music, Popular

MAKING JUDGMENTS AND DRAWING CONCLUSIONS (Slides/Cassettes). Center for Humanities, 1977 (How to Get the Most from What You Read). 240 slides in 3 Carousel cartridges; 3 cassettes or discs; Teacher's Guide; 30 Student Activity Books; Teacher's Desk Copy. ($189.50). Gr. 9-A. BKL (6/1/77), PRV (2/78). 372.4
This set is designed to help students develop critical reading skills. Part one helps students distinguish between fact and opinion. Part two shows how faulty thinking can lead to absurd conclusions through the use of false premises, stereotypes, or overgeneralizations. Part three stresses the importance of carefully considering the source of information.

MEN BEHIND THE BRIGHT LIGHTS (Kit). *See* Musicians

THE MIGHTY MIDGETS (Kit). *See* Automobile racing

PHONETIC RULES IN READING (Kit). Society for Visual Education, 1978 (Life Skills Program). 4 color, sound filmstrips, 50-55 fr. ea.; 4 cassettes, 8-9 min. ea.; 8 Tests; 24 Activity Sheets; Teacher's Guide. #LG329-SATC ($99). Gr. 7-12. BKL (1978). 372.4
Contents: 1. Review of Consonants. 2. Review of Blends, Diagraphs, and Clusters. 3. Review of the Vowel Family. 4. Irregular Patterns in Phonics and Syllabication.
This program reviews phonetic rules and shows how phonics can help expand reading skills.

READING COMPREHENSION SKILLS (Kit). Society for Visual Education, 1978. 4 filmstrips; 4 cassettes; 19 Worksheets; 1 Guide. #330-SATC ($99). Gr. 7-10. BKL (11/1/78). 372.4
Contents: 1. Contextual and Structural Clues. 2. Getting the Main Idea. 3. Making Inferences. 4. Critical Reading Skills.
Demonstrates methods for improving comprehension skills: examining words, style, and sentence structure; determining fact and opinion; drawing inferences about time, place, characters, action.

READING IN THE CONTENT AREAS (Kit). Society for Visual Education, 1978. 4 filmstrips; 4 cassettes; 31 worksheets; 1 Guide. #328-SATC ($99). Gr. 9-C. BKL (9/15/78), M&M (4/19/78). 372.4
Contents: 1. Reading Skills in Science. 2. Reading Skills in Social Studies. 3. Reading Skills in Mathematics. 4. Reading Skills in Literature.
Defines characteristics that distinguish four areas of subject matter, and shows students how to use context, structure, and visual aids within the text to improve comprehension in these areas.

ROCK (Kit). *See* Rock music—History

UNDERSTANDING THE MAIN IDEA AND MAKING INFERENCES (Slides/Cassettes). *See* Communication

USING CLUE WORDS TO UNLOCK MEANING (Slides/Cassettes). *See* Communication

RECYCLING (WASTE, ETC.)

JUNK ECOLOGY (Filmstrip—Sound). *See* Handicraft

REDUCING

PSYCHOLOGY OF WEIGHT LOSS (Filmstrip—Sound). Multi-Media Productions, 1975. 1 sound filmstrip, 47 fr.; 1 cassette or disc, 10 min.; Teacher's Manual. #7131 C or R ($12.95). Gr. 9–12. MDU (1975). 613.25
Offers insights into specific problems of obesity and suggests a number of psychological devices that can be employed to eliminate unwanted pounds.

REFERENCE BOOKS

BIOGRAPHY: BACKGROUND FOR INSPIRATION (Filmstrip—Sound). *See* Biography

REFERENCE SERVICE (LIBRARIES). *See* Libraries—Reference service

REFORMATION

THE REFORMATION: AGE OF REVOLT (Motion Picture—16mm—Sound). Signet Productions/EBEC. 1973. 16mm color, sound film, 24 min. #47789 ($330). Gr. 10–C. AVI (1973), CPR (1973), LFR (1973). 270.6
The 16th century was an era shaken by radical changes in technology, split by social conflict, and torn by the bloody clash of ideologies. The passion of the age was loosened by Martin Luther, whose grievances against the Church produced a chain of events that left a profound impact on the modern world, with religious, moral, and political effects.

REFORMERS

PROGRESSIVE ERA: REFORM WORKS IN AMERICA (Motion Picture—16mm—Sound). *See* U.S.—History—1898–1919

RELATIVITY (PHYSICS)

THE SPECIAL THEORY OF RELATIVITY (Filmstrip—Sound). Audio Visual Narrative Arts, 1977. 2 color, sound filmstrips, 89–96 fr. ea.; 2 cassettes or discs, av. 17 min. ea.; Teacher's Guide. Cassette ($52.50), Disc ($48.50). Gr. 7–A. BKL (12/15/77), PRV (5/78). 530.11.
Contents: 1. Relativity Is Not Difficult to Understand. 2. Space and Time, Energy and Mass.
The history of common sense and Newtonian physics is traced to the 19th century.

Problems that could not be explained before the Relativity Theory, and unification of space and time and a unification of mass and energy are discussed. Familiar objects are used to clarify concepts with a minimum of math.

RELIGION AND SCIENCE

AN INQUIRY INTO THE ORIGIN OF MAN: SCIENCE AND RELIGION (Slides/Cassettes). *See* Man—Origin and antiquity

REMBRANDT

REMBRANDT (Motion Picture—16mm—Sound). National Gallery of Art & Visual Images/WETA-TV/EBEC, 1974 (The Art Awareness Collection, National Gallery of Art). 16mm color, sound film, 7 min. #3534 ($105). Gr. 7–C. EGT (1978), LFR (1978). 759.3
Opulence and restraint, light and dark—all brought together by the creative genius of Rembrandt Van Rijn. Glowing colors and dramatic contrasts reveal the inner drama of his personal life in his portraits of rich clients and penniless beggars.

REMINGTON, FREDERIC

AMERICANS WHO CHANGED THINGS (Filmstrip—Sound). *See* U.S.—Civilization—Biography

RENAISSANCE

THE ART OF THE RENAISSANCE (Filmstrip—Sound). *See* Art, Renaissance

SPIRIT OF THE RENAISSANCE (Motion Picture—16mm—Sound). EBEC, 1971. 16mm color, sound film, 31 min. #47737 ($390). Gr. 9–C. CGE (1972), FLN (1972), MER (1973). 940.21
This film explores the intellectual and artistic climate of Florence during the 14th and 15th centuries, framed against scenes from the daily life of a contemporary Florentine. The lives of three historical figures, Petrarch, Alberti, and Leonardo da Vinci, illustrate the several facets of the Renaissance that made it unique in religion, education, discovery, art, literature, and politics. The invasion of Tuscany by French armies brought the Renaissance to an end.

RENOIR, PIERRE AUGUSTE

RENOIR (Motion Picture—16mm—Sound). National Gallery of Art & Visual Images/WETA-TV/EBEC, 1974 (The Art Awareness Collection, National Gallery of Art). 16mm color, sound film, 7 min. #3533 ($125). Gr. 7–C. EGT (1978), LFR (1978), M&M (1977). 759.4

RENOIR, PIERRE AUGUSTE (cont.)

"There are enough boring things in life without having to create still more." Lovely women and children gave Renoir material for his fresh, colorful impressionistic work, and narration adapted from his own words record the flavor of a happy France.

REPORT WRITING

HOW TO WRITE A RESEARCH PAPER (Filmstrip—Sound). McGraw-Hill Films, 1976. 4 color, sound filmstrips, 45–51 fr. ea.; 4 cassettes, 7–8 min. ea.; Guide. #102861–7 Cassettes ($72), #106874–0 Discs ($72). Gr. 9–C. BKL (7/15/77), PRV (4/78). 808
Contents: 1. Getting Started. 2. Selecting and Restricting a Topic. 3. Gathering and Organizing Information. 4. Writing the Research Paper.
An introduction to research and writing skills, this program leads viewers through each phase from choosing a topic to note-taking and outlining.

HOW TO WRITE A TERM PAPER (Filmstrip—Sound). United Learning, 1974. 4 color, sound filmstrips, approx. 60 fr.; 2 discs or cassettes, 12 min. ea.; Script/Guide. #77701 Cassettes ($60), #77702 Discs ($60). Gr. 5–10. PRV (5/74). 808
Contents: 1. Getting Ready. 2. Taking Notes. 3. Organizing and Writing. 4. Preparing a Final Draft.
The research, organization, and writing methods of three students assigned a term paper are analyzed.

THE RESEARCH PAPER (Kit). Society for Visual Education, 1978. 4 filmstrips; 4 cassettes; 25 Worksheets; 1 Guide. #323–SATC ($99). Gr. 9–C. BKL (10/15/78). 808
Contents: 1. Getting Started. 2. Follow-Through. 3. Thesis and Organization. 4. Final Draft.
Thorough overview of planning, researching, organizing, and writing a paper; examines library sources, note taking, outlining, proofreading.

RESEARCH PAPER MADE EASY: FROM ASSIGNMENT TO COMPLETION (Slides/Cassettes). Center for Humanities, 1976. 240 color slides in 3 carousel cartridges; 3 cassettes or discs; Teacher's Guide. ($179.50). Gr. 9–C. PRV (1/78), TSS (5/6/77). 808
A systematic approach to the research paper. Part One defines, gives examples of, and explains the reasons for writing research papers. It stresses choice, narrowing the topic, and stating objectives. Part Two concentrates on doing the research in support of the thesis with the use of library tools. Instructions for using working outlines and taking notes are given. Part Three covers writing the paper—presenting the thesis with supporting research. At this point, the paper is outlined, following a review and revision of the statement of objectives. Notes are in order and materials are in logical sequence. Extra information on bibliographies and footnotes is included.

REPRODUCTION

BIRTH AND CARE OF BABY CHICKS (Film Loop—8mm—Captioned). See Chickens

THE BROWN TROUT (Film Loop—8mm—Captioned). See Trout

THE GROWING TRIP (Filmstrip—Sound). Marshfilm, 1975 (The Human Growth Series). 1 color, sound filmstrip, 65 fr.; 1 cassette or disc, 15 min.; Teacher's Guide. #1121 ($22). Gr. 5–8. PRV (12/75), PVB (4/76), TCT (9/75). 612.6
This filmstrip illustrates male and female anatomy and functions, human reproduction and birth, heredity and genetics.

MIRACLE OF LIFE (Motion Picture—16mm—Sound). See Embryology

REPTILES

ENDANGERED SPECIES: REPTILES AND AMPHIBIANS (Study Print). See Rare reptiles

RATTLESNAKE (Film Loop—8mm—Silent). See Snakes

SNAKES OF THE AMAZON (Film Loop—8mm—Silent). See Snakes

RESCUE WORK

RESCUE SQUAD (Motion Picture—16mm—Sound). EBEC, 1972. 16mm color, sound film, 14 min. #3112 ($185). Gr. 5–12. FLN (1973), PRV (1972), STE (1974). 614.8
Documents the work of a fire department rescue squad through a close-up look. Rescue work may involve racing to aid a heart attack victim or stopping a suicide attempt, assisting at a fire, or giving emergency first aid at an accident. Viewers become aware of the dedication and alertness needed for this exciting, often dangerous career.

RESEARCH

BIOGRAPHY: BACKGROUND FOR INSPIRATION (Filmstrip—Sound). See Biography

INDEXES (Filmstrip—Sound). See Indexes

RESEARCH PAPER. See Report writing

RESPIRATION

LUNGS AND THE RESPIRATORY SYSTEM (Motion Picture—16mm—Sound).

EBEC, 1975. 16mm color, sound film, 17 min. #3395 ($240). Gr. 6–9. INS (3/77), LFR (11/12/75), PRV (4/76). 612
This film uses diagrams, demonstrations, and microphotography to show how oxygen—the vital element in air—gets from lungs to blood to each and every tiny body cell. It reveals the body's intricate system for filtering and purifying the air we breathe—and how smoking and breathing industrial air pollutants can break that marvelous system down.

RESTAURANTS, BARS, ETC.

CAREERS IN FOOD SERVICE (Filmstrip—Sound). *See* Cookery, Quantity

REVERE, PAUL

PAUL REVERE'S RIDE AND HIS OWN STORY (Cassette). *See* American poetry

RHETORIC

RHETORIC IN EFFECTIVE COMMUNICATION (Kit). Society for Visual Education, 1978. 4 filmstrips; 4 cassettes; 21 Worksheets; 1 Guide. #321–SATC ($99). Gr. 9–12. BKL (11/1/78). 808
Contents: 1. Introduction to Rhetoric. 2. Rhetoric: Substance. 3. Rhetoric: Structure. 4. Rhetoric: Style.
Defines rhetoric, investigates its uses and methods for developing rhetorical styles. Demonstrates how rhetoric is vital to effective communication.

RICHARDS, BEAH

BEAH RICHARDS: A BLACK WOMAN SPEAKS (Videocassette). KCET-TV; Los Angeles/Public Television Library, 1975. 1 color videocassette, 30 min. ($175). Gr. 9–A. BKL (2/1/79). 920
Beah Richards, a black actress and playwright from Mississippi, is an interesting woman of wide experience. This series of dramatic monologues is entertaining and conveys the advantages and disadvantages of being a black woman.

RICHARDS, DOROTHY

MY FORTY YEARS WITH BEAVERS (Filmstrip—Sound). *See* Beavers

RICHARDSON, ELLIOT

TALKING WITH THOREAU (Motion Picture—16mm—Sound). *See* Philosophy, American

RICHTER, CONRAD

THE LIGHT IN THE FOREST (Phonodisc or Cassette). *See* Indians of North America—Fiction

RIEGER, SHAY

THE BRONZE ZOO (Motion Picture—16mm—Sound). *See* Bronzes

RIGHT TO LIVE. *See* Abortion

RIVERS

ALL ABOUT RIVERS (Filmstrip—Sound). EBEC, 1974. 5 color, sound filmstrips, av. 64 fr. ea.; 5 discs or cassettes, 8 min. ea.; Teacher's Guide. #6494K Cassettes ($72.50), #6494 Discs ($72.50). Gr. 4–8. BKL (12/15/74), ESL (1977), PRV (11/75–76). 551.4
Contents: 1. What is a River? 2. Rivers Work for Us. 3. Rivers Are Highways. 4. Rivers Work Against Us. 5. Rivers Need Our Help.
This series helps children to explore. How rivers benefit people as suppliers of water, power, and recreation becomes vividly clear. Children also learn that rivers can be destructive and, conversely, that people can destroy rivers.

ROBERTSON, CLIFF

THE KENNEDYS (Videocassette). *See* Kennedy (family)

MEN BEHIND THE BRIGHT LIGHTS (Kit). *See* Musicians

ROCK MUSIC—HISTORY

ROCK (Kit). Troll Associates, 1976 (Troll Jam Sessions). 1 color, sound filmstrip, 34 fr.; 1 cassette, 14 min.; 10 Books, 4 Duplicating Masters; 1 Teacher's Guide. ($42.95). Gr. 5–8. BKL (1/15/77). 781.5
This kit covers the development of rock music from its beginnings in the early 1950s to its present forms, emphasizing that rock music is constantly changing. Rock music provides the background for the narration and photos of different rock performers are used in the visuals. A high-interest/low-reading-ability kit.

RODGERS, RICHARD

MARCHES (Phonodiscs). *See* Marches (music)

ROGERS, WILL

WILL ROGERS' NINETEEN TWENTIES (Motion Picture—16mm—Sound). *See* U.S.—History—1919-1933

ROMANTICISM

THE SPIRIT OF ROMANTICISM (Motion Picture—16mm—Sound). EBEC, 1977. 16mm color, sound film, 16 min. #47829 ($410). Gr. 9–A. BKL (5/15/78), LRF (1978), PRV (1978). 709.034

Dramatizations of key events and personalities during the Romantic movement in literature, music, and art (1789–1838). Highlights include: Carlyle describing the French Revolution, the impact of the revolution on Wordsworth, Delacroix painting "Liberty Leading the People," Shelley as a young rebel, the friendship of Shelley and Byron, the unique contribution to music of Beethoven and Chopin. This film is based on actual letters and documents of the period.

ROOSEVELT, FRANKLIN DELANO

PRESIDENTS AND PRECEDENTS (Kit). *See* Presidents—U.S.

ROOSEVELT, THEODORE

PRESIDENTS AND PRECEDENTS (Kit). *See* Presidents—U.S.

ROSE, GEORGE

THROUGH THE LOOKING GLASS (Phonodiscs). *See* Fantasy—Fiction

ROSENMANN, DOROTHY

PRESIDENTIAL STYLES: SOME GIANTS AND A PYGMY (Cassette). *See* Presidents—U.S.

ROUAULT, GEORGES

GEORGES ROUAULT (Motion Picture—16mm—Sound). Texture Films, 1972. 16mm color, sound film, 30 min. ($350). Gr. 10–A. AFF (1973). 921

A definitive film on this 20th-century master painter: Rouault's childhood, the atelier of Gustave Moreau, the scandalous Fauves, violent social protest, the Miserere, and the last glowing biblical landscapes. Directed by Isabelle Rouault, the painter's daughter and curator, the film is unusually rich in documentation and insight.

RUGS, HOOKED

RUG HOOKING: A MODERN APPROACH (Filmstrip—Sound). Warner Educational Productions, 1974. 2 color, sound filmstrips, 62–71 fr.; 1 cassette, approx. 12½ min.; Teaching Guide. #510 ($47.50). Gr. 7–A. PRV (9/19/75). 746.7

The latch hook and punch needle techniques of rug hooking are illustrated in this set. Designing, transferring of designs to fabric, and the consideration of materials and supplies are included.

RURAL LIFE. *See* Agriculture; Farm life—U.S.; Outdoor life

RUSSIA

INSIDE THE USSR (Filmstrip—Sound). Educational Enrichment Materials, 1977. 8 color filmstrips; 8 cassettes; Teacher's Guide. #51037 ($144). Gr. 6–12. BKL (4/1/78). 914.7

Contents: 1. An Historical Overview, Part I. 2. An Historical Overview, Part II. 3. Artistic Heritage. 4. Party History & Governmental Structure, Part I. 5. Party History & Governmental Structure, Part II. 6. Industry & Agriculture. 7. The Many Contrasts Within, Part I. 8. The Many Contrasts Within, Part II.

Introduces students to Russian history, culture, government, politics, geography, agriculture, and industry to encourage an understanding of the Soviet Union today. Discusses the impact of modern life-styles and technology on this nation of numerous contrasts.

RUSSIA (Kit). Nystrom, 1974. 5 color, sound filmstrips, 70–92 fr. ea.; 1 cassette, 8–12 min.; 10 Student Readers; 1 Transparency; Set of Activity Sheets; Teacher's Guide. ($165). Gr. 7–12. BKL (6/15/75), LGB (1974–75), PRV (11/75). 914.7

Contents: 1. Daily Life. 2. The Communist Revolution. 3. Stalin. 4. War and Its Aftermath. 5. Prospects for the Future.

This program utilizes a variety of instructional materials to show the complex transformations that have taken place in Russian society under Communist leadership.

SIBERIA NOW (Kit). EMC, 1974. 4 color, sound filmstrips, 70–111 fr.; 4 discs or cassettes, 11–19 min.; 4 Paperback Books; Political Map; Teacher's Guide. #SS–214000 ($89). Gr. 8–C. BKL (6/74), PRV (3/75). 915.7

Contents: 1. Background for Change. 2. Energy: Key to Development. 3. From Resources to Products. 4. Living on a New Frontier.

An up-to-date approach to Siberia's geography, history, varied cultures, resources, and technological development.

SOVIET UNION: EPIC LAND (Motion Picture—16mm—Sound). EBEC, 1972. 16mm color, sound film, 29 min. #3096 ($360). Gr. 7–12. BKL (1972), LFR (1972), STE (1972). 914.7

Unique footage gives panorama of the Russian land and people, capturing the vast size, political structure, and potential of the Soviet Union. Film explores the cities and industry of European Russia, the Caucasus, Central Asia, and sparsely populated but mineral-rich Siberia. Songs and instrumental music give viewers an understanding of the people and the varying cultures.

THE SOVIET WORLD (Kit). EMC, 1976. 4 color, sound filmstrips, 103–179 fr.; 4 cassettes, 12–18 min.; 8 Student Books; Wall Map; Teacher's Guide. #SS–218000 ($98). Gr. 7–12. BKL (3/77). 914.7

Contents: 1. Setting the Stage. 2. Resource Use: Land and Machines. 3. A Better Life. 4. Culture: Preserving, Creating.

This study of the Soviet Union, reinforced by firsthand observations, focuses on the vast area west of the Urals. Government structure and function, agriculture and industrial growth under Communism, contrasting life-styles, historic and contemporary contribution to the arts are all presented on an open-ended inquiry basis.

RUSSIA—HISTORY—FICTION

THE CROCODILE (Motion Picture—16mm—Sound). See Fantasy—Fiction

A DISCUSSION OF THE CROCODILE (Motion Picture—16mm—Sound). See Motion pictures—History and criticism

RUSSIA—SOCIAL LIFE AND CUSTOMS

THE SOVIET UNION: A STUDENT'S LIFE (Motion Picture—16mm—Sound). See Student life—Russia

RUSSIAN LITERATURE—HISTORY AND CRITICISM

A DISCUSSION OF THE CROCODILE (Motion Picture—16mm—Sound). See Motion pictures—History and criticism

SAFETY EDUCATION

ALL ABOUT FIRE (Motion Picture—16mm—Sound). See Fire prevention

BICYCLE SAFELY (Motion Picture—16mm—Sound). See Bicycles and bicycling

BICYCLING ON THE SAFE SIDE (Motion Picture—16mm—Sound). See Bicycles and bicycling

THE D.W.I. (DRIVING WHILE INTOXICATED) DECISION (Motion Picture—16mm—Sound). Lawren Productions, 1977. 16mm color, sound film, 25 min. ($380). Gr. 9–12. FLN (9/10/78), LFR (11/12/78). 614.8

The viewer is brought face to face with the DWI decision. The importance of making that decision with a clear mind and knowledge is emphasized. An added feature is a description of transactional analysis and its applications to the problems connected with drinking.

EFFECTIVE TRAILERING (Filmstrip—Sound). See Automobiles—Trailers

FIRE EXTINGUISHERS AND SMOKE DETECTORS (Motion Picture—16mm—Sound). See Fire prevention

HEALTH AND SAFETY: KEEPING FIT (Kit). See Hygiene

MOTORCYCLE SAFETY (Kit). See Motorcycles

NOBODY'S VICTIM (Motion Picture—16mm—Sound). See Self-defense

NOT SO EASY (MOTORCYCLE SAFETY) Motion Picture—16mm—Sound). See Motorcycles

PLAYING IT SAFE (Filmstrip—Sound). Current Affairs Films, 1977. 2 color, sound filmstrips, 74 fr.; 2 cassettes, 16 min.; Teacher's Guide. #576–581 ($48). Gr. 5–12. PRV (4/78). 614.8

Contents: Part I: 1. In the Home. 2. En Route. 3. In the Water. Part II: 1. Babysitting/Child Care. 2. With Food. 3. In School.

The safety needs of six areas of everyday life are depicted with emphasis on preventing problems and handling emergencies.

SAFE IN THE WATER (Motion Picture—16mm—Sound). FilmFair Communications, 1972. 16mm color, sound film, 15¼ min. ($190). Gr. 4–12. LFR (1972), PRV 5/74). 614.8

Shows the most frequent causes of water accidents and demonstrates various rescue techniques that can save a distressed victim.

SKATEBOARD SAFETY (Motion Picture—16mm—Sound). See Skateboards

SKIN DIVING (Filmstrip—Sound). See Skin diving

SPEEDING? (Motion Picture—16mm—Sound). See Automobiles—Law and legislation

THUMBS DOWN (HITCHHIKING) (Motion Picture—16mm—Sound). See Hitchhiking

WHATEVER HAPPENED TO LINDA? (Filmstrip—Sound). Marshfilm, 1975. 1 color, sound filmstrip, 50 fr.; 1 cassette or disc; Teacher's Guide. #1131 ($21). Gr. 4–8. INS (1978), NYF (1976). 371.77

Dramatizes potentially dangerous situations students face in everyday life. Action moves from how to avoid trouble to what to say and do in response to unusual demands made by strangers, friends, or relatives.

YOU AND OFFICE SAFETY (Motion Picture—16mm—Sound). See Office management

SALEMME, ANTONIO

INTERVIEW WITH ANTONIO SALEMME (Phonodiscs). Folkways Records, 1978. 2 discs, 101 min., with notes. #FX 6004 ($17.96). Gr. 11–A. BKL (1/15/79). 921

An 85-year-old Italian-born artist, Antonio Salemme talks about his sculpture and drawing. He discusses how he approaches his subject, how he models to suggest the subject, and his reliefs and "environments," which are scenes in sculptural form.

SALES PERSONNEL

A CAREER IN SALES (Filmstrip—Sound). Associated Press/Pathescope Educational Media, 1973 (Careers In). 2 color, sound filmstrips, 68–80 fr. ea.; 2 cassettes; Teacher's Guide. #706 ($50). Gr. 7–12. PRV (5/74). 658.85023

Discusses the role of the salesman in the history of our country. It recounts the hardships of the Yankee peddler and the qualities of perseverance and endurance necessary for survival and success in this field. Interviews with sales people from a variety of areas present a broad background for this field.

SALMON

SALMON RUN (Film Loop—8mm—Silent). Walt Disney Educational Media, 1966. 8mm color, silent film loop, approx. 4 min. #62–5020L ($30). Gr. K–12. MDU (1974). 597.5

Follows salmon on their spawning run through rapids and falls, and as they lay eggs.

SALOMON, HAYM

FAMOUS PATRIOTS OF THE AMERICAN REVOLUTION (Filmstrip—Sound). *See* U.S.—History—Biography

SAMPSON, DEBORAH

DEBORAH SAMPSON: A WOMAN IN THE REVOLUTION (Motion Picture—16mm—Sound). BFA Educational Media, 1975. 16mm color, sound film, 15¼ min. #11639 ($225). Gr. 4–A. PRV (11/76). 921

Deborah Sampson was committed to the causes of the American Revolution and unwilling to adopt one of the traditional female roles. She served in Washington's army under an assumed name and identity, taking an active part in the Battle of Tarrytown and Yorktown. The film shows reenactments of battles, ambushes and dramatic encounters.

SAND PAINTING

NAVAJO SAND PAINTING CEREMONY (Film Loop—8mm—Silent). *See* Indians of North America—Art

SANDBURG, CARL

DAYBREAK (Motion Picture—16mm—Sound). *See* Nature in poetry

FOG (Motion Picture—16mm—Sound). *See* Art and nature

POETS OF THE TWENTIETH CENTURY (Filmstrip—Sound). *See* American literature—History and criticism

SATIRE

DE FACTO (Motion Picture—16mm—Sound). *See* Decision making

SATIRICAL JOURNALISM (Cassette). *See* Cartoons and caricatures

SATIRE, AMERICAN

ANIMATED WOMEN (Motion Picture—16mm—Sound). *See* Women—U.S.

BILL COSBY ON PREJUDICE (Motion Picture—16mm—Sound). *See* Prejudices and antipathies

HARDWARE WARS (Motion Picture—16mm—Sound). Pyramid Films, 1978. 16mm color, sound film, 13 min.; ¾-in. videocassette also available. ($225). Gr. 7–A. AFF (1978), FLQ (1978), LFR (1978). 817

A parody of the spectacular space epic "Star Wars," in which the special effects are created with household appliances available in any hardware store. "Star Wars" look-alikes and a "coming attractions" format contribute to the effectiveness of this example of satire.

SCANDINAVIA

NORTHERN EUROPE: SCANDINAVIA (Filmstrip—Sound). Encyclopaedia Britannica Educational Corporation, 1977. 4 color, sound filmstrips, 108–116 fr.; 4 cassettes, 15½–19 min.; Teacher's Guide. #17006/K ($57.95). Gr. 7–12. PRV (2/79). 914.8

Contents: 1. Norway: New Wealth Confronts Tradition. 2. Finland: Prosperity Facing the East. 3. Sweden: Experiment in Socialized Democracy. 4. Denmark: A Farmer Moves to Town.

This series shows how the four tiny nations that make up Scandinavia continue to prosper in the age of superpowers. Despite a common Viking heritage, there are surprising differences. Norway is shown as a country of rugged individualists—but with a social conscience. Industrious Finland shows how to live with its Russian neighbor. Realities of Sweden's socialized democracy come alive and Denmark's vigorous economic life is described.

SCHLIEMANN, HEINRICH

HEROES OF THE ILIAD (Cassette). *See* Mythology, Classical

SCHOLARSHIPS, FELLOWSHIPS, ETC.

HOW CAN I PAY FOR COLLEGE? (Filmstrip—Sound). *See* Student loan funds

SCHUBERT, FRANZ

MARCHES (Phonodiscs). *See* Marches (music)

SCIENCE

AIR, EARTH, FIRE AND WATER (Filmstrip—Sound). Hawkhill Associates, 1975. 5 color, sound filmstrips, 78–94 fr.; 5 cassettes, 9–17 min.; Teacher's Guide; 96 Real Gold Idea-Bank "coins." ($110). Gr. 5–10. BKL (1/15/76). 500

Contents: 1. Air. 2. Earth. 3. Fire. 4. Water. 5. Air, Earth, Fire and Water.

Unusual presentation of vital elements of air, earth, fire, and water, which stresses their aesthetic as well as their scientific and environmental value. The narration and photography examine each element, its role in creation of our planet, its changes under various influences, its continuing function in life as we know it, and our abuse of the element.

SCIENCE AND YOU (Filmstrip—Sound). Pix Production/Barr Films, 1977. 6 color, sound filmstrips, 62–77 fr.; 6 cassettes, 6–10 min.; Teacher's Guide. #50300 ($125). Gr. 4–8. PRV (12/78). 500

Contents: 1. You and Energy. 2. You and Communication. 3. You and Transportation. 4. You and Health. 5. You and Shelter. 6. You and Food.

These filmstrips are intended to create awareness of the world and environment around us, of the inventions and discoveries people have made, and of the problems brought by these changes.

SPACESHIP EARTH (Filmstrip—Sound). *See* Natural history

SCIENCE—HISTORY

THE HOUSE OF SCIENCE (Motion Picture—16mm—Sound). EBEC, 1973. 16mm color, sound film, 15 min. #3185 ($185). Gr. 7–C. BKL (1974), LFR (1975), MMT (1974). 509

Science is shown expanding and subdividing from its early history of navigation and calendar making to its explosive growth today. Following this animated prologue, the film presents a variety of contemporary scientists, their laboratories and their work. Emphasis is on science as an artistic or philosophical enterprise.

SCIENCE—STUDY AND TEACHING

ENERGY (Filmstrip—Sound). *See* Science and the humanities

HUNGER (Filmstrip—Sound). *See* Science and the humanities

OUR CHANGING EARTH (Filmstrip—Sound). *See* Science and the humanities

OURSELVES (Filmstrip—Sound). *See* Science and the humanities

SCIENCE AND CIVILIZATION

INVENTIONS AND TECHNOLOGY THAT SHAPED AMERICA, SET ONE (Filmstrip—Sound). *See* Technology and civilization

SCIENCE AND SOCIETY: RECONCILING TWO PERSPECTIVES (Filmstrip—Sound). Harper and Row Publishers/Sunburst Communications, 1975. 6 color, sound filmstrips, 63–70 fr. ea.; 6 cassettes or discs, 10½–12 min. ea.; Teacher's Guide. #73 ($145). Gr. 9–12. BKL (2/1/76), TSS (1/78). 301.24

Contents: 1. Science and Social Values. 2. Science and Religion. 3. Science and the Humanities. 4. Science and the Law. 5. Science and Economics. 6. Science and Government.

Analyzes the effects of today's scientific advances on six key areas of our lives. Stresses the close interrelationship between science and critical social issues. Incisive comments of outstanding scientists provide a challenging framework for students to weigh the impact of future scientific advances before they happen, to keep abreast of scientific developments, and to examine and question the influence of science.

SCIENCE AND RELIGION. *See* Religion and science

SCIENCE AND THE HUMANITIES

ENERGY (Filmstrip—Sound). Scholastic Book Services, 1975 (Human Issues in Science). 4 color, sound filmstrips, 88–95 fr. ea.; 4 cassettes or discs, 15–17 min. ea.; Guide. #010015 Cassettes ($79.50), #010014 Discs ($79.50). Gr. 7–12. BKL (10/1/75). 001.3

Contents: 1. Energy and the Land. 2. Energy and the Sea. 3. Using Energy. 4. The Future of Energy.

Photographs and drawings illustrate this set, which is narrated by several voices. The strips examine the problems of strip mining, oil spills, junk and garbage, nuclear and solar energy.

HUNGER (Filmstrip—Sound). Scholastic Book Services, 1975 (Human Issues in Science). 4 color, sound filmstrips, 75–77 fr. ea.; 4 cassettes of discs, 15–17 min. ea.;

SCIENCE AND THE HUMANITIES (cont.)

Guide. #010019 Cassettes ($79.50), #010018 Discs ($79.50). Gr. 9–12. BKL (6/1/76). 001.3

Contents: 1. Too Many People. 2. Growing Enough Food. 3. Artificial Foods.

A grim set that explores the population explosion, the green revolution, the negative and positive effects of pesticides, and the concerns with artificial food coloring.

OUR CHANGING EARTH (Filmstrip—Sound). Scholastic Book Services, 1975 (Human Issues in Science). 4 color, sound filmstrips, 76–87 fr. ea.; 4 cassettes or discs, 15–17 min. ea.; Guide. #010017 Cassettes ($79.50), #010016 Discs ($79.50). Gr. 7–12. BKL (6/1/76). 001.3

Contents: 1. Changing Local Weather. 2. Changing Global Climate. 3. Changing the Land. 4. Changing the Sea.

This set discusses how pollution has effected changes in weather conditions and how scientists have attempted to control rain, snow, hail, fog, and hurricanes. The dangers of even slight changes in global climate are explained. Other concerns regarding our use and misuse of land and sea are explored.

OURSELVES (Filmstrip—Sound). Scholastic Book Services, 1975 (Human Issues in Science). 4 color, sound filmstrips, 88–95 fr. ea.; 4 cassettes of discs, 15–17 min. ea.; Guide. #010011 Cassettes ($79.50), #010012 Discs ($79.50). Gr. 7–12. BKL (10/1/75), PRV (1/76). 001.3

Contents: 1. Experimenting on People . . .? 2. Behavior Control . . .? 3. People Made-to-Order . . .? 4. Living Longer . . .?

Drawings, paintings, color and black-and-white photographs are used to show the recent discoveries in science that pose moral questions for people today. The pros and cons of performing experiments on human and animal subjects, controlling human behavior, producing replicas of human beings, and extending the lifetime of men and women are discussed.

SCIENCE FICTION

A DISCUSSION OF DR. HEIDEGGER'S EXPERIMENT (Motion Picture—16mm—Sound). See Motion pictures—History and criticism

DOCTOR HEIDEGGER'S EXPERIMENT (Motion Picture—16mm—Sound). See American fiction

THE FOGHORN BY RAY BRADBURY (Filmstrip—Sound). Listening Library, 1975 (Visu-Literature). 1 color, sound filmstrip, 31 fr.; 1 cassette, 11½ min.; Teacher's Guide. ($19.95). Gr. 8–C. LNG (1976). Fic

An ancient sea monster is lured by the sound of a foghorn in an annual pilgrimage in search for a mate in this macabre and unusual, yet touching, tale by Ray Bradbury.

THE ILLUSTRATED MAN (Filmstrip—Sound). Listening Library, 1976. 1 color, sound filmstrip; 1 cassette. #SS156 CFX ($19.95). Gr. 5–10. LGB (12/76), PRV (1976). Fic

One of Ray Bradbury's most memorable short stories has been adapted into a sound filmstrip production. It explores the theme of reality versus illusion. The tattoos on a man's body come to life, revealing the future.

THE MARTIAN CHRONICLES (Phonodiscs). Caedmon Records, 1 disc or cassette. #CDL51466 Cassette ($7.95), #TC1466 Disc ($6.98). Gr. 5–9. PRV (4/76). Fic

Contents: 1. There Will Come Soft Rains. 2. Usher II.

These are chilling masterpieces of science fiction by Ray Bradbury. They are read by Leonard Nimoy, Mr. Spock of "Star Trek."

THE PORTABLE PHONOGRAPH (Motion Picture—16mm—Sound). See American fiction

SCIENCE FICTION (Filmstrip—Sound/Captioned). Society For Visual Education, 1977. 4 color, captioned, sound filmstrips, av. 55–62 fr. ea.; 4 cassettes or discs, av. 8–10 min. #Y454–SAR Discs ($80), #Y454–SATC Cassettes ($80), Individual Titles ($20). Gr. 4–12. BKL (12/15/78). Fic

Contents: 1. Off on a Comet. 2. The People of the Pit. 3. The Time Traveler. 4. The Mortal Immortal.

Science fiction classics adapted to captioned sound filmstrips by Jules Verne, H. G. Wells, Mary Shelley, and A. Merritt. Four main themes of science fiction writing are represented: man's journey through space, the future vision, the battle of man against technology, and the alien world saga.

STORY INTO FILM: CLARK'S THE PORTABLE PHONOGRAPH (Motion Picture—16mm—Sound). See Motion pictures—History and criticism

THE TIME MACHINE (Filmstrip—Sound). Listening Library, 1975 (Visu-Literature). 1 color, sound filmstrip, 68 fr.; 1 cassette, 17 min.; Teacher's Guide. ($19.95). Gr. 7–12. LNG (1976). Fic

The Time Traveler recounts his amazing adventures into the land of the future in this adaptation from the classic science fiction book.

THE VELDT (Filmstrip—Sound). Listening Library, 1976. 1 color, sound filmstrip; 1 cassette. #SS159 CFX ($19.95). Gr. 5–10. LGB (1976). Fic

Relationships between parents and children are considered in this Ray Bradbury story. Set in a nursery, the children's needs are met by futuristic, mechanized contraptions.

SCIENCE FICTION—HISTORY AND CRITICISM

A DISCUSSION OF THE FALL OF THE HOUSE OF USHER (Motion Picture—16mm—Sound). *See* Motion pictures—History and criticism

ENCOUNTERS WITH TOMORROW: SCIENCE FICTION AND HUMAN VALUES (Filmstrip—Sound). Harper and Row Publishers/Sunburst Communications, 1976. 6 color, sound filmstrips, 85–108 fr. ea.; 6 cassettes, 14–18 min. ea.; Teacher's Guide. #76 ($145). Gr. 7–C. BKL (11/77), LGB (12/76), PRV (9/77). 809.3876

Contents: 1. From the Odyssey to the New Wave. 2. The Cult and Its Creators. 3. Facing the Biomedical Future. 4. Mythology of the Technical Age. 5. Scenarios for Tomorrow. 6. Science Fiction or Science Fact?

Examines the literary origins and development of science fiction, from its mythological underpinnings to its emergence as a mainstream genre. Analyzes science fiction's social function as a predictor of things to come. Scientist and writer Dr. Isaac Asimov describes the creative process he goes through in developing a novel.

FAHRENHEIT 451 (Kit). Current Affairs Films, 1978. 1 color, sound filmstrip, 70 fr.; 1 cassette, 15 min; 1 Book; 1 Guide. #625 ($30). Gr. 7–12. BKL (1/1/79). 809.3876

The title of this science fiction novel by Ray Bradbury is taken from the temperature at which paper burns. It portrays a future society in which book burning is an accepted government policy. The narrator offers thoughts on the novel's themes and stylistic elements.

READING SCIENCE FICTION (Filmstrip—Sound). Coronet Instructional Media, 1978. 4 color, sound filmstrips, 51–54 fr.; 2 discs or 4 cassettes, 12–14 min.; Teacher's Guide. #M301 Cassette ($69.50), #S301 Disc ($69.50). Gr. 9–12. PRV (12/19/78). 809.3876

Contents: 1. Voyages to Imaginary Worlds. 2. Visions of the Future. 3. Heroes, Robots and Aliens. 4. Tomorrow's Discoveries Today.

An overview of the history of science fiction from the early Greeks to tomorrow's dreams. Ethical and moral dilemmas of the future, based on current scientific experiments such as cloning, are discussed.

TALES OF TIME AND SPACE (Filmstrip—Sound). Educational Dimensions Group, 1978. 4 color, sound filmstrips, 80 fr.; 4 cassettes, 16 min. ea.; Teacher's Guide. #754 ($98). Gr. 9–12. PRV (2/79). 809.3876

Contents: 1. "Future Past Revisited: Vintage Season", by C. L. Moore and H. Kuttner. 2. "Science Puzzles: Neutron Star," by Larry Niven. 3. "What Happens Now? First Contact," by Murray Leinster. 4. "Perceptions of Other Realities: Nightfall," by I. Asimov.

How science fiction writers make profound human statements using fictional and scientific principles is explored. These excellent stories are enhanced by the reading aloud, the sound effects, and the drawings.

SCIENTISTS

CHARLES DARWIN (Cassette). *See* Darwin, Charles

CONVERSATION WITH LOREN EISELEY (Cassette). *See* Eiseley, Loren

THE HOUSE OF SCIENCE (Motion Picture—16mm—Sound). *See* Science—History

ISAAC NEWTON (Cassette). *See* Newton, Isaac

LINUS PAULING: SCIENTISTS AND RESPONSIBILITY (Cassette). *See* Pauling, Linus

THOMAS ALVA EDISON (Cassette). *See* Edison, Thomas Alva

SCOTLAND

THE BRITISH ISLES: SCOTLAND AND IRELAND (Filmstrip—Sound). University Films/McGraw-Hill Films, 1972 (European Geography, Set Four). 4 color, sound filmstrips, av. 65 fr. ea.; 4 cassettes or discs, 14 min. ea.; Teacher's Guide. Cassette #102419-0 ($65), Discs #103519-2 ($65). Gr. 5–9 BKL (1973), PRV (5/73). 914.1

Contents: 1. Scotland: The Highlands. 2. Scotland: Central Lowlands and Southern Uplands. 3. Ireland: The Land. 4. Ireland: Past and Present.

The series explores the history, geography, economics, and culture of Ireland and Scotland through photographs, maps, and some reproductions. Discussion questions are included at the end of each filmstrip.

EUROPEAN GEOGRAPHY, SET FOUR (Filmstrip—Sound). McGraw-Hill Films, 1972. 4 color, sound filmstrips, 55–70 fr.; 4 cassettes or discs, 12–15 min.; Teacher's Guide. #103519-2 Discs ($65), #102419-0 Cassettes ($65). Gr. 6–10. BKL (9/1/73), PRV (1/73), PVB (5/73). 914.1

Contents: 1. Ireland: The Land. 2. Ireland: Past and Present. 3. Scotland: The Central Lowlands and the Southern Uplands. 4. Scotland: The Highlands.

These filmstrips explore the beautiful and rugged environments of Scotland and Ireland, emphasizing how geography and climate have shaped living conditions in the various regions.

SCOTT, DRED

DRED SCOTT: BLACK MAN IN A WHITE COURT (Filmstrip—Sound). Current Af-

SCOTT, DRED (cont.)

fairs Films, 1977. 1 color, sound filmstrip, 74 fr.; 2 cassettes—"Pro-and-Con", 16 min. ea.; Teacher's Guide. #586 ($30). Gr. 7–A. PRV (3/78). 973.7

The trial of Dred Scott in 1846 marked the first time that a black slave appeared in a white court suing for his freedom. Scott contended that he had become free because of his stay in Illinois and Wisconsin. The case eventually reached the Supreme Court of the United States, and the decision handed down by this court helped to set up a chain of events that culminated in the Civil War. Through the use of two cassettes, both sides of the argument over the Dred Scott Decision are presented.

SCOTT, WALTER

GREAT WRITERS OF THE BRITISH ISLES, SET I (Filmstrip—Sound). *See* Authors, English

SCUBA DIVING

SCUBA (Filmstrip—Sound). Great American Film Factory, 1976 (Outdoor Education). 1 color, sound filmstrip, 80 fr.; 1 cassette, 18 min. #FS–12 ($35). Gr. 5–A. BKL (3/15/77), PRV (1/78). 797.23

This filmstrip pursues the interest in the ocean's depths shared by young students who might be interested in scuba diving as a sport or career. The principles of scuba are introduced and the class joins an expedition in search of a mysterious sunken ship. Beautiful underwater shots gives the strip pictorial and esthetic value.

SCULPTORS

THE BRONZE ZOO (Motion Picture—16mm—Sound). *See* Bronzes

INTERVIEW WITH ANTONIO SALEMME (Phonodiscs). *See* Salemme, Antonio

SCULPTURE

THE BRONZE ZOO (Motion Picture—16mm—Sound). *See* Bronzes

HOW TO DO: CARDBOARD SCULPTURE (Filmstrip—Sound). Educational Dimensions Group, 1978. 2 color, sound filmstrips, 60 fr.; 2 cassettes, 18 min. ea.; Teacher's Guide. #618 ($49). Gr. 9–A. PRV (1/79). 731

This set deals with low relief—3-dimensional types of sculpture. Directions and photographs of the creations are easy to follow, but projects take care and patience to produce.

MOBILES: ARTISTRY IN MOTION (Filmstrip—Sound). *See* Mobiles (sculpture)

SCULPTURE WITH STRING (Filmstrip—Sound). *See* String art

SOFT SCULPTURE (Filmstrip—Sound). Warner Educational Productions, 1978. 3 color, sound filmstrips, 93–116 fr.; 2 cassettes, 10–11 min.; Teacher's Guide. #180 ($61.50). Gr. 4–A. BKL (2/1/79). 731

Shows several examples of soft sculpture, then each set demonstrates one piece being constructed. The viewer sees the sculptures' developing characteristics, while the narrative explains the steps in their creation.

SEA HORSES

SEA HORSE (Film Loop—8mm—Silent). Walt Disney Educational Media, 1966. 8mm color, silent film loop, approx. 4 min. #62-5456L ($30). Gr. K–12. MDU (1974). 597.53

Male sea horse is shown ejecting the young from his pouch. The young congregate on a nearby plant while the exhausted father rests on the sea floor.

SEA STORIES

A DISCUSSION OF THE SECRET SHARER (Motion Picture—16mm—Sound). *See* Motion pictures—History and criticism

THE SECRET SHARER (Motion Picture—16mm—Sound). *See* American fiction

SEAFARING LIFE—HISTORY

THERE SHE BLOWS (Filmstrip—Sound). Hawkhill Associates, 1977. 1 color, sound filmstrip, 123 fr.; 1 cassette, 22 min.; 1 Guide. ($22.50). Gr. 3–A. BKL (7/15/77). 910.4

Old recordings of whaling folk songs provide background for the informational narration and the dramatic readings of life at sea aboard a whaler 150 years ago. Drawings, photographs, and classical paintings are used in the presentation of details of the daily life of whalers.

SEASHORE

SEA, SAND AND SHORE (Filmstrip—Sound). *See* Ocean

SEASONS

WEATHER, SEASONS AND CLIMATE (Filmstrip—Sound). *See* Meteorology

SEASONS—POETRY

POETRY OF THE SEASONS (Filmstrip—Sound). Centron Films, 1972. 4 color, sound filmstrips; 4 discs or cassettes. Discs ($52.50), Cassettes ($60.50). Gr. K–A. FLN (2/3/74), INS (4/74), SCT (2/74). 808.81

Contents: 1. Spring. 2. Summer. 3. Autumn. 4. Winter.

Four filmstrips that capture the visual splendor of the seasons. Sparse narrative consists of blank verse poetry.

SEEDS

HOW FLOWERS REPRODUCE: THE CALIFORNIA POPPY (Study Print). *See* Plant propagation

SEED DISPERSAL (Film Loop—8mm—Silent). Walt Disney Educational Media, 1966. 8mm color, silent film loop, approx. 4 min. #62–5501L ($30). Gr. K–12. MDU (1974). 582

Shows the various manners by which seeds are dispersed: by the wind, by pods bursting, by cones dropping, and by shooting pods.

SEEDS SPROUTING (Film Loop—8mm—Silent). Walt Disney Educational Media, 1966. 8mm color, silent film loop, approx. 4 min. #62–5503L ($30). Gr. K–12. MDU (1974). 582

Time-lapse photography shows the germination of seeds from the sprouting of roots to the growth of stems and leaves.

SELF PLANTING SEEDS (Film Loop—8mm—Silent). Walt Disney Educational Media, 1966. 8mm color, silent film loop, approx. 4 min. #62–5502L ($30). Gr. K–12. MDU (1974). 582

Shows the way that seeds are released from the plant and move across the ground in search of a suitable bit of soil.

SEEGER, PETE

FOLK SONGS IN AMERICAN HISTORY, SET ONE: 1700–1864 (Filmstrip—Sound). *See* Folk songs—U.S.

FOLK SONGS IN AMERICAN HISTORY, SET TWO: 1865–1967 (Filmstrip—Sound). *See* Folk songs—U.S.

SELF

DIVIDED MAN, COMMITMENT OR COMPROMISE? (Motion Picture—16mm—Sound). *See* Decision making

THE INDIVIDUAL (Kit). Scholastic Book Services, 1975 (Family Living Program). 3 color, sound filmstrips, 93–103 fr.; 3 cassettes or discs, 10–12 min.; 9 Ditto Masters; Student Worksheet; Teacher's Guide. #01124 Cassette ($69.50), #1114 Disc ($69.50). Gr. 8–A. BKL (10/1/75). 155.2

Contents: 1. Who Am I? 2. Growing through Changes. 3. Your Choice.

This program shows how young people handle problems of identity, frustration, peer pressure and the constant state of change.

PERSONALITY: ROLES YOU PLAY (Filmstrip—Sound). Sunburst Communications, 1974. 2 color, sound filmstrips, 64–74 fr.; 2 cassettes or discs, 10–13 min.; Teacher's Guide. #214 ($59). gr. 7–12. BKL (6/1/75), LGB (12/75), RMF (1975). 155.2

Contents: 1. Roles You Play. 2. Style and Popularity.

Explores individuality, looks at the concept of personality and the roles we play in different situations. Using firsthand accounts and examples, explores techniques, goals and results of role playing. Young people's attitudes about popularity are presented and lead into a discussion of the relative importance of popularity in later life.

SQUARE PEGS—ROUND HOLES (Motion Picture—16mm—Sound). *See* Individuality

TEENAGE RELATIONSHIPS: VENESSA AND HER FRIENDS (Motion Picture—16mm—Sound). *See* Adolescence

SELF—FICTION

REALLY ME (Kit). *See* Girls—Fiction

SELF-CONTROL

CONTROL: WHO PUSHES YOUR BUTTON? (Filmstrip—Sound). Harper and Row Publishers/Sunburst Communications, 1974. 2 color, sound filmstrips, 66–68 fr.; 2 cassettes or discs, 13$\frac{1}{2}$ min. ea.; Teacher's Guide. #82 ($59). Gr. 7–12. PRV (5/76), PVB (1977). 153.8

Contents: 1. Society's View. 2. The Personal View.

Uses literary excerpts, interviews with social scientists, and interviews with students to explore key questions: To what extent should individuals exert self-control and assume responsibility for their actions? How much control should they accept from society? Who decides what's "good?"

COPING WITH LIFE: THE ROLE OF SELF-CONTROL (Filmstrip—Sound). Center for Humanities/Sunburst Communications, 1975. 2 color, sound filmstrips, 80 fr. ea.; 2 cassettes or discs, 12–13 min.; Teacher's Guide. #602-81 ($59). Gr. 7–12. BKL (5/15/75). 153.8

Contents: 1. Self-Control and Society.

2. Self-Control and the Family.

Combines examples from relevant works of literature with movie stills, photographs, graphics, and masterpieces of art to give young people a broader perspective on the role of self-control in their lives. Stresses the functions of dialogue and compromise when emotions clash. Outlines ways of channeling emotions into constructive action.

SELF-DEFENSE

NOBODY'S VICTIM (Motion Picture—16mm—Sound). FilmFair Communications,

SELF-DEFENSE (cont.)

1972. 16mm color, sound film, 20 min. ($230). Gr. 7–A. BKL (1/73), FHE (9/73), PRV (12/73). 613.66

The film covers the two basics of personal safety—avoiding danger and dealing with danger. The first section covers working alone, thwarting purse snatchers, using a dog for protection, driving alone, car trouble, home security, etc. The second section emphasizes escape as the object of self-defense and describes what to do if confronted. It covers the use of weapons such as purse, keys, and magazines, etc. The film then demonstrates easily learned physical self-defense techniques.

SELF-PERCEPTION

THE CHILD AND THE FAMILY (Kit). *See* Child development

GRAVITY IS MY ENEMY (Motion Picture—16mm—Sound). *See* Hicks, Mark

THE GROWTH OF INTELLIGENCE (Kit). *See* Child development

I THINK (Motion Picture—16mm—Sound). *See* Values

PHYSICAL DEVELOPMENT (Kit). *See* Child development

THE SCHOOL EXPERIENCE (Kit). *See* Child development

A SENSE OF SELF (Kit). *See* Child development

SELF-REALIZATION

ALL MY FAMILIES (Kit). *See* Humanities

THE ARTIST INSIDE ME (Kit). *See* Humanities

THE BEST I CAN (Motion Picture—16mm—Sound). Films, 1976 (Best of Zoom). 16mm color, sound film, 12 min. #393–0019 ($200). Gr. 4–12. AFF (1978). 155.2

Demonstrating the importance of "trying again" after failure, rather than giving up, are a gymnast and a diver. Marcie, during the Junior Olympics competitions, slips from the balance beam, but gets up and continues. Diver Laura has a similar experience.

CLIMB (Motion Picture—16mm—Sound). *See* Mountaineering

DEALING WITH STRESS (Filmstrip—Sound). *See* Stress (psychology)

THE HUMAN-I-TIES OF LANGUAGE (Kit). *See* Humanities

THE I IN IDENTITY (Kit). *See* Humanities

LIFE GOALS: SETTING PERSONAL PRIORITIES (Filmstrip—Sound). Human Rela-

tions Media Center, 1976. 3 color, sound filmstrips, 70–77 fr.; 3 cassettes or discs, approx. 11½ min. ea.; Teacher's Guide. #619 ($90). Gr. 7–C. BKL (5/15/77), M&M (4/77), PRV (11/77). 155.2

Contents: 1. Values. 2. Establishing Objectives. 3. Risks and Strategy.

Gives students practical suggestions on making decisions about their goals in life. Stresses the importance of clarifying their values, establishing clear and practical objectives, and developing strategies. Explains that most decisions involve some risks. Encourages students to determine what they would like most from life and provides guidelines for realizing their goals.

LIVING IN THE FUTURE: NOW—INDIVIDUAL CHOICES (Filmstrip—Sound). BFA Educational Media, 1974. 4 color, sound filmstrips; 4 discs or cassettes. #VEP000 Cassettes ($70), #VEN000 Discs ($70). Gr. 9–12. BKL (5/15/75), PRV (1975). 155.2

Contents: 1. Our Changing Values. 2. Lifestyles In and Out of the Family. 3. Work and Leisure. 4. Liberty and Individualism.

This set isolates many of the decisions young people will have to make regarding personal values, family, and work. It gives them a social perspective on the issues, but implies that they can have an effect on the course of the future. This open-ended set combines student's comments, narrator's questions, and documentary facts.

SALLY GARCIA AND FAMILY (Motion Picture—16mm—Sound). *See* Family

TAKING RESPONSIBILITY (Filmstrip—Sound). BFA Educational Media, 1975 (Understanding Myself and Others). 4 color, sound filmstrips, 80–95 fr.; 4 discs or cassettes, 9–13 min.; Guide. #VFR000 Cassettes ($70), #VFP000 Discs ($70). Gr. 9–12. BKL (1/15/76). 155.2

Contents: 1. Being Who You Are. 2. Making Your Life. 3. Beyond Roles. 4. A Family.

Rapid social and cultural change have placed an increasing importance on taking responsibility for our own actions and identity. This set assists young people in examining their roles. Several individuals, with differing opinions, discuss their lives and values as they relate to the concept presented in each strip.

THE WORLD OUTSIDE ME (Kit). *See* Humanities

SELF-REALIZATION—FICTION

THE RAZOR'S EDGE (Cassette). Books on Tape, 1978. 7 cassettes #1087 ($65). Gr. 9–A. PRV (1/79). Fic

W. Somerset Maugham's novel of the World War I returning hero who wants to "loaf" and his journey to Paris, Germany, and India. It is in the latter that he becomes a student of Vedanta. The theme of exploration of mysticism and Eastern philosophy is of interest to adolescents and adults today.

SENSES AND SENSATION

SENSES AND PERCEPTION: LINKS TO THE OUTSIDE WORLD (Motion Picture—16mm—Sound). Media Four Productions/EBEC, 1975. 16mm color, sound film, 18 min. #3475 ($255). Gr. 6–9. AAS (12/76), LFR (3/4/76), SCT (1/77). 152.1

This film explores the world of the five senses, among both humans and animals, with emphasis on the role of the human eye and brain. Optical illusions help students appreciate the role of the brain in sorting sensory messages.

SEWING

FINISHING TOUCHES (Kit). Butterick Publishing, 1978 (Butterick Sewing Series). 2 color, sound filmstrips, av. 76 fr.; 2 cassettes or discs, 7–9 min.; 12 Transparencies; 8 Spirit Masters; Wall Poster; Paperback Book; Teacher's Guide. #C–138–XE Cassette ($79). #137–IE Disc ($79). Gr. 6–10. BKL (2/15/79). 646.4

Shows the variety of choices available for allowing imagination and creativity to achieve the finishing touch that makes a project unique.

PLANNING TO SEW (Kit). Butterick Publishing, 1978 (The Butterick Sewing Series). 2 color, sound filmstrips, 76–85 fr.; 2 discs or cassettes, 8–11 min.; 8 Spirit Masters; 1 Wall Chart; 1 Book; 12 Overhead Transparencies; Teacher's Guide, #C–128–2E Cassettes ($79), #R–127–4E Discs ($79). Gr. 6–10. BKL (1/15/79). 646.4

Shows how to plan, step-by-step, a sewing project.

SETTING UP TO SEW (Kit). Butterick Publishing, 1978 (The Butterick Sewing Series). 2 color, sound filmstrips, 74–80 fr.; 2 discs or cassettes, 7–10 min.; 12 Transparencies; 8 Spirit Masters; Wall Poster; 1 Paperback Book; Teacher's Guide. #C–132–OE Cassettes ($79), #R–131–2E Discs ($79). Gr. 6–10. AUR (1978). 646.4

The techniques of laying out a pattern, cutting, and marking are illustrated in filmstrips, spelled out in overhead transparencies, and reinforced by exercises on the spirit masters.

STARTING TO SEW (Kit). Butterick Publishing, 1977 (The Butterick Sewing Series). 2 color, sound filmstrips; 2 discs or cassettes; 8 Transparencies; 8 Spirit Masters; Wall Chart; 1 Paperback Book; Teacher's Guide. #C–137E Cassettes ($79.50), #R–133–9E Discs ($79.50). Gr. 6–10. BKL (12/15/78). 646.4

A quick overview of how the sewing machine works; what it can and cannot do; how to cope with thread jams, skipped stitches, and other problems.

TAKING SHAPE (Kit). Butterick Publishing, 1978 (The Butterick Sewing Series). 2 color, sound filmstrips, av 76 fr.; 2 discs or cassettes, 7–9 min; 12 Transparencies; 8 Spirit Masters; Wall Poster; 1 Book; Teacher's Guide. #C–136–3E Cassettes ($79), #R–135–5E Discs ($79). Gr. 6–10. BKL (2/15/79). 646.4

With this set, students learn "tricks of the trade" for completing just about any shaped construction assignments. Easing, gathering, interfacing, grading, trimming, clipping, and notching are all demonstrated.

A TRIP TO THE FABRIC STORE (Kit). Butterick Publishing, 1978 (The Butterick Sewing Series). 2 color, sound filmstrips, 76–85 fr.; 2 discs or cassettes, 8–11 min.; 8 Spirit Masters; 8 Overhead Transparencies; 1 Game; Teacher's Guide. #C–130–4E Cassettes ($79), #R–129–OE Discs ($79). Gr. 6–10. BKL (1/15/79). 646.4

This set demonstrates how to use the pattern catalogs and the pattern envelope as shopping guides and lists. Helps to decide on the right fabric pattern, trims, buttons, interfacing, lining.

SEX INSTRUCTION

ABOUT SEX (Motion Picture—16mm—Sound). Herman N. Engel/Texture Films, 1972. 16mm color, sound film, 23 min. ($315). Gr. 9–A. AFF (1973), PRV (2/73). 612.6

This film gives teenagers information about sexual fantasies, body growth, homosexuality, masturbation, birth control, venereal disease, and mutuality between the sexes.

THE GROWING TRIP (Filmstrip—Sound). *See* Reproduction

THERE'S A NEW YOU COMIN' FOR BOYS (Filmstrip—Sound). *See* Boys

THERE'S A NEW YOU COMIN' FOR GIRLS (Filmstrip—Sound). *See* Girls

SEX ROLES

ANYTHING YOU WANT TO BE (Motion Picture—16mm—Sound). Liane Brandon/New Day Films, 1975. 16mm b/w, sound film, 8 min. ($117). Gr. 9–A. AFF (1976), BKL (1976), FLQ (1976). 301.41

A film about sex-role stereotyping for women. It is the funny/not funny story of a high school girl who gets conflicting verbal (anything you want to be) and nonverbal messages.

MALE AND FEMALE ROLES (Filmstrip—Sound). Coronet Instructional Media, 1975.

SEX ROLES (cont.)

6 color, sound filmstrips. av. 74 fr.; 6 discs or cassettes, av. 11^1/$_2$ min. #M714 Cassettes (102), #S714 Discs ($102). Gr. 7–12. PRV (1976). 301.41

Contents: 1. The Stereotypes. 2. How Stereotypes Evolved. 3. How They Are Learned. 4. Emerging Dissatisfactions. 5. New Perspectives. 6. How Aware Are You?

These strips, their photographs, lithographs, and interviews, help clarify views on male and female roles.

REEXAMINING SEX ROLES: EVOLU-TION OR REVOLUTION (Filmstrip—Sound). Harper and Row Publishers/Sunburst Communications, 1975, 6 color, sound filmstrips, 71–84 fr. ea.; 6 cassettes or discs, 15^1/$_2$–17^1/$_2$ min. ea.; Teacher's Guide. #75 ($145). Gr. 7–C. BKL (4/15/76), TSS (5/77). 301.41

Contents: 1. The New Feminism. 2. The Individual and Sexual Identity. 3. Family Roles Challenges. 4. The Economic World. 5. Images, Messages and Media. 6. Institutions in Transition.

Scrutinizes the radical changes in today's sex roles. Identifies the forces behind these changes, then examines their impact on the individual and on society at large.

SHAKESPEARE, WILLIAM

GREAT WRITERS OF THE BRITISH ISLES, SET I (Filmstrip—Sound). *See* Authors, English

HIS INFINITE VARIETY: A SHAKE-SPEAREAN ANTHOLOGY (Cassette). *See* English drama

INTERVIEWS WITH PLAYWRIGHTS: SOPHOCLES, SHAKESPEARE, O'NEILL (Cassette). *See* Dramatists

SHAKESPEARE (Filmstrip—Sound). Coronet Instructional Media, 1974. 8 color, sound filmstrips, av. 50 fr.; 4 discs or 8 cassettes, av. 16 min. #M260 Cassettes ($120), #S260 Discs ($120). Gr. 9–12. PRV (1975). 822.33

Contents: 1. His Life. 2. His Theatre. 3. His Comedies. 4. His Tragedies. 5. His Historical Plays. 6. His Poems. 7. Macbeth, Part I. 8. Macbeth, Part II.

Shakespeare's major literary forms plus two-part dramatiziation of *Macbeth* are presented. Researched artwork depicts the costumes and settings of Elizabethan England.

TALES FROM SHAKESPEARE (Phonodiscs). *See* English literature

THEATER IN SHAKESPEARE'S ENG-LAND (Filmstrip—Sound). *See* Theater—England

WILLIAM SHAKESPEARE (Cassette). Ivan Berg Associates/Jeffrey Norton Publishers,

1976 (History Makers). 1 cassette, approx. 81 min. #41009 ($11.95). Gr. 10–A. BKL (1/15/78), PRV (11/19/78). 921

The story of Shakespeare's life in the days of Queen Elizabeth and the early years of James the First. Shakespeare managed in a comparatively brief life—fifty-two years—to produce literary masterpieces that have enriched not only his native England but the world.

SHANNON, TERRY

THE WORLD OF INNERSPACE (Filmstrip—Sound). *See* Oceanography

SHAPE. *See* Size and shape

SHAPIRO, KARL

KARL SHAPIRO'S AMERICA (Motion Picture—16mm—Sound). *See* American poetry

SHAW, GEORGE BERNARD

THE SEVEN AGES OF GEORGE BERNARD SHAW (Cassette). *See* English drama

SHEEP

LIFE CYCLE OF COMMON ANIMALS, GROUP 2 (Filmstrip—Sound). Imperial Learning Corp., 1973. 2 color, sound filmstrips, av. 37 fr.; 2 cassettes or discs, av. 10 min. #3RG 44400 Discs ($30), #3KG 444000 Cassettes ($33). Gr. 4–9 TEA (2/73). 591.5

Contents: 1. The Life Cycle of a Bighorn Sheep, Part One. 2. The Life Cycle of a Bighorn Sheep, Part Two.

These strips trace the evolution of the bighorn's habitat showing how they came to depend on a special grassland environment.

SHELLEY, MARY WOLLSTONECRAFT

SCIENCE FICTION (Filmstrip—Sound/Captioned). *See* Science fiction

SHEPARD, ERNEST

MR. SHEPARD AND MR. MILNE (Motion Picture—16mm—Sound). *See* Authors, English

SHOPPERS GUIDES. *See* Consumer education

SHOPPING

LET'S GO SHOPPING (Kit). Changing Times Education Service/EMC, 1973. 2 color, sound filmstrips, 238 fr.; 2 discs or cassettes; 10 Linemasters; Reading and Resources List; Exercises for Review; Guide to Inquiry. #2345 Cassettes ($49.50), #2340 Discs ($49.50). Gr. 7–12. IFT (1973). 640.73

Contents: 1. Comparison Shopping. 2. Hunting Bargains. 3. The Supermarket Maze. 4. Evaluating Services. 5. Settling a Grievance.

Provides information to help get the most and best for your money. Helps formulate values and develop decision-making skills that can be applied to all kinds of buying. Guides inquiry and discussion based on 5 real-life case studies involving young people.

PLAY THE SHOPPING GAME (Motion Picture—16mm—Sound). *See* Consumer education

SHORT STORIES

BARTLEBY (Motion Picture—16mm—Sound). *See* American fiction

A DISCUSSION OF BARTLEBY (Motion Picture—16mm—Sound). *See* Motion pictures—History and criticism

A DISCUSSION OF DR. HEIDEGGER'S EXPERIMENT (Motion Picture—16mm—Sound). *See* Motion pictures—History and criticism

A DISCUSSION OF MY OLD MAN (Motion Picture—16mm—Sound). *See* Motion pictures—History and criticism

A DISCUSSION OF THE CROCODILE (Motion Picture—16mm—Sound). *See* Motion pictures—History and criticism

A DISCUSSION OF THE HUNT (Motion Picture—16mm—Sound). *See* Motion pictures—History and criticism

A DISCUSSION OF THE LADY OR THE TIGER? (Motion Picture—16mm—Sound). *See* Motion pictures—History and criticism

A DISCUSSION OF THE LOTTERY (Motion Picture—16mm—Sound). *See* Motion pictures—History and criticism

A DISCUSSION OF THE SECRET SHARER (Motion Picture—16mm—Sound). *See* Motion pictures—History and criticism

DOCTOR HEIDEGGER'S EXPERIMENT (Motion Picture—16mm—Sound). *See* American fiction

JACK LONDON CASSETTE LIBRARY (Cassette). Listening Library, 1976. 6 cassettes; Teacher's Guide. #CXL 517 ($44.95). Gr. 6–A. BKL (12/15/76), PRV (11/76). Fic

Contents: 1. "To Build a Fire." 2. "A Piece of Steak." 3. "Lost Face." 4. "Told in the Drooling Ward." 5. Selections from the "Call of the Wild." 6. "The Sea Wolf." 7. "Martin Eden." 8. "The Poker Game."

Some of Jack London's best stories are included in this album as well as some lesser-known short works.

THE LADY OR THE TIGER? (Motion Picture—16mm—Sound). *See* American fiction

THE LEGEND OF SLEEPY HOLLOW AND ICHABOD CRANE (Phonodisc or Cassette). *See* Legends—U.S.

THE LOTTERY BY SHIRLEY JACKSON (Motion Picture—16mm—Sound). *See* American fiction

THE MANY FACES OF TERROR (Cassette). *See* Horror—Fiction

THE PORTABLE PHONOGRAPH (Motion Picture—16mm—Sound). *See* American fiction.

THE SECRET SHARER (Motion Picture—16mm—sound). *See* American fiction

STORY INTO FILM: CLARK'S THE PORTABLE PHONOGRAPH (Motion Picture—16mm—Sound). *See* Motion pictures—History and criticism

SHORT STORY SHOWCASE

THE CROCODILE (Motion Picture—16mm—Sound). *See* Fantasy—Fiction

A DISCUSSION OF BARTLEBY (Motion Picture—16mm—Sound). *See* Motion pictures—History and criticism

A DISCUSSION OF DR. HEIDEGGER'S EXPERIMENT (Motion Picture—16mm—Sound). *See* Motion pictures—History and criticism

A DISCUSSION OF MY OLD MAN (Motion Picture—16mm—Sound). *See* Motion picture—History and criticism

A DISCUSSION OF THE CROCODILE (Motion Picture—16mm—Sound). *See* Motion pictures—History and criticism

A DISCUSSION OF THE HUNT (Motion Picture—16mm—Sound). *See* Motion picture—History and criticism

A DISCUSSION OF THE LADY OR THE TIGER? (Motion Picture—16mm—Sound). *See* Motion pictures—History and criticism

A DISCUSSION OF THE LOTTERY (Motion Picture—16mm—Sound). *See* Motion pictures—History and criticism

A DISCUSSION OF THE SECRET SHARER (Motion Picture—16mm—Sound). *See* Motion pictures—History and criticism

DOCTOR HEIDEGGER'S EXPERIMENT (Motion Picture—16mm—Sound). *See* American fiction

THE HUNT (Motion Picture—16mm—Sound). *See* American fiction

THE LADY OR THE TIGER? (Motion Picture—16mm—Sound). *See* American fiction

SHORT STORY SHOWCASE (cont.)

THE LOTTERY BY SHIRLEY JACKSON (Motion Picture—16mm—Sound). *See* American fiction

MY OLD MAN (Motion Picture—16mm—Sound). *See* American fiction

THE PORTABLE PHONOGRAPH (Motion Picture—16mm—sound). *See* American fiction

THE SECRET SHARER (Motion Picture—16mm—Sound). *See* American fiction

STORY INTO FILM: CLARK'S THE PORTABLE PHONOGRAPH (Motion Picture—16mm—Sound). *See* Motion pictures—History and criticism

SHYNESS. *See* Bashfulness

SIBELIUS, JEAN

SYMPHONIC MOVEMENTS NO. 2 (Phonodiscs). *See* Symphonies

SIBERIA

SIBERIA NOW (Kit). *See* Russia

SIERRA NEVADA

JOHN MUIR'S HIGH SIERRA (Motion Picture—16mm—Sound). *See* Muir, John

SILVERSTEIN, SHEL

THE GIVING TREE (Motion Picture—16mm—Sound). *See* Friendship—Fiction

SINGER, ISAAC BASHEVIS

ZLATEH THE GOAT (Motion Picture—16mm—Sound). *See* Folklore—Poland

SIZE AND SHAPE

SHAPES AND STRUCTURES IN NATURE (Filmstrip—Captioned). *See* Matter

SKATEBOARDS

THE MAGIC ROLLING BOARD (Motion Picture—16mm—Sound). MacGillivray/Freeman Films/Pyramid Films, 1976. 16mm color, sound film, 15 min.; ³/₄ in. videocassette also available. ($225). Gr. K–A. CFF (1976), LFR (1976), WLB (1976). 796.21
A joyous, lyrical skateboarding extravaganza, this film captures every expression of the sport. Graceful, poetic slow motion shots; exhibitions of skill, stunts, and mishaps combine to form a compelling mixture of humor and excitement.

SKATEBOARD MANIA (Kit). Children's Press, 1976 (Ready, Get-Set, Go). 1 cassette; 4 Paperback Books, 2 Task Cards; Ditto Masters; packaged in plastic bag. #27275–6 ($16.95). Gr. 1–8. PRV (9/15/77). 796.21
High-interest, low reading ability material. The cassette contains a word-for-word reading of the book on a subject of great interest.

SKATEBOARD SAFETY (Motion Picture—16mm—Sound). Pyramid Films, 1976. 16mm color, sound film, 13 min. #2908 ($200). Gr. 5–12. BKL (4/15/77). 796.21
This film shows the exhilaration and excitement of skateboarding while advising of its dangers and the ways in which accidents and injuries can be minimized.

SKIN, COLOR OF. *See* Color of people

SKIN DIVING

SKIN DIVING (Filmstrip—Sound). Great American Film Factory, 1976 (Outdoor Education). 1 color, sound filmstrip; 1 cassette, 17 min. #FS–11 ($35). Gr. 5–A. BKL (3/15/77), PRV (1/19/77), SCT (12/19/76). 797.23
Developed in close cooperation with the Brooks Institute of Underwater Photography, Jacques Cousteau's U.S. Diver's Company filmed this strip on location on both the East and West Coasts. Discusses the things any beginner needs to know about skin diving and exploring the underwater world for the first time. Shares important rules of water safety, equipment, and selection of diving location. Makes the first experience more enjoyable by knowing what to look out for and how to prepare.

SKINNER, B. F.

TALKING WITH THOREAU (Motion Picture—16mm—Sound). *See* Philosophy, American

SKIS AND SKIING

SKI FLYING (Motion Picture—16mm—Sound). Jugoslavija Films/EBEC, 1975. 16mm color, sound film, 6 min. #3419 ($80). Gr. 5–9. EGT (1976), PRV (1976). 796.9
A European sports meet depicts the skills and thrills of ski jumping. Without narration, the film transcends the typical sports documentary by using the camera in unusual ways to tell a real story. The skiers seem actually to fly and the spectators become a part of the action.

SLAVERY IN THE U.S.

DRED SCOTT: BLACK MAN IN A WHITE COURT (Filmstrip—Sound). *See* Scott, Dred

SLIDES (PHOTOGRAPHY)

HOW TO DO: SLIDE COLLAGE (Filmstrip—Sound). *See* Collage

SMITH, DICK

DICK SMITH, MAKE-UP ARTIST (Motion Picture—16mm—Sound). *See* Makeup, Theatrical

SMITHSONIAN INSTITUTION

SMITHSONIAN INSTITUTION (Motion Picture—16mm—Sound). *See* Museums

SMOKING

HOW TO STOP SMOKING (Filmstrip—Sound). *See* Tobacco habit

TOBACCO PROBLEM: WHAT DO YOU THINK? (Motion Picture—16mm—Sound). *See* Tobacco habit

SNAKES

RATTLESNAKE (Film Loop—8mm—Silent). Walt Disney Educational Media, 1966. 8mm color, silent film loop, approx. 4 min. #62-5110L ($30). Gr. K–12. MDU (1974). 598.12
The rattlesnake is shown hunting a pocket mouse, which escapes while the snake is diverted by a tarantula. A study is made of the snake's fangs, tongue, and rattle.

THE SNAKE: VILLAIN OR VICTIM? (Motion Picture—16mm—Sound). Avatar Learning/EBEC, 1976 (The Wide World of Adventure). 16mm color, sound film, 24 min. #3393 ($320). Gr. 10–A. AVI (2/19/77), BKL (7/15/77), LFR (3/19/77). 598.12
Historically veiled by myth and legend, the snake is actually a valuable member of the ecosystem. Film explores many of the misconceptions surrounding snakes.

SNAKES OF THE AMAZON (Film Loop—8mm—Silent). Walt Disney Educational Media, 1966. 8mm color, silent film loop, approx. 4 min. #62-5317L ($30). Gr. K–12. MDU (1974). 598.12
Shows boa constrictors as they live in the jungle. A marmoset is seen annoying an emerald tree boa until it is engulfed in the boa's coils.

SOAP BOX DERBIES

DEAR KURT (Motion Picture—16mm—Sound). Weston Woods Studios, 1973. 16mm color, sound film, 24 min. #418 ($305). Gr. K–A. MMT (4/74). 796.6
Told in the form of a letter from a father to his son, the film is a portrayal of the actual experiences of one thirteen-year-old boy who took part in the 1972 All-American Soap Box Derby in Akron, Ohio. The film follows the boy, Kurt, as he constructs his soap box racer, wins the elimination "heats" in New Rochelle, N.Y. and goes on to represent his home town in the international finals in Akron. It is a film illustrating the competitiveness of modern life, the virtues and rewards of work, and the spirit of giving.

RACING NUMBERS (Kit). *See* Automobile racing

THE SOAP BOX DERBY SCANDAL (Motion Picture—16mm—Sound). Weston Woods Studios, 1975. 16mm color, sound film, 24 min; captioned film available. #422 ($305). Gr. K–A. BKL (6/1/77), PRV (3/76). 796.6
This documentary records the events of the important 1973 competition during which the venerable Soap Box Derby was rocked by scandal. It is a penetrating look at a sophisticated annual event that began 36 years ago and raises questions about winning, losing, and cheating.

SOBOL, DONALD

THE BEST OF ENCYCLOPEDIA BROWN (Filmstrip—Sound). *See* Mystery and detective stories

SOCCER

DRIBBLING (Film Loop—8mm—Silent). Athletic Institute (Soccer). 8mm color, silent film loop, approx. 4 min.; Guide. #F–5 (22.95). Gr. 1–12. AUR (1978). 796.33
A single-concept loop demonstrates the basic technique.

GOAL KEEPER, CLEARING (Film Loop—8mm—Silent). Athletic Institute (Soccer). 8mm color, silent film loop, approx. 4 min.; Guide. #F–11 ($22.95). Gr. 1–12. AUR (1978). 796.334
A single-concept loop demonstrates the basic technique.

GOAL KEEPER, PART 1 (Film Loop—8mm—Silent). Athletic Institute (Soccer). 8mm color, silent film loop, approx. 4 min.; Guide. #F–8 ($22.95). Gr. 1–12. AUR (1978). 796.334
A single concept loop demonstrates the basic technique.

GOAL KEEPER, PART 2 (Film Loop—8mm—Silent). Athletic Institute (Soccer). 8mm color, silent film loop, approx. 4 min.; Guide. #F–9 ($22.95). Gr. 1–12. AUR (1978). 796.334
A single-concept loop demonstrates the basic technique.

GOAL KEEPER, PART 3 (Film Loop—8mm—Silent). Athletic Institute (Soccer). 8mm color, silent film loop, approx. 4 min.; Guide. #F–10 ($22.95). Gr. 1–12. AUR (1978). 796.334
A single-concept loop demonstrates the basic technique.

HEADING AND BACK-HEADING (Film Loop—8mm—Silent). Athletic Institute

SOCCER (cont.)

(Soccer). 8mm color, silent film loop, approx. 4 min.; Guide. #F-4 ($22.95). Gr. 1–12. AUR (1978). 796.334

A single-concept loop demonstrates the basic technique.

KICKING (Film Loop—8mm—Silent). Athletic Institute (Soccer). 8mm color, silent film loop, approx. 4 min.; Guide. #F-1 ($22.95). Gr. 1–12. AUR (1978). 796.334

A single concept loop demonstrates the basic technique.

NO. 1 (Motion Picture—16mm—Sound). Texture Films, 1976. 16mm color, sound film, 10 min. ($160). Gr. 8–A. BKL (1977), FLN (1977). 796.334

In soccer, No. 1 is the position of the goalkeeper, a human body in continuous, unbelievable motion—diving, leaping, bouncing, sprinting, hurtling through space.

TACKLING (Film Loop—8mm—Silent). Athletic Institute (Soccer). 8mm, color, silent film loop, approx. 4 min.; Guide. #F-6 ($22.95). Gr. 1–12. AUR (1978). 796.334

A single concept loop demonstrates the basic technique.

THROW IN (Film Loop—8mm—Silent). Athletic Institute (Soccer). 8mm color, silent film loop, approx. 4 min.; Guide. #F-7 ($22.95). Gr. 1–12. AUR (1978). 796.334

A single concept loop demonstrates the basic technique.

TRAPPING—BALL IN AIR (Film Loop—8mm—Silent). Athletic Institute (Soccer). 8mm, color, silent film loop, approx. 4 min.; Guide. #F-3 ($22.95). Gr. 1–12. AUR (1978). 796.334

A single concept loop demonstrates the basic technique.

TRAPPING—GROUND BALL (Film Loop—8mm—Silent). Athletic Institute (Soccer). 8mm, color, silent film loop, approx. 4 min.; Guide. #F-3 ($22.95). Gr. 1–12. AUR (1978). 796.334

A single concept loop demonstrates the basic technique.

SOCIAL ACTION—CITIZEN PARTICIPATION

COMMITMENT TO CHANGE: REBIRTH OF A CITY (Filmstrip—Sound). *See* Urban renewal

THE WOODLAWN ORGANIZATION (Filmstrip—Sound). *See* Chicago—Social conditions

SOCIAL CHANGE

DIMENSIONS OF CHANGE (Filmstrip—Sound). Westport Communications/Sun-

burst Communications, 1974. 6 color, sound filmstrips; 6 cassettes or discs; Teacher's Guide. #40 ($145). Gr. 7–12. M&M (1973), PVB (1975). 301.24

Contents: 1. The Man-Made Planet. 2. Shelter. 3. Energy. 4. Food. 5. Mobility. 6. Communications.

Challenges students to examine new ways of looking at shelter, energy, food, mobility, and communication. Emphasizes the array of possible solutions and presents some of the most creative ones as guideposts for future planning.

THE FAMILY IN CHANGING SOCIETY (Filmstrip—Sound). *See* Family

FORECASTING THE FUTURE: CAN WE MAKE TOMORROW WORK? (Filmstrip—Sound). *See* Forecasting

LIVING WITH TECHNOLOGY: CAN WE CONTROL APPLIED SCIENCE? (Filmstrip—Sound). *See* Technology and civilization

PROGRESSIVE ERA: REFORM WORKS IN AMERICA (Motion Picture—16mm—Sound). *See* U.S.—History—1898–1919

SCIENCE, TECHNOLOGY AND MODERN MAN (Filmstrip—Sound). *See* Technology and civilization

TECHNOLOGY AND CHANGE (Kit). *See* Technology and civilization

SOCIAL CONFLICT

BALABLOK (Motion Picture—16mm—Sound). *See* Prejudices and antipathies

COPING WITH CONFLICT: AS EXPRESSED IN LITERATURE (Filmstrip—Sound). *See* American literature—Study and teaching

URBAN WORLD—VALUES IN CONFLICT (Filmstrip—Sound). *See* Metropolitan areas

SOCIAL PROBLEMS

FAMILIES IN CRISIS (Filmstrip—Sound). S. A. Films/Coronet Instructional Media, 1973. 8 color, sound filmstrips, av. 65 fr. ea.; 4 discs or 8 cassettes, av. 13 min. ea. #M300 Cassettes ($120), #S300 Discs ($120). Gr. 7–12. BKL (1974), IDF (1974), M&M (1974), 362.8

Contents: 1. Divorce. 2. Occupational Stress. 3. A Handicapped Child. 4. Coping with Death. 5. Care of the Aged. 6. Financial Reverses. 7. Pulling Up Roots. 8. A Brush with the Law.

Case studies of problems in family loving offer alternatives and open up questions. Interviews with real families explore the effects of crisis on careers and family relationships.

OUR CHANGING CITIES: CAN THEY BE SAVED? (Motion Picture—16mm—Sound). *See* City life

UNDERSTAND INSTITUTIONAL RACISM (Kit). *See* Prejudices and antipathies

See also Alcoholism; Crime; Divorce; Drug abuse; Juvenile delinquency; Migrant labor—Fiction

SOCIAL PSYCHOLOGY

MAN ALONE AND LONELINESS: THE DILEMMA OF MODERN SOCIETY (Slides/Cassettes). *See* Human relations

SOCIAL SCIENCES—STUDY AND TEACHING

WHY DO WE HAVE TO TAKE SOCIAL STUDIES (Filmstrip—Sound). *See* Educational guidance

SOCIAL STUDIES. *See* Social sciences

SOCIAL WORK

CAREERS IN SOCIAL WORK (Filmstrip—Sound). Pathescope Educational Media, 1974. 2 color, sound filmstrips, 75–90 fr. ea.; 2 cassettes, 11–15 min. ea.; Teacher's Guide. #721 ($50). Gr. 7–12. PRV (5/75). 361
Presents a career of helping others who have capabilities of making their own decisions by providing them with technical support and guidance. A general look at the varied types of jobs and a more detailed look at education, working conditions, opportunities, and compensations are presented in the two parts. Interviews with social workers bring out the variety of activities performed in this field.

SOFTBALL

BATTING (Film Loop—8mm—Silent). Athletic Institute (Soft Ball—Women's). 8 mm color, silent film loop, approx. 4 min.; Guide. #SB-4 ($22.95). Gr. 4–12. AUR (1978). 796.357
A single-concept loop demonstrates the basic technique.

THE CATCHER (Film Loop—8mm—Silent). Athletic Institute (Soft Ball—Women's). 8 mm color, silent film loop, approx. 4 min.; Guide. #SB-13 ($22.95). Gr. 4–12. AUR (1978). 796.357
A single-concept loop demonstrates the basic technique.

CATCHING ABOVE WAIST/BELOW WAIST (Film Loop—8mm—Silent). Athletic Institute (Soft Ball—Women's). 8mm color, silent film loop, approx. 4 min.; Guide. #SB-2 ($22.95). Gr. 4–12. AUR (1978). 796.357
A single-concept loop demonstrates the basic technique.

DEFENSIVE RUN DOWN (Film Loop—8mm—Silent). Athletic Institute (Soft Ball—Women's). 8mm color, silent film loop, approx. 4 min.; Guide. #SB-7 ($22.95). Gr. 4–12. AUR (1978). 796.357
A single-concept loop demonstrates the basic technique.

DOUBLE PLAY BY SHORT STOP/SECOND BASE WOMAN (Film Loop—8mm—Silent). Athletic Institute (Soft Ball—Women's). 8mm color, silent film loop, approx. 4 min.; Guide. #SB-12 ($22.95). Gr. 4–12. AUR (1978). 796.357
A single-concept loop demonstrates the basic technique.

FIELDING LONG HIT FLY BALL/GROUND BALL (Film Loop—8mm—Silent). Athletic Institute (Soft Ball—Women's). 8mm color, silent, film loop, approx. 4 min.; Guide. #SB-3 ($22.95). Gr. 4–12. AUR (1978). 796.357
A single-concept loop demonstrates the basic technique.

HOOK SLIDE/STRAIGHT IN SLIDE (Film Loop—8mm—Silent). Athletic Institute (Soft Ball—Women's). 8mm color, silent film loop, approx. 4 min.; Guide #SB-8 ($22.95). Gr. 4–12. AUR (1978). 796.357
A single-concept loop demonstrates the basic technique.

OVERHAND THROW—SIDE ARM THROW (Film Loop—8mm—Silent). Athletic Institute (Soft Ball—Women's). 8mm color, silent film loop, approx. 4 min.; Guide. #SB-1 ($22.95). Gr. 4–12. AUR (1978). 796.357
A single concept loop demonstrates the basic technique.

PITCHING/WINDMILL STYLE/SLINGSHOT STYLE (Film Loop—8mm—Silent). Athletic Institute (Soft Ball—Women's). 8mm color, silent film loop, approx. 4 min.; Guide. #SB-9 ($22.95). Gr. 4–12. AUR (1978). 796.357
A single-concept loop demonstrates the basic technique.

RUNNING TO FIRST BASE/RUNNING EXTRA BASES/RUNNER'S HEADOFF (Film Loop—8mm—Silent). Athletic Institute (Soft Ball—Women's). 8mm color, silent film loop, approx. 4 min.; Guide. #SB-6 ($22.95). Gr. 4–12. AUR (1978). 796.357
A single-concept loop demonstrates the basic technique.

SACRIFICE BUNT/BUNT FOR BASE HIT (Film Loop—8mm—Silent). Athletic Institute (Soft Ball—Women's). 8mm color, silent film loop, approx. 4 min.; Guide. #SB-5 ($22.95). Gr. 4–12. AUR (1978). 796.357
A single-concept loop demonstrates the basic technique.

SOFTBALL (cont.)

TAG OUTS/FORCE OUTS (Film Loop—
8mm—Silent). Athletic Institute (Soft Ball—
Women's). 8mm color, silent film loop, approx. 4 min,; Guide. #SB–11 ($22.95). Gr.
4–12. AUR (1978). 796.357

A single-concept loop demonstrates the basic technique.

SOLAR ENERGY

THE SUN: ITS POWER AND PROMISE
(Motion Picture—16mm—Sound). *See* Sun

SOLAR SYSTEM

THE SUN AND THE MOON (Slides). Educational Dimensions Group, 1976. 20 slides;
cardboard mount, in plastic sheet; Notes.
#9129 ($25). Gr. 6–12. BKL (4/1/77). 523.2

The set identifies the sun, the moon, the
earth, and various astronomical phases of
these bodies.

SONGS

AMERICA SINGS (Phonodisc or Cassette).
See Folk songs

SOPHOCLES

INTERVIEWS WITH PLAYWRIGHTS:
SOPHOCLES, SHAKESPEARE,
O'NEILL (Cassette). *See* Dramatists

SOUND

MUSIC APPRECIATION: TRAVELING
SOUND TO SOUND (Filmstrip—Sound).
See Music—Analysis, Appreciation

SOUTH AFRICA—RACE RELATIONS

WHITE ROOTS IN BLACK AFRICA (Filmstrip—Sound). Current Affairs Films, 1978.
2 color, sound filmstrips, 74 fr.; 2 cassettes,
16 min. ea. #606–607 ($52). Gr. 7–12. BKL
(6/15/78), PRV (12/19/78). 320.5

This two-part presentation gives a moderate
view of the racial problem in African nations
south of the Sahara. Part I traces the heritage of white Rhodesians and South Africans
and provides insights into how they see their
role in relation to the black population. Part
II focuses on the black majority and its attitude toward the white minority in Kenya,
Rhodesia, and South Africa. Very little attention is given to apartheid or the Mau
Mau.

SOUTH AMERICA

THE ANDEAN LANDS (Filmstrip—Sound).
EBEC, 1973. 5 color, sound filmstrips, av.
81 fr.; 5 discs or cassettes, av. 8 min. ea;
Spanish-English cassettes available. Teach-
er's Guides. #6463K Cassettes ($72.50),
#6463 Discs ($72.50), #6491K Biligual ed.
($72.50). Gr. 4–9. BKL (11/73); PRV (4/75),
TEA (5/6/74). 918

Contents: 1. Life in the Highlands. 2. Life in
the Lowlands. 3. Venezuela: Sowing the Oil.
4. A Highland Indian Village. 5. Coffee
Farmer of Colombia.

Along the Andes lies a group of countries in
which striking contrasts abound. In this series students travel through these contrasts,
and see the influences of geography and history on today's economic and cultural life.

THE MANY FACES OF SOUTH AMERICA
(Filmstrip—Sound). Current Affairs Films,
1976. 4 color, sound filmstrips, 74 fr.; 4 cassettes, 16 min.; Discussion Guide. #YW–15
($88). Gr. 3–6. PRV (11/78). 918

Contents: 1. People and Resources of South
America. 2. Life in a South American City.
3. Life in a South American Mountain Community. 4. Life in a South American Indian
Tribe.

A unit of study that provides an introduction
to how people live in South America, whether it be in one of the many rapidly growing
city areas, the countryside, a mountain village, or an Indian tribe. Depicts the tremendous progress and crises that exist on this
great continent, and provides an overview of
the human and physical resources.

SOUTH AMERICA: A REGIONAL STUDY
(Filmstrip—Sound). Skylight Productions/
Educational Enrichment Materials, 1975. 5
color, sound filmstrips, 53–66 fr.; 5 cassettes
or discs, 13–15 min.; Teacher's Guide.
#41080 C or R ($90). Gr. 5–10. BKL (7/15/
76). 918

Contents: 1. Perspectives on a Continent. 2.
Forms of Cultural Expression. 3. The Bitter
Legacy of the Inca Indians. 4. Politics and
the Economy.

An interdisciplinary approach to the study of
South America that focuses on the geographical, cultural, social, anthropological, historical, and political aspects of the continent.

SOUTHERN STATES

KUDZU (Motion Picture—16mm—Sound).
See Weeds

SOUTHERN STATES—FICTION

TO KILL A MOCKINGBIRD (Kit). Current
Affairs Films, 1978. 1 color, sound filmstrip,
70 fr.; 1 cassette, 15 min.; 1 Book; 1 Guide.
#646 ($30). Gr. 9–C. BKL (2/1/79). Fic

This Pulitzer Prize–winning novel is set in a
small rural town of the Deep South during
the years 1933–1935. It is the story both of a
little girl's life and personal problems and of
her lawyer father's personal courage in de-
fending a black man wrongly accused of rap-
ing a white woman in the town.

SOUTHWEST, NEW

CULTURE OF REGIONS: THE SOUTH-
WEST (Filmstrip—Sound). Filmstrip
House, 1971. 4 color, sound filmstrips, 48–
59 fr.; 2 discs; Script; 16 Spirit Masters.
($45). Gr. 4–12. PRV (9/72), PVB (5/73).
917.9

Contents: 1. Early Indian Culture. 2. Spanish Heritage. 3. Early Anglo Culture. 4. The
Modern Southwest.

This set on the Southwest traces the influence of traders, miners, cowboys, and farmers and presents the Indian culture and the
influence of the Indian, Spanish and Anglo
on this part of the country.

SOVIET UNION. *See* Russia; Siberia

SOYBEANS

SOYBEANS—THE MAGIC BEANSTALK
(Motion Picture—16mm—Sound). Centron
Films, 1975. 16mm color, sound film, 11-¹/₂
min.; Leader's Guide. ($185). Gr. 4–9. AAS
(5/76), AFF, CGE. 633.34

History of soybeans and their use, including
how soybeans are playing a major role in
meeting world food shortages. Illustrates
planting, tending, and harvesting of soybeans.

SPACE AND TIME

THE SPECIAL THEORY OF RELATIVITY
(Filmstrip—Sound). *See* Relativity (physics)

SPACE FLIGHT TO THE MOON

THE MOON: A GIANT STEP IN GEOLOGY
(Motion Picture—16mm—Sound). *See* Lunar geology

SPACEBORNE (Motion Picture—16mm—
Sound). Pyramid Films, 1977. 16mm color,
sound film, 14 min.; ³/₄ in. videocassette also
available. ($250). Gr. K–A. BKL (3/1/78),
FLN (1978). 629.45

In an exhilarating tribute to our exploration of space, NASA film photography of actual space voyages transforms space flight
into one symbolic journey to the moon.

SPACE PHOTOGRAPHY

MARS MINUS MYTH (REV.) Motion Picture—16mm—Sound). *See* Mars (planet)—
Exploration

SPACE PROBES

MARS MINUS MYTH (REV.) (Motion Picture—16mm—Sound). *See* Mars (planet)—
Exploration

SPACE SCIENCES

SPACE EXPLORATION (Filmstrip—Sound).
Educational Dimensions Group, 1978. 2 col-
or, sound filmstrips, 60 fr.; 2 cassettes;
Teacher's Guide. #1215 ($37). Gr. 7–12.
PRV (2/79). 629.4

Contents: 1. Man and Machine in Space. 2.
Skylab and Space Shuttle.

An overview of the space program that includes the discovery of the Van Allen belts,
the development of tracking networks, the
use of satellites and unmanned probes.

SPAIN

SOUTHERN EUROPE: MEDITERRA-
NEAN LANDS (Filmstrip—Sound). *See*
Europe

SPAIN—DESCRIPTION AND TRAVEL

MI HERMANA SE CASO EN ESPANA (MY
SISTER WAS MARRIED IN SPAIN) (Kit).
Kevin Donovan Films, 1976. 4 color, sound
filmstrips, 85–90 fr.; 4 English/Spanish cassettes, 15–20 min.; 4 Reproducible Spanish
Scripts. ($95). Gr. 7–C. BKL (9/77), FLN (9/
78), PRV (1/76). 916

Contents: 1. Espana Es Diferente (Spain Is
Different). 2. Hay Moros En la Costa (There
Are Moors Along the Coast). 3. Llegada en
Nuestro Toledo (Arrival in Our Toledo). 4.
La boda (The Wedding).

True family experience told by teenage sister
comparing and contrasting life and customs
as she travels in Spain with her family and
meets Emilio and his family. The section on
the Moors gives insight into the influence of
these people on the culture of Spain. Original art and music.

PASSPORT TO SPAIN (Filmstrip—Sound).
EMC, 1978. 5 color, sound filmstrips, 93–108
fr. ea.; 5 cassettes or reel-to-reel tapes;
Teacher's Guide. #SP116000 ($98). Gr. 9–A.
BKL (1/15/79). 916

Contents: 1. Preparation and Flight. 2. At
the Bank, At the Drugstore. . . . 3. At the
Tourist Office, Means of Transportation. . . . 4. At the Fleamarket, At the Department Store. . . . 5. Coping with the
Metric System.

A multimedia cultural program presented in
English and Spanish, designed to acquaint
students with day-to-day experiences of
traveling in Spain.

SPAIN—HISTORY—1936–1939—CIVIL WAR

PRELUDE TO WORLD WAR II—THE
SPANISH CIVIL WAR (Filmstrip—
Sound). Multi-Media Productions, 1976. 2
sound filmstrips, color & b/w, 38–53 fr.; 1
cassette or disc, 10–14 min.; Teacher's Manual. #7180 R or C ($17.95). Gr. 7–12. PVB
(5/74). 946.081

An analysis of the forces behind the Spanish
Civil War, including the reaction of the major world powers to the struggle between fas-

SPAIN—HISTORY—1936-1939—CIVIL WAR (cont.)

cism and democracy that culminated in World War II.

SPANISH LANGUAGE—READING MATERIALS

FROM HOME TO SCHOOL (Filmstrip— Sound). *See* Family life

LA LECHUZA—CUENTOS DE MI BAR- REO (Cassette). *See* Owls—Fiction

LEARNING BEGINS AT HOME (Film- strip—Sound). *See* Family life

MOTOCROSS RACING (Kit). Educational Activities, 1973. 1 color, sound filmstrip, 1 disc or cassette; 10 Paperback Books; Teacher's Guide. Kit with Disc #FSR182 ($35), Kit with Cassette #FSC182 ($36). Gr. 3-9. BKL (8/15/77). 468.6
The challenges, hazards, qualifications, and vocabulary of motocross racing are shown in vivid color in Spanish or English.

OUR LANGUAGE, OUR CULTURE, OUR- SELVES (Filmstrip—Sound). *See* Family life

PARENT-SCHOOL RELATIONSHIPS (Filmstrip—Sound). *See* Family life

PASSPORT TO MEXICO (Filmstrip— Sound). *See* Mexico—Description and travel

PRESENTANDO MEDIDAS (IN- TRODUCING MEASURING) (Film- strip—Sound). *See* Measurement

SPANISH LANGUAGE—STUDY AND TEACHING

MI HERMANA SE CASO EN ESPANA (MY SISTER WAS MARRIED IN SPAIN). (Kit). *See* Spain—Description and travel

MI HERMANO SE CASO EN MEXICO (MY BROTHER WAS MARRIED IN MEXI- CO). (Kit). *See* Mexico—Description and travel

SPELLING. *See* English language—Spelling

SPIDERS

TARANTULA (Film Loop—8mm—Silent). Walt Disney Educational Media, 1966. 8mm color, silent film loop, approx. 4 min. #62- 5117L ($30). Gr K-12. MDU (1974). 595.4
The tarantula is seen removing soil from its burrow and attacking several beetles. In- cludes views of the courtship rituals of this spider.

SPIER, PETER

THE ERIE CANAL (Filmstrip—Sound). *See* Erie Canal—Songs

SPORTS

BASIC DRIBBLE/CONTROL DRIBBLE/ SPEED (Film Loop—8mm—Silent). *See* Basketball

BATTING (Film Loop—8mm—Silent). *See* Softball

THE CATCHER (Film Loop—8mm—Silent). *See* Softball

CATCHING ABOVE WAIST/BELOW WAIST (Film Loop—8mm—Silent). *See* Softball

CENTER FOR A FIELD GOAL (Film Loop—8mm—Silent). *See* Football

CENTER SNAP FOR PUNT (Film Loop— 8mm—Silent). *See* Football

CENTER TO QUARTERBACK EX- CHANGE (Film Loop—8mm—Silent). *See* Football

CHEST PASS/BOUNCE PASS (Film Loop— 8mm—Silent). *See* Basketball

CHEST PASS/OVERHEAD PASS (Film Loop—8mm—Silent). *See* Basketball

COMPETITIVE VALUES: WINNING AND LOSING (Filmstrip—Sound). Human Rela- tions Media Center/Sunburst Communica- tions, Inc., 1975. 2 color, sound filmstrips, 91-95 fr.; 2 cassettes or discs, 17 min. ea.; Teacher's Guide. #610 ($75). Gr 7-12. BKL (7/15/75), LNG (12/75), PVB (4/77). 796.01
Contents: 1. Sports and Society. 2. Winning
Examines the dynamics of sports in our so- ciety, by outlining the benefits of sports ac- tivity, studying the impact it has on our personal values, and questioning the heavy emphasis placed upon winning, aggressive- ness, and competition.

CROSSOVER CHANGE/REVERSE PIVOT CHANGE (Film Loop—8mm—Silent). *See* Basketball

CROSSOVER DRIBBLE/REVERSE DRIBBLE (Film Loop—8mm—Silent). *See* Basketball

DEFENSIVE RUN DOWN (Film Loop— 8mm—Silent). *See* Softball

DOUBLE PLAY BY SHORT STOP/SEC- OND BASE WOMAN (Film Loop—8mm— Silent). *See* Softball

DRIBBLING (Film Loop—8mm—Silent). *See* Soccer

DRIVE/CROSSOVER DRIVE (Film Loop— 8mm—Silent). *See* Basketball

FIELD GOAL AND EXTRA POINTS (Film Loop—8mm—Silent). *See* Football

FIELD GOAL AND KICKOFF (SOCCER STYLE) (Film Loop—8mm—Silent). *See* Football

FIELDING LONG HIT FLY BALL/ GROUND BALL (Film Loop—8mm—Silent). *See* Softball

GOAL KEEPER, CLEARING (Film Loop—8mm—Silent). *See* Soccer

GOAL KEEPER, PART 1 (Film Loop—8mm—Silent). *See* Soccer

GOAL KEEPER, PART 2 (Film Loop—8mm—Silent). *See* Soccer

GOAL KEEPER, PART 3 (Film Loop—8mm—Silent). *See* Soccer

HAND OFF (Film Loop—8mm—Silent). *See* Football

HEADING AND BACK-HEADING (Film Loop—8mm—Silent). *See* Soccer

HOOK SLIDE/STRAIGHT IN SLIDE (Film Loop—8mm—Silent). *See* Softball

INSIDE POWER SHOT (Film Loop—8mm—Silent). *See* Basketball

JUMP SHOT (Film Loop—8mm—Silent). *See* Basketball

JUMP SHOT/ONE HAND SET/TURN-AROUND JUMP SHOT (Film Loop—8mm—Silent). *See* Basketball

KICKING (Film Loop—8mm—Silent). *See* Soccer

KICKOFF AND ONSIDE KICK (Film Loop—8mm—Silent). *See* Football

LAY UP SHOT (Film Loop—8mm—Silent). *See* Basketball

MIDDLE GUARD PLAY (Film Loop—8mm—Silent). *See* Football

OFFENSIVE BACKS (Film Loop—8mm—Silent). *See* Football

OFFENSIVE LINE BLOCKING (Film Loop—8mm—Silent). *See* Football

ONE-ON-ONE DRIVE (Film Loop—8mm—Silent). *See* Basketball

OVERARM PASS/OVERHEAD PASS/UNDERHAND PASS (Film Loop—8mm—Silent). *See* Basketball

OVERHAND THROW—SIDE ARM THROW (Film Loop—8mm—Silent). *See* Softball

PASS PROTECTION (Film Loop—8mm—Silent). *See* Football

PASSING SKILLS, PART 1(Film Loop—8mm—Silent). *See* Football

PASSING SKILLS, PART 2(Film Loop—8mm—Silent). *See* Football

PITCHING/WINDMILL STYLE/SLINGSHOT STYLE (Film Loop—8mm—Silent). *See* Softball

PUNTING (Film Loop—8mm—Silent). *See* Football

REBOUNDING (Film Loop—8mm—Silent). *See* Basektball

REBOUNDING/BLOCKING OUT (Film Loop—8mm—Silent). *See* Basketball

RUNNING TO FIRST BASE/RUNNING EXTRA BASES/RUNNER'S HEADOFF (Film Loop—8mm—Silent). *See* Softball

SACRIFICE BUNT/BUNT FOR BASE HIT (Film Loop—8mm—Silent). *See* Softball

SPEED DRIBBLE/CONTROL DRIBBLE (Film Loop—8mm—Silent). *See* Basketball

SPORTS ACTION 1 (Filmstrip—Sound). *See* Sports

TAG OUTS/FORCE OUTS (Film Loop—8mm—Silent). *See* Softball

THROW IN (Film Loop—8mm—Silent). *See* Soccer

TRAPPING—BALL IN AIR (Film Loop—8mm—Silent). *See* Soccer

TRAPPING—GROUND BALL (Film Loop—8mm—Silent). *See* Soccer

TURN AROUND JUMP SHOT (Film Loop—8mm—Silent). *See* Basketball

SPORTS—BIOGRAPHY

SUPER THINK PROGRAM (Cassette). Troll Associates, 1975. 4 modules each containing 6 cassettes, 10 min.; 12 Duplicating Masters; Teacher's Guide. Individual Module ($48), Series ($192). Gr. 5–8. PRV (10/76), TEA (3/77). 796

Contents: 1. Football. 2. Basketball. 3. Baseball. 4. Fast Action.

A listening program dealing with notable sports figures in a variety of sports fields.

SPYRI, JOHANNA

EPISODES FROM FAMOUS STORIES (Filmstrip—Sound). *See* Literature—Collections

LIBRARY 3 (Cassette). *See* Literature—Collections

SQUIRRELS

INVESTIGATING HIBERNATION: THE GOLDEN-MANTLED SQUIRREL (Motion Picture—16mm—Sound). *See* Animals—Hibernation

STAINED GLASS. *See* Glass painting and staining

STAMPS, POSTAGE. *See* Postage stamps

STATE GOVERNMENTS

STATE GOVERNMENT—RESURGENCE OF POWER (Motion Picture—16mm—Sound). Concept Media/EBEC, 1976. 16mm color, sound film, 21 min. #3490 ($290). Gr. 7–12. LFR (1977). 359.9

State governments, all over the country, are taking new initiatives once reserved for the federal level. Viewers look in on legislative sessions in two key states, see evidence of judicial reform in action, and become aware of the growing problem of state financing.

STEFFENS, LINCOLN

A MISERABLE MERRY CHRISTMAS (Motion Picture—16mm—Sound). *See* Christmas

STEINBECK, JOHN

CANNERY ROW, LIFE AND DEATH OF AN INDUSTRY (Filmstrip—Sound). *See* Monterey, California—History

FIVE MODERN NOVELISTS (Filmstrip—Sound). *See* American literature—History and criticism

THE PEARL BY JOHN STEINBECK (Filmstrip—Sound). *See* Mexico—Fiction

STEVENSON, ROBERT LOUIS

DOCTOR JEKYLL AND MR. HYDE (Filmstrip—Sound). *See* Fantastic fiction

EPISODES FROM FAMOUS STORIES (Filmstrip—Sound). *See* Literature—Collections

GREAT WRITERS OF THE BRITISH ISLES, SET II (Filmstrip—Sound). *See* Authors, English

TREASURE ISLAND (Cassette). *See* Pirates—Fiction

TREASURE ISLAND (Filmstrip—Sound). *See* Pirates—Fiction

STOCKTON, FRANK

A DISCUSSION OF THE LADY OR THE TIGER? (Motion Picture—16mm—Sound) *See* Motion pictures—History and criticism

THE LADY OR THE TIGER (Motion Picture—16mm—Sound). *See* American fiction

STORIES—COLLECTIONS. *See* Short stories

STORIES WITHOUT WORDS

TCHOU TCHOU (Motion Picture—16mm—Sound). *See* Fantasy—Fiction

STORMS

MOUNTAIN STORM (Film Loop—8mm—Silent). *See* Weather

STORMS: THE RESTLESS ATMOSPHERE (Motion Picture—16mm—Sound). EBEC, 1974. 16mm color, sound film, 22 min. #3331 ($310). Gr. 5–A. AAS (1976), CFF (1975), LGB (1975). 551.5

In this study of thunderstorms, tornadoes, and hurricanes, the film examines the nature, structure, incidence, and consequences of these storms; then examines the elaborate systems of detection, data collection, and interpretation that meteorologists use to investigate storms. Highlights include a time-lapse buildup of cumulonimbus clouds, a weather satellite's view of the eye of a hurricane, and a miniature tornado whirlwind demonstration.

STOUT, RUTH

RUTH STOUT'S GARDEN (Motion Picture—16mm—Sound). Arthur Mokin Productions, 1976. 16mm color, sound film, 23 min. ($365). Gr. 7–A. BKL (6/1/76), CGE (1976), PRV (11/76). 301.43

Ruth Stout, now more than ninety years old, illustrates that age need not be a barrier to continued productivity and creativity. She remains active writing books, contributing regularly to periodicals, lecturing and perfecting her "no dig/no work" method of gardening.

STOUTENBURG, ADRIEN

JOHN HENRY AND JOE MAGARAC (Phonodisc or Cassette). *See* Folklore—U.S.

JOHNNY APPLESEED AND PAUL BUNYAN (Phonodisc or Cassette). *See* Folklore—U.S.

STRAVINSKY, IGOR FEDOROVICH

STRAVINSKI GREETING PRELUDE, ETC. (Phonodiscs). *See* Orchestral music

STRESS (PSYCHOLOGY)

DEALING WITH STRESS (Filmstrip—Sound). Human Relations Media Center, 1975. 2 color, sound filmstrips, 76–81 fr.; 2 cassettes or discs, 12 min. ea.; Teacher's Guide. #616 ($60). Gr. 7–A. BKL (2/1/76), MNL (4/76). 131.33

Contents: 1. What is Stress? 2. Coping.

Examines causes, effects, and handling of stress. Demonstrates that stress is a normal, necessary part of life, but is unnatural when it becomes an emotional or physical crippler. Case studies illustrate that many of life's stressful experiences offer opportunities for personal growth and development. Specific guidelines and examples help the student assess his or her reactions to life's tensions and modify his or her behavior, if necessary.

MIND POWER (Filmstrip—Sound). Society for Visual Education, 1977. 4 color, sound filmstrips with captions, av. 55 fr.; 4 discs or cassettes, av. 6¼ min.; Guide. #LG456–SATC Cassettes ($84), #LG456–SAR Discs ($84). Gr. 7–12. BKL (4/15/78). 131.11

Contents: 1. Yoga: The Union of Body and Mind. 2. Transcendental Meditation: The Search for Self. 3. Biofeedback: The Healing Will. 4. Behavior Modification: The Habit Breaker.

Four techniques for coping with mental stress and anxiety are examined in this set of high-interest controlled-vocabulary filmstrips.

STRESS, SANITY, AND SURVIVAL (Filmstrip—Sound). Current Affairs Films, 1978. 1 color, sound filmstrip, 74 fr.; 1 cassette, 16 min.; Discussion Guide. #609 ($24). Gr. 8–C. BKL (7/15/78). 131.33

This filmstrip discusses, from several points of view, the problem of stress in modern society, the reasons for it, and the possible cures.

STRING ART

SCULPTURE WITH STRING (Filmstrip—Sound). Warner Educational Productions, 1975. 2 color, sound filmstrips, 93–104 fr.; 1 cassette, 9–12 min.; Teaching guide. ($47.50). Gr. 6–A. BKL (2/1/76), PRV (11/76), PVB (4/77). 731

Close-ups illustrate the techniques of wrapping string around nails to form a pattern/design. The narrator explains the basic procedures in string sculpture.

STUDENT LIFE—RUSSIA

THE SOVIET UNION: A STUDENT'S LIFE (Motion Picture—16mm—Sound). EBEC, 1972. 16mm color, sound film, 21 min. #3145 ($290). Gr. 7–12. BKL (1972), CGE (1972), EFL (1974). 914.7

Filmed in the Soviet Union is the "story" of Nikolai Diachenko, a chemical engineering student. As he traces his life from nursery school to the university in Moscow, viewers see the vital part Communist youth movements play during school years. Nikolai spends one summer at a work camp in Siberia, an example of the dual emphasis on intellectual achievement and manual labor to help develop the vast land.

STUDENT LOAN FUNDS

HOW CAN I PAY FOR COLLEGE? (Filmstrip—Sound). Eye Gate Media, 1977. 4 color, sound filmstrips, 46–63 fr. ea.; 2 cassettes, 10–12 min.; Guide. #TV727 ($63.60). Gr. 10–A. BKL (4/1/77). 378.3

Contents: 1. Available Help for College. 2. Making the System Respond. 3. Maximizing Your Assets. 4. Doing It by Day-Hopping.

This program provides valuable explanations on the types of aid available for college students and the various sources of such help. It introduces three students who relate their experiences in search for assistance.

STUDY, METHOD OF

DEVELOPING EFFECTIVE STUDY SKILLS (Filmstrip—Sound). Knowledge Aid/United Learning, 1976. 6 color, sound filmstrips, av. 65–75 fr. ea.; 6 cassettes, av. 12–16 min. ea.; Teacher's Guide, Duplicating Masters. #1310 ($85). Gr. 5–8. PRV (4/77). 028.7

Contents: 1. How to Take Notes in Class. 2. Taking Notes at Home. 3. Writing Reports. 4. Following Directions—Or Else. 5. Developing Your Listening Skills. 6. How to Remember.

The essential skills of note taking, report writing, and listening to and interpreting instructions, and aids for effective retention of information are all logically and memorably brought together in this unit of study.

DEVELOPING STUDY SKILLS (Filmstrip—Sound). Coronet Instructional Media, 1974. 8 color, sound filmstrips, 48–54 fr.; 4 discs or 8 cassettes, 9–12 min.; Teacher's Guide. #S175 Discs ($120), #M175 Cassettes ($120). Gr. 4–8. BKL (1975), PRV (4/76). 028.7

Contents: 1. Why We Study. 2. Planning Your Work. 3. Learning to Listen. 4. Taking Notes. 5. Using a Textbook. 6. Doing Homework. 7. How to Review. 8. Using the Library.

This set has as its objective motivating young people to want to study while teaching them how to study. The strength of the series rests on its clear, down-to-earth analytical approach to a task. Examples to motivate students to improve study habits include an airline pilot, a football coach, a television newsman, a publisher, and a magician as they demonstrate the value of learning and using good study habits.

DICTIONARY SKILL BOX (Cassette). *See* English language—Dictionaries

HOW TO SURVIVE IN SCHOOL: NOTE-TAKING AND OUTLINING SKILLS (Filmstrip—Sound). Center for Humanities, 1976. 6 filmstrips, av. 60 fr. ea.; 6 cassettes; 6 discs; Teacher's Guide. #2308–1200 ($160). Gr. 7–12. BKL (3/1/78), PRV (2/78). 371.302

Contents: 1. Note-taking Techniques. 2. Tips for Outlining Materials. 3. Synthesize What You've Learned.

This program teaches students the important steps involved in note-taking and outlining skills.

STUDY, METHOD OF (cont.)

HOW TO WRITE A RESEARCH PAPER (Filmstrip—Sound). *See* Report writing

HOW TO WRITE A TERM PAPER (Filmstrip—Sound). *See* Report writing

LET'S LEARN TO STUDY (Filmstrip—Sound). Guidance Associates, 1974. 2 color, sound filmstrips, 66–72 fr.; 2 discs or cassettes, 9 min. ea.; Teacher's Guide. #9A–301 141 Discs ($52.50), #9A–301 158 Cassettes ($52.50). Gr. 4–8. PRV (2/76), PVB (4/76). 028.7
Spotlights tips for grasp and retention. Covers organizing study time, setting objectives, ordering priorities, mastering material, asking questions; suggests way to choose topics for independent projects, do research, and present reports to class.

LIBRARY SKILL BOX (Cassette). *See* Library skills

MEDIA: RESOURCES FOR DISCOVERY (Filmstrip—Sound). *See* Audiovisual materials

RESEARCH PAPER MADE EASY: FROM ASSIGNMENT TO COMPLETION (Slides/Cassettes). *See* Report writing

SCHOOL SURVIVAL SKILLS: HOW TO STUDY EFFECTIVELY (Slides). Center for Humanities, 1977. 3 units with 80 slides ea.; 3 cassettes and discs of the same sound track; Teacher's Guide. #0322 ($179.50). Gr. 10–12. PRV (1979). 371.302
Parts I and II present the study method of survey, question, read, write, review. Part III addresses management of time.

SUCCESS

THE BEST I CAN (Motion Picture—16mm—Sound). *See* Self-realization

JERRY'S RESTAURANT (Motion Picture—16mm—Sound). *See* Individuality

SUICIDE

RONNIE'S TUNE (Motion Picture—16mm—Sound). *See* Death

SUICIDE: CAUSES AND PREVENTION (Filmstrip—Sound). Human Relations Media Center, 1976. 2 color, sound filmstrips, approx 89 fr.; 2 cassettes or discs, approx. 16 min. ea.; Teacher's Guide with Script. ($60). Gr. 7–C. NCF (1977). 157.744
Contents: 1. Causes. 2. Prevention.
Examines the problem of suicide in our society and discusses ways in which suicide can be prevented. Points out that suicide is a serious problem that can be alleviated only if more people recognize it and try to understand it; that suicide is one of the leading causes of death among high school and college students. Describes the typical warning signs of a potential suicide and discusses good and bad ways of handling such a person.

SULLIVAN, ANNE

CHILDHOOD OF FAMOUS WOMEN, VOLUME THREE (Cassette). *See* Women —Biography

SUN

THE SUN: ITS POWER AND PROMISE (Motion Picture—16mm—Sound). Avatar Learning/EBEC, 1976. 16mm color, sound film, 24 min. #3512 ($320). Gr. 7–12. AVI (1977), PRV (1978), SCT (1978). 523.7
Spectacular photography of the sun and lively animation combine to explore ways in which the sun's energy might be used to help replace our ever-dwindling supplies of fossil fuels.

SUPERSTITION

THE SNAKE: VILLAIN OR VICTIM? (Motion Picture—16mm—Sound). *See* Snakes

SURFING

SURFING, THE BIG WAVE (Kit). Troll Associates, 1976 (Troll Reading Program). 1 color, sound filmstrip, 34 fr.; 1 cassette, 14 min.; 10 Soft-cover Books, 1 Library Edition; 4 Duplicating Masters, Teacher's Guide. ($48). Gr. 5–9. BKL (1/15/77). 797.1
A high-interest kit features an easy-to-read book on surfing, a sound filmstrip of the book, and practice in word building, contest clues, vocabulary, and comprehension.

SURINAM—HISTORY

I SHALL MOULDER BEFORE I SHALL BE TAKEN (Motion Picture—16mm—Sound). Educational Development, 1978. 16mm color, sound film, two reels, 58 min. ($795). Gr. 7–A. BKL (2/1/79), CGE (8/19/78), CRA (1978). 988.3
A documentary film on the Djuka of Surinam; descendants of West Africans who were brought to South America in the 1600s as slaves. Soon after their arrival, the slaves rebelled, escaping into the jungle and establishing a black tribal society in the New World. The film serves as a record of a black people who refused to submit to slavery, fought for their freedom, and have a positive self-image as a free people.

SURREALISM

SURREALISM (AND DADA) (Motion Picture—16mm—Sound). Texture Films, 1970. 16mm color, sound film, 18 min. ($250). Gr. 9–A. AFF (1971), CRH (1971). 759.06

"A film about the radical attack Surrealism and Dada wrought on the familiar and traditional art values of the past. Through the perceptive choice of paintings and sculptures—from the work of de Chirico, Duchamp, Man Ray, Picabia, Ernst, Miro, Magritte and others—we become exposed to the philosophy, psychology, poetry, and politics of these fascinating art movements."

SURVIVAL

SAFE IN NATURE (Motion Picture—16mm—Sound). *See* Wilderness survival

WEATHER IN THE WILDERNESS (Filmstrip—Sound). Great American Film Factory, 1976 (Outdoor Education). 1 color and b/w, sound filmstrip, 58 fr.; 1 cassette, 19 min. #FS-4 ($35). Gr. 6-A. PRV (2/19/77). 613.6

The importance of and procedures for checking weather forecasts are detailed. Survival and safety filmstrip explains the dangers and signs of hypothermia (exposure). Its prevention, identification, and on-the-spot treatment are discussed. Pictures range from U.S. Weather Bureau maps and charts to scenery.

SURVIVAL—FICTION

SWISS FAMILY ROBINSON (Phonodisc or Cassette). *See* Adventure and adventurers—Fiction

SWAMPS. *See* Marshes

SYMPHONIES

LEONARD BERNSTEIN CONDUCTS FOR YOUNG PEOPLE (Phonodiscs). Columbia Records. 1 disc, 12 in., 33$\frac{1}{3}$ rpm. #3ML5841 ($7.98). Gr. 4-A MOV (1978). 785.1

Contents: 1. "Afternoon of a Fawn." 2. "Nutcracker Suite."

The New York Philharmonic plays these well-known selections.

SYMPHONIC MOVEMENTS NO. 2 (Phonodiscs). Bowmar Publishing 1 disc; #087 ($10). Lesson Guides; Theme Chart. Gr. 3-10. AUR (1978). 785.1

Focuses students' attention on the music while they are listening and relates the sound of music to music concepts and terminology. The music can also be enjoyed with no follow-up. Some of the symphonic movements included are Symphony No. 5 by Beethoven, Symphony No. 2 by Sibelius.

SYNGE, JOHN MILLINGTON

THE WELL OF THE SAINTS (Motion Picture—16mm—Sound). *See* Irish drama

SYSTEM ANALYSIS

LOOKING AT THE FUTURE: SYSTEMS THINKING (Filmstrip—Sound). *See* Thought and thinking

TALL TALES. *See* Folklore; Legends

TAZEWELL, CHARLES

THE LITTLEST ANGEL (Phonodisc or Cassette). *See* Christmas—Fiction

TEACHERS

PARENT INVOLVEMENT: A PROGRAM FOR TEACHERS AND EDUCATORS (Filmstrip—Sound). *See* Home and school

TECHNOLOGY AND CIVILIZATION

COMPUTERS AND HUMAN SOCIETY (Filmstrip—Sound). *See* Computers

FORECASTING THE FUTURE: CAN WE MAKE TOMORROW WORK? (Filmstrip—Sound). *See* Forecasting

INVENTIONS AND TECHNOLOGY THAT SHAPED AMERICA, SET ONE (Filmstrip—Sound). International Cinemedia Center/Learning Corporation of America, 1971 (Colonial Times to Civil War). 6 color, sound filmstrips, 48–68 fr. ea.; 3 cassettes, 8-1/2–14 min.; Guides. ($74.50). Gr. 6–10. BKL (2/15/73), LFR (4/72), STE (12/72). 608

Contents: 1. The Second American Revolution: Technology. 2. Eli Whitney Changes America: The Cotton Gin and Interchangeable Parts. 3. The Agricultural Revolution: Plow and Reaper. 4. The Steam Engine: From Riverboat to Iron Horse. 5. Man Makes His Climate: Refrigeration. 6. The Communications Breakthrough: Telegraph and Telephone.

This set shows how Eli Whitney influenced the Civil War; how the steam engine led to the railroad; how refrigeration changed medical and household practices; how the telephone and telegraph revolutionized communications. The emphasis is historical and social, as well as scientific.

LIVING WITH TECHNOLOGY: CAN WE CONTROL APPLIED SCIENCE? (Filmstrip—Sound). Harper and Row Publishers/Sunburst Communications, 1975. 5 color, sound filmstrips, 82–98 fr.; 5 cassettes or discs, 16–17½ min.; Teacher's Guide. #74 ($125). Gr. 7–12. BKL (2/1/76), TSS (1/78). 608

Contents: 1. The Transformation of Society. 2. The American Dream. 3. Implications for the World System. 4. Vulnerability: The System Tested. 5. Visions of the Future.

Probes the historical and contemporary impact of technology on our lives and values. Challenges students to question the direction

TECHNOLOGY AND CIVILIZATION (cont.)

in which technology has taken us, and to measure its relative worth against its cost in human terms.

MEDIA AND MEANING: HUMAN EXPRESSION AND TECHNOLOGY (Slides/Cassettes). *See* Mass media

REDESIGNING MAN: SCIENCE AND HUMAN VALUES (Filmstrip—Sound). *See* Forecasting

SCIENCE, TECHNOLOGY AND MODERN MAN (Filmstrip—Sound). Educational Dimensions Group, 1974. 2 color, sound filmstrips, 79–83 fr.; 2 cassettes, 17 min.; Teacher's Guide. #1034 ($49). Gr. 6–12. PRV (4/76). 608

A survey of man's development and use of technology from prehistoric times to the 1970s shows that advancing technology both solves and creates problems.

TECHNOLOGY AND CHANGE (Kit). Nystrom, 1975 (The American Experience). 5 color, sound filmstrips, 70–90 fr.; 5 cassettes, 8–10 min.; 10 Student Readers, 1 Set Activity Sheets; 2 Teacher's Guides. #AE300 ($160). Gr. 7–12. BKL (12/76). 608

Contents: 1. Supergrow (Resources and Growth). 2. Machines of Plenty (Agriculture and Mechanization). 3. Interchangeable Parts (Mass Production). 4. Workingmen's Blues (Workers in Industry). 5. Cities (Technology and the Environment).

This kit shows how national growth and the progressive mechanization of American lifestyles have resulted in tremendous waste.

TEENAGERS. *See* Adolescence; Youth as parents

TELEVISION

HOW TO WATCH TV (Filmstrip—Sound). *See* Mass media

TELEVISION: AN INSIDE VIEW (Filmstrip—Sound). Educational Audio Visual, 1976. 3 color, sound filmstrips, 89–127 fr.; 3 discs or cassettes, 15½–18 min.; Teacher's Guide. #7KF0015 Cassettes ($70), #7RF0015 Discs ($70). Gr. 7–C. PRV (2/77), PVB (4/77). 384.55

Contents: 1. On the Air. 2. The History of Television. 3. The Business of Networks.

The development of the technology and business of television is the basis for this set. Behind-the-scenes tour of a production, the complexities of daily taping, the frequent changes of text, the role of the director and staff, live sports events filming, are all shared with the viewer.

THE THIRTY SECOND DREAM (Motion Picture—16mm—Sound). *See* Advertising

TELEVISION—PRODUCTION AND DIRECTION

TV NEWS: BEHIND THE SCENES (Motion Picture—16mm—Sound). EBEC, 1973 (World of Work). 16mm color, sound film, 17 min. #3205 ($360). Gr. 7–C. BKL (1/15/74), CFF (1974), STE (1974). 791.45

A documentary on New York's ABC Eyewitness News reveals the tight editorial and technical teamwork responsible for a local television news program.

TELEVISION—PSYCHOLOGICAL ASPECTS

THE THIRTY SECOND DREAM (Motion Picture—16mm—Sound). *See* Advertising

TELEVISION ADVERTISING

SIXTY SECOND SPOT (Motion Picture—16mm—Sound). Pyramid Films, 1974. 16mm color, sound film, 25 min.; ¾-in. videocassette also available ($350). Gr. 7–A. BKL (10/1/74), NEF (1974), STE (1975). 659.14

A behind-the-scenes account of the epic, at times humorous, struggles inherent in the making of a one-minute television commercial. All phases of the production are covered, from storyboard to casting, budgeting to finished product.

TENNIS

TENNIS: HOW TO PLAY THE GAME (Filmstrip—Sound). EBEC, 1978. 5 color, sound filmstrips, av. 98 fr. ea.; 5 cassettes, 14 min. ea.; Teacher's Guide. #17043K ($82.88). Gr. 7–A. PRV (10/78). 796.34

Contents: 1. Tennis: The Fundamentals. 2. Tennis: Groundstrokes. 3. Tennis: The Serve and Overhead. 4. Tennis: Supplementary Strokes. 5. Tennis: Strategy and Tactics.

Using a combination of set shots and sequential photography, former touring pro Wes Tenney teaches and demonstrates the basics of tennis. The fundamentals of equipment, scoring, and terminology, and strategy and tactics for both singles and doubles play are covered.

TERRARIUMS

AMPHIBIANS (Filmstrip—Sound). *See* Amphibians

TERRORISM

TERRORISM WORLDWIDE: IS ANYBODY SAFE? (Filmstrip—Sound). Current Affairs Films, 1977. 1 color, sound filmstrip, 74 fr.; 1 cassette, 16 min.; Teacher's Guide. #567 ($24). Gr. 7–A. PRV (5/78). 332.4

An introduction to terrorism and its many complexities: terrorists are presented as individuals or groups who merge their psychological motives with a political cause to gain recognition for that cause.

TEXTILE DESIGN

CREATIVE BATIK (Filmstrip—Sound). *See* Batik

CREATIVE TIE/DYE (Filmstrip—Sound). *See* Batik

TEXTILES AND ORNAMENTAL ARTS OF INDIA (Motion Picture—16mm—Sound). *See* India—Handicraft

TEXTILE INDUSTRY

THE CLOTHES WE WEAR (Filmstrip—Sound). *See* Clothing and dress

THEATER. *See* Drama

THEATER—ENGLAND

THEATER IN SHAKESPEARE'S ENGLAND (Filmstrip—Sound). EBEC, 1974. 4 color, sound filmstrips, 56 fr. ea.; 4 discs or cassettes, 9 min. ea.; Teacher's Guide. #6903K Cassettes ($57.95), #6903 Discs ($57.95). Gr. 9–12. EGT (12/75). 792
Contents: 1. Origins of English Drama. 2. Theater in Elizabethan London. 3. The Globe: Design and Construction. 4. The Globe: A Day at Shakespeare's Theater.
Photographs and color drawings trace the development of English theater from the religious drama of the Middle Ages to the plays of rowdy Elizabethan London. The series shows how English political and economic supremacy injected power into a young and growing language, and how the language, in turn, shaped British drama. A model depicts the famous Globe Theater and shows how the theater design influenced the form of the plays themselves.

THEATER—HISTORY

MEDIEVAL THEATER: THE PLAY OF ABRAHAM AND ISAAC (Motion Picture—16mm—Sound). EBEC, 1974. 16mm color, sound film, 26 min. #47796 ($360). Gr. 10–A. BKL (11/15/74), CFF (1975), EGT (1975). 792.09
This film shows how a family of traveling players produced the play at an English estate in 1482. Attitudes of actors and peasants are contrasted with those of the aristocracy and church. Dialogue for the play portion of the film was taken from the original Brome manuscript. Social status of audience and actors, speech, costumes, sets, and props are based on 15th-century records.

THEATER IN SHAKESPEARE'S ENGLAND (Filmstrip—Sound). *See* Theater—England

THEATER—PRODUCTION AND DIRECTION

TENNESSEE WILLIAMS: THEATER IN PROCESS (Motion Picture—16mm—Sound). Signet Productions/EBEC, 1976. 16mm color, sound film, 29 min. #47825 ($430). Gr. 10–A. EFL (1977), LFR (1977), M&M (1977). 792.9
This film takes viewers behind the scenes for a glimpse of the theater playgoers seldom see: the creation of the play itself. The play in process is Tennessee Williams's *The Red Devil Battery Sign*. Williams narrates as viewers watch the play progress from first rehearsals through opening performances.

THEATER—U.S.

AMERICAN THEATRE (Filmstrip—Sound). Aids of Cape Cod, 1978. 4 color, sound filmstrips, 56–69 fr.; 4 cassettes, 5–5$^1/_2$ min.; Teacher's Guide. #073 ($84.50). Gr. 7–12. PRV (2/79). 792
Contents: 1. American Theatre Comes Alive. 2. Sound and Splendor. 3. 20th-Century Playwrights. 4. Enter Stage Right.
A brief history of American theater emphasizing the European influences. An actor shares his thoughts as he prepares for a performance. A general appraisal of acting as a career and the idealism and hard facts of employment are stated.

THERMAL WATERS. *See* Geothermal resources

THOMAS, VIOLA SCOTT

THE GOOD OLD DAYS—THEY WERE TERRIBLE (Cassette). *See* History—Philosophy

THOREAU, HENRY DAVID

THE ROMANTIC AGE (Filmstrip—Sound). *See* American literature—History and criticism

TALKING WITH THOREAU (Motion Picture—16mm—Sound). *See* Philosophy, American

THOREAU ON THE RIVER: PERSPECTIVE ON CHANGE (Filmstrip—Sound). *See* Ecology

THOUGHT AND THINKING

IDIOT . . . GENIUS? (Motion Picture—16mm—Sound). *See* Intellect

LET'S LOOK AT LOGIC (Filmstrip—Sound). *See* Logic

THOUGHT AND THINKING (cont.)

LISTENING BETWEEN THE LINES (Motion Picture—16mm—Sound). *See* Listening

LOOKING AT THE FUTURE: SYSTEMS THINKING (Filmstrip—Sound). BFA Educational Media, 1975. 4 color, sound filmstrips, 55-74 fr. ea.; 4 discs or cassettes, 7-11 min.; Guide. #VFT000 Cassettes ($70), #VFS000 Discs ($70). Gr. 7-9. BKL (1/1/76). 001.5

Contents: 1. Tunnel-Vision: A Great Invention. 2. What is a System? 3. What Kind of Change? Revolution or Evolution. 4. Interdependence.

People create systems in their technological and social lives and these are delicately balanced. When there is a disturbance in one part, other parts are usually affected. This set is designed to help students to think of a whole system when dealing with social change.

MAKING JUDGMENTS AND DRAWING CONCLUSIONS (Slides/Cassettes). *See* Reading—Study and teaching

UNDERSTANDING THE MAIN IDEA AND MAKING INFERENCES (Slides/Cassettes). *See* Communication

USING CLUE WORDS TO UNLOCK MEANING (Slides/Cassettes). *See* Communication

VIEW FROM THE PEOPLE WALL (Motion Picture—16mm—Sound). Eames film/EBEC, 1973. 16mm color, sound film, 14 min. #3184 ($185). Gr. 7-A. BKL (3/15/74), LFR (11/75). 153.4

Both humans and computers employ abstract models to solve problems. This film demonstrates physical, mathematical, and spatial models; traces the steps in the problem-solving process; and illustrates the computer's role in solving complex problems. Examples include city planning and weather simulation.

THRIFT. *See* Consumer education

THURBER, JAMES

THE GREAT QUILLOW (Phonodisc or Cassette). *See* Giants—Fiction

THE GRIZZLY AND THE GADGETS AND FURTHER FABLES FOR OUR TIME (Phonodisc or Cassette). *See* Fables

TIME

TIME: FROM MOONS TO MICROSECONDS (Filmstrip—Sound). Guidance Associates, 1975 (Math Matters). 2 color, sound filmstrips, av. 57 fr.; 2 discs or cassettes, av. 12 min.; Teacher's Guide. #1B-301 331 Cassettes ($52.50), #1B-301 349 Discs ($52.50). Gr. 5-8. ATE. 529

Shows early use of natural cycles to mark time. Discusses ancient and modern calendars. Traces development of timepieces and explains global time zones and world standard times.

TIME STUDY

THE MANAGEMENT OF TIME (Motion Picture—16mm—Sound). *See* Management

TOBACCO—PHYSIOLOGICAL EFFECT

PHYSIOLOGY OF SMOKING AND DRINKING (Filmstrip—Sound). *See* Alcohol—Physiological effect

TOBACCO HABIT

HOW TO STOP SMOKING (Filmstrip—Sound). Sunburst Communications, 1977. 2 color, sound filmstrips, 80 fr.; 2 cassettes, 15 min. ea.; Teacher's Guide. #238-50 ($59). Gr. 7-12. BKL (4/15/78), PRV (1/79). 613.8

Contents: 1. Why Do I Smoke? 2. Six Ways to Quit.

Encourages viewers to analyze their smoking habits and reviews psychological need patterns, including relaxation, stimulation, rebellion, and conformity. The six methods of quitting are described, ranging from "cold turkey" to positive reinforcement.

TOBACCO PROBLEM: WHAT DO YOU THINK? (Motion Picture—16mm—Sound). EBEC, 1972. 16mm color, sound film, 17 min. #3091 ($220). Gr. 7-12. CGE (1972), PRV (1973), SCN (8/73). 613.8

Presents historical perspective and timely data on smoking habit and health risks. Footage shows congressional hearings concerning warning labels, rebuttal from consumers and tobacco companies, and debate by teenagers on the pros and cons of smoking. The question is asked: How would prohibition of smoking affect the national economy?

TOKYO

THE ISLAND NATION CAPITALS (Filmstrip—Sound). *See* Cities and towns

TOLKIEN, J. R. R.

THE HOBBIT (Kit). *See* Fantasy—Fiction

TOPOLOGY

MATHEMATICAL PEEP SHOW (Motion Picture—16 mm—Sound). Eames Films/EBEC, 1973. 16mm color, sound film, 12 min. #3187 ($150), Gr. 5-9. AAS (1976), ATE (1974), BKL (1974). 514

The film is divided into five segments, each illustrating a concept of sophisticated mathematics in a clear, simplified way. Using clever animation and photography, the film conveys basic principles of Eratosthenes' method of abstract measurement of the earth, of topology, symmetry, functions, and exponents.

TOTALITARIANISM—FICTION

ANIMAL FARM BY GEORGE ORWELL (Filmstrip—Sound). *See* Fantasy—Fiction

TOWNS. *See* Cities and towns

TOYS

TOPS (Motion Picture—16mm—Sound). Eames Film/EBEC, 1973. 16mm color, sound film, 8 min. #3190 ($115). Gr. K–A. BKL (3/15/74), LFR (9/19/75), PRV (4/19/74). 790.13

Tops are among our oldest and most popular toys. This unique and beautiful non-narrated film features a cast of 123 spinning tops from different lands. There are tops of carved wood, metal, and plastic; tops simple and sophisticated; tops antique and modern. The camera reveals how tops are wound or prepared; how they are launched; how they spin, wobble, and die.

TOYS—FICTION

THE VELVETEEN RABBIT (Filmstrip—Sound). *See* Fantasy—Fiction

TOYS—HISTORY

THIS TINY WORLD (Motion Picture—16mm—Sound). Phoenix Films, 1973. 16mm color, sound film, 15 min. #0025 ($250). Gr. K–C. FLN (3/74). 649.55

Offers a glimpse into the tiny world of toys in former days, when children were supposed to be little grown-ups, and so their toy world was a meticulous copy of the world of adults.

TRADE. *See* Business

TRADE UNIONS. *See* Labor unions

TRAFFIC REGULATIONS

MOTORCYCLE SAFETY (Kit). *See* Motorcycles

SPEEDING? (Motion Picture—16mm—Sound). *See* Automobiles—Law and legislation

TRAINS. *See* Railroads—History

TRAMPS

HOBO: AT THE END OF THE LINE (Motion Picture—16mm—Sound). Altus Films Productions/EBEC, 1977. 16mm color, sound film, 24 min. #3566 ($350). Gr. 7–C. BKL (9/1/78), EGT (1978), M&M (1978). 301.44

The hobo symbolizes an American romantic tradition. He lives on the road, hitches rides on freight trains, cooks mulligan stew over an open fire, and plays the harmonica. He "rides the rails." He frequents the skid-row missions for a little preaching and a free meal. This film portrayal of hobo life reveals an element of Americana approaching extinction, and considers what values and standards dictate any person's choice of lifestyle.

TRANSCENDENTAL MEDITATION

MIND POWER (Filmstrip—Sound). *See* Stress (psychology)

TRANSPORTATION—HISTORY

MOVIN' ON (Motion Picture—16mm—Sound). Films, 1977. 16mm color, sound, animated film, 4 min. #394–0003 ($100). Gr. 4–A. AFF (1978), BKL (9/1/78). 380.509

An animated history of travel on land, on sea, and in the air points out the creative leaps necessary between successive inventions.

TRAVEL

PASSPORT TO MEXICO (Filmstrip—Sound). *See* Mexico—Description and travel

TRAVEL TRAILERS AND CAMPERS

EFFECTIVE TRAILERING (Filmstrip—Sound). *See* Automobiles—Trailers

TREASON

THE DREYFUS AFFAIR (Motion Picture—16mm—Sound). *See* Dreyfus, Alfred

TREASURE-TROVE

SCUBA (Filmstrip—Sound). *See* Scuba diving

TREES

THE GODS WERE TALL AND GREEN (Filmstrip—Sound). *See* Forests and forestry

TRIALS

DRED SCOTT: BLACK MAN IN A WHITE COURT (Filmstrip—Sound). *See* Scott, Dred

THE JOHN PETER ZENGER TRIAL (Filmstrip—Sound). *See* Freedom of the press

TROPICS

CULTURE AND ENVIRONMENT: LIVING IN THE TROPICS (Filmstrip—Sound). BFA Educational Media, 1974. 8 color, sound filmstrips, 88–120 fr.; 8 cassettes or discs, 8–10$^{1}/_{2}$ min.; Teacher's Guide. #VW 4000 Cassettes ($138), #VW 3000 Discs ($138). Gr. 4–8. PRV (4/76), PVB (4/77). 301.3

Contents: 1. Shelter. 2. Clothing. 3. Obtaining Food. 4. Food Preparation. 5. Transportation. 6. Technology. 7. Family Life. 8. Community and Traditions.

Both the music and photography support the minimal narrative to assist in understanding the Cuna culture's fascinating transition as the Indian life-style reacts to modern practices.

TROUT

THE BROWN TROUT (Film Loop—8mm—Captioned). BFA Educational Media, 1973 (Animal Behavior). 8mm color, captioned film loop, approx. 4 min. #481417 ($30). Gr. 4–9. BKL (5/1/76). 597.55

Demonstrates experiments, behavior, and characteristics of the brown trout.

TRUCK DRIVERS

THE ROAD NEVER ENDS (Motion Picture—16mm—Sound). Texture Films, 1977. 16mm color, sound film, 15 min. ($225). Gr. 7–A. AFF (1977), EFL (1977), PRV (1977). 629.22

Joe Fusaro was determined to be a truck driver. The truck driver is an almost mythological figure in our culture: rugged, footloose, rootless, he is a kind of cowboy of the auto era. This film is a portrait of Joe as a truck driver that shows his work and his life as they really are, which is not entirely what he imagined they would be.

TRUDEAU, GARRY

THE DOONESBURY SPECIAL (Motion Picture—16mm—Sound). See Comic books, strips, etc.

TRUMAN, HARRY S

PRESIDENTS AND PRECEDENTS (Kit). See Presidents—U.S.

TRUMBULL, JOHN

AMERICAN CIVILIZATION: 1738–1840 (Filmstrip—Sound). See Art, American

TCHAIKOWSKY, PETER ILYICH

STORIES IN BALLET AND OPERA (Phonodiscs). See Music—Analysis, Appreciation

TURNER, JOSEPH M. W.

JOSEPH MALLORD WILLIAM TURNER, R.A. (Motion Picture—16mm—Sound). National Gallery of Art & Visual Images/ WETA-TV/EBEC, 1974 (The Art Awareness Collection, National Gallery of Art). 16mm color, sound film, 8 min. #3532 ($145). Gr. 7–C. EFL (1977), EGT (1978), LFR (1978). 759.2

Turner confounds the artistic establishment of nineteenth-century England with his landscapes and seascapes. His forms, dissolving into luminous color and glowing light, earn harsh criticism from his contemporaries.

TUTANKHAMUN

KING TUTANKHAMUN: HIS TOMB AND HIS TREASURE (Kit). See Egypt—Antiquities

TREASURES OF KING TUT (Videocassette). Educational Dimensions Group, 1978. videocassette, 30 min. #CV720 Beta ($95), #CV720-3/4-U ($190). Gr. 9–A. BKL (1/15/79), CIF (1978). 913.32

This program explores the relics of the ever-fascinating boy king and describes what is known of the pharaoh's life. Photographs illustrate the discovery and excavation of his tomb.

TWAIN, MARK

EPISODES FROM FAMOUS STORIES (Filmstrip—Sound). See Literature—Collections

HUCKLEBERRY FINN (Cassette). See U.S.—Social life and customs—Fiction

HUCKLEBERRY FINN (Phonodisc or Cassette). See U.S.—Social life and customs—Fiction

MARK TWAIN (Filmstrip—Sound). Coronet Instructional Media, 1974. 4 color, sound filmstrips, av. 60 fr.; 2 discs or 4 cassettes, av. 15 min. #M2704 Cassettes ($65), #S2704 Discs ($65). Gr. 7–12. BKL (10/1/74), MMT (1975). 921

Contents: 1. The Man. 2. The Humorist. 3. The Pessimist. 4. The Social Critic.

This set explores the life and writings of Samuel Clemens as humorist and social critic of the nineteenth century. It portrays his years as a miner, journalist, printer, author, and lecturer, and includes some costumed scenes from *Tom Sawyer, Huckleberry Finn,* "The Celebrated Jumping Frog of Calaveras County," and other stories.

MARK TWAIN: MISSISSIPPI RENAISSANCE MAN (Filmstrip—Sound). Aids of Cape Cod, 1977 (Americans Who Changed Things). 1 color, sound filmstrip, 84 fr.; 1 cassette, 17 min.; Teacher's Guide. #DM011Z ($26.50). Gr. 7–12. PRV (2/79). 921

Twain's unusual, rich, fascinating life is portrayed. His interest with the river, the origin of his creative ideas, his wit, and his boyhood memories that became classics are all presented. A number of quotes from Twain's humor are included.

NOVELISTS AND THEIR TIMES (Kit). *See* Authors

THE PRINCE AND THE PAUPER (Phonodisc or Cassette). *See* Fantasy—Fiction

U.S.—ANTIQUITIES

MESA VERDE (Study Print). *See* Cliff dwellers and cliff dwellings

U.S.—ARMED FORCES

CAREER TRAINING THROUGH THE ARMED FORCES (Filmstrip—Sound). Pathescope Educational Media, 1975. 2 color, sound filmstrips, av. 86 fr. ea.; 2 cassettes, av. 11–13 min.; Teacher's Manual packaged in a library storage unit. #737 ($50). Gr. 9–12. BKL (5/76). 355.0023
Contents: Part 1. Part 2.
This set describes the different jobs available within the armed forces by using interviews to determine necessary training and job satisfactions.

U.S.—CIVILIZATION

AMERICAN CIVILIZATION: 1783–1840 (Filmstrip—Sound). *See* Art, American

AMERICAN CIVILIZATION: 1840–1876 (Filmstrip—Sound). *See* Art, American

AMERICANS ON AMERICA: OUR IDENTITY AND SELF IMAGE (Slides/Cassettes). *See* American literature—Study and teaching

RISE OF THE AMERICAN CITY (Motion Picture—16mm—Sound). *See* Cities and towns—U.S.

TECHNOLOGY AND CHANGE (Kit). *See* Technology and civilization

U.S.—CIVILIZATION—BIOGRAPHY

AMERICANS WHO CHANGED THINGS (Filmstrip—Sound). Aids, 1977. 6 color—b/w, sound filmstrips, 64–81 fr.; 6 cassettes, 10–18 min.; Script. #919–928 ($135.50). Gr. 7–12. PRV (12/19/78). 917.3
Contents: 1. Buckminster Fuller: Geodesic Man. 2. Frank Lloyd Wright: Organic Architect. 3. Frederick Law Olmsted: Space in the City. 4. Frederic Remington: Artist of the Wild West. 5. George Gershwin: "Rhapsody in Blue." 6. Marian Anderson: "Once in a Hundred Years . . . ."
Chronological facts are coupled with critical events or achievements for each person presented in this set.

U.S. CONGRESS

FOR ALL THE PEOPLE (Kit). EMC, 1974. 3 color, sound filmstrips, 87–96 fr. ea.; 3 cassettes, 10–15 min. ea.; 2 Newsletters; Teacher's Guide. #SS–213000 ($65). Gr. 9–12. HST (11/75), LGB (1975), PVB (5/75). 328.973
Contents: 1. Facing It Squarely. 2. If I Am Elected. 3. Making It a Two Way Street.
A study of the legislative process that provides a historical perspective on Congress from its conception to contemporary problems. The set dramatizes the value conflicts faced by a first-term Congressman, and follows a student helping to defeat a bill that would destroy a wildlife area but is favored by powerful interests.

THE UNITED STATES CONGRESS: OF, BY, AND FOR THE PEOPLE (Motion Picture—16mm—Sound). Concept Media/EBEC, 1972. 16mm color, sound film, 26 min.; 2nd edition. #3158 ($325). Gr. 9–C. CGE (1974), EFL (1975), PRV (2/74). 328.973
The development of the U.S. congressional system is surveyed from earliest times to the present. The film examines the day-to-day work of a congressman and a senator. Through the numerous activities—staff meetings, committee hearings, meetings with constituents, briefings with students— viewers learn of the multitudinous responsibilities of these lawmakers. Former Vice-President Hubert H. Humphrey comments on historical and constitutional developments and the philosophy that has led to the evolution of Congress to its present state.

U.S. CONSTITUTION

THE CONSTITUTION—A LIVING DOCUMENT (Filmstrip—Sound). Pathescope Educational Media, 1971. 6 color, sound filmstrips, 49–62 fr. ea.; 6 cassettes, 15–20 min. ea.; Teacher's Guide. #511 ($100). Gr. 7–A. SLJ (4/15/72), TSS (1973). 342.73
Contents: 1. Formation. 2. Ratification and the Bill of Rights. 3. The Executive Branch. 4. The Legislative Branch. 5. Judicial Branch—The Supreme Court. 6. Checks and Balances.
Explores the nature of the Constitution and examines its past, present, and future. Focuses on the continual shaping and modifications of the document under stresses of economic changes, wars, civil rights restrictions, awakening human conscience, and the changing needs of the country.

U.S. CONSTITUTION—AMENDMENTS

YOUR FREEDOM AND THE FIRST AMENDMENT (Filmstrip—Sound). Educational Enrichment Materials, 1976. 6 color, sound filmstrips, 79–113 fr.; 6 cassettes or discs, 15–20 min.; Teacher's Guide.

U.S. CONSTITUTION—AMENDMENTS (cont.)

#51005 C or R ($108). Gr. 7–12. PRV (4/77). 342.2

Contents: 1. A History of Liberty. 2. Freedom of the Press. 3. Freedom of Assembly. 4. Freedom of Speech. 5. Freedom of Religion. 6. The Continuing Struggle for Freedom.

Traces the substance and evolution of egalitarian philosophies from Socrates to the American Revolution. The program then examines the struggle to establish each freedom in the First Amendment, as well as the judicial decisions that altered or strengthened this important part of the Bill of Rights.

U.S. DECLARATION OF INDEPENDENCE

THE DECLARATION OF INDEPENDENCE (Cassette). Heritage Productions, 1974. 1 cassette, 20 min. ea. side. #103 ($14.50). Gr. 5–A. PRV (1/19/75). 973.3

Introduced with U.S. Military Academy Glee Club's recording of ''America, The Beautiful.'' Also includes the causes of the Revolutionary War and events that led to the writing of this declaration.

DECLARATION OF INDEPENDENCE AND PROGRAM OF PATRIOTIC MUSIC (Phonodisc or Cassette). Liberty Tree Productions/Children's Classics on Tape, 1976. 1 cassette or disc. #201. ($8). Gr. 10–A. ESL (1977), PRV (12/75). 973.3

Narration of Declaration of Independence with background music with program of patriotic music.

U.S.—DESCRIPTION AND TRAVEL

AN AMERICAN SAMPLER (Filmstrip—Sound). See U.S.—Social life and customs

CITIES OF AMERICA, PART ONE (Filmstrip—Sound). Teaching Resources Films, 1972. 6 color, sound filmstrips, 53–76 fr.; 3 discs or cassettes, 5–6$^1/_2$ min.; Teacher's Guide. ($75). Gr. 5–12. PRV (2/74), PVB (5/74). 917.3

Contents: 1. New York. 2. Chicago. 3. Boston. 4. Detroit. 5. Washington, D.C. 6. Miami/Atlanta.

A factual exploration of major American cities encompasses history, geography, landmarks, transportation, commerce, mercantile establishments, culture, education, ethnic groups and contemporary problems.

CITIES OF AMERICA, PART TWO (Filmstrip—Sound). Teaching Resources Films, 1972. 6 color, sound filmstrips, 53–76 fr.; 3 discs or cassettes, 5–6$^1/_2$ min.; Teacher's Guide. ($75). Gr. 5–12. PRV (2/74), PVB (5/74). 917.3

Contents: 1. San Francisco. 2. Denver. 3. Houston. 4. Seattle. 5. Los Angeles. 6. New Orleans/St. Louis.

A factual exploration of major American cities encompassing history, geography, landmarks, transportation, commerce, mercantile establishments, culture, education, recreation, ethnic groups and contemporary problems.

TALKING ENCYCLOPEDIA OF THE NEW AMERICAN NATION (Cassette). See U.S.—History

U.S.—ECONOMIC CONDITIONS

THE AMERICAN ECONOMY, SET 1 (Filmstrip—Sound). See Economics

THE AMERICAN ECONOMY, SET 2 (Filmstrip—Sound). See Economics

THE AMERICAN ECONOMY, SET 3 (Filmstrip—Sound). See Economics

ECONOMIC ISSUES IN AMERICAN DEMOCRACY (Kit). Teaching Resources Films, 1973. 4 color, sound filmstrips, 41–50 fr. ea.; 2 cassettes or discs, 7–8 min. ea.; 4 Paperbound Texts; Teacher's Guide. ($60). Gr. 9–12. PRV (10/74), PVB (5/19/75). 330.973

Contents: 1. Unemployment and Inflation. 2. The Profit System. 3. Government and Our Economic System. 4. The World Economy.

A commentary, produced in cooperation with the Joint Council on Economic Education, exploring the important aspects of economics and government in society, this kit probes the interaction of productivity and distribution, world trade and economics.

ECONOMICS AND THE AMERICAN DREAM (Kit). See Economics

ECONOMICS IN AMERICAN HISTORY (Slides). See U.S.—History

FOCUS ON AMERICA, ALASKA AND HAWAII (Filmstrip—Sound). Society for Visual Education, 1973. 4 color, sound filmstrips, 71–82.; 2 discs or cassettes, 15$^1/_2$–18 min. #250–SFR Discs ($56), #250–SFTC Cassettes ($56). Gr. 4–9. BKL (7/74), PRV (5/74). 917

Contents: 1. Alaska: Wealth or Wilderness? 2. Alaska: The Old Frontier Meets the New. 3. Honolulu: The Tourist Explosion. 4. Lanai: The Pineapple Island.

This set focuses on several issue-oriented, sociological studies on the chief regional problems of these two states. Although the issues are unrelated, they provide challenging, thought-provoking topics.

FOCUS ON AMERICA, THE MIDWEST (Filmstrip—Sound). Society for Visual Education, 1972. 6 color, sound filmstrips, 71–87 fr.; 3 discs or cassettes, 15–18 min.; 6 Teacher's Guides. #A250SDR Discs ($83), #A250SDTC Cassette ($83). Gr. 5–9. BKL (4/1/73), PRV (5/1/73). 917.7

Contents: 1. North Dakota: Are Subsidies Necessary? 2. The Great Lakes: America's "Inland Sea." 3. Howard, Kansas: A Struggle to Survive. 4. Akron, Ohio: The Rubber City. 5. Chicago: The Airport Conflict. 6. Detroit: A City Rebuilds.

The cities seem to have little in common, yet all these problems are interdependent. The strips show some of the people in conflicts that exist in the vast and varied midwest region. Prints, maps, and diagrams as well as photographs are used.

FOCUS ON AMERICA, THE NEAR WEST REGION (Filmstrip—Sound). Society For Visual Education, 1972. 6 color, sound filmstrips, av. 82 fr.; 3 discs or cassettes; 1 Guide. #250–SBR Discs ($83), #250–SBTC Cassettes ($83). Gr. 4–9. BKL (4/73). 917

Contents: 1. The Cherokee Nation of Oklahoma. 2. Texas: Land of Cattle and Oil. 3. The Spanish Americans of New Mexico. 4. The Mormons of Utah. 5. Butte, Montana: City in Transition. 6. Colorado: Agriculture Technology.

Three groups of early settlers—the Indians, Spanish-Americans, and the Mormons—share their past and present and look to the future in this set. The wealth of this area comes from the land, as portrayed in visits to oil wells and oil boom towns in Texas, cattle feed lots in Colorado, and copper mines in Butte, Montana.

HAS BIG BUSINESS GOTTEN TOO BIG? (Filmstrip—Sound). *See* Economics

THE INEQUALITY OF WEALTH IN AMERICA (Filmstrip—Sound). Multi-Media Productions, 1976. 1 color, sound filmstrip, 59 fr.; 1 cassette or disc, 14 min.; Teacher's Manual. #7179C Cassette ($12.95), #7179R Disc ($12.95). Gr. 9–12. PVB (4/78). 330.973

Explores the issue of basic inequality in the United States and the disparity between lower-, middle-, and upper-income groups in America. The circular effect of economy and social conditions, each reacting on the other, is presented along with proposals for equalizing wealth in order to ease social tensions.

LIFE IN RURAL AMERICA (Filmstrip—Sound). *See* U.S.—Social life and customs

WORKING IN U.S. COMMUNITIES, GROUP ONE (Filmstrip—Sound). *See* Occupations

U.S.—FOREIGN POPULATION

AMERICA'S ETHNIC HERITAGE—GROWTH AND EXPANSION (Filmstrip—Sound). *See* Minorities

IMMIGRANT AMERICA (Filmstrip—Sound). *See* Immigration and emigration

IMMIGRATION AND EMIGRATION (Motion Picture—16mm—Sound). *See* U.S.—History

JEWISH IMMIGRANTS TO AMERICA (Filmstrip—Sound). *See* Jews in the U.S.

U.S.—FOREIGN RELATIONS

CHINA (Cassette). Cinema Sound Ltd./Jeffrey Norton Publishers, 1975 (Avid Reader). 1 cassette, approx. 55 min. #40105 ($11.95). Gr. 10–A. BKL (3/15/77). 327.73

O. Edmund Clubb, author of *The Witness and I*, was the director of the state department's Office of Chinese Affairs and an expert on China for some twenty years before Joseph McCarthy called him a security risk in the fifties. He describes the Kafkaesque results of that smear campaign in the light of our recent friendly relations with China.

U.S.—GEOGRAPHY

REGIONS OF THE UNITED STATES (Filmstrip—Sound). Troll Associates, 1977. 8 color, sound filmstrips, av. 67 fr. ea.; 8 cassettes, av. 12 min. ea.; Teacher Guide. ($112). Gr. 4–9. BKL (1/15/78). 917.3

Contents: 1. New England. 2. Middle Atlantic States. 3. Southwest. 4. Midwest. 5. North Central States. 6. South Central States. 7. Rocky Mountain States. 8. Western and Pacific States.

Provides basic overview of regions of the United States. Students learn about lifestyles, working conditions, geography, and resources that have built a nation and the special character of the American people.

U.S.—HISTORY

AMERICA IN THE 19TH CENTURY: THE PLURALISTIC SOCIETY (Filmstrip—Sound). Learning Corporation of America, 1975 (The Social History of the United States). 4 color, sound filmstrips, 78–88 fr.; 4 cassettes, 10–14 min.; Teacher's Guide. ($79.50). Gr. 7–C. BKL (10/15/76), M&M (4/77), PRV (12/76). 973

Contents: 1. The City in America. 2. Race in America. 3. The Immigrant Stream. 4. The Business Ethic.

During the nineteenth century our cities "boomed," and today over half of our population lives in cities. The effects of attitudes toward race are shown as they relate to our history and social structure. We see the emotional and social aspects of the "immigration stream" and watch American commerce and industry evolve from free enterprise competition to monopolistic capitalism. The historical roots of our consumer-oriented life-style are examined.

THE AMERICAN CITY (Kit). *See* Cities and towns—U.S.

AMERICAN HISTORY ON STAMPS (Kit). *See* Postage stamps

U.S. HISTORY (cont.)

THE AMERICAN SPIRIT (Motion Picture—16mm—Sound). Fred A. Niles Communications Centers/Best Films, 1976. 16mm color, sound film, 25 min., or videocassette #BF-1 ($295). Gr. 5-A. IFT (1976), LFR (9/76). 973

A "super-positive" film that presents the historic development of the spirits that, over the years, have helped the United States to become and remain a great Nation: discovery, cooperation, volunteerism, generosity, laughter, opportunity, invention.

THE AMERICAN WOMAN: A SOCIAL CHRONICLE (Filmstrip—Sound). See Women—U.S.

AMERICAN WOMEN (Kit). See Women—U.S.

CHANGING VALUES IN AMERICA: THE 20TH CENTURY (Filmstrip—Sound). Learning Corporation of America, 1975 (The Social History of the United States). 4 color, sound filmstrips, 75–84 fr.; 4 cassettes, 9–13 min.; Teacher's Guide. ($79.50). Gr. 7-C. BKL (10/15/76), M&M (4/19/77), PRV (12/19/76). 973

Contents: 1. The Changing Place of Religion. 2. The Changing American Family. 3. The Changing Role of Women. 4. The Changing Faith in Technology.

Many basic values have evolved into new modes of living during this century. This set looks at some of the key influences affecting these values.

ECONOMICS IN AMERICAN HISTORY (Slides). Educational Enrichment Materials, 1973. 120 slides in plastic holders; 1 Teacher's Guide. #480002 ($95). Gr. 7-C. PRV (4/76). 973

Designed to point out U.S. historical events that can be enriched by a consideration of relevant economic topics. Suggested groupings of slides are indicated to complement a range of pertinent social studies courses. The 120 slides include diagrams, pie charts, bar graphs, and pictograms.

IMMIGRATION AND MIGRATION (Motion Picture—16mm—Sound). Nystrom, 1975 (The American Experience). 5 color, sound filmstrips, 69–84 fr.; 5 cassettes, 10–12 min.; 10 Student Readers, 1 Set of Activity Sheets; 2 Teacher's Programs. #AE-100 ($125). Gr. 6–11. BKL (2/15/77). 973

Contents: 1. The Golden Land. 2. The First Wave. 3. Frontiers. 4. A Flood of Peoples. 5. Strangers in the City.

Intended to present "history without politics" but with a special focus on the role of the common person in building and shaping America.

INVENTIONS AND TECHNOLOGY THAT SHAPED AMERICA, SET ONE (Filmstrip—Sound). See Technology and civilization

A LETTER TO CETI (Filmstrip—Sound). Hawkhill Associates, 1976. 1 color sound filmstrip, approx. 100 fr.; 1 cassette, approx. 18 min.; Teacher's Guide. ($26.50). Gr. 7-12. PRV (3/77). 973

The sound filmstrip is an imaginary letter to our nearest star about the meaning of our country, how it has been, how it is, and the American Dream. U.S. paintings and photographs of persons, places, and events that have shaped America are presented with excerpts from folk music.

THE ROOTS OF THE AMERICAN CHARACTER (Filmstrip—Sound). International Cinemedia Center/Learning Corporation of America, 1975 (The Social History of the United States). 4 color, sound filmstrips, 79–88 fr.; 4 cassettes, 9–12 min.; 1 Guide. ($79.50). Gr. 9-C. BKL (10/15/76), M&M (4/77), PRV (12/76). 973

Contents: 1. The American Character: What Is It? 2. The Democratic Ideal. 3. The Puritan Ethic. 4. The Frontier Influence.

This set explores the forces that have contributed to the image that many people have of the "typical American."

SETTLERS OF NORTH AMERICA (Filmstrip—Sound). See Frontier and pioneer life

TALKING ENCYCLOPEDIA OF THE NEW AMERICAN NATION (Cassette). Troll Associates, 1973. 51 cassette tapes, av. 15–20 min. ea. (8 volumes). #973–980 ($299). Gr. 5-A. PRV (5/75), TEA (12/73). 973

Contents: 1. New England. 2. Middle States. 3. Southeast. 4. Prairie States. 5. Plains States. 6. South Central. 7. Rocky Mountain States. 8. Western and Pacific.

The emphasis is on events and people that played an important part in the state's history. There is also material on geography, industry, and points of interest. Included is a cross-referenced subject guide.

UNIVERSAL VALUES IN AMERICAN HISTORY (Kit). Guidance Associates, 1977. 5 sets of 2 color, sound filmstrips, ea., 114–144 fr. ea. set; cassettes or discs, 17–23 min. ea. set; Spirit Duplicating Masters; Guide. #9A–401–859 Cassettes ($187), #9A–401–842 Discs ($187). Gr. 9-C. BKL (12/15/77), PRV (2/78), PVB (4/78). 973

Contents: 1. Personal Conflicts in the Revolutionary Era. 2. Personal Conflicts in a Divided Nation. 3. Personal Conflicts on the Western Frontier. 4. Personal Conflicts at the Turn of the Century. 5. Personal Conflicts in the Modern Era.

Original, authentically detailed photo-essays portray costumes, life-styles, and cultural details of different eras. Historical vignettes recreate events and issues. Students are involved in personalized moral dilemmas illustrating historical problems, which help them see parallels between past and present issues.

U.S.—HISTORY—1600–1775—COLONIAL PERIOD

AMERICAN FOLK ARTS (Filmstrip—Sound). *See* Folk art, American

AMERICAN REVOLUTION—ROOTS OF REBELLION (Filmstrip—Sound). Coronet Instructional Media, 1974. 6 color, sound filmstrips, av. 53 fr. ea.; 6 cassettes or discs, av. 11 min. ea. #M705 Cassettes ($99), #S705 Discs ($99). Gr. 6–12. M&M (1975), PVB (1975), TEA (1975). 973.2

Contents: 1. The Search Warrant Dispute (1761). 2. The Stamp Act Riots (1765). 3. The Boston Massacre (1770). 4. The Boston Tea Party (1773). 5. The Battle of Lexington and Concord (1775). 6. The Declaration of Independence (1776).

This set explores six key events leading to the American Revolution. Photos, artwork, and historical prints are used in the strips.

COLONIAL AMERICA (Filmstrip—Sound). McGraw-Hill Films, 1971 (American Heritage). 5 color, sound filmstrips, av. 50 fr. ea.; 5 cassettes or discs, 7 min. #101906–5 Cassettes ($95), #101895–6 Discs ($95). Gr. 7–A. SLJ (2/72). 973.2

Contents: 1. England Stakes a Claim. 2. Religious Havens. 3. The American Melting Pot. 4. Expansion and Conflict. 5. War for Empire.

Woodcuts and paintings by early American artists, a slave sale advertisement, official portraits, and other period illustrations provide insights into colonial life.

THE JOHN PETER ZENGER TRIAL (Filmstrip—Sound). *See* Freedom of the press

U.S.—HISTORY—1775–1783—REVOLUTION

AMERICA IN ART: THE AMERICAN REVOLUTION (Filmstrip—Sound). Miller-Brody Productions, 1974. 2 color, sound filmstrips, 66–71 fr.; 2 cassettes or discs, av. 14¹/₂ min; Teacher's Guide. #A211C Cassettes ($44), #A211 Discs ($44). Gr. 6–12. BKL (12/15/74). 973.3

Contents: 1. Part I. 2. Part II.

More about the American Revolution than about art, this set provides insight into the personalities and significant events in and the course of the struggle for independence. It features both familiar and lesser-known paintings, cartoons and, engravings of the period. The narrator includes incidental information about the artist, but the teacher guide indicates the artist and year (when available) of each work used, plus information on art styles and the artists.

AMERICAN REVOLUTION—ROOTS OF REBELLION (Filmstrip—Sound). *See* U.S.—History—1600–1775—Colonial period

BENEDICT ARNOLD: TRAITOR OR PATRIOT? (Filmstrip—Sound). *See* Arnold, Benedict

DEBORAH SAMPSON: A WOMAN IN THE REVOLUTION (Motion Picture—16mm—Sound). *See* Sampson, Deborah

THE DECLARATION OF INDEPENDENCE (Cassette). *See* U.S. Declaration of Independence

FAMOUS PATRIOTS OF THE AMERICAN REVOLUTION (Filmstrip—Sound). *See* U.S.—History—Biography

GEORGE WASHINGTON (Cassette). *See* Washington, George

PAUL REVERE'S RIDE AND HIS OWN STORY (Cassette). *See* American poetry

PRELUDE TO REVOLUTION (Motion Picture—16mm—Sound). EBEC, 1975, 16mm color, sound film, 12 min. #3426 ($150). Gr. 7–12. LFR (1976), PRV (1975). 973.3

Featuring original watercolors by the artist A. N. Wyeth, this animated film portrays the tug-of-wag between radicals and conservatives as the colonies struggled first to keep their ties with England and, finally, reluctantly, to break them.

THE REVOLUTION (Filmstrip—Sound). McGraw-Hill Films, 1971 (American Heritage). 5 color, sound filmstrips, av. 50 fr. ea; 5 cassettes or discs, 7 min. ea. #101889–1 Cassettes ($95), #101878–6 Discs ($95). Gr. 7–A. SLJ (2/72). 973.3

Contents: 1. Eve of Revolt. 2. Toward Independence. 3. The Times that Try Men's Souls. 4. Frontiers Aflame. 5. Independence Won.

Paul Revere's famous engraving of the Boston Massacre, and works by other American, British, and French artists of the period, plus statements and speeches of Washington, Jefferson, John Adams, Thomas Paine, and others lend reality to the program as it presents the Revolutionary War period.

THOMAS PAINE (Motion Picture—16mm—Sound). *See* Paine, Thomas

VOICES OF THE AMERICAN REVOLUTION (Filmstrip—Sound). Coronet Instructional Media, 1975. 6 color, sound filmstrips, av. 66 fr.; 3 discs or 6 cassettes, av. 13¹/₂ min. ea.; #M294 Cassettes ($95), #S294 Discs ($95). Gr. 7–12. BKL (1976), M&M (1976), PRV (1976). 973.3

Contents: 1. Josiah Thatcher, Farmer Recruit. 2. Samuel Stark, Merchant. 3. Constance Ware, Wartime Civilian. 4. Daniel Pittman, Wilderness Warrior. 5. Edward Marshall, Loyalist. 6. Josiah Thatcher, Yorktown Veteran.

First-person narratives focusing on different times, areas, and viewpoints in the eight-year struggle of the American Revolution.

U.S.—HISTORY—1775–1783—REVOLUTION (cont.)

The problems and reactions of ordinary people are based on journals, diaries, and letters of the period.

WOMEN IN THE AMERICAN REVOLUTION (Filmstrip—Sound). Multi-Media Productions, 1975. 2 color and b/w, sound filmstrips, 45–47 fr.; 1 cassette or disc, 20 min.; Teacher's Manual. #7142 C or R ($17.95). Gr. 8–12. PVB (10/76). 973.3

Discusses why there were so few known heroines of the American Revolution. Shows how the move to America from Europe and how the War for Independence broadened the traditional role of all colonial women. They participated in varied and responsible labors analogous to the image of Rosie the Riveter of WW II.

U.S.—HISTORY—1775–1783—REVOLUTION—FICTION

MY BROTHER SAM IS DEAD (Phonodisc or Cassette). Miller-Brody Productions, 1976. 1 disc or cassette, 41 min. #NAR3089 Disc ($6.95), #NAC3089 Cassette ($7.95). Gr. 5–8. BKL (9/15/77). Fic

A dramatization based on a Newbery Honor Book by J. L. Collier. A Revolutionary War story about Sam, who volunteers in the rebel cause leaving behind his brother, who tells the story.

U.S.—HISTORY—1783–1865

THE AMERICA OF CURRIER AND IVES (Filmstrip—Sound). See Prints, American

AMERICAN CIVILIZATION: 1783–1840 (Filmstrip—Sound). See Art, American

AMERICAN CIVILIZATION: 1840–1876 (Filmstrip—Sound). See Art, American

GHOST TOWNS OF THE WESTWARD MARCH (Motion Picture—16mm—Sound). See The West (U.S.)—History

GOLD TO BUILD A NATION: THE UNITED STATES BEFORE THE CIVIL WAR (Filmstrip—Sound). See California—Gold discoveries

U.S.—HISTORY—1787–1865

DRED SCOTT: BLACK MAN IN A WHITE COURT (Filmstrip—Sound). See Scott, Dred

U.S.—HISTORY—1801–1805—TRIPOLITAN WAR

AMERICA REACTS TO A WORLD IN CONFLICT, 1800–1815 (Filmstrip—Sound). Multi-Media Productions, 1976. 3 color, sound filmstrips, 54–79 fr.; 3 cassettes or discs, 9–14 min.; Teacher's Manual. #7167C

Cassettes ($36), #7167R Discs ($36). Gr. 9–12. BKL (1/15/77), PRV (3/77). 973.5

Contents: 1. The Barbary Wars: ". . . to the shores of Tripoli." 2. Jefferson's Embargo: A Search for an Alternative to War. 3. 1812: Mr. Madison's War.

Presentation of three international situations occurring between 1800 and 1815 that challenged the moral, economic, and political values of the American people and their leaders.

U.S.—HISTORY—1805—NONIMPORTATION EMBARGO

AMERICA REACTS TO A WORLD IN CONFLICT, 1800–1815 (Filmstrip—Sound). See U.S.—History—1801–1805—Tripolitan War

U.S.—HISTORY—1812, WAR OF

AMERICA REACTS TO A WORLD IN CONFLICT, 1800–1815 (Filmstrip—Sound). See U.S.—History—1801–1805—Tripolitan War

U.S.—HISTORY—1861–1865—CIVIL WAR

THE CIVIL WAR (Filmstrip—Sound). McGraw-Hill Films, 1971 (American Heritage). 5 color, sound filmstrips, av. 50 fr. ea.; 5 cassettes or discs, 7 min. ea. #101872–7 Cassettes ($95), #101861–1 Discs ($95). Gr. 7–A. BKL (1/72), CFF (1972), IFT (1972). 973.7

Contents: 1. A Nation Divided. 2. The Clash of Amateur Armies. 3. The Iron Vise Is Forged. 4. Gettysburg. 5. An Ending and a Beginning.

The photographs of Mathew Brady and others, newspaper sketches, and battlefield paintings depict various phases of the Civil War struggle.

MATHEW BRADY: PHOTOGRAPHER OF AN ERA (Motion Picture—16mm—Sound). See Photography—History

U.S.—HISTORY—1861–1865—CIVIL WAR—FICTION

ACROSS FIVE APRILS (Filmstrip—Sound). Miller-Brody Productions. 1974 (Newbery Filmstrips). 2-part color, sound filmstrip, 165 fr.; cassette or disc, 38 min.; Teacher's Guide. #NAC 3034 ($32). Gr. 5–8. BKL (9/15/74), LGB (1974). Fic

Growing up is the theme of this popular story about the Civil War. The issue is the commitment of the people to a free or a slave society. Based on the book by Irene Hunt.

THE RED BADGE OF COURAGE (Kit). Listening Library, 1975. 4 cassettes, 290 min. total; 1 Book. #YRA 66 CX ($29.95). Gr. 7–12. PRV (4/76). Fic

The complete novel by Stephen Crane is read with no sound effects or musical background.

U.S.—HISTORY—1865–1898

THE AMERICA OF CURRIER AND IVES (Filmstrip—Sound). *See* Prints, American

ANDREW JOHNSON COMES TO TRIAL (Filmstrip—Sound). Current Affairs Films, 1977. 1 color, sound filmstrip, 74 fr.; 2 cassettes—"Pro-and-Con," 16 min. ea.; Teacher's Guide. #587 ($30). Gr. 9–C. PRV (1/78). 973.8

The clash between President Andrew Johnson and his Congress after the Civil War led to a great trial in American History, one that took place not in a courtroom with a judge and jury, but in the Senate of the United States. The impeachment of Andrew Johnson is closely analyzed in this filmstrip program from the political, judicial, and human points of view. Using two cassettes, students are able to listen to the arguments for the prosecution and the defense.

RAILROADS WEST (Filmstrip—Sound). Associated Press/Prentice-Hall Media, 1974 (America Comes of Age 1870–1917). 3 color, sound filmstrips, 80–94 fr.; 3 discs or cassettes, 14–20 min.; Teacher's Guide. #KAC6850 Cassettes ($72), #KAR6850 Discs ($72). Gr. 5–A. PRV (1/75), PVB (5/75). 973.8

Contents: 1. The Golden Spike. 2. The Iron Horse. 3. The End of the Wild West.

This set focuses on our country's development and the role the railroad played. It uses a documentary approach with each strip composed of black-and-white photographs and prints. On-the-spot reports, authentic sound effects, and music bring alive this era of American history.

U.S.—HISTORY—1898–1919

PROGRESSIVE ERA: REFORM WORKS IN AMERICA (Motion Picture—16mm—Sound). EBEC, 1971. 16mm color, sound film, 23 min. #3043 ($150). Gr. 7–12. CIF (1972), M&M (1974), SLJ (1972). 973.9

Describes the years spanning 1890–1915 and how they marked this period of American history as one of great turmoil and violence. Though a Progressive candidate was elected in 1912, major legislation for improved living conditions was not enacted until the depression years of the thirties. Original film clips bring sharp perspective to this historical era.

RAILROADS WEST (Filmstrip—Sound). *See* U.S.—History—1865–1898

U.S.—HISTORY—1919–1933

THE DECADES: NINETEEN 30'S (Cassette). Visual Education, 1976. 3 cassettes,

40–55 min.; Guide. #5203–00 ($34). Gr. 9–A. BKL (5/15/77). 973.91

Contents: 1. Politics. 2. Society. 3. Technology.

This set records the leadership and New Deal legislation of President Roosevelt, a nation of unemployed and fearful people fighting to retain their optimism and using radio and motion pictures as an escape from reality, and the laying of the groundwork necessary for the war effort.

THE LONG THIRST: PROHIBITION IN AMERICA (Cassette). *See* Prohibition

THE TWENTIES AND THIRTIES (Filmstrip—Sound). McGraw-Hill Films, 1972 (American Heritage). 5 color, sound filmstrips, av. 65 fr. ea.; 5 cassettes or discs, 7 min. ea. #102454–9 Cassettes ($95), #103693–8 Discs ($95). Gr. 7–A. AIF (1973), PVB (4/73). 973.91

Contents: 1. "What This Country Needs Is A Good Five-Cent Cigar." 2. "My Friends Are Keeping Me Awake Nights." 3. "Everyone Ought to Be Rich." 4. "Many Persons Left Their Jobs to Sell Apples." 5. "Due to Circumstances Beyond Our Control . . . ."

A contrast in words, images, and music between the prosperous, roaring 1920s and the depression-ridden, socially conscious 1930s. The voices of Franklin D. Roosevelt, Charles A. Lindbergh, and Will Rogers are heard and quotes from F. Scott Fitzgerald, Edna St. Vincent Millay, Henry Ford, H. L. Mencken, and others are included. Newspaper photographs, cartoons, and paintings of the time illustrate the set.

WILL ROGERS' NINETEEN TWENTIES (Motion Picture—16mm—Sound). Oklahoma State University/Churchill Films, 1976. 16mm color, sound film, 41 min.; Guide. ($425). Gr. 9–A. AUR (1979). 973.91

A compilation of archive materials with Will Rogers' funny, penetrating comments provides an engrossing review of the period from the end of the war to the start of the Depression.

U.S.—HISTORY—1933–1945

THE DECADES: NINETEEN 40'S (Cassette). Visual Education, 1976. 3 cassettes, 45–60 min.; Guide. #5204–00 ($34). Gr. 9–A. BKL (5/15/77), PRV (10/76). 973.91

Contents: 1. Politics. 2. Society. 3. Technology.

This set conveys the drama, tension, and sacrifices of the war years; the chartering of the United Nations; the closing of the Iron Curtain; and the effects of the great social and technological changes that followed the war.

WORLD WAR TWO (Filmstrip—Sound). McGraw-Hill Films, 1972 (American Heri-

U.S.—HISTORY—1933–1945 (cont.)

tage). 5 color, sound filmstrips, av. 70 fr. ea.; 5 cassettes or discs, 7 min. ea. #102460–3 Cassettes ($95), #103704–7 Discs ($95). Gr. 7–A. M&M (9/73). 973.91

Contents: 1. Nightmare. 2. Blitzkrieg. 3. Counterattack. 4. Invasion. 5. Inferno.

These strips cover the highlights of World War II: the personalities, battles, politics, treaties. Selections of music from the period and the voices of Roosevelt, Hitler, Mussolini, Churchill, Eisenhower, and others are heard.

U.S.—HISTORY—1945–1953

THE DECADES: NINETEEN 50'S (Cassette). Visual Education, 1975 (The Decades). 3 cassettes, 35–65 min. #5205–00 ($30). Gr. 5–10. BKL (7/15/77). 973.918

Contents: 1. Political Machinations of the 1950s. 2. American Society in the 1950s. 3. Technology of the 1950s.

The hopes, aspirations, dreams, and disappointments as well as the fears and frustrations of America in the period between Korea and Vietnam is portrayed using primary sources such as news reports and radio programs.

U.S.—HISTORY—1953–1961

THE DECADES: NINETEEN 50'S (Cassette). *See* U.S.—History—1945–1953

U.S.—HISTORY—BIOGRAPHY

AUNT ARIE (Motion Picture—16mm—Sound). *See* Carpenter, Aunt Arie

BENJAMIN FRANKLIN—SCIENTIST, STATESMAN, SCHOLAR AND SAGE (Motion Picture—16mm—Sound). *See* Franklin, Benjamin

DEBORAH SAMPSON: A WOMAN IN THE REVOLUTION (Motion Picture—16mm—Sound). *See* Sampson, Deborah

FAMOUS PATRIOTS OF THE AMERICAN REVOLUTION (Filmstrip—Sound). Coronet Instructional Media, 1973. 6 color, sound filmstrips, av. 45 fr.; 6 discs or cassettes, av. 11 min.; Teacher's Guide. #M249 Cassettes ($87), #S249 Discs ($79). Gr. 4–9. BKL (1974), PRV (1974). 920

Contents: 1. Patrick Henry. 2. Crispus Attucks. 3. Nathanael Greene. 4. Haym Salomon. 5. Molly Pitcher. 6. John Paul Jones.

Artist-illustrated biographies of familiar or little-known patriots, filled with personal and ethnic characteristics.

THE FOUNDING FATHERS IN PERSON (Cassette). Cinema Sound/Jeffrey Norton Publishers, 1976 (Avid Reader). 1 cassette, approx. 55 min. #40222 ($11.95). Gr. 7–A. BKL (7/15/77). 920

Heywood Hale Broun finds that in this program on the founding fathers, based on Claude-Ann Lopez's *The Private Franklin: The Man and His Family* (Norton, 1975) and on *The Book of Abigail and John: Selected Letters of the Adams Family* (Harvard, 1975), comparisons are made of the relationship of each man to his family, focusing on the intensity or lack of feeling each expressed to his wife and children.

GEORGE WASHINGTON—THE COURAGE THAT MADE A NATION (Motion Picture—16mm—Sound). *See* Washington, George

LEADERS, DREAMERS AND HEROES (Cassette). *See* Biography—Collections

THE PICTORIAL LIFE-STORY OF GEORGE WASHINGTON (Kit). *See* Washington, George

THE PICTORIAL LIFE-STORY OF THOMAS JEFFERSON (Kit). *See* Jefferson, Thomas

THOMAS JEFFERSON (Motion Picture—16mm—Sound). *See* Jefferson, Thomas

WOMEN: AN AMERICAN HISTORY (Filmstrip—Sound). *See* Women—U.S.

WOMEN IN AMERICAN HISTORY (Filmstrip—Sound). *See* Women—U.S.

U.S.—HISTORY—MUSIC

AMERICAN HISTORY: IN BALLAD AND SONG, VOLUME ONE (Phonodiscs). *See* Ballads, American

AMERICAN HISTORY: IN BALLAD AND SONG, VOLUME TWO (Phonodiscs). *See* Ballads, American

FOLK SONGS IN AMERICAN HISTORY, SET ONE: 1700–1864 (Filmstrip—Sound). *See* Folk songs—U.S.

FOLK SONGS IN AMERICAN HISTORY, SET TWO: 1865–1967 (Filmstrip—Sound). *See* Folk songs—U.S.

FOLK SONGS IN AMERICAN HISTORY: THE AMERICAN FLAG (Filmstrip—Sound). *See* Folk songs—U.S.

FOLK SONGS IN AMERICA'S HISTORY SINCE 1865 (Filmstrip—Sound). *See* Folk songs—U.S.

U.S.—HISTORY—SOURCES

VOICES OF THE AMERICAN REVOLUTION (Filmstrip—Sound). *See* U.S.—History—1755–1783—Revolution

U.S.—POLITICS AND GOVERNMENT

THE AMERICAN PRESIDENCY (Kit). *See* Presidents—U.S.

GOVERNMENT AND YOU (Filmstrip—Sound). *See* Political science

LOBBYING: A CASE HISTORY (2ND EDITION) (Motion Picture—16mm—Sound). *See* Lobbying

THE PRESIDENCY (Kit). *See* Presidents—U.S.

PRESIDENTS AND PRECEDENTS (Kit). *See* Presidents—U.S.

U.S.—RACE RELATIONS

KING: A FILMED RECORD, MONTGOMERY TO MEMPHIS (Videocassette). *See* King, Martin Luther

U.S.—SOCIAL CONDITIONS

THE MOUNTAIN PEOPLE (Motion Picture—16mm—Sound). *See* Appalachian Mountains

U.S.—SOCIAL LIFE AND CUSTOMS

AMERICA IN THE 19TH CENTURY: THE PLURALISTIC SOCIETY (Filmstrip—Sound). *See* U.S.—History

AMERICAN MAN: TWO HUNDRED YEARS OF AUTHENTIC FASHION (Kit). *See* Clothing and dress

AN AMERICAN SAMPLER (Filmstrip—Sound). Joshua Tree Productions, 1972. 6 color, sound filmstrips, 63–97 fr.; 6 discs or cassettes, 15 min. Teacher's Guide. #1R-6900 Discs ($95.50), #1T-6900 Cassettes ($106.50). Gr. 6-A. PRV (9/72), PVB (5/74). 917.3

Contents: 1. America Celebrates Tradition. 2. American Variety and Individualism. 3. America Changing and Unchanged. 4. America on the Go. 5. Americans and Their Land. 6. America and a Job Well Done.

An exploration of grass-roots America, its heritage, and both familiar and unfamiliar life-styles. Based on Charles Kuralt's CBS news program "On the Road," it is narrated by him.

THE AMERICAN WOMAN: A SOCIAL CHRONICLE (Filmstrip—Sound). *See* Women—U.S.

CHANGING VALUES IN AMERICA: THE 20TH CENTURY (Filmstrip—Sound). *See* U.S.—History

FOCUS ON AMERICA, ALASKA AND HAWAII (Filmstrip—Sound). *See* U.S.—Economic conditions

FOCUS ON AMERICA, THE MIDWEST (Filmstrip—Sound). *See* U.S.—Economic conditions

FOCUS ON AMERICA, THE NEAR WEST REGION (Filmstrip—Sound). *See* U.S.—Economic conditions

A LETTER TO CETI (Filmstrip—Sound). *See* U.S.—History

LIFE IN RURAL AMERICA (Filmstrip—Sound). National Geographic Educational Services, 1973. 5 color, sound filmstrips, av. 60 fr.; 5 cassettes or discs, 13 min. ea.; Teacher's Guide. #03738 ($74.50), #03739 ($74.50). Gr. 5-A. BKL (2/1/74), PRV (1/75), PVB (5/75), 917.3

Contents: 1. The Family Farm. 2. Cowboys. 3. Coal Miners of Appalachia. 4. Harvester of the Golden Plains. 5. Settlers on Alaska's Frontier.

This set discusses the problems and rewards of living on the land. It depicts the rigors and satisfaction of physical toil, the dangers and risks in producing some of our most common necessities, and the life-styles of the people involved.

THE QUALITY OF LIFE IN THE UNITED STATES (Kit). Educational Enrichment Materials, 1977. 5 color, sound filmstrips; 5 cassettes; 12 Spirit Duplicating Masters; Teacher's Guide. #51031 ($102). Gr. 7-12. PRV (5/78). 301.44

Contents: 1. Our Quality of Life—Can We Measure It? Can We Improve It? 2. Law/Government and the Quality of Life. 3. Your Career and the Quality of Life. 4. The Quality of Life in the Future.

Discusses the various criteria used to measure the quality of life: economic status, aesthetic appeal, ethical/moral climate, etc. Provides students with a structure within which they can investigate the elements of "the good life," for society and for themselves.

THE ROOTS OF THE AMERICAN CHARACTER (Filmstrip—Sound). *See* U.S.—History

U.S.—SOCIAL LIFE AND CUSTOMS—FICTION

THE ADVENTURES OF HUCKLEBERRY FINN BY MARK TWAIN (Phonodisc or Cassette). *See* Adventure and adventurers—Fiction

HUCKLEBERRY FINN (Cassette). Jabberwocky Cassette Classics, 1972. 3 cassettes, av. 60 min. #1071-1091 ($23.94).Gr. 5-A. BKL (3/15/74), JOR (11/74). Fic

Contents: 1. Part One. 2. Part Two. 3. Part Three.

A colorful dramatization of Mark Twain's portrait of a real boy. Musical interludes are used to indicate changes of scenes and pas-

U.S.—SOCIAL LIFE AND CUSTOMS—FICTION (cont.)

sage of time in this novel, which has been adapted and condensed. Twain's satire and irony come through clearly. Can be used to introduce Twain's book or to bring the novel to life after it has been read.

HUCKLEBERRY FINN (Phonodisc or Cassette). Caedmon Records, 1968. 2 cassettes or discs, approx. 60 min. ea. #TC2038 Discs ($13.96), #CDL 52038 Cassettes ($15.90). Gr. 5–9. LBJ. Fic

Contents: 1. "Sivilizing" Huck. 2. Huck and Pap. 3. Huck and Jim. 4. Emmeline Grangerford. 5. Life on the Raft. 6. Hamlet's Soliloquy. 7. The Shooting of Boggs. 8. Jim Gets Homesick. 9. You Can't Pray a Lie. 10. Escape Plan.

Seemingly a rambling tale for boys, this novel is in reality a subtly crafted and carefully structured masterpiece. Careful selection from the various chapters carries the story line and presents the flavor and excellence of Twain's writing.

LITTLE WOMEN (Phonodisc or Cassette). *See* Family life—Fiction

U.S. SUPREME COURT

A CONVERSATION WITH EARL WARREN (Videocassette). *See* Warren, Earl

JUSTICE AND THE LAW (Filmstrip—Sound). *See* Law

UNIVERSE

THE SCOPE OF THE UNIVERSE (Videocassette). Educational Dimensions Group, 1978 (The Cosmos: The Study of the Universe, Part 1). videocassette, 30 min. #V716 ($380). Gr. 7–12. BKL (1/15/79), IFT (1978). 523

What is the Universe? Time? How do cosmologists work? These and more vital questions are answered or explored.

URBAN AREAS. *See* Metropolitan areas

URBAN RENEWAL

THE AMERICAN URBANIZATION (Filmstrip—Sound). Pathescope Educational Media, 1974 (Great Cities in Transition). 2 color, sound filmstrips, 81–84 fr. ea.; 2 cassettes, 11-1/2–12 min. ea.; Guide. #754 ($50). Gr. 7–12. BKL (9/74). 301.36

Contents: 1. Megalopolis East. 2. Megalopolis West.

This set studies the coastal areas of New York and Los Angeles in terms of transportation, education, taxes, housing, and pollution. The urban spread has seen these two communities develop into supercities.

COMMITMENT TO CHANGE: REBIRTH OF A CITY (Filmstrip—Sound). Current Affairs Films, 1976. 2 color, sound filmstrips, approx 115 fr. ea.; 2 cassettes, 20 min. ea.; Teacher's Guide. #546 ($48). Gr. 7–12. PRV (3/77). 309.2

This true story tells of the determination of minority groups who believe it is not too late to change the grim conditions of ghetto life. They are persistent people who continually face ponderous bureaucratic stumbling blocks, political and social resistance, and financial difficulties.

VALUES

CHANGING VALUES IN AMERICA: THE 20TH CENTURY (Filmstrip—Sound). *See* U.S.—History

COMPETITIVE VALUES: WINNING AND LOSING (Filmstrip—Sound). *See* Sports

CONTROL: WHO PUSHES YOUR BUTTON? (Filmstrip—Sound). *See* Self-control

COPING WITH LIFE: THE ROLE OF SELF-CONTROL (Filmstrip—Sound). *See* Self-control

CORRUPTION IN AMERICA: WHERE DO YOU STAND? (Filmstrip—Sound). *See* Corruption in politics

ESSAY ON WAR (Motion Picture—16mm—Sound). *See* War and civilization

I THINK (Motion Picture—16mm—Sound). Wombat Productions, 1970. 16mm color, sound film, 19 min.; available in Spanish. ($285). Gr. 4–A. CRA (1971), LFR (1972), SOC (3/74). 170.2

A young girl is under many influences. But where does she make her stand and assert what she believes, even in opposition to her peers?

IT'S A MATTER OF PRIDE (Motion Picture—16mm—Sound). *See* Work

LAW AND JUSTICE (Filmstrip—Sound). *See* Law

LAW AND JUSTICE: MAKING VALUE DECISIONS (Filmstrip—Sound). *See* Law

LEISURE (Motion Picture—16mm—Sound). *See* Man

LIFE GOALS: SETTING PERSONAL PRIORITIES (Filmstrip—Sound). *See* Self-realization

LIFESTYLES: OPTIONS FOR LIVING (Kit). *See* Life-styles—U.S.

LIVING IN THE FUTURE: NOW—INDIVIDUAL CHOICES (Filmstrip—Sound). *See* Self-realization

LIVING WITH TECHNOLOGY: CAN WE CONTROL APPLIED SCIENCE? (Filmstrip—Sound). *See* Technology and civilization

MATURITY: OPTIONS AND CON-
SEQUENCES (Filmstrip—Sound). *See*
Adolescence

MEET MARGIE (Motion Picture—16mm—
Sound). *See* Finance, Personal

SQUARE PEGS—ROUND HOLES (Motion
Picture—16mm—Sound). *See* Individuality

URBAN WORLD—VALUES IN CON-
FLICT (Filmstrip—Sound). *See* Metropolitan areas

VALUES (Filmstrip—Sound). Davidson
Films/Xerox Educational Publications, 1975.
7 color, sound filmstrips, av. 65 fr. ea.; 7 cassettes or discs, av. 7 min. ea.; 7 Teaching
Guides. #SC011 Cassettes ($140), #SR011
Discs ($140). Gr. 5–9. M&M (12/76), PRV (9/
76). 170.2

Contents: 1. Idioms Delight. 2. If You're a
Horse. 3. Smiles. 4. The Truth about Horse-
feather. 5. Where Did Leonard Harry Go? 6.
Who Knows? 7. Yes and No.

Important value questions are explored
through cartoon fables. The amusing stories
are underplayed, but raise basic value questions, such as learning from our mistakes,
personal freedom limits, place of new or
original ideas in society, accepting consequences of decisions, dealing with success
and happiness, facing ignorance, anonymity.

WHAT ARE YOU GOING TO DO ABOUT
ALCOHOL (Filmstrip—Sound). *See* Alcohol

THE WITNESS (Motion Picture—16mm—
Sound). Vitascope Film/FilmFair Communications, 1976. 16mm color, sound film, 14
min. ($210). Gr. 7–A. CFF (1977), LFR (3/
77), PRV (3/78). 170.2

Charles, a high school sophomore collides
with a much larger boy who is running from
an alley. The boys recognize each other.
When Charles see a woman lying unconscious he realizes what has happened. Although he can provide information regarding
the crime, he slips away from the crowd and
runs home. The purse snatcher is one of the
toughest kids in school and Charles knows
he will be in jeopardy if he goes to the police.
The question is, what to do?

VALUES—FICTION

ANGEL AND BIG JOE (Motion Picture—
16mm—Sound). Learning Corporation of
America, 1975 (Learning to Be Human).
16mm color, sound film, 27 min. ($355). Gr.
3–A. AAW (1975), BKL (6/1/77), NEF
(1976). Fic

Depicts the deep friendship between Big
Joe, the phone lineman, and Angel, a young
migrant worker. The climax comes when
Angel must choose between moving on with
his family or staying and developing a business with Big Joe.

VALUES IN LITERATURE

THE DILEMMA OF PROGRESS (Filmstrip—Sound). *See* American literature—
Study and teaching

ENCOUNTERS WITH TOMORROW: SCI-
ENCE FICTION AND HUMAN VALUES
(Filmstrip—Sound). *See* Science fiction—
History and criticism

QUESTIONING THE WORK ETHIC (Filmstrip—Sound). *See* American literature—
Study and teaching

SEEKING THE GOOD LIFE (Filmstrip—
Sound). *See* American literature—Study and
teaching

VELIKOVSKY, IMMANUEL

WORLDS IN COLLISION (Motion Picture—
16mm—Sound). *See* Astronomy

VENEREAL DISEASES

HOW YOU GET VD (Filmstrip—Sound).
Sunburst Communications, 1974. 2 color,
sound filmstrips, 57–80 fr.; 2 cassettes or
discs, 10–14 min.; Teacher's Guide. #208
($59). Gr. 9–12. PRV (5/76), PVB (5/75).
616.9

Contents: 1. Gonorrhea. 2. Syphilis.

Uses eye-catching graphics and a fact-filled
story line to explode the many myths that
surround these diseases, and to make a
strong case for periodic medical examinations. Describes symptoms and treatment
for cure.

VD: TWENTIETH CENTURY PLAGUE
(Filmstrip—Sound). Marshfilm, 1975 (The
Human Growth Series). 1 color, sound film-
strip, 50 fr., 1 cassette or disc, 15 min.;
Teacher's Guide. #1122 ($21). Gr. 5–9. CIF
(1975), PRV (4/76). 616.9

Venereal diseases are presented as serious
communicable diseases with which modern
man must cope. The emphasis is on medical,
rather than social, aspect.

VENEREAL DISEASE: THE HIDDEN EPI-
DEMIC (Motion Picture—16mm—Sound).
EBEC, 1972. 16mm color, sound film, 23
min.; also in Spanish. #3197 ($325). Gr. 7–C.
BKL (2/1/73), FHE (1973), NEF (1973). 616.9

A scientifically accurate film is organized
from three perspectives: history of venereal
diseases and attitudes toward them; clinical
analysis of the two major venereal diseases—gonorrhea and syphilis; frank discussions between doctors and patients in a
treatment center. Cases are illustrated and
diagnosed in order to inform and aid in the
prevention of the diseases and to know what
to do if one contracts VD.

VERNE, JULES

AROUND THE WORLD IN EIGHTY DAYS (Phonodisc or Cassette). *See* Fantastic fiction

SCIENCE FICTION (Filmstrip—Sound/Captioned). *See* Science fiction

VINES. *See* Climbing plants

VISION

GLASSES FOR SUSAN (Motion Picture—16mm—Sound). EBEC, 1972. 16mm color, sound film, 13 min. #3166 ($185). Gr. 3–8. BKL (12/15/73). 617.7

Susan has poor vision, but she does not know it. Like many other youngsters who see the world out of focus, she blunders through a series of daily mishaps at school and at play. This film impresses students with the importance of good vision and of wearing glasses, if they are needed, with a positive attitude.

HOW WE SEE (Filmstrip—Captioned). BFA Educational Media, 1971. 4 color, sound filmstrips, captioned. #VL7000 ($32). Gr. 7–9. BKL (4/15/72). 617.7

Contents: 1. Eyes in Nature. 2. Human Vision. 3. What Can You See? 4. How Do We See?

This set describes the mechanics of eyes, particularly the human eye. Viewers are led to investigate what kinds of stimuli the eyes can receive, and how messages are received and interpreted by the eyes.

VOCABULARY

EXPANDING YOUR VOCABULARY (Kit). *See* English language—Usage

MAKING WORDS WORK (Filmstrip—Sound). Coronet Instructional Media, 1974. 6 color, sound filmstrips, 50–58 fr.; 6 cassettes or 3 discs, 11–13 min. #M251 Cassettes ($95), #S251 Discs ($95). Gr. 5–9. BKL (5/1/75), PRV. 372.4

Contents: 1. Using Pointer Words. 2. Using Basket Words. 3. Exploring Word Meanings. 4. Using Fact and Opinion Pointers. 5. Using Figurative Language. 6. Sound and Word Order.

Combining cartoons with photography, this set describes various ways in which one can improve one's writing and speaking. Television sportcasts, conversations, lyrics, and speeches stress communication, rather than grammar, to show language as a tool.

SUPER THINK PROGRAM (Cassette). *See* Sports—Biography

WORDS, MEDIA AND YOU (Filmstrip—Sound). *See* Mass media

VOCATIONAL EDUCATION

METRICS FOR CAREER EDUCATION (Filmstrip—sound). *See* Metric system

VOCATIONAL OPPORTUNITIES IN HIGH SCHOOL (Motion Picture—16mm—Sound). EBEC, 1972. 16mm color, sound film, 14 min. #3167 ($185). Gr. 7–12. FLN (1973). 370.11

This film offers suggestions in the area of vocational education. A sampling of vocational shops and classes is included, as the narrator points out the advantages of early exposure to different trades and fields of employment. Stressed is the idea that a student needs job knowledge and experience to compete in the labor market after graduation.

THE VOCATIONAL TECHNICAL SCHOOL: GETTING ACQUAINTED (Filmstrip—Sound). Current Affairs Films, 1976. 1 color, sound filmstrip, 65 fr.; 1 cassette, 12 min.; Teacher's Guide. #547 ($24). Gr. 7–12. PRV (2/78). 370.11

Examines today's vocational/technical school as an extensive alternative education for students of high school age through adulthood. Includes.: General Organization; Cocurricular Activities; Differences between the Traditional High School and the Vocational/Technical School; Occupational Choices and Curricula; History of Vocational/Technical Schools; and Philosophy and Methods.

VOCATIONAL GUIDANCE

ADVENTURES IN THE WORLD OF WORK, SET ONE (Kit). Visual Education/Random House, 1976. 6 color, sound filmstrips, 92–106 fr.; 6 cassettes or discs, 10–13 min.; 36 Paperback Books (6 titles); 6 Duplicating Masters; 1 Discussion Guide; Teacher's Guide. #06148–9 Cassettes ($108), #06149–7 Discs ($108). Gr. 5–10. BKL (11/1/77), LGB (1976), TEA. 371.425

Contents: 1. Who Puts the Light in the Bulb? 2. Who Puts the Prints on the Page? 3. Who Puts the Ice in the Cream? 4. Who Puts the Blue in the Jeans? 5. Who Puts the Room in the House? 6. Who Puts the Grooves in the Record?

Personable, surprisingly interesting tours of some basic industries about which many people have never thought. Developed in harmony with the U.S. Office of Education career cluster chart.

ADVENTURES IN THE WORLD OF WORK, SET TWO (Kit). Visual Education/Random House, 1976. 6 color, sound filmstrips, 92–106 fr.; 6 cassettes or discs, 10–13 min.; 36 paperback books (6 titles); 6 Duplicating Masters; 1 Discussion Guide; 1 Teacher's Guide. #06151–9 Cassettes ($108), #06150–0 Discs ($108). Gr. 5–10. BKL (11/1/77), LGB (1976), TEA. 371.425

Contents: 1. Who Works for You? 2. Who Puts the Plane in the Air? 3. Who Puts the News on Television? 4. Who Puts the Care in Health Care? 5. Who Puts the Fun in Free Time? 6. Who Keeps America Clean?

A continuation of Set One, visiting some basic, interesting industries and occupations not often considered.

ART CAREERS (Motion Picture—16mm—Sound). *See* Art, American

BEGINNING CONCEPTS: PEOPLE WHO WORK, UNIT ONE (Kit). Scholastic Magazine, 1975. 5 color, sound filmstrips, 56–63 fr.; 5 cassettes or discs, av. 7 min.; 1 Teacher's Guide/Activity Book; 5 Paper Hats; 5 Punch-out Finger puppets. #4218 Cassettes ($79.50), #4230 Discs ($79.50). Gr. 3–8. INS (4/77), LAM (2/77), LGB (1975). 371.425

Contents: 1. Say, Ah! 2. Bake a Batch of Bread. 3. Pick a Pattern, Pick a Patch. 4. Park Ranger. 5. Stitch and Stuff.

This set emphasizes the variety of types of work, how the various kinds of work relate to the life and family of the person who does the work, some of the satisfactions and problems in working, and how the work is important. Unit one shows a pediatrician, a baker, a quilt maker, a naturalist, and a toy maker.

BEYOND HIGH SCHOOL (Filmstrip—Sound). Sunburst Communications, 1977. 2 color, sound filmstrips, 81–91 fr.; 2 cassettes or discs, 13–16 min.; Teacher's Guide. #237 ($59). Gr. 9–12. BKL (11/19/77), FHE (2/19/78), TSS (1/19/78). 371.425

Contents: 1. School after High School. 2. Learning on the Job.

Provides noncollege-bound students with the tolls for making realistic and intelligent choices of career and career training. Provides students with practical strategies for first choosing and then preparing for an occupation. Practical points are offered concerning junior and community colleges, vocational training through the armed services, vocational schools, and apprenticeships. Examines methods for matching capabilities and occupations.

CAREER CHOICE: A LIFELONG PROCESS (Filmstrip—Sound). Guidance Associates, 1977. 2 color, sound filmstrips, av. 85 fr. ea.; 2 cassettes or discs, av. 15 min. ea.; 1 Guide. #9A–103–232 Cassettes ($52.50), #9A–103–224 Discs ($52.50). Gr. 9–12. PRV (1977), PVB (4/77). 371.425

Part I examines factors that determine the relative importance of work for each individual and explores the career phases most individuals experience. Part II is made up of four documentary interviews. Case histories probe aspects of job satisfaction and examine changing career goods.

CAREERS IN HOME ECONOMICS (Filmstrip—Sound). *See* Home economics

CAREERS IN THE FASHION INDUSTRY (Kit). *See* Clothing trade

ECONOMICS (Kit). *See* Economics

GETTING READY UNIT ONE (Motion Picture—16mm—Sound). *See* Occupations

IT'S A MATTER OF PRIDE (Motion Picture—16mm—Sound). *See* Work

THE MOST IMPORTANT THING (Motion Picture—16mm—Sound). WNVT (Northern Virginia Educational TV)/EBEC, 1973 (Whatcha Gonna Do?). 16mm color, sound film, 15 min. #3354 ($200). Gr. 5–9. BKL (1975), LFR (1975). 371.425

A comparison is made of the reasons why two different workers—a youth leader and a mechanic—chose their occupations. The point is not to draw comparisons about the life styles of youth leaders as opposed to those of mechanics, but to show that a person's life-style (which includes his or her work) is a reflection of values.

OPPORTUNITY (Kit). Scholastic Book Services, 1976. 8 color, sound filmstrips; 8 cassettes or discs, 10–20 min.; 30 Student Logbooks; Teacher's Guide. #03690 Cassettes ($169.50), #03689 Discs ($169.50). Gr. 9–12. BKL (12/15/77), PRV (3/77). 371.425

Students get actual practice in employment activities and procedures, beginning with a job interview. Later, they explore 15 careers including a customs agent, a photographer, a dry cleaner, and others. More on-the-job knowledge is provided regarding particular jobs.

STARTING TO THINK ABOUT WORK (Filmstrip—Sound). Insight Productions/Denoyer-Geppert, 1977. 2 color, sound filmstrips, 70–73 fr. ea.; 2 cassettes, 9–10 min. ea.; Teacher's Guide. ($44). Gr. 7–12. BKL (3/1/78), PRV (5/78). 371.425

An introduction to an overview of working with a brief history of the industrial revolution, and a discussion of the ways in which attitudes are formed at home and at school. Interviews with four individuals from diverse educational backgrounds present the advantages, disadvantages, and benefits of their careers. The conclusion that all jobs have an element of tedium is presented. It is the individual's responsibility to search for the job that presents the greatest satisfaction for him or her.

TRY OUT: UNIT ONE (Motion Picture—16mm—Sound). WNVT (Northern Virginia Education TV)/EBEC, 1973 (Whatcha Gonna Do?). 16mm color, sound film, 15 min. #3353 ($200). Gr. 5–A. BKL (5/15/75), LFR (12/74), PRV (10/76). 371.425

This film presents children in six different activities that are career related: nature studies, television broadcasting, building, tutoring, raising livestock and a classroom manufacturing simulation. Throughout the

VOCATIONAL GUIDANCE (cont.)

film, children comment on their work, what they are learning about themselves, and what they might want to do. The film motivates students to reflect on how their own experiences can guide them in career choices and to take an active interest in career opportunities around them.

UNIONS AND YOU (Kit). *See* Labor unions

VOCATIONAL SKILLS FOR TOMORROW (Filmstrip—Sound). Coronet Instructional Media, 1977. 6 color, sound filmstrips, 34–54 fr.; 3 discs or 6 cassettes, 5–11 min.; Teacher's Guide. #340 ($95). Gr. 4–12. PRV (4/78). 371.425
Contents: 1. A New World of Jobs. 2. Business and Clerical Vocations. 3. Commercial and Graphic Arts. 4. Communication and Information Trades. 5. Construction and Manufacturing Trades. 6. Health and Medical Vocations.
Surveys many fields of vocational opportunity, with emphasis on the need to train for the job. Shows how rapid advances in technology creates new jobs.

WHAT DO YOU THINK? UNIT ONE (Motion Picture—16mm—Sound). *See* Occupations

WHAT'S THE LIMIT? UNIT TWO (Motion Picture—16mm—Sound). *See* Occupations

WHO WORKS FOR YOU? (Kit). Visual Education/Random House, 1976. 6 color, sound filmstrips, av. 106–126 fr.; 6 cassettes or discs, av. 11–13 min.; 36 Softcover Books (6 Titles), 6 Duplicating Masters, Teacher's Guide. #04754–0 Cassettes ($158.16), #04781–8 Discs ($158.16). Gr. 5–8. BKL (11/15/77). 371.425
This kit covers a wide variety of service occupations. The first module explores occupations in city, state, and national government. The second covers behind-the-scene work in television news. Health service professions are explored as well as careers with the airlines. Expanding leisure industries and environmental management and protection are also examined.

WORK/WORKING/WORKER (Filmstrip—Sound). Guidance Associates, 1976. 3 color, sound filmstrips, 69–100 fr. ea.; 3 cassettes or discs, 10–19 min. ea.; Teacher's Guide. #104–099 Cassettes ($70), #104–081 Discs ($70). Gr. 9–A. PRV (2/78). 371.425
Contents: 1. The Business of Staying Alive. 2. What Do You Want to Be When You Grow Up? 3. Workers Talk about Work.
Examines work and society in terms of historical background and present social realities. Reviews work attitudes, factors of status, materialism, school experience, the Puritan work ethic, and the myth of equal opportunity. A cross section of American workers talks about reasons for working, job rewards, conflicts, and choices.

WORKING (Filmstrip—Sound). *See* Work

VOLCANOES

FIRE IN THE SEA (Motion Picture—16mm—Sound). EBEC, 1974. 16mm color, sound film, 10 min. #3191 ($150). Gr. 4–12. BKL (10/1/73), CGE (10/73), PRV (10/74). 551.4
When the Mt. Kilauea volcano erupted, molten lava swept through the Hawaiian countryside and into the sea. In this dangerous environment, photographers captured dramatic and rare close-up views of the awesome lava flow. The absence of narration will inspire individual interpretation of the film.

FIRE MOUNTAIN (Motion Picture—16mm—Sound). EBEC, 1972. 16mm color, sound film, 9 min. #2991 ($115). Gr. 4–12. CGE (1973). 551.21
Rare, close-up views of the great eruption of Mt. Kilauea in Hawaii. Color, music and, natural sound enhance this narrationless film allowing the viewer to "experience" the eruption.

VOLCANOES: EXPLORING THE REST-LESS EARTH (Motion Picture—16mm—Sound). EBEC, 1973. 16mm color, sound film, 18 min. #3208 ($255). Gr. 5–12. BKL (2/1/74), LFR (2/74), LGB (1974). 551.4
On-location photography and animated drawings illustrate volcanic phenomena in Hawaii, Mexico, Italy, and Iceland. The film distinguishes shield, cinder cone, and strato-volcano by showing how each type of volcano is formed and how each erupts.

VOLLEYBALL

SPORTS IN ACTION (Filmstrip—Sound). EBEC, 1976, 3 color, sound filmstrips, av. 77 fr.; 3 cassettes, 7 min. ea. #6942 ($43.50). Gr. 6–12. MDU (4/77). 796.32
Contents: 1. Before You Play. 2. How to Play. 3. Play to Win.
Discusses everything players need to know about volleyball.

SPORTS IN ACTION: VOLLEYBALL (Filmstrip—Sound). EBEC, 1976. 3 color, sound filmstrips, av. 77 fr. ea.; 3 discs or cassettes, 7 min. ea.; Teacher's Guide. #6942K Cassettes ($43.50), #6942 Discs ($43.50). Gr. 7–12. LFR (9/10/77), MDU (4/77), PRV (2/78). 796.32
Contents: 1. Before You Play. 2. How to Play. 3. Play to Win.
Volleyball—the ideal team sport for both sexes, all ages, beginners or serious athletes. Filmstrips progress from basic rules and beginning skills to advanced strategies.

VOLTAIRE

VOLTAIRE PRESENTS CANDIDE: AN IN-
TRODUCTION TO THE AGE OF EN-
LIGHTENMENT (Motion Picture—
16mm—Sound). *See* French fiction

WAR—FICTION

FLOWER STORM (Motion Picture—16mm—
Sound). *See* Iran—Fiction

WAR AND CIVILIZATION

ESSAY ON WAR (Motion Picture—16mm—
Sound). Essay Productions/EBEC, 1971.
16mm color, sound film, 23 min. #3109
($290). Gr. 7–12. CGE (1972), EFL (5/19/73),
PRV (2/19/73). 301.6
This film explores the puzzle of war: Why
we wage it, how we justify it, how we are
affected as nations and as individuals. Flashbacks provide perspective and stimulate
thought.

WARREN, EARL

A CONVERSATION WITH EARL WAR-
REN (Videocassette). WGBH Educational
Foundation, 1976. 1 videocassette, 60 min.
($250). Gr. 11–A. BKL (9/1/78). 921
In this interview, former Chief Justice Earl
Warren discusses the Supreme Court and
how it influences the justices and their decisions. The segregation cases and cases from
the McCarthy indictments are described.
The opinions of Warren had a tremendous
influence on the social history of this century
and are recorded on this tape.

WASHINGTON, D.C.

THE IDEOLOGICAL CAPITALS OF THE
WORLD (Filmstrip—Sound). *See* Cities and
towns

WASHINGTON, GEORGE

GEORGE WASHINGTON (Cassette). Ivan
Berg Associates/Jeffrey Norton Publishers,
1977 (History Makers). 1 cassette, approx.
58 min. #41026 ($11.95). Gr. 7–A. BKL (4/
15/78), PRV (5/19/78). 921
The biography of George Washington is presented by a narrator with interludes of music
and dialogue by individual character voices.
Key events in his life are dramatized with
enough information to be the basis for history or biography reports.

GEORGE WASHINGTON—THE COUR-
AGE THAT MADE A NATION (Motion
Picture—16mm—Sound). Handel Film Corporation, 1968 (Americana Series #4). 16mm
color, sound film, 30 min.; Film Guide.
($360), Rental 7 day ($36). Gr. 4–A. LFR (9/
68). 921

Portrays the powerful personality of George
Washington and some of the events that led
to his decisive role in history. Describes his
education, social life, military experience,
and political activities against the background of the authentic locale.

THE PICTORIAL LIFE-STORY OF
GEORGE WASHINGTON (Kit). Davco
Publishers, 1976 (Life Stories of Great Presidents). 4 color, sound filmstrips, 68–74 fr.; 4
cassettes, av. 13 min. ea.; 1 Book, 1 Teacher's Guide. ($89). Gr. 5–9. BKL (12/15/77).
921
Well organized, accurate presentation of the
myriad aspects of George Washington's life
and career.

PRESIDENTS AND PRECEDENTS (Kit).
See Presidents—U.S.

WATER

EARTH AND UNIVERSE SERIES, SET 2
(Filmstrip—Sound). *See* Geology

WATER: WHY IT IS WHAT IT IS (Motion
Picture—16mm—Sound). Moody Institute
of Science, 1975. 16mm color, sound film, 11
min. ($170). Gr. 7–12. SCT (11/76). 551.4
A survey of water properties and their relationship to molecular structure. Physical
properties (rather than chemical ones) are
stressed.

WATER ANIMALS. *See* Fresh-water biology;
Marine animals

WATER BIRDS

AUDUBON'S SHORE BIRDS (Motion Picture—16mm—Sound). Fenwick Productions, 1977 (America's Wildlife Heritage).
16mm color, sound film, 18^1/$_2$ min. ($225).
Gr. 5–12. LGB (1977). 598.2
Recreates what Audubon first saw and painted, using his own observations to describe
ornithological adaptations and behavior.
Filmed by the late Roy Wilcox.

WATER SPORTS

SAFE IN THE WATER (Motion Picture—
16mm—Sound). *See* Safety education

WATER SUPPLY

THE THIRSTY CITY (Filmstrip—Sound).
Multi-Media Productions, 1976. 1 color,
sound filmstrip, 87 fr.; 1 cassette or disc, 13
min.; Teacher's Manual. #7182 C or R
($12.95). Gr. 9–12. PVB (4/78). 352.911
A look at the problem of supplying fresh water to our cities. Los Angeles is used as a
case study in surveying not only the quest
for water but also a range of problems the
city imposes on the environment.

WATSON, JANE WERNER

HEROES OF THE ILIAD (Cassette). *See* Mythology, classical

WAVES

STANDING WAVES AND THE PRINCIPLES OF SUPERPOSITION (Motion Picture—16mm—Sound). EBEC, 1971. 16mm color, sound film, 11 min. #3035 ($150). Gr. 9–C. AFF (1972), MER (1973), SCT (1974). 531.113

Explains the formation and characteristics of standing waves, a physics principle basic to understanding the behavior of matter. Computer animation and simple experiments demonstrate how standing waves are produced by the superposition of two identical wave patterns traveling in opposite directions. The principle is then extended to an explanation of atomic structure.

WEALTH

THE INEQUALITY OF WEALTH IN AMERICA (Filmstrip—Sound). *See* U.S.—Economic conditions

WEATHER

MOUNTAIN STORM (Film Loop—8mm—Silent). Walt Disney Educational Media, 1966. 8mm color, silent film loop, approx. 4 min. #62–5023L ($30). Gr. K–12. MDU (1974). 551.6

The spectacle of a mountain storm is viewed from start to finish. A bolt of lightning strikes a huge tree, causing it to topple.

WEATHER (Kit). *See* Meteorology

WEATHER FORECASTING

WEATHER FORECASTING (Motion Picture—16mm—Sound). EBEC, 1975. 16mm color, sound film, 22 min. #3372 ($290). Gr. 4–10. CPR (3/76), LFR (9/10/75), NST (2/77). 551.6

This film traces the development of meteorology to its present state a sophisticated and increasingly valuable science. Focusing on the work of a meteorologist in an actual weather station, the film shows how data from satellites, reports from weather stations all over the world, and trained around-the-clock observations are combined to give the most accurate possible prediction of a region's weather.

WEAVING

LOOMLESS WEAVING (Filmstrip—Sound). Warner Educational Productions, 1974. 2 color, sound filmstrips, 73–74 fr.; 1 cassette, 15–17½ min.; Teaching Guide. #530 ($47.50). Gr. 7–A. PRV (9/75). 746

Introduces the skill of loomless weaving and teaches five techniques of weaving without the use of looms. It also introduces objects woven into designs. Texture in design and suggested materials are shown.

WEBSTER, MARGARET

HIS INFINITE VARIETY: A SHAKESPEAREAN ANTHOLOGY (Cassette). *See* English drama

NO COWARD SOUL: A PORTRAIT OF THE BRONTES (Cassette). *See* English drama

THE SEVEN AGES OF GEORGE BERNARD SHAW (Cassette). *See* English drama

WEEDS

KUDZU (Motion Picture—16mm—Sound). Pyramid Films, 1976. 16mm color, sound film, 16 min.; ³/₄-in. videocassette also available. ($250). Gr. 7–A. AFF (1978), BKL (11/1/77). 632

An off-beat, witty documentary about Kudzu—a vine originally imported to the South as a means of erosion control, which has become a botanical problem of menacing proportions. James Dickey, Jimmy Carter, and a series of real-life characters comment on the role Kudzu plays in developing Southern traditions.

A WEEK ON THE CONCORD AND MERRIMACK RIVERS

THOREAU ON THE RIVER: PERSPECTIVE ON CHANGE (Filmstrip—Sound). *See* Ecology

WELLES, ORSON

IS IT ALWAYS RIGHT TO BE RIGHT? (Motion Picture—16mm—Sound). *See* Human relations

WELLS, H. G.

SCIENCE FICTION (Motion Picture—16mm—Sound). *See* Science fiction

THE TIME MACHINE (Filmstrip—Sound). *See* Science fiction

THE WEST (U.S.)

FOCUS ON AMERICA, THE NEAR WEST REGION (Filmstrip—Sound). *See* U.S.—Economic conditions

THE WEST (U.S.)—HISTORY

GHOST TOWNS OF THE WESTWARD MARCH (Motion Picture—16mm—Sound).

Alfred Higgins Productions, 1973. 16mm color, sound film, 18 min. ($255). Gr. 5–A. BKL (11/1/73), CFF (1973), LFR (10/73). 978

This film traces our westward expansion against a background of ghost towns and mines once rich with priceless ore.

WEST, BENJAMIN

AMERICAN CIVILIZATION: 1783–1840 (Filmstrip—Sound). *See* Art, American

WEST GERMANY. *See* Germany

WHALES

WHALES: CAN THEY BE SAVED? (Motion Picture—16mm—Sound). Avatar Learning/ EBEC, 1976. 16mm color, sound film, 24 min. #3506 ($320). Gr. 5–12. CGE (1977), LFR (1977), PRV (1977). 599.5

Examines the behavior of many types of whales and explains how whales can be trained to perform. Film shows how modern technology has brought whales close to extinction and explains the steps people must take to save them.

WHALES SURFACING (Film Loop—8mm— Silent). Walt Disney Educational Media, 1966. 8mm color, silent film loop, approx. 4 min. #62–5462L ($30). Gr. K–12. MDU (1974). 599.5

Whales are shown as they surface and dive. These huge mammals are shown spouting not water but moist air from their lungs.

WHALING

THERE SHE BLOWS (Filmstrip—Sound). *See* Seafaring life—History

WHARTON, EDITH

WOMEN WRITERS: VOICES OF DISSENT (Filmstrip—Sound). *See* Women authors

WHEELOCK, WARREN

HISPANIC HEROES OF THE U.S.A. (Kit). *See* Biography—Collections

WHITE, JOSH

FOLK SONGS IN AMERICAN HISTORY, SET ONE: 1700–1864 (Filmstrip—Sound). *See* Folk songs—U.S.

FOLK SONGS IN AMERICAN HISTORY, SET TWO: 1865–1967 (Filmstrip—Sound). *See* Folk songs—U.S.

WHITMAN, WALT

NINETEENTH CENTURY POETS (Filmstrip—Sound). *See* American literature— History and criticism

WALT WHITMAN (Filmstrip—Sound). Coronet Instructional Media, 1978. 4 color, sound filmstrips, 68–82 fr.; 4 cassettes, 14–17 min.; 1 Guide. #MO299 ($75). Gr. 10–12. BKL (10/78), IDF (1978). 921

Contents: 1. Country Boy and City Journalist. 2. Prophet of American Democracy. 3. Spokesman for the Universal Self. 4. A Poet of the Civil War.

The set traces Whitman's years as a farm boy, office boy, typesetter, and journalist as well as his Civil War experiences. His life experiences are interwoven with his poetry and other writings, giving special insight into his concepts of nature, democracy and self-identity and his dismay at the horrors of war.

WALT WHITMAN: POET FOR A NEW AGE (Motion Picture—16mm—Sound). EBEC, 1972. 16mm color, sound film, 29 min. #47783 ($390). Gr. 9–A. CGE (1972), EFL (4/74), M&M (1977). 921

This is a study of poet Walt Whitman—his beliefs and his conflicts with contemporaries. The film reveals the "cosmic consciousness" of Whitman: his strong belief in democracy; the oneness and sacredness of all living things, and the mystical truths of life and death; his distaste for war; and his concern for the primacy of personality and love. Historical settings and special cinematographic effects are used.

WILD ANIMALS. *See* Animals; also names of individual animals, e.g., Lions

WILD LIFE. *See* Wildlife conservation

WILDE, OSCAR

THE HAPPY PRINCE AND OTHER OSCAR WILDE FAIRY TALES (Phonodisc or Cassette). *See* Fairy tales

WILDEBEEST. *See* Gnu

WILDER, LAURA INGALLS

LAURA: LITTLE HOUSE, BIG PRAIRIE (Filmstrip—Sound). Perfection Form Company, 1976. 1 color, sound filmstrip, 140 fr.; 1 cassette or disc, 18 min.; 1 Guide. #KH95267 Cassette ($22.95), #KH95266 Disc ($21.45). Gr. 3–A. BKL (6/1/77). 921

Biographical facts are combined with personal impressions in this production that intertwines the plots of the books with their author's life.

WILDER, THORNTON

ACTING FOR FILM (LONG CHRISTMAS DINNER) (Motion Picture—16mm— Sound). *See* Acting

THE LONG CHRISTMAS DINNER (Motion Picture—16mm—Sound). *See* American drama

WILDERNESS SURVIVAL

SAFE IN NATURE (Motion Picture—16mm—Sound). FilmFair Communications, 1975. 16mm color, sound film, 20 min. ($265). Gr. 3–12. BKL (12/75), (5/75). 796.5

Dramatizes basic techniques of survival when lost in wilderness areas—in the woods and in the desert. Emphasized are three basics: stay put, make shelter, and give distress signals, with detailed coverage of each.

WILDERNESS SURVIVAL—FICTION

THE GRIZZLY (Kit). *See* Bears—Fiction

WILDLIFE CONSERVATION

BUFFALO: AN ECOLOGICAL SUCCESS STORY (Motion Picture—16mm—Sound). *See* Bison

ECOLOGY AND THE ROLE OF MAN (Filmstrip—Sound). EBEC, 1975. 6 color, sound filmstrips, av. 58 fr. ea.; 6 discs or cassettes, 9 min. ea.; Teacher's Guide. Discs #6913 ($86.95), Cassettes #6913K ($86.95). Gr. 5–8. PRV (11/76). ESL (1977). 639

Contents: 1. The Passenger Pigeon: Caring Too Late. 2. The Buffalo: Caring in Time. 3. The Bald Eagle: An Endangered Symbol. 4. The Kaibab Deer: A Lesson in Management. 5. The American Elm: A Fatal Journey. 6. The Redwood: Why Save a Tree?

Six animated filmstrips make a powerful case for conservation by depicting the life histories of four endangered species—plus one that has been saved and a sixth for which it is too late.

ENDANGERED ANIMALS: WILL THEY SURVIVE? (Motion Picture—16mm—Sound). *See* Rare animals

ENDANGERED SPECIES: BIRDS (Study Print). *See* Rare birds

ENDANGERED SPECIES: MAMMALS (Study Print). *See* Rare animals

ENDANGERED SPECIES: REPTILES AND AMPHIBIANS (Study Print). *See* Rare reptiles

WILLIAMS, MARGERY

THE VELVETEEN RABBIT (Filmstrip—Sound). *See* Fantasy—Fiction

WILLIAMS, TENNESSEE

TENNESSEE WILLIAMS: THEATER IN PROCESS (Motion Picture—16mm—Sound). *See* Theater—Production and direction

WILSON, WOODROW

WOODROW WILSON AND THE SEARCH FOR PEACE (Filmstrip—Sound). Multi-Media Productions, 1976. 1 color, sound filmstrip, 58 fr.; 1 cassette, 10 min.; Teacher's Guide. #7199C ($14.95). Gr. 8–12. PRV (5/78). 921

President Wilson had definite proposals for a lasting peace at the end of World War I, including the establishment of an international organization, the League of Nations. The vengeful European leaders emasculated his proposals and the U.S. Senate refused to approve this nation's membership. Finally, Wilson took his case to the people and suffered a breakdown, leaving the nation with a physically incapacitated chief executive.

WINTER

WINTER IN THE FOREST (Filmstrip—Sound). Aids, 1974. 4 color, sound filmstrips, av. 50 fr. ea.; 4 cassettes or 4 discs, av. 12 min. ea.; Teacher's Guide. ($72). Gr. 4–10. BKL (11/74). 525

Contents: 1. Enjoying the Woodlands. 2. Looking Closely at Forest Trees. 3. A School and a Sawmill. 4. Sweets from the Forests.

The inquiry method is used to take the viewer into the forest in the winter to increase awareness of problems, adventure, concepts, and relationships in nature.

WINTER SPORTS

BOBSLEDDING: DOWN THE CHUTE (Kit). Troll Associates, 1976 (Troll Reading Program). 1 color, sound filmstrip, 34 fr.; 1 cassette, 14 min.; 10 Softcover Books; 1 Library Edition; 4 Duplicating Masters; 1 Teacher's Guide. ($48). Gr. 4–9. BKL (2/15/77). 796.95

This filmstrip and cassette, based on the high-interest/low-reading level of Ed Radlauer's material, exactly duplicate the text of the book.

WIT AND HUMOR

THE CONCERT (Motion Picture—16mm—Sound). *See* Pantomimes

THE DOONESBURY SPECIAL (Motion Picture—16mm—Sound). *See* Comic books, strips etc.

GERTRUDE, THE GOVERNESS, AND OTHER WORKS (Phonodisc or Cassette). *See* Canadian fiction

HARDWARE WARS (Motion Picture—16mm—Sound). *See* Satire, American

HUCKLEBERRY FINN (Cassette). *See* U.S.—Social life and customs—Fiction

HUCKLEBERRY FINN (Phonodisc or Cassette). *See* U.S.—Social life and customs—Fiction

THE PETERKIN PAPERS (Phonodisc or Cassette). *See* Family life—Fiction

WOODY ALLEN (Cassette). *See* Allen, Woody

WOLVES

WOLF FAMILY (Film Loop—8mm—Silent). Walt Disney Educational Media, 1966. 8mm color, silent film loop, approx. 4 min. #62–5356L ($30). Gr. K–12. MDU (1974). 599.74
A wolf family is studied as it hunts for food in the Arctic.

THE WOLVES OF ISLE ROYALE (Kit). Classroom Complements/EBEC, 1976 (Animal Life Stories). 1 color, sound filmstrip; 1 cassette; 5 Identical Storybooks; Teacher's Guide. #6969K ($27.95). Gr. K–6. TEA (11/77). 599.74
Young readers can learn how animals live, move, build homes, and fight for food and survival. A scientifically accurate story describes the life of wolves in their natural habitat. A 56-page storybook reproduces all of the narration and pictures from the filmstrip.

WOLVES—FICTION

JULIE OF THE WOLVES (Phonodisc or Cassette). *See* Eskimos—Fiction

WOMEN

GROWING UP FEMALE (Videocassette). Videotape Network/Jeffrey Norton Publishers, 1977. Videocassette, ³/₄-in. U-Matic, ¹/₂-in. Betamax, or ¹/₂-in. EIAJ (b/w); 50 min. #71089 ($375). Gr. 10–A. BKL (7/1/79). 301.41
Six individual portraits illustrate various stages of growing into womanhood. The program shows how women are conditioned to assume their roles in society.

WOMEN AUTHORS

WOMEN WRITERS: VOICES OF DISSENT (Filmstrip—Sound). Educational Enrichment Materials, 1975. 3 color, sound filmstrips, 60–70 fr.; 3 cassettes or discs, 14–16 min.; Teacher's Guide. #41087 C or R ($60). Gr. 9–12. BKL (9/1/76), PRV (12/77). 920
Contents: 1. Edith Wharton: The Decadent New York Society. 2. Ellen Glasgow: The Southern Myth. 3. Willa Cather: The Pioneer West.
Utilizes historical prints and dramatized quotations to paint a portrait of three talented women writers. Discusses the life and work of each against a backdrop of a male-dominated literary world.

WOMEN—BIOGRAPHY

BEAH RICHARDS: A BLACK WOMAN SPEAKS (Videocassette). *See* Richards, Beah

CHILDHOOD OF FAMOUS WOMEN, VOLUME THREE (Cassette). H. Wilson, 1975. 2 cassettes; Student Response Sheets; Teacher's Guide. #S31-CT ($17.50). Gr. 4–8. PRV (11/70). 920
A collection of dramatized biographies featuring Edna St. Vincent Millay, Marie Curie, Louisa May Alcott, and Anne Sullivan. An attempt is made to describe the motivation behind the later successes of these women.

WOMEN BEHIND THE BRIGHT LIGHTS (Kit). *See* Entertainers

WOMEN WHO WIN, SET 1 (Kit). *See* Athletes

WOMEN WHO WIN, SET 2 (Kit). *See* Athletes

WOMEN WHO WIN, SET 3 (Kit). *See* Athletes

WOMEN WHO WIN, SET 4 (Kit). *See* Athletes

WOMEN—CANADA

GREAT GRANDMOTHER (Motion Picture—16mm—Sound). *See* Canada—History

WOMEN—EMPLOYMENT

NON-TRADITIONAL CAREERS FOR WOMEN (Filmstrip—Sound). Pathescope Educational Media, 1974 (Careers In). 2 color, sound filmstrips, 84–89 fr. ea.; 2 cassettes, 12¹/₂–13 min. ea.; Teacher's Manual. #736 ($50). Gr. 9–12. BKL (3/75), NVG (1975). 331.4
Part 1 describes different types of jobs now available for women. Part 2 uses interviews to determine the necessary training and opportunities. The Teacher's Manual deals with background information, topics for discussion, activities, and scripts.

WHY MOTHERS WORK (Motion Picture—16mm—Sound). EBEC, 1976. 16mm color, sound film, 19 min. #3485 ($255). Gr. K–8. LFR (5/6/77), PRV (5/77). 331.4
Two working mothers tell why they hold jobs and reveal their hopes for the future as viewers share a long, busy day in the life of each woman.

WOMEN AT WORK: CHANGE, CHOICE, CHALLENGE (Motion Picture—16mm—Sound). EBEC, 1977. 16mm color, sound film, 19 min. #3520 ($270). Gr. 9–A. AAS (1978), NEF (1978), PRV (1978). 331.4
A dialogue with seven women, expressed in counterpoint with actual on-the-job scenes, reveals their attitudes about work, training,

WOMEN—EMPLOYMENT (cont.)

and their personal roles. All have strong reasons for their career choices and different views of their work-family-community relationships.

WOMEN—U.S.

THE AMERICAN WOMAN: A SOCIAL CHRONICLE (Filmstrip—Sound). Educational Enrichment Materials, 1976. 6 color, sound filmstrips, 78–93 fr.; 6 cassettes or discs, 12–15 min.; Teacher's Guide, #51022C Cassettes ($108), #51002R Discs ($108). Gr. 9–A. PRV (1/78). 301.41
Contents: 1. Puritans & Patriots (1628–1776). 2. Mill "Girls," Intellectuals & the Southern Myth (1776–1860). 3. Pioneer Women & Belles of the Wild West (1876–1886). 4. The Suffragist, Working Women, & Flappers (1890–1929). 5. Breadlines, Assembly lines, and Togetherness (1929–1960). 6. Liberation Now! (1960–   ).
A survey of the social and economic history of women in the United States: who they were, how they lived, and what they accomplished.

AMERICAN WOMEN (Kit). Proof Press, 1978. 12 duotone prints, 10 in. × 13 in. on poster-weight cardboard; 6 Booklets; 1 Chronological Chart; Teacher's Guide. ($35). Gr. 9–A. PRV (1/79). 301.41
An introduction to the history of American women. The booklets are divided into chronological periods, but each covers the major economic, political, and social influences on women during each particular period.

ANIMATED WOMEN (Motion Picture—16mm—Sound). Texture Films, 1978. 16mm color, sound film, 15 min. ($235). Gr. 11–A. FLN (1978). 301.41
Contents: 1. The Ballad of Lucy Jordan. 2. Later That Night. 3. Brews & Potions. 4. Made for Each Other. 5. A La Votre
Animation is the sharp tool of the satirist in these provocative comments on today's female of the species. Women's aspirations, relationships, and power (real and imaginary) are explored in this five-part film, which is alternately humorous, tragic, witty, and sardonic.

ANYTHING YOU WANT TO BE (Motion Picture—16mm—Sound). See Sex roles

MALE AND FEMALE ROLES (Filmstrip—Sound). See Sex roles

WOMEN: AN AMERICAN HISTORY (Filmstrip—Sound). EBEC, 1976. 6 color, sound filmstrips, av. 100 fr. ea.; 6 cassettes or discs, 17 min. ea.; Teacher's Guide. #6916 Discs ($86.95), #6916K Cassettes ($86.95). Gr. 5–A. LGB (75/76), LNG (12/76), MDU (1/77). 301.41

Contents: 1. Women of the New World. 2. The Mill Girl and the Lady. 3. The Fight for Equality. 4. A Combination of Work and Hope. 5. Beyond the Vote. 6. The Modern Women's Movement.
Six filmstrips trace 350 years in the history of the American woman from her paradoxical colonial role as a valued helpmate—who was legally a "nonperson"—to the many-faceted woman of the 1970s. In between, viewers see how it all came about. Throughout the series, outstanding women of history and of today's complex world come dramatically to life.

WOMEN IN AMERICAN HISTORY (Filmstrip—Sound). Activity Records/Educational Activities, 1974. 6 color, sound filmstrips, 50–65 fr.; 3 discs or cassettes, 12–15 min.; Teacher's Guide. #FSR 460 Discs ($61.95), #FSC 460 Cassettes ($64.95). Gr. 5–12. PRV (10/75), PVB (4/76). 301.41
Contents: 1. The Colonies. 2. After the Revolution. 3. Slavery and Suffrage. 4. Reformers. 5. The Artists. 6. Crisis of Identity.
These strips present an accurate history of women's experiences in America. Women's struggle for justice and equality and their contributions to American life are presented through memorable vignettes from the lives of outstanding women and brief excerpts from their speeches and writings. Discrimination in law, politics, religion, education, work, etc., is vividly revealed.

WOMEN IN THE AMERICAN REVOLUTION (Filmstrip—Sound). See U.S.—History—1775–1783—Revolution

WONDER, STEVIE

MEN BEHIND THE BRIGHT LIGHTS (Kit). See Musicians

WOODWORK

CIRCULAR SAW, SET FOUR (Film Loop—8mm—Silent). Raybar Technical Films/McGraw-Hill Films, 1968 (Woodworking Series). 14 silent, 8mm technicolor loops, 4 min. #698746–9 ($309). Gr. 7–A. AFF (1970). 694
Contents: 1. Parts and Functions of the Circular Saw. 2. Changing Saw Blades. 3. Ripping a Board. 4. Crosscutting Long Uniform Lengths. 5. Crosscutting Short Uniform Pieces. 6. Squaring a Board. 7. Chamfering. 8. Mitering. 9. Sawing Rabbets. 10. Sawing with a Pattern. 11. Tapering with a Tapering Jig. 12. Sawing Inside Cut-Outs. 13. Setting Up a Dado Head. 14. Cutting with a Dado Head.
This set covers the circular saw, acquainting students with the characteristics of the material in use and demonstrating the proper use of a particular tool. Emphasis is placed on safety and proper care and upkeep of tools.

HAND TOOL OPERATIONS, SET 1 (Film Loop—8mm—Silent). Raybar Technical Films/McGraw-Hill Films, 1968 (Woodworking Series). 14 silent, 8mm technicolor loops, 4 min. #698701–9 ($309). Gr. 7–A. AFF (1970). 694

Contents: 1. Wood: Terminology and Measurement. 2. The Combination Square and Its Uses. 3. Using the Marker Gauge. 4. Laying Out Corner Radii. 5. Driving and Setting Nails. 6. Drawing Nails. 7. Methods of Driving Screws. 8. Drilling: Anchor Holes, Clearance Holes, and Countersinking. 9. Uses of Handscrews. 10. Smoothing with Abrasive Paper. 11. Using a Crosscut Saw. 12. The Back Saw. 13. Using a Coping Saw. 14. Using a Coping Saw to Cut Thin Material.

This set covers basic hand tools, acquainting students with the characteristics of the material in use and demonstrating the proper use of a particular tool. Emphasis is placed on safety and proper care and upkeep of tools.

HAND TOOL OPERATIONS, SET 2 (Film Loop—8mm—Silent). Raybar Technical Films/McGraw-Hill Films, 1968 (Woodworking Series). 14 silent, 8mm technicolor loops, 4 min. #6998716–7 ($309). Gr. 7–A. AFF (1970). 694

Contents: 1. The Compass Saw. 2. Boring with an Auger Bit. 3. Drilling. 4. Boring with the Forstner Bit. 5. Setting a Plane. 6. Planing a Chamfer. 7. Methods of Planing End Grain. 8. Squaring Up Stock. 9. Filing Wood. 10. Scraping Wood. 11. Using Spoke Shaves. 12. Using a Gouge. 13. Chiseling to a Finished Line. 14. Gluing Edge to Edge.

This set covers basic hand tools, acquainting students with the characteristics of the material in use and demonstrating the proper use of a particular tool. Emphasis is placed on safety and proper care and upkeep of tools.

HAND TOOL OPERATIONS, SET 3 (Film Loop—8mm—Silent). Raybar Technical Films/McGraw-Hill Films, 1968 (Woodworking Series). 14 silent, 8mm technicolor loops, 4 min. #698731–0 ($309). Gr. 7–A. AFF (1970). 694

Contents: 1. Preparing a Dowel. 2. Using a Doweling Jig. 3. Gluing a Dowel Joint. 4. Mortising with an Auger Bit. 5. Mitering a Picture Frame. 6. Gluing and Fastening a Miter Joint with a Spline. 7. Cutting a Dado. 8. Laying Out and Cutting a Gain. 9. Cutting a Tenon. 10. Sharpening a Hand Scraper. 11. Laying Out and Cutting an End Lap Joint. 12. Grinding a Chisel. 13. Whetting a Chisel. 14. Hardware Commonly Used in Woodworking.

This set covers basic hand tools, acquainting students with the characteristics of the material in use and demonstrating the proper use of a particular tool. Emphasis is placed on safety and proper care and upkeep of tools.

LATHE (Film Loop—8mm—Silent). Raybar Technical Films/McGraw-Hill Films, 1968 (Woodworking Series). 14 silent, 8mm technicolor loops, 4 min. #698761–2 ($309). Gr. 7–A. AFF (1970). 694

Contents: 1. Introduction to the Woodworking Lathe. 2. Centering and Mounting for Spindle Turning. 3. Roughing Stock between Centers. 4. Turning Work to a Finished Diameter. 5. Facing and Parting between Centers. 6. Cutting a Taper. 7. Cutting a Shoulder. 8. Cutting a Cove. 9. Turning Bees and Beads. 10. Turning Square Sections and Rounding Corners. 11. Preparing Stock for Faceplate Turning. 12. Facing and Squaring an Edge on a Faceplate. 13. Convex and Concave Turning on a Faceplate. 14. Sanding on a Faceplate.

This set covers the lathe, acquainting students with the characteristics of the material in use and demonstrating the proper use of a particular tool. Emphasis is placed on safety and proper care and upkeep of tools.

WOOLF, VIRGINIA

THE LETTERS OF VIRGINIA WOOLF (Cassette). Cinema Sound/Jeffrey Norton Publishers, 1976 (Avid Reader). 1 cassette, approx. 55 min. #40219 ($11.95). Gr. 7–A. BKL (7/15/77). 921

Heywood Hale Broun's discussion with Nigel Nicholson, author of *The Letters of Virginia Woolf* (Harcourt Brace, 1975), reveals not an eccentric woman, but one with purpose and robust intensity of life.

WORDS. *See* Vocabulary

WORK

IT'S A MATTER OF PRIDE (Motion Picture—16mm—Sound). FilmFair Communications, 1975. 16mm color, sound film, 17 min. ($230). Gr. 7–A. BKL (7/75), LFR (9/75), M&M (12/75). 331.3

Dramatized vignettes of several job situations are humorously portrayed to reveal the rewards of having pride in one's work and how lack of it harms ourselves and others. These situations encourage pride in work by helping viewers recognize the human needs that pride in work fulfills and the negative effects on ourselves and others of carelessness, boredom, and selfishness.

STARTING TO THINK ABOUT WORK (Filmstrip—Sound). *See* Vocational guidance

WHY WE WORK (Filmstrip—Sound). Educational Direction/Learning Tree Filmstrips, 1977. 4 color, sound filmstrips, av. 49 fr. ea.; 4 cassettes, av. 7–8 min. ea.; Teacher's Guide. ($58). Gr. 3–8. FLN (9/10/77), IFT (1977), LGB (1977). 331.1

Contents: 1. Things We Need. 2. Things We Want. 3. Liking a Job. 4. Everyone Works.

Introduction to the various aspects of work. Emphasizes different reasons to work—sat-

WORK (cont.)

isfaction of basic needs and desires, personal development, achievement, peer consideration, etc.—and also discusses kinds of work we all do at different stages in our lives.

WORK/WORKING/WORKER (Filmstrip—Sound). *See* Vocational guidance

WORKING (Filmstrip—Sound). BFA Educational Media, 1974. 4 color, sound filmstrips; 4 cassettes or discs; 20 Activity Master Sheets. Cassettes #VEK000 ($74.50), Discs #VEJ000 ($74.50). Gr. 4–8. BKL (5/1/75). 331.1
Contents: 1. Why Work? 2. What You're Worth. 3. On-The-Job. 4. Forecast for the Future.
This series examines the reasons people work, the economic structure of work, the training and experience needed for different jobs, and the future of work.

WORLD WAR, 1914–1918

HAYWIRE MAC (Phonodiscs). *See* Folk songs

WORLD WAR, 1939–1945

VOICES OF WORLD WAR TWO (Cassette). Visual Education, 1975. 12 cassettes, approx. 60 min. ea. #48799 ($119). Gr. 8–A. BKL (9/15/76), PRV (9/76). 940.53
Contents: 1. Peace for Our Time. 2. Blitzkrieg. 3. Britain Stands Alone. 4. America Enters the War. 5. Turning Points. 6. Counteroffensive. 7. The Air and Sea War. 8. D-Day and the Battle for Europe. 9. Victory in Europe. 10. Victory in the Pacific. 11. The Lessons of Nuremberg. 12. Media in Wartime.
Actual eyewitness accounts, newscasts, speeches, and oral history interviews recreate how it was in London during the blitz, in Pearl Harbor on December 7, 1941, and in Hiroshima when the atomic bomb was dropped. The voices include Churchill, Hitler, Roosevelt, Tokyo Rose, Axis Sally, Lord Haw Haw, Chamberlain, and many private individuals—both military and civilian. The cassettes are arranged in chronological order from 1933–1947.

WORLD WAR TWO (Filmstrip—Sound). *See* U.S.—History—1933–1945

WORLD WAR, 1939–1945—CAUSES

PRELUDE TO WORLD WAR II—THE SPANISH CIVIL WAR (Filmstrip—Sound). *See* Spain—History—1936–1939—Civil War

WORLD WAR, 1939–1945—JEWS

THE DIARY OF ANNE FRANK (Filmstrip—Sound). *See* Autobiographies

WORLD WAR, 1939–1945—PRISONERS & PRISONS—FICTION

SUMMER OF MY GERMAN SOLDIER (Phonodisc or Cassette). Miller-Brody Productions, 1976 (Young Adult Recordings and Sound Filmstrips). 1 cassette or disc, 40 min.; Teacher's Notes. #YA 401C Cassette ($7.95), #YA 401 Disc ($7.95). Gr. 7–C. BKL (2/15/78), CRC (6/19/76). Fic
Betty Greene's *Summer of My German Soldier* (Dial, 1973) is the story of 12-year-old Patty Bergen's friendship with an escaped German POW and is dramatized in this recording of an abridgment of the book.

WORMS

NEMATODE (Motion Picture—16mm—Sound). *See* Nematodes

WRIGHT, FRANK LLOYD

AMERICANS WHO CHANGED THINGS (Filmstrip—Sound). *See* U.S.—Civilization—Biography

WRIGHT, ORVILLE

THE WRIGHT BROTHERS (Cassette). *See* Aeronautics—Biography

WRIGHT, WILBUR

THE WRIGHT BROTHERS (Cassette). *See* Aeronautics—Biography

WRITING (AUTHORSHIP). *See* Authorship; English language—Composition and exercises; Journalism; Language arts

WYETH, ANDREW N.

PRELUDE TO REVOLUTION (Motion Picture—16mm—Sound). *See* U.S.—History—1775–1783—Revolution

WYSS, JOHANN

SWISS FAMILY ROBINSON (Phonodisc or Cassette). *See* Adventure and adventurers—Fiction

YATES, ELIZABETH

ELIZABETH YATES (Filmstrip—Sound). Miller-Brody Productions, 1976 (Meet the Newbery Author). 1 color, sound filmstrip, 113 fr.; 1 cassette or disc, 19 min. Cassette #MNA–1009–C ($32), Disc #MNA–1009–R ($32). Gr. 5–12. BKL (12/15/76). 921

Intimate family photos from Elizabeth Yates' childhood and adult adventures document the life of this vibrant individual and provide an authentic note to the story of a well-known author.

YELLOWSTONE NATIONAL PARK

ECOLOGY OF A HOT SPRING: LIFE AT HIGH TEMPERATURE (Motion Picture—16mm—Sound). *See* Ecology

GEYSER VALLEY (Motion Picture—16mm—Sound). *See* Geysers

YOGA

MIND POWER (Filmstrip—Sound). *See* Stress (psychology)

YOUNG, SCOTT

HOCKEY HEROES (Kit). *See* Athletes

YOUTH—ATTITUDES

THE AGE OF SENSATION (Cassette). Cinema Sound/Jeffrey Norton Publishers, 1978 (Avid Reader). 1 cassette, approx. 55 min. #40242 ($11.95). Gr. 9–C. PRV (2/79). 301.43

Heywood Hale Broun discusses the book *The Age of Sensation* with its author, Herbert Hendin. Hendin has attempted to learn what is changing the lives of young people. He explores their flight from deep involvement, their envy of other people's lives and experiences, the meaninglessness that is a protection against the risk of making choices, the drug culture, and the wandering culture.

YOUTH—LAW AND LEGISLATION

YOUTH AND THE LAW (Filmstrip—Sound). Barr Films, 1977. 6 color, sound filmstrips, 70–83 fr.; 6 cassettes, 12–15 min.; Teacher's Guide. #74202 ($125). Gr. 7–A. PRV (2/79). 346.013

Contents: 1. Law and the Judge. 2. Law and the Police. 3. Law and the Dissenter. 4. Law and the Youthful Offender. 5. Law and the Individual. 6. Law and the Accused.

The viewpoint of a young person is presented as well as that of the establishment.

YOUTH AS PARENTS

BECOMING A PARENT: THE EMOTIONAL IMPACT (Filmstrip—Sound). Parents' Magazine Films, 1977 (A Life Begins—A Life Changes—The School-Age Parent). 5 color, sound filmstrips, approx 60 fr.; 3 cassettes or discs, 8 min.; 5 Scripts; 1 Guide. ($65). Gr. 7–A. BKL (12/19/77), PRV (5/19/78). 301.42

Introduces several families with school-age parents dramatizing the common feelings, problems, and adjustments these young people make. Shows how attitudes of family and friends influence their decisions. Options such as going back to school or preparing for a career are discussed.

BUILDING A FUTURE (Filmstrip—Sound). Parents' Magazine Films, 1977 (A Life Begins—A Life Changes—The School-Age Parent). 5 color, sound filmstrips, approx. 60 fr. ea.; 3 cassettes or discs, 8 min.; 5 Scripts; 1 Guide. ($65). Gr. 7–A. BKL (12/19/77), PRV (5/19/78). 301.42

School-age parents have not only themselves to consider, but their young children as well. In this set, viewers learn how to guide a child's growth and development while developing their own maturity and establishing an independent adult life for themselves.

DEALING WITH PRACTICAL PROBLEMS OF PARENTHOOD (Filmstrip—Sound). Parents' Magazine Films, 1977 (A Life Begins—A Life Changes—The School-Age Parent). 5 color, sound filmstrips, approx. 60 fr. ea.; 3 cassettes or discs, 8 min.; 5 Scripts; Teacher's Guide. ($65). Gr. 7–A. BKL (12/19/77), PRV (5/19/78). 301.42

Dramatizations illustrate the value of counseling for new (or about to be) parents trying to cope with problems such as prenatal care, day care, and family planning. Proper nutrition; effects of smoking, drinking, and drugs on the newborn; and fears surrounding childbirth are discussed.

A LIFE BEGINS . . . LIFE CHANGES . . . THE SCHOOL-AGE PARENT (Filmstrip—Sound). *See* Pregnancy

MATURITY: OPTIONS AND CONSEQUENCES (Filmstrip—Sound). *See* Adolescence

RIGHTS AND OPPORTUNITIES (Filmstrip—Sound). Parents' Magazine Films, 1977 (A Life Begins—A Life Changes—A School-Age Parent). 5 color, sound filmstrips, approx. 60 fr. ea.; 3 discs or cassettes, 8 min.; 5 Scripts; Teacher's Guide. ($65). Gr. 7–12. PRV (5/78), BKL (12/77). 301.42

This set discusses financial assistance and opportunities for education and careers and explains legal limitations of a minor person.

ZEBRAS

ZEBRA (Motion Picture—16mm—Sound). EBEC, 1971 (Silent Safari Series). 16mm color, sound, narrationless film, 10 min.; Teacher's Guide. #3122 ($150). Gr. K–9. BKL (10/1/76). 599.7

Zebras travel in large herds, following edible grasses and available water. Chief among their striking physical traits are the striped

ZEBRAS (cont.)

markings that help camouflage them. Scenes show zebras as sociable and playful, greeting each other with conversational barking noises, enjoying running games, and grooming each other with their teeth while they depend on friendly birds to signal approaching danger.

ZENGER, JOHN PETER

THE JOHN PETER ZENGER TRIAL (Filmstrip—Sound). *See* Freedom of the press

ZINDEL, PAUL

THE PIGMAN (Phonodisc or Cassette). *See* Friendship—Fiction

ZOOLOGICAL GARDENS

MUSEUMS AND MAN (Filmstrip—Sound). *See* Museums

ZOOFARI (Filmstrip—Sound). Lyceum/Mook & Blanchard, 1972. 2 color, sound filmstrips, av. 53 fr.; 2 cassettes or discs, $8^{1}/_{2}$–$14^{1}/_{2}$ min.; Teacher Guide. #LY35573SC Cassettes ($46), #LY35573SR Discs ($37). Gr. 3–8. FLN (2/73), PRV (9/72), PVB (5/73). 590.74

Contents: 1. New at the Zoo. 2. Safari: North American Style.

The objective of this set is to explore the varieties of animal life available in animal parks and to understand the attempts that are made to provide natural surroundings for the animals.

ZOOLOGY

ANIMAL LIFE SERIES, SET ONE (Film Loop—8mm—Silent). *See* Growth

ANIMAL LIFE SERIES, SET TWO (Film Loop—8mm—Silent). *See* Adaptation (biology)

BABOONS AND THEIR YOUNG (Film Loop—8mm—Silent). *See* Baboons

AN INSIDE LOOK AT ANIMALS (Study Print). *See* Anatomy, Comparative

ZOOS. *See* Zoological gardens

MEDIA INDEXED BY TITLE

A. J. Miller's West: The Plains Indian—1837—Slide Set (Slides) *Indians of North America —Paintings*

A. Lincoln (Filmstrip—Sound) *Lincoln, Abraham*

About Sex (Motion Picture—16mm—Sound) *Sex instruction*

Acid (Motion Picture—16mm—Sound) *Hallucinogens*

Across Five Aprils (Filmstrip—Sound) *U.S.—History —1861-1865—Civil War —Fiction*

Acting for Film (Long Christmas Dinner) (Motion Picture—16mm—Sound) *Acting*

The Action Process (Filmstrip—Sound) *Consumer education*

Acupuncture: An Exploration (Motion Picture—16mm—Sound) *Acupuncture*

Adobe Oven Building (Film Loop—8mm—Silent) *Indians of North America*

Adolescence, Love and Dating (Kit) *Adolescence*

Adolescent Conflict: Parents vs. Teens (Filmstrip—Sound) *Adolescence*

Adolescent Responsibilities: Craig and Mark (Motion Picture—16mm—Sound) *Adolescence*

Adolescent to Adulthood: Rites of Passage (Filmstrip—Sound) *Adolescence*

Adolf Hitler (Cassette) *Hitler, Adolf*

Adventure and Suspense (Filmstrip—Sound) *Literature —Study and teaching*

The Adventure of the Speckled Band by Doyle (Cassette) *Mystery and detective stories*

Adventures in Imagination Series (Filmstrip—Sound) *English language —Composition and exercises*

Adventures in the World of Work, Set One (Kit) *Vocational guidance*

Adventures in the World of Work, Set Two (Kit) *Vocational guidance*

The Adventures of Huckleberry Finn by Mark Twain (Phonodisc or Cassette) *Adventure and adventurers —Fiction*

Africa: Learning about the Continent (Filmstrip—Sound) *Africa*

Africa: Portrait of a Continent (Kit) *Africa*

African Cliff Dwellers: The Dogon People of Mali, Part One (Kit) *Africa*

African Cliff Dwellers: The Dogon People of Mali, Part Two (Kit) *Africa*

Age of Exploration and Discovery (Filmstrip—Sound) *Explorers*

The Age of Sensation (Cassette) *Youth—Attitudes*

Agricultural Reform in India: A Case Study (Filmstrip—Sound) *Agriculture —India*

Agua Salada (Motion Picture—16mm—Sound) *Passion plays*

Air Conditioning Serviceman, Set 7 (Film Loop—8mm—Silent) *Automobiles—Maintenance and repair*

Air Conditioning Serviceman, Set 8 (Film Loop—8mm—Silent) *Automobiles—Maintenance and repair*

Air, Earth, Fire and Water (Filmstrip—Sound) *Science*

Aircraft: Their Power and Control (Filmstrip—Sound) *Aeronautics*

Alaska: The Big Land and Its People (Filmstrip—Sound) *Alaska*

Alcohol and Alcoholism (Filmstrip—Sound) *Alcoholism*

Alcohol and Alcoholism: The Drug and the Disease (Filmstrip—Sound) *Alcoholism*

Alcohol, Drugs or Alternatives (Motion Picture—16mm—Sound) *Alcoholism*

Alcohol: Facts, Myths and Decisions (Filmstrip—Sound) *Alcohol*

Animal Life Series, Set One (Film Loop—8mm—Silent) *Growth*

Animal Life Series, Set Two (Film Loop—8mm—Silent) *Adaptation (biology)*

Animals, Animals (Filmstrip—Sound) *Animals—Habits and behavior*

Animated Women (Motion Picture—16mm—Sound) *Women—U.S.*

Antarctic Penguins (Film Loop—8mm—Silent) *Penguins*

Antarctica: The White Continent (Filmstrip—Sound) *Antarctic regions*

Anthropologist at Work (Filmstrip—Sound) *Anthropology*

Anything You Want To Be (Motion Picture—16mm—Sound) *Sex roles*

The Arab Civilization (Filmstrip—Sound) *Arab countries—Civilization*

The Arab World (Kit) *Arab countries*

Archeological Dating: Retracing Time (Motion Picture—16mm—Sound) *Archeology*

Around the World in Eighty Days (Phonodisc or Cassette) *Fantastic fiction*

Arrow to the Sun (Motion Picture—16mm—Sound) *Indians of North American—Legends*

Art by Talented Teenagers (Filmstrip) *Art—Exhibitions*

Art Careers (Motion Picture—16mm—Sound) *Art, American*

Art of Dave Brubeck: The Fantasy Years (Phonodiscs) *Jazz music*

The Art of Jewelry (Art Prints) *Jewelry—History*

The Art of Mosaics (Filmstrip—Sound) *Mosaics*

The Art of Motivation (Motion Picture—16mm—Sound) *Motivation (psychology)*

Art of Persepolis (Slides) *Iran—Antiquities*

The Art of Seeing (Filmstrip—Sound) *Art appreciation*

The Art of the Middle Ages (Filmstrip—Sound) *Art, medieval*

The Art of the Renaissance (Filmstrip—Sound) *Art, renaissance*

The Artist inside Me (Kit) *Humanities*

The Arts of Japan—Slide Set (Slides) *Art, Japanese*

As Long as the Grass Shall Grow (Phonodiscs) *Folk songs, Indian*

Asian Man: China (Kit) *China—History*

Atmosphere in Motion (Motion Picture—16mm—Sound) *Atmosphere*

The Atom Bomb (Filmstrip—Sound) *Atomic bomb*

An Audio Visual History of American Folk Music (Filmstrip—Sound) *Folksongs—U.S.*

An Audio Visual History of European Literature (Filmstrip—Sound) *Literature*

Audubon's Shore Birds (Motion Picture—16mm—Sound) *Water Birds*

Aunt Arie (Motion Picture—16mm—Sound) *Carpenter, Aunt Arie*

Australia and New Zealand (Filmstrip—Sound) *Australia*

Auto-Body Sheet Metal Man, Set One (Film Loop—8mm—Silent) *Automobiles—Maintenance and repair*

Auto-Body Sheet Metal Man, Set Two (Film Loop—8mm—Silent) *Automobiles—Maintenance and repair*

Auto Painter Helper, Set 3 (Film Loop—8mm—Silent) *Automobiles—Maintenance and repair*

Auto Painter, Set 4 (Film Loop—8mm—Silent) *Automobiles—Maintenance and repair*

Auto Racing: Something for Everyone (Kit) *Automobile Racing*

Automobile Glass Man, Set Nine (Film Loop—8mm—Silent) *Automobiles—Maintenance and repair*

Automobile Glass Man, Set Ten (Film Loop—8mm—Silent) *Automobiles—Maintenance and repair*

Automobile Upholstery Repairman, Set Eleven (Film Loop—8mm—Silent) *Automobiles—Maintenance and repair*

Automobile Upholstery Repairman, Set Twelve (Film Loop—8mm—Silent) *Automobiles—Maintenance repair*

Ayn Rand—Interview (Videocassette) *Rand, Ayn*

B. Kliban (Cassette) *Kliban, B.*

Baboons and Their Young (Film Loop—8mm—Silent) *Baboons*

Backpacking Basics (Slides/Cassettes) *Backpacking*

Backpacking: Revised (Filmstrip—Sound) *Backpacking*

Balablok (Motion Picture—16mm—Sound) *Prejudices and antipathies*

Bartleby (Motion Picture—16mm—Sound) *American fiction*

Basic Dribble/Control Dribble/Speed (Film Loop—8mm—Silent) *Basketball*

Basic Photography (Filmstrip—Sound) *Photography*

The Bat Poet (Phonodisc or Cassette) *Individuality—Fiction*

Bate's Car (Motion Picture—16mm—Sound) *Fuel*

Bats (Film Loop—8mm—Silent) *Bats*

Batting (Film Loop—8mm—Silent) *Softball*

Bauhaus (Filmstrip—Sound) *Art—Study and teaching*

Beah Richards: A Black Woman Speaks (Videocassette) *Richards, Beah*

Bear Country and Beaver Valley (Filmstrip—Sound) *Natural history*

Beaver (Film Loop—8mm—Silent) *Beavers*

Beaver Dam and Lodge (Film Loop—8mm—Silent) *Beavers*

Because It's Just Me Unit Two (Motion Picture—16mm—Sound) *Occupations*

Becoming a Parent: The Emotional Impact (Filmstrip—Sound) *Youth as parents*

Becoming an Adult: The Psychological Tasks of Adolescence (Filmstrip—Sound) *Adolescence*

Beekeeping (Filmstrip—Sound) *Bees*

Before You Take That Bite (Motion Picture—16mm—Sound) *Nutrition*

The Beginning (Motion Picture—16mm—Sound) *Creativity—Fiction*

Beginning Concepts: People Who Work, Unit One (Kit) *Vocational Guidance*

Benedict Arnold: Traitor or Patriot? (Filmstrip—Sound) *Arnold, Benedict*

Benjamin Franklin—Scientist, Statesman, Scholar and Sage (Motion Picture—16mm—Sound) *Franklin, Benjamin*

Beowulf and the Monsters (Cassette) *Beowulf*

The Best I Can (Motion Picture—16mm—Sound) *Self-realization*

The Best of Encyclopedia Brown (Filmstrip—Sound) *Mystery and detective stories*

Better Choice, Better Chance: Selecting a High School Program (Filmstrip—Sound) *Educational guidance*

A Better Train of Thought (Motion Picture—16mm—Sound) *Creativity—Fiction*

Beyond High School (Filmstrip—Sound) *Vocational guidance*

Bicycle Safely (Motion Picture—16mm—Sound) *Bicycles and bicycling*

Bicycling on the Safe Side (Motion Picture—16mm—Sound) *Bicycles and bicycling*

The Big Dig (Motion Picture—16mm—Sound) *Israel—Antiquities*

Bill Cosby on Prejudice (Motion Picture—16mm—Sound) *Prejudices and antipathies*

Biography: Background for Inspiration (Filmstrip—Sound) *Biography*

Biological Catastrophes: When Nature Becomes Unbalanced (Slides/Cassettes) *Ecology*

Biological Dissection (Filmstrip—Sound) *Dissection*

Biological Rhythms: Studies in Chronobiology (Motion Picture—16mm—Sound) *Biology—Periodicity*

Bip as a Skater (Motion Picture—16mm—Sound) *Pantomimes*

Bird of Freedom (Motion Picture—16mm—Sound) *Eagles*

Birds Feeding Their Young (Film Loop—8mm—Silent) *Birds*

Birds of a Feather (Motion Picutre—16mm—Sound) *Birds—Fiction*

Birds of Prey (Filmstrip—Sound) *Birds of Prey*

Birth and Care of Baby Chicks (Film Loop—8mm—Captioned) *Chickens*

Bison and Their Young (Film Loop—8mm—Silent) *Bison*

Black American Athletes (Kit) *Athletes*

Black Holes of Gravity (Motion Picture—16mm—Sound) *Astronomy*

Blessingway: Tales of a Navajo Family (Kit) *Navajo*

The Blue Ridge: America's First Frontier (Filmstrip—Sound) *Blue Ridge Mountains*

Bobsledding: Down the Chute (Kit) *Winter sports*

The Body (Filmstrip—Sound) *Physiology*

Boomsville (Motion Picture—16mm—Sound) *Cities and Towns*

Bottlenose Dolphin (Film Loop—8mm—Silent) *Dolphins*

Boxing (Kit) *Boxing*

Brazil (Filmstrip—Sound) *Brazil*

Breads and Cereals (Kit) *Cookery*

The British Isles: Scotland and Ireland (Filmstrip—Sound) *Scotland*

The Bronze Zoo (Motion Picture—16mm—Sound) *Bronzes*

The Brown Trout (Film Loop—8mm—Captioned) *Trout*

Buffalo: An Ecological Success Story (Motion Picture—16mm—Sound) *Bison*

Building a Future (Filmstrip—Sound) *Youth as parents*

The Building Blocks of Language (Filmstrip—Sound) *English language—Study and teaching*

The Business of Motion Pictures (Cassette) *Motion pictures—Economic aspects*

But Is It Art? (Filmstrip—Sound) *Art, modern*

Butterfly: The Monarch's Life Cycle (Motion Picture—16mm—Sound) *Butterflies*

The Butterick Interior Design Series (Kit) *Interior decoration*

The Buy Line (Motion Picture—16mm—Sound) *Advertising*

Buying Health Care (Kit) *Insurance, health*

C P R: To Save a Life (Motion Picture—16mm—Sound) *First aid*

The Cactus: Adaptations for Survival (Motion Picture—16mm—Sound) *Cactus*

Call of the Wild (Filmstrip—Sound) *Dogs—Fiction*

The Call of the Wild (Kit) *Dogs—Fiction*

Camping (Filmstrip—Sound) *Camping*

Canada (Filmstrip—Sound) *Canada*

Canada: Land of New Wealth (Filmstrip—Sound) *Canada*

The Canadians (Filmstrip—Sound) *Canada*

Cancer (Filmstrip—Sound) *Cancer*

Cannery Row, Life and Death of an Industry (Filmstrip—Sound) *Monterey, California—History*

Canyon De Chelly Cliff Dwellings (Film Loop—8mm—Silent) *Cliff dwellers and cliff dwellings*

Career Choice: A Lifelong Process (Filmstrip—Sound) *Vocational guidance*

Career English: Communicating on the job (Slides/Cassettes) *English language —Business English*

A Career in Computers (Filmstrip—Sound) *Computers*

A Career in Sales (Filmstrip—Sound) *Sales personnel*

Career Training through the Armed Forces (Filmstrip—Sound) *U.S. —Armed forces*

Careers in Banking and Insurance (Filmstrip—Sound) *Banks and banking*

Careers in Business Administration (Filmstrip—Sound) *Business*

Careers in Engineering (Filmstrip—Sound) *Engineering*

Careers in Food Service (Filmstrip—Sound) *Cookery, quantity*

Careers in Graphic Arts (Filmstrip—Sound) *Graphic arts*

Careers in Health Services (Filmstrip—Sound) *Medical care*

Careers in Home Economics (Filmstrip—Sound) *Home economics*

Careers in Law Enforcement (Filmstrip—Sound) *Law enforcement*

Careers in Nursing (Filmstrip—Sound) *Nursing*

Careers in Social Work (Filmstrip—Sound) *Social work*

Careers in the Fashion Industry (Kit) *Clothing trade*

Carnivorous Plants (Motion Picture—16mm—Sound) *Insectivorous plants*

The Catcher (Film Loop—8mm—Silent) *Softball*

Catching Above Waist/Below Waist (Film Loop—8mm—Silent) *Softball*

Catherine de'Medici (Cassette) *Catherine de Medici*

Caves: The Dark Wilderness (Motion Picture—16mm—Sound) *Caves*

Center for a Field Goal (Film Loop—8mm—Silent) *Football*

Center Snap for Punt (Film Loop—8mm—Silent) *Football*

Center to Quarterback Exchange (Film Loop—8mm—Silent) *Football*

Central America: Finding New Ways (Motion Picture—16mm—Sound) *Central America*

Changes (Motion Picture—16mm—Sound) *Physically handicapped*

Changing Human Behavior (Filmstrip—Sound) *Behavior modification*

Changing the Face of Things (Filmstrip) *Design, decorative*

Changing Values in America: The 20th Century (Filmstrip—Sound) *U.S.—History*

Changing Views on Capital Punishment (Filmstrip—Sound) *Capital punishment*

Chaplin—A Character Is Born (Motion Picture—16mm—Sound) *Chaplin, Charles Spencer*

Charles Darwin (Cassette) *Darwin, Charles*

Charles Dickens (Cassette) *Dickens, Charles*

Cheetah (Motion Picture—16mm—Sound) *Cheetahs*

Chemistry: Dissecting the Atom (Filmstrip—Sound) *Chemistry*

Chemistry in Nature (Motion Picture—16mm—Sound) *Chemistry*

Chest Pass/Bounce Pass (Film Loop—8mm—Silent) *Basketball*

Chest Pass/Overhead Pass (Film Loop—8mm—Silent) *Basketball*

Child Abuse: America's Hidden Epidemic (Filmstrip—Sound) *Child abuse*

The Child and the Family (Kit) *Child development*

Child Care and Development, Set One (Filmstrip—Sound) *Children —Care and hygiene*

Child Care and Development, Set Two (Filmstrip—Sound) *Children —Care and hygiene*

Child of Fire (Phonodisc or Cassette) *Mexicans in the U.S. —Fiction*

Childhood of Famous Women, Volume Three (Cassette) *Women —Biography*

Children of Alcoholic Parents (Filmstrip—Sound) *Alcoholism*

Children of the Northlights (Motion Picture—16mm—Sound) *Authors, American*

China (Cassette) *U.S. —Foreign relations*

China: A Network of Communes (Motion Picture—16mm—Sound) *China —Civilization*

China: Education for a New Society (Motion Picture—16mm—Sound) *China —Civilization*

China Multimedia Program (Kit) *China*

Chinese Food (Kit) *Cookery, Chinese*

Choking: To Save a Life (Motion Picture—16mm—Sound) *First aid*

A Christmas Carol (Phonodisc or Cassette) *Christmas —Fiction*

A Christmas Carol (Kit) *Christmas —Fiction*

Chuparosas: The Hummingbird Twins (Filmstrip—Sound) *Hummingbirds*

Circular Saw, Set Four (Film Loop—8mm—Silent) *Woodwork*

Cities of America, Part One (Filmstrip—Sound) *U.S. —Description and travel*

Cities of America, Part Two (Filmstrip—Sound) *U.S. —Description and travel*

City and Town (Filmstrip—Sound) *Cities and towns*

The City at the End of the Century (Motion Picture—16mm—Sound) *Cities and towns —U.S.*

The Civil War (Filmstrip—Sound) *U.S. —History —1861-1865 —Civil War*

Classical Music for People Who Hate Classical Music (Phonodiscs) *Orchestral music*

Climb (Motion Picture—16mm—Sound) *Mountaineering*

Closed Mondays (Motion Picture—16mm—Sound) *Art —Exhibitions —Fiction*

Four Biomes (Filmstrip—Sound) *Ecology*

Four Families (Filmstrip—Sound) *Family life*

Four Families of Kenya (Filmstrip—Sound) *Kenya*

Four Wheels (Filmstrip—Sound) *Camping*

Fragonard (Motion Picture—16mm—Sound) *Fragonard, Jean Honore*

Frank Film (Motion Picture—16mm—Sound) *Mouris, Frank*

Free Press: A Need to Know the News (Kit) *Freedom of the press*

Freedom River (Motion Picture—16mm—Sound) *Freedom*

From Cave to City (Motion Picture—16mm—Sound) *Civilization*

From Home to School (Filmstrip—Sound) *Family life*

Fruits and Vegetables (Kit) *Cookery*

Fur, Fins, Teeth, and Tails (Filmstrip—Sound) *Animals —Habits and behavior*

The Future (Filmstrip—Sound) *Literature — Study and teaching*

Futurism (Motion Picture—16mm—Sound) *Painters, Italian*

Gail E. Haley: Wood and Linoleum Illustration (Filmstrip—Sound) *Haley, Gail E.*

Galapagos: Darwin's World within Itself (Motion Picture—16mm—Sound) *Galapagos Islands*

The Gammage Cup (Filmstrip—Sound) *Fantasy—Fiction*

Gene Deitch: The Picture Book Animated (Motion Picture—16mm—Sound) *Deitch, Gene*

Geology: Our Dynamic Earth (Filmstrip—Sound) *Geology*

George Washington (Cassette) *Washington, George*

George Washington—The Courage That Made a Nation (Motion Picture—16mm—Sound) *Washington, George*

Georges Rouault (Motion Picture—16mm—Sound) *Rouault, Georges*

German Food (Kit) *Cookery, German*

Gertrude the Governess and Other Works (Phonodisc or Cassette) *Canadian fiction*

Get It Together (Motion Picture—16mm—Sound) *Handicapped*

Getting Ready Unit One (Motion Picture—16mm—Sound) *Occupations*

Geyser Valley (Motion Picture—16mm—Sound) *Geysers*

Ghost of Captain Peale (Motion Picture—16mm—Sound) *Metric system*

Ghost Towns of the Westward March (Motion Picture—16mm—Sound) *The West (U.S.)—History*

Giraffe (Motion Picture—16mm—Sound) *Giraffes*

Girl Stuff (Kit) *Girls—Fiction*

The Giving Tree (Motion Picture—16mm—Sound) *Friendship—Fiction*

Glacier on the Move (Motion Picture—16mm—Sound) *Glaciers*

Glasses for Susan (Motion Picture—16mm—Sound) *Vision*

Goal Keeper, Clearing (Film Loop—8mm—Silent) *Soccer*

Goal Keeper, Part 1 (Film Loop—8mm—Silent) *Soccer*

Goal Keeper, Part 2 (Film Loop—8mm—Silent) *Soccer*

Goal Keeper, Part 3 (Film Loop—8mm—Silent) *Soccer*

The Gods Were Tall and Green (Filmstrip—Sound) *Forests and forestry*

Going Back Unit Three (Motion Picture—16mm—Sound) *Occupations*

Gold to Build a Nation: The United States Before the Civil War (Filmstrip—Sound) *California—Gold discoveries*

Golden Lizard: A Folk Tale from Mexico (Motion Picture—16mm—Sound) *Folklore — Mexico*

The Good Old Days—They Were Terrible (Cassette) *History—Philosophy*

Goodnight Miss Ann (Motion Picture—16mm—Sound) *Boxing*

Got Something to Tell You (Slides/Cassettes) *Blues (songs, etc.)*

Government and You (Filmstrip—Sound) *Political science*

Government: How Much Is Enough? (Filmstrip—Sound) *Political science*

Goya (Motion Picture—16mm—Sound) *Goya, Francisco*

Grammar (Slides/Cassettes) *English language —Grammar*

Grammar: Everything You Wanted to Know about Usage (Slides/Cassettes) *English language —Grammar*

Gramp: A Man Ages and Dies (Filmstrip—Sound) *Old age*

La Grande Breteche (Motion Picture—16mm—Sound) *Horror—Fiction*

The Graphic Arts: An Introduction (Filmstrip—Sound) *Graphic arts*

Grassland Ecology—Habitats and Change (Motion Picture—16mm—Sound) *Grasslands*

Gravity Is My Enemy (Motion Picture—16mm—Sound) *Hicks, Mark*

The Great Cover Up (Motion Picture—16mm—Sound) *Clothing and dress*

Great Grandmother (Motion Picture—16mm—Sound) *Canada—History*

Great Myths of Greece (Filmstrip—Sound) *Mythology, classical*

The Great Quillow (Phonodisc or Cassette) *Giants—Fiction*

Great Writers of the British Isles, Set I (Filmstrip—Sound) *Authors, English*

Great Writers of the British Isles, Set II (Filmstrip—Sound) *Authors, English*

The Grizzly (Kit) *Bears —Fiction*

The Grizzly and the Gadgets and Further Fables for Our Time (Phonodisc or Cassette) *Fables*

Growing Old with Grace (Cassette) *Old age*

The Growing Trip (Filmstrip—Sound) *Reproduction*

Growing Up Female (Videocassette) *Women*

The Growth of Intelligence (Kit) *Child development*

Haiku: The Hidden Glimmering (Filmstrip—Sound) *Haiku*

Haiku: The Mood of Earth (Filmstrip—Sound) *Haiku*

Hand Off (Film Loop—8mm—Silent) *Football*

Hand Tool Operations, Set 1 (Film Loop—8mm—Silent) *Woodwork*

Hand Tool Operations, Set 2 (Film Loop—8mm—Silent) *Woodwork*

Hand Tool Operations, Set 3 (Film Loop—8mm—Silent) *Woodwork*

Handling, Transferring and Filtering of Chemicals (Motion Picture—16mm—Sound) *Chemical apparatus*

The Hands (Motion Picture—16mm—Sound) *Pantomimes*

Hang Gliding: Riding the Wind (Kit) *Gliding and soaring*

The Happy Lion Series, Set One (Filmstrip—Sound) *Lions —Fiction*

The Happy Lion Series, Set Two (Filmstrip—Sound) *Lions —Fiction*

The Happy Prince and Other Oscar Wilde Fairy Tales (Phonodisc or Cassette) *Fairy tales*

Hardware Wars (Motion Picture—16mm—Sound) *Satire, American*

Harpsichord Builder (Motion Picture—16mm—Sound) *Harpsichord*

Harvest (Motion Picture—16mm—Sound) *Agriculture —U.S.*

Has Big Business Gotten Too Big? (Filmstrip—Sound) *Economics*

Hawaii: The Fiftieth State (Filmstrip—Sound) *Hawaii*

Haywire Mac (Phonodiscs) *Folk songs*

Heading and Back-Heading (Film Loop—8mm—Silent) *Soccer*

Health and Safety: Keeping Fit (Kit) *Hygiene*

The Heart and the Circulatory System (Motion Picture—16mm—Sound) *Blood —Circulation*

Heroes of the Iliad (Cassette) *Mythology, classical*

Heroic Adventures (Filmstrip—Sound) *Legends*

The Higher Invertebrates (Filmstrip—Sound) *Invertebrates*

His Infinite Variety: A Shakespearean Anthology (Cassette) *English drama*

Hispanic Heroes of the U.S.A. (Kit) *Biography —Collections*

History, Poetry and Drama in the Old Testament (Motion Picture—16mm—Sound) *Bible —Old Testament*

The Hoarder (Motion Picture—16mm—Sound) *Fables*

The Hobbit (Kit) *Fantasy —Fiction*

Hobo: At the End of the Line (Motion Picture—16mm—Sound) *Tramps*

Hockey Heroes (Kit) *Athletes*

Holidays: Set One (Filmstrip—Sound) *Holidays*

Home Decoration Series, Set One (Filmstrip—Sound) *Interior decoration*

Home Decoration Series, Set Two (Filmstrip—Sound) *Interior decoration*

Homer's Mythology: Tracing a Tradition (Filmstrip—Sound) *Greek poetry*

The Honey Bee (Film Loop—8mm—Captioned) *Bees*

Hook Slide/Straight in Slide (Film Loop—8mm—Silent) *Softball*

Hopis—Guardians of the Land (Filmstrip—Sound) *Hopi*

Horizontal Belt Loom (Film Loop—8mm—Silent) *Looms*

Horse Flickers (Motion Picture—16mm—Sound) *Collage*

Horses . . . to Care Is to Love (Motion Picture—16mm—Sound) *Horses*

The House of Science (Motion Picture—16mm—Sound) *Science —History*

Housing and Home Furnishings: Your Personal Environment (Kit) *Houses*

How a Picture Book Is Made (Filmstrip—Sound) *Books*

How Beaver Stole Fire (Motion Picture—16mm—Sound) *Indians of North America —Legends*

How Can I Pay for College? (Filmstrip—Sound) *Student loan funds*

How Flowers Reproduce: The California Poppy (Study Print) *Plant propagation*

How Our Continent Was Made (Filmstrip—Sound) *North America*

How to Build an Igloo—Slide Set (Slides) *Igloos*

How to Do: Cardboard Sculpture (Filmstrip—Sound) *Sculpture*

How to Do: Slide Collage (Filmstrip—Sound) *Collage*

How to Get Elected President (Filmstrip—Sound) *Elections*

How to Stop Smoking (Filmstrip—Sound) *Tobacco habit*

How to Survive in School: Note-Taking and Outlining Skills (Filmstrip—Sound) *Study, method of*

How to Use Maps and Globes (Filmstrip—Sound) *Maps*

How to Watch TV (Filmstrip—Sound) *Mass media*

Koala Bear (Film Loop—8mm—Silent) *Koala Bears*

Kudzu (Motion Picture—16mm—Sound) *Weeds*

La Lechuza—Cuentos de mi Barreo (Cassette) *Owls—Fiction*

Labor Unions: Power to the People (Filmstrip—Sound) *Labor unions*

Labor Unions: What You Should Know (Filmstrip—Sound) *Labor unions*

The Lady or the Tiger (Motion Picture—16mm—Sound) *American fiction*

Landmarks in Psychology (Filmstrip—Sound) *Psychology—History*

Language—The Mirror of Man's Growth (Filmstrip—Sound) *English Language—History*

Lassie Come Home (Phonodisc or Cassette) *Dogs—Fiction*

Lathe (Film Loop—8mm—Silent) *Woodwork*

Laura: Little House, Big Prairie (Filmstrip—Sound) *Wilder, Laura Ingalls*

Law and Crime (Kit) *Criminal justice, administration of*

Law and Justice: Making Value Decisions (Filmstrip—Sound) *Law*

Law and Lawmakers (Kit) *Law*

Law and Society: Law and Crime (Kit) *Law enforcement*

Law and the Environment (Kit) *Environment—Law and legislation*

Lay Up Shot (Film Loop—8mm—Silent) *Basketball*

Leaders, Dreamers and Heroes (Cassette) *Biography—Collections*

Learning about Heat: Second Edition (Motion Picture—16mm—Sound) *Heat*

Learning about Light: Second Edition (Motion Picture—16mm—Sound) *Light*

Learning about Magnetism: Second Edition (Motion Picture—16mm—Sound) *Magnetism*

Learning Away from Home (Kit) *Family life*

Learning Begins at Home (Filmstrip—Sound) *Family life*

Learning: Conditioning or Growth? (Filmstrip—Sound) *Education—Aims and objectives*

Learning in the Home (Kit) *Family life*

Learning Through Play (Kit) *Family life*

The Leather-Crafter (Filmstrip—Sound) *Leather work*

Lee Baltimore: Ninety Nine Years (Motion Picture—16mm—Sound) *Baltimore, Lee*

The Legend of John Henry (Motion Picture—16mm—Sound) *Folklore—U.S.*

The Legend of Paul Bunyan (Motion Picture—16mm—Sound) *Folklore—U.S.*

The Legend of Sleepy Hollow (Motion Picture—16mm—Sound) *Legends—U.S.*

The Legend of Sleepy Hollow and Ichabod Crane (Phonodisc or Cassette) *Legends—U.S.*

Legend of Sleepy Hollow and Other Stories (Cassette) *Legends—U.S.*

The Legend of the Magic Knives (Motion Picture—16mm—Sound) *Indians of North America—Legends*

Leisure (Motion Picture—16mm—Sound) *Man*

Lemming Migration (Film Loop—8mm—Silent) *Lemmings*

Leo Beuerman (Motion Picture—16mm—Sound) *Beuerman, Leo*

Leonard Bernstein Conducts for Young People (Phonodiscs) *Symphonies*

Let No Man Regret (Motion Picture—16mm—Sound) *Man—Influence on nature*

Let's Go Shopping (Kit) *Shopping*

Let's Learn to Study (Filmstrip—Sound) *Study, method of*

Let's Look at Logic (Filmstrip—Sound) *Logic*

Let's Sing a Round (Phonodiscs) *Choruses and part songs*

A Letter to Ceti (Filmstrip—Sound) *U.S.—History*

The Letters of Virginia Woolf (Cassette) *Woolf, Virginia*

Library of Congress (Motion Picture—16mm—Sound) *Library of Congress*

Library Skill Box (Cassette) *Library skills*

Library 3 (Cassette) *Literature—Collections*

A Life Begins . . . Life Changes . . . The School-Age Parent (Filmstrip—Sound) *Pregnancy*

Life Cycle (Filmstrip—Sound) *Animals—Habits and behavior*

Life Cycle of Common Animals, Group 2 (Filmstrip—Sound) *Sheep*

Life Cycle of Common Animals, Group 3 (Filmstrip—Sound) *Butterflies*

Life Goals: Setting Personal Priorities (Filmstrip—Sound) *Self-realization*

Life in Rural America (Filmstrip—Sound) *U.S.—Social life and customs*

Life in the Sea (Filmstrip—Sound) *Marine biology*

Life of a Worker Bee (Film Loop—8mm—Silent) *Bees*

Life Times Nine (Motion Picture—16mm—Sound) *Life*

Lifestyles: Options for Living (Kit) *Life styles—U.S.*

A Light Beam Named Ray (Motion Picture—16mm—Sound) *Color*

The Light in the Forest (Phonodisc or Cassette) *Indians of North America—Fiction*

Linus Pauling: Scientists and Responsibility (Cassette) *Pauling, Linus*

Lion (Motion Picture—16mm—Sound) *Lions*

Lion: Mother and Cubs (Film Loop—8mm—Silent) *Lions*

Measuring Volumes of Liquids (Motion Picture—16mm—Sound) *Chemical apparatus*

Mechanics of Phototropism (Film Loop—8mm—Captioned) *Phototropism*

Media and Meaning: Human Expression and Technology (Slides/Cassettes) *Mass media*

Media Classic Adaptations (Filmstrip—Sound) *Literature —Collections*

Media: Resources for Discovery (Filmstrip—Sound) *Audio-Visual materials*

Medieval Europe (Filmstrip—Sound) *Middle ages*

Medieval Theater: The Play of Abraham and Isaac (Motion Picture—16mm—Sound) *Theater —History*

Meet Margie (Motion Picture—16mm—Sound) *Finance, personal*

Men Behind the Bright Lights (Kit) *Musicians*

The Mental/Social Me (Kit) *Anatomy*

Mesa Verde (Study Print) *Cliff dwellers and cliff dwellings*

Metamorphosis (Motion Picture—16mm—Sound) *Butterflies*

Metric Meets the Inchworm (Motion Picture—16mm—Sound) *Metric system*

The Metric Move (Motion Picture—16mm—Sound) *Metric system*

Metric System (Filmstrip—Sound) *Metric system*

Metric System of Measurement (Filmstrip—Sound) *Metric system*

The Metric System of Measurement (Filmstrip—Sound) *Metric system*

Metrics for Career Education (Filmstrip—Sound) *Metric system*

Metroliner (Motion Picture—16mm—Sound) *Railroads —History*

The Mexican-American Speaks: Heritage in Bronze (Motion Picture—16mm—Sound) *Mexicans in the U.S.*

Mexican Food (Kit) *Cookery, Mexican*

Mexican Indian Legends (Motion Picture—16mm—Sound) *Folklore —Mexico*

Mexico: Images and Empires (Filmstrip—Sound) *Mexico*

Mexico in the Twentieth Century (Filmstrip—Sound) *Mexico*

Mi Hermana Se Caso en Espana (My Sister Was Married in Spain) (Kit) *Spain —Description and travel*

Mi Hermano Se Caso en Mexico (My Brother Was Married in Mexico) (Kit) *Mexico —Description and travel*

The Middle East: A Unit of Study (Kit) *Middle East*

The Middle East: Facing a New World Role (Filmstrip—Sound) *Middle East*

The Middle East: Lands in Transition (Kit) *Middle East*

Middle Guard Play (Film Loop—8mm—Silent) *Football*

The Mighty Midgets (Kit) *Automobile racing*

Milk and Dairy Products (Kit) *Cookery*

Mind Power (Filmstrip—Sound) *Stress (psychology)*

Mindscape (Motion Picture—16mm—Sound) *Painters —Fiction*

Minorities—USA (Filmstrip—Sound) *Minorities*

The Miracle of All Life: Values in Biology (Slides/Cassettes) *Biology*

Miracle of Life (Motion Picture—16mm—Sound) *Embryology*

A Miserable Merry Christmas (Motion Picture—16mm—Sound) *Christmas*

Mr. Shepard and Mr. Milne (Motion Picture—16mm—Sound) *Authors, English*

Mobiles: Artistry in Motion (Filmstrip—Sound) *Mobiles (sculpture)*

Model Airplanes (Kit) *Airplanes —Models*

Modern Morality: Old Values in New Settings (Filmstrip—Sound) *Ethics, American*

Monarch: Story of a Butterfly (Film Loop—8mm—Silent) *Butterflies*

Money: From Barter to Banking (Filmstrip—Sound) *Money*

Money Management (Kit) *Finance, personal*

Monsters and Other Science Mysteries (Filmstrip—Sound) *Curiosities and wonders*

Monuments to Erosion (Motion Picture—16mm—Sound) *Erosion*

The Moon: A Giant Step in Geology (Motion Picture—16mm—Sound) *Lunar geology*

More Silver Pennies (Phonodisc or Cassette) *Poetry —Collections*

The Most Important Thing (Motion Picture—16mm—Sound) *Vocational guidance*

Moths—Slide Set (Slides) *Moths*

Motocross Racing (Kit) *Spanish Language —Reading materials*

Motorcycle Safety (Kit) *Motorcycles*

The Mountain People (Motion Picture—16mm—Sound) *Appalachian Mountains*

Mountain Storm (Film Loop—8mm—Silent) *Weather*

Mountaineering (Filmstrip—Sound) *Mountaineering*

Movin' On (Motion Picture—16mm—Sound) *Transportation —History*

The Moving Earth: New Theories of Plate Tectonics (Filmstrip—Sound) *Plate tectonics*

Mozart, the Marriage of Figaro (Filmstrip—Sound) *Operas*

Museums and Man (Filmstrip—Sound) *Museums*

Music Appreciation: Traveling Sound to Sound (Filmstrip—Sound) *Music —Analysis, appreciation*

The Music Child (Motion Picture—16mm—Sound) *Handicapped children —Education*

The Music Makers (Phonodiscs) *Musicians*

Musical Visions of America (Filmstrip—Sound) *Music, American*

This Tiny World (Motion Picture—16mm—Sound) *Toys —History*

Thomas Alva Edison (Cassette) *Edison, Thomas Alva*

Thomas Jefferson (Motion Picture—16mm—Sound) *Jefferson, Thomas*

Thomas Paine (Motion Picture—16mm—Sound) *Paine, Thomas*

Thoreau on the River: Perspective on Change (Filmstrip—Sound) *Ecology*

Through the Looking Glass (Phonodiscs) *Fantasy —Fiction*

Throw In (Film Loop—8mm—Silent) *Soccer*

Thumbs Down (Hitchhiking) (Motion Picture—16mm—Sound) *Hitchhiking*

Tiger, Tiger, Burning Bright (Filmstrip—Sound) *Fire prevention*

Time: From Moons to Microseconds (Filmstrip—Sound) *Time*

The Time Machine (Filmstrip—Sound) *Science fiction*

A Time of Changes Unit Three (Motion Picture—16mm—Sound) *Occupations*

To Kill a Mockingbird (Kit) *Southern states —Fiction*

To Lead a Nation (Motion Picture—16mm—Sound) *Presidents —U.S.*

Tobacco Problem: What Do You Think? (Motion Picture—16mm—Sound) *Tobacco habit*

Today's Family: A Changing Concept (Filmstrip—Sound) *Family*

The Toddler (Kit) *Child development*

Tommy's First Car (Buying a Used Car) (Motion Picture—16mm—Sound) *Consumer education*

Tomorrow's Cities Today (Filmstrip—Sound) *Cities and towns*

Too Much of a Good Thing (Filmstrip—Sound) *Nutrition*

Tops (Motion Picture—16mm—Sound) *Toys*

Trapping—Ball in Air (Film Loop—8mm—Silent) *Soccer*

Trapping—Ground Ball (Film Loop—8mm—Silent) *Soccer*

Treasure Island #1 (Filmstrip—Sound) *Pirates —Fiction*

Treasure Island #2 (Cassette) *Pirates — Fiction*

Treasures of King Tut (Videocassette) *Tutankhamun*

A Trip to the Fabric Store (Kit) *Sewing*

Try Out: Unit One (Motion Picture—16mm—Sound) *Vocational guidance*

Turn Around Jump Shot (Film Loop—8mm—Silent) *Basketball*

TV News: Behind the Scenes (Motion Picture—16mm—Sound) *Television —Production and direction*

The Twenties and Thirties (Filmstrip—Sound) *U.S. —History —1919–1933*

Two for Adventure (Kit) *Animals —Fiction*

Underground Wilderness (Filmstrip—Sound) *Caves*

Understand Institutional Racism (Kit) *Prejudices and antipathies*

Understanding and Using Decimals (Filmstrip—Sound) *Arithmetic*

Understanding and Using Percent (Filmstrip—Sound) *Arithmetic*

Understanding and Using Whole Numbers (Kit) *Arithmetic —Study and teaching*

Understanding Fiction (Filmstrip—Sound) *Fiction*

Understanding Natural Environments: Swamps and Deserts (Slides) *Ecology*

Understanding the Main Idea and Making Inferences (Slides/Cassettes) *Communication*

Understanding the Responsibilities of Child Care (Kit) *Children —Care and hygiene*

The Underwater Environment, Group One (Filmstrip—Sound) *Marine animals*

The Underwater Environment, Group Two (Filmstrip—Sound) *Marine plants*

Unions and You (Kit) *Labor unions*

The United States Congress: Of, By, and For the People (Motion Picture—16mm—Sound) *U.S. Congress*

Universal Values in American History (Kit) *U.S. —History*

Up Is Down (Motion Picture—16mm—Sound) *Individuality —Fiction*

Upright Loom (Film Loop—8mm—Silent) *Looms*

Urban World—Values in Conflict (Filmstrip—Sound) *Metropolitan areas*

Use of Triple Beam Balance (Motion Picture—16mm—Sound) *Chemical apparatus*

Using Clue Words to Unlock Meaning (Slides/Cassettes) *Communication*

Using the Bunsen Burner and Working with Glass (Motion Picture—16mm—Sound) *Chemical apparatus*

Values (Filmstrip—Sound) *Values*

Values for Dating (Filmstrip—Sound) *Dating (social customs)*

VD: Twentieth Century Plague (Filmstrip—Sound) *Venereal diseases*

The Veldt (Filmstrip—Sound) *Science fiction*

The Velveteen Rabbit (Filmstrip—Sound) *Fantasy —Fiction*

Venereal Disease: The Hidden Epidemic (Motion Picture—16mm—Sound) *Venereal diseases*

View from the People Wall (Motion Picture—16mm—Sound) *Thought and thinking*

Virginia Hamilton (Filmstrip—Sound) *Hamilton, Virginia*

Vocational Opportunities in High School (Motion Picture—16mm—Sound) *Vocational education*

Vocational Skills for Tomorrow (Filmstrip—Sound) *Vocational guidance*

The Vocational Technical School: Getting Acquainted (Filmstrip—Sound) *Vocational education*

Voices of the American Revolution (Filmstrip—Sound) *U.S. —History —1775-1783 — Revolution*

Voices of World War Two (Cassette) *World War, 1939-1945*

Volcanoes: Exploring the Restless Earth (Motion Picture—16mm—Sound) *Volcanoes*

Voltaire Presents Candide: An Introduction to the Age of Enlightenment (Motion Picture—16mm—Sound) *French fiction*

A Walk in the Forest (Motion Picture—16mm—Sound) *Forests and forestry*

Walls and Walls (Motion Picture—16mm—Sound) *Prejudices and antipathies*

Walt Whitman (Filmstrip—Sound) *Whitman, Walt*

Walt Whitman: Poet for a New Age (Motion Picture—16mm—Sound) *Whitman, Walt*

War Whoops and Medicine Songs (Phonodiscs) *Folk songs, Indian*

Watch Your Language: Usage (and Abusage) (Slides/Cassettes) *English language —Usage*

Watching Artists at Work (Filmstrip—Sound) *Art—Technique*

Water: Why It Is What It Is (Motion Picture—16mm—Sound) *Water*

Weather (Kit) *Meteorology*

Weather Forecasting (Motion Picture—16mm—Sound) *Weather forecasting*

Weather in the Wilderness (Filmstrip—Sound) *Survival*

Weather, Seasons, and Climate (Filmstrip—Sound) *Meteorology*

Weed (Marijuana) (Motion Picture—16mm—Sound) *Marijuana*

The Well of the Saints (Motion Picture—16mm—Sound) *Irish drama*

West African Artists and Their Art (Filmstrip—Sound) *Art, African*

Western Europe: France (Filmstrip—Sound) *France*

Western Europe: Germany (Filmstrip—Sound) *Germany*

Western Europe, Group One (Filmstrip—Sound) *Europe*

Western Europe, Group Two (Filmstrip—Sound) *Europe*

Whales: Can They Be Saved? (Motion Picture—16mm—Sound) *Whales*

Whales Surfacing (Film Loop—8mm—Silent) *Whales*

What about Marriage? (Filmstrip—Sound) *Marriage*

What Are You Going to Do about Alcohol (Filmstrip—Sound) *Alcohol*

What Do You Think? Unit One (Motion Picture—16mm—Sound) *Occupations*

What Does It Mean to Be Human? (Slides/Cassettes) *Anthropology*

What Is a Cat? (Motion Picture—16mm—Sound) *Cats*

What Is Ecology? (Second edition) (Motion Picture—16mm—Sound) *Ecology*

What Is Journalism? (Filmstrip—Sound) *Journalism*

What Makes Rain? (Motion Picture—16mm—Sound) *Rain and rainfall*

Whatever Happened to Linda? (Filmstrip—Sound) *Safety education*

What's Going on Here? (Filmstrip—Sound) *Propaganda*

What's the Limit? Unit Two (Motion Picture—16mm—Sound) *Occupations*

Where Is Dead? (Motion Picture—16mm—Sound) *Death*

White Roots in Black Africa (Filmstrip—Sound) *South Africa —Race relations*

Who Owns the Oceans? (Filmstrip—Sound) *Maritime law*

Who Stole The Quiet Day? (Motion Picture—16mm—Sound) *Noise pollution*

Who Works for You? (Kit) *Vocational guidance*

Who's OK, Who's Not OK: An Introduction to Abnormal Psychology (Filmstrip—Sound) *Psychology, pathological*

Who's Running the Show? (Filmstrip—Sound) *Leadership*

Why Am I Studying This? (Filmstrip—Sound) *Educational guidance*

Why Cities? (Filmstrip—Sound) *Cities and towns*

Why Cultures Are Different (Filmstrip—Sound) *Culture*

Why Do We Have to Take Social Studies (Filmstrip—Sound) *Educational guidance*

Why Man Creates (Motion Picture—16mm—Sound) *Creation (literary, artistic, etc.)*

Why Mothers Work (Motion Picture—16mm—Sound) *Women —Employment*

Why Skin Has Many Colors (Filmstrip—Sound) *Color of people*

Why We Do What We Do! Human Motivation (Filmstrip—Sound) *Motivation (psychology)*

Why We Work (Filmstrip—Sound) *Work*

The Wild Young Desert Series (Filmstrip—Sound) *Deserts*

Will Rogers' Nineteen Twenties (Motion Picture—16mm—Sound) *U.S. —History — 1919-1933*

William H. Armstrong (Filmstrip—Sound) *Armstrong, William*

William Shakespeare (Cassette) *Shakespeare, William*

Winter in the Forest (Filmstrip—Sound) *Winter*

With Justice for All (Kit) *Justice, administration of*

The Witness (Motion Picture—16mm—Sound) *Values*

Wolf Family (Film Loop—8mm—Silent) *Wolves*

The Wolves of Isle Royale (Kit) *Wolves*

A Woman's Place Is in the House (Motion Picture—16mm—Sound) *Noble, Elaine*

Women: An American History (Filmstrip—Sound) *Women—U.S.*

Women at Work: Change, Choice, Challenge (Motion Picture—16mm—Sound) *Women—Employment*

Women Behind the Bright Lights (Kit) *Entertainers*

Women in American History (Filmstrip—Sound) *Women—U.S.*

Women in the American Revolution (Filmstrip—Sound) *U.S.—History—1775-1783—Revolution*

Women Who Win, Set 1 (Kit) *Athletes*

Women Who Win, Set 2 (Kit) *Athletes*

Women Who Win, Set 3 (Kit) *Athletes*

Women Who Win, Set 4 (Kit) *Athletes*

Women Writers: Voices of Dissent (Filmstrip—Sound) *Women authors*

Women's Gymnastics, Beginning Level (Motion Picture—16mm—Sound) *Gymnastics*

The Woodlawn Organization (Filmstrip—Sound) *Chicago—Social conditions*

Woodrow Wilson and the Search for Peace (Filmstrip—Sound) *Wilson, Woodrow*

Woody Allen (Cassette) *Allen, Woody*

Words, Media and You (Filmstrip—Sound) *Mass media*

Work of the Kidneys (Second edition) (Motion Picture—16mm—Sound) *Kidneys*

Work/Working/Worker (Filmstrip—Sound) *Vocational guidance*

Working (Filmstrip—Sound) *Work*

Working in U.S. Communities, Group One (Filmstrip—Sound) *Occupations*

World Myths and Folktales (Filmstrip—Sound) *Folklore*

The World of Innerspace (Filmstrip—Sound) *Oceanography*

The World of Jungle Books, Set One (Kit) *Animals—Fiction*

The World of Just So Stories, Set One (Kit) *Animals—Fiction*

The World Outside Me (Kit) *Humanities*

World War Two (Filmstrip—Sound) *U.S.—History—1933-1945*

Worlds in Collision (Motion Picture—16mm—Sound) *Astronomy*

The Wright Brothers (Cassette) *Aeronautics—Biography*

Writing: From Assignment to Composition (Filmstrip—Sound) *English Language—Study and teaching*

Writing Skills—The Final Touch: Editing, Rewriting & Polishing (Slides/Cassettes) *English language—Composition and exercises*

Writing the Expository Essay (Kit) *Essay*

The Year of the Wildebeest (Motion Picture—16mm—Sound) *Gnu*

You and Office Safety (Motion Picture—16mm—Sound) *Office management*

You and the Group (Filmstrip—Sound) *Human relations*

Your Freedom and the First Amendment (Filmstrip—Sound) *U.S. Constitution—Amendments*

Your Newspaper (Filmstrip—Sound) *Newspapers*

Youth and the Law (Filmstrip—Sound) *Youth—Law and legislation*

Youth, Maturity, Old Age and Death (Filmstrip—Sound) *Pantomimes*

Zebra (Motion Picture—16mm—Sound) *Zebras*

Zlateh the Goat (Motion Picture—16mm—Sound) *Folklore—Poland*

Zoofari (Filmstrip—Sound) *Zoological gardens*

PRODUCER/ DISTRIBUTOR DIRECTORY

Where it is known that the producer differs from the distributor, both names are given in the entry; however, only the distributor's address is listed in this directory. One source in the entry assumes that the producer/distributor is the same.

ACI MEDIA, INC.
see Paramount Communications

AIDS (AUDIOVISUAL INSTRUCTIONAL DEVICES INC.)
see Aids of Cape Cod

AIDS OF CAPE COD
110 Old Town House Rd.
South Yarmouth, MA 02664

AIMS INSTRUCTIONAL MEDIA SERVICE
626 Justin Ave.
Glendale, CA 91201

ACORN FILMS, INC.
33 Union Sq. W.
New York, NY 10003

ACTIVITY RECORDS
see Educational Activities, Inc.

AESOP FILMS, INC.
3701 Buchanan St.
San Francisco, CA 94123

ARGO RECORDS, INC.
539 W. 25 St.
New York, NY 10001

ASSOCIATED EDUCATIONAL MATERI-ALS CO.
Box 2087
14 Glenwood Ave.
Raleigh, NC 27602

ASSOCIATED PRESS
see Pathescope Educational Media, Inc.

ATHLETIC INSTITUTE
200 Castlewood Dr.
North Palm Beach, FL 33408

ATLANTIC RECORDS, INC.
75 Rockefeller Plaza
New York, NY 10020

ATLANTIS PRODUCTIONS, INC.
1252 La Granada Dr.
Thousand Oaks, CA 91360

AUDIO VISUAL NARRATIVE ARTS
Pleasantville, NY 10570

AUDIO VISUAL ASSOCIATES, INC.
180 E. California Blvd.
Pasadena, CA 91105

AVATAR LEARNING, INC.
see Encyclopaedia Britannica Educational Corp.

BFA EDUCATIONAL MEDIA
Box 1795
2211 Michigan Ave.
Santa Monica, CA 90406

BACKPACKERS MAGAZINE
65 Adams Rd.
Bedford Hills, NY 10507

BAILEY FILM ASSOCIATES
see BFA Educational Media

BARR FILMS
Box 5667
Pasadena, CA 91107

DAVE BELL ASSOCIATES, INC.
3211 Cahuenga Blvd. W.
Los Angeles, CA 90038

BENCHMARK FILMS, INC.
145 Scarborough Rd.
Briarcliff Mnr., NY 10510

BENDICK ASSOCIATES, INC.
360 Grace Church St.
Rye, NY 10580

BERGWALL PRODUCTIONS
839 Stewart Ave.
Garden City, NY 11530

BEST FILM
Box 725
Delmar, CA 92014

BOOKS ON TAPE
Box 71405
Los Angeles, CA 90071

NICK BOSUSTOW
see Bosustow Productions

BOSUSTOW PRODUCTIONS
Stephen Bosustow
1649 11th St.
Santa Monica, CA 90404

BOWMAR PUBLISHING CORP.
4563 Colorado Blvd.
Los Angeles, CA 90039

BUTTERICK PUBLISHING
161 Ave. of the Americas
New York, NY 10003

CMS RECORDS
14 Warren St.
New York, NY 10007

CAEDMON RECORDS, INC.
505 Eighth Ave.
New York, NY 10018

CAL INDUSTRIES, INC.
76 Madison Ave.
New York, NY 10016

CENTER FOR HUMANITIES
2 Holland Ave.
White Plains, NY 10603

CENTER FOR SOUTHERN FOLKLORE
Box 4081
1216 Peabody Ave.
Memphis, TN 38104

CENTRON EDUCATIONAL FILMS
1621 W. Ninth St.
Box 687
Lawrence, KS 66044

CHILDREN'S CLASSICS ON TAPE
6722 Bostwick Dr.
Springfield, VA 22151

CHILDREN'S PRESS
1224 W. Van Buren St.
Chicago, IL 60607

CHURCHILL FILMS
662 N. Robertson Blvd.
Los Angeles, CA 90069

CLEARVUE, INC.
6666 N. Oliphant
Chicago, IL 60631

KENNETH E. CLOUSE
333 Quail Hollow Rd.
Felton, CA 95018

COLUMBIA RECORDS (SPECIAL PROD-
    UCTS)
51 W. 52 St.
New York, NY 10019

COMMUNICATIONS GROUP WEST
6066 Sunset Blvd.
Hollywood, CA 90028

CONCEPT MEDIA, INC.
Box 19542
Irvine, CA 92714

CONTEMPORARY DRAMA SERVICE
Box 457
Downers Grove, IL 60515

CORONET INSTRUCTIONAL MEDIA
65 E. South Water St.
Chicago, IL 60601

JEAN-MICHEL COUSTEAU
see BFA Educational Media

CURRENT AFFAIRS FILMS
24 Danbury Rd.
Wilton, CT 06897

CYPRESS PUBLISHING
1763 Gardenia Ave.
Glendale, CA 91204

DAVCO PUBLISHERS, INC.
8154 Ridgeway
Skokie, IL 60073

DAVIDSON FILMS
3701 Buchanan
San Francisco, CA 94123

DENOYER-GEPPERT
5235 Ravenswood Ave.
Chicago, IL 60640

DEUTSCHE GRAMMOPHON
810 Seventh Ave.
New York, NY 10019

DIRECT CINEMA
Box 315
Franklin Lakes, NJ 07417

WALT DISNEY EDUCATIONAL MEDIA
CO.
500 S. Buena Vista St.
Burbank, CA 91521

DONARS PRODUCTIONS
Box 24
Loveland, CO 80537

KEVIN DONOVAN FILMS
44 Treat Rd.
Glastonbury, CT 06033

EBEC
see Encyclopaedia Britannica Educational
Corp.

EMC CORP.
180 W. Sixth St.
St. Paul, MN 55101

EDUCATIONAL ACTIVITIES, INC.
Box 392
Freeport, NY 11520

EDUCATIONAL AUDIO VISUAL, INC.
Pleasantville, NY 10570

EDUCATIONAL DESIGN, INC.
47 W. 13 St.
New York, NY 10011

EDUCATIONAL DEVELOPMENT CEN-
TER
39 Chapel St.
Newton, MA 12160

EDUCATIONAL DEVELOPMENT CORP.
202 Lake Miriam Dr.
Lakeland, FL 33803

EDUCATIONAL DIMENSIONS GROUP
Box 126
Stamford, CT 06904

EDUCATIONAL DIRECTION, INC.
181 W. State St.
Westport, CT 06880

EDUCATIONAL ENRICHMENT MATERI-
ALS
110 S. Bedford Rd.
Mt. Kisco, NY 10549

EDUCATIONAL RECORDS, INC.
157 Chambers St.
New York, NY 10007

ENCORE VISUAL EDUCATION
1235 S. Victory Blvd.
Burbank, CA 91502

ENCYCLOPAEDIA BRITANNICA EDU-
CATIONAL CORP.
425 North Michigan Ave.
Chicago, IL 60611

EYE GATE MEDIA
146–01 Archer Ave.
Jamaica, NY 11435

FARMHOUSE FILMS
425 N. Michigan Ave.
Chicago, IL 60611

FENWICK PRODUCTIONS, INC.
134 Steele Rd.
West Hartford, CT 06119

FILM COMMUNICATORS
11136 Weddington St.
North Hollywood, CA 91601

FILM POLSKI
see Encyclopaedia Britannica

FILMFAIR COMMUNICATIONS
10820 Ventura Blvd.
Studio City, CA 91604

FILMS, INC.
1144 Wilmette Ave.
Wilmette, IL 60091

FILMSTRIP HOUSE
432 Park Ave. S.
New York, NY 10016

FLIGHT PLAN I
see Encyclopaedia Britannica

FOLKWAYS RECORDS
43 W. 63 St.
New York, NY 10023

FRANKLIN WATTS, INC.
730 Fifth Ave.
New York, NY 10019

GAKKEN CO., LTD.
Tokyo, Japan

GLOBE FILMSTRIPS, INC.
175 Fifth Ave.
New York, NY 10010

GRANADA INTERNATIONAL TV
1221 Ave. of the Americas
New York, NY 10020

GREAT AMERICAN FILM FACTORY,
INC.
Box 9195
Sacramento, CA 95816

GUIDANCE ASSOCIATES
757 Third Ave.
New York, NY 10017

HBJ FILMS
see Harcourt Brace Jovanovich

HANDEL FILM CORP.
8730 Sunset Blvd.
Los Angeles, CA 90069

HARCOURT BRACE JOVANOVICH, INC.
757 Third Ave.
New York, NY 10017

HARPER AND ROW PUBLISHERS, INC.
10 E 53 St.
New York, NY 10022

HAWKHILL ASSOCIATES, INC.
125 Gilman St.
Madison, WI 53703

HERITAGE PRODUCTIONS, INC.
1437 Central Ave.
Memphis, TN 38104

ALFRED HIGGINS PRODUCTIONS, INC.
9100 Sunset Blvd.
Los Angeles, CA 90064

HUMAN RELATIONS MEDIA CENTER
175 Thompkins Ave.
Pleasantville, NY 10570

IFB
see International Film Bureau, Inc.

IHS ASSOCIATES, INC.
see Film Communicators

IMPERIAL EDUCATIONAL RESOURCES,
    INC.
202 Lake Miriam Dr.
Lakeland, FL 33803

IMPERIAL INTERNATIONAL LEARN-
    ING, INC.
Box 548
Kankakee, IL 60901

INSIGHT MEDIA PROGRAMS, INC.
13900 Panay Way M–120
Marina del Rey, CA 90291

INSIGHT!, INC.
100 E. Ohio St.
Chicago, IL 60611

INSTRUCTIONAL MATERIALS LABORA-
    TORIES
200 Madison Ave.
New York, NY 10016

INTERNATIONAL FILM BUREAU,
    INC.
332 S. Michigan Ave.
Chicago, IL 60604

JABBERWOCKY CASSETTE CLASSICS
Box 6727
San Francisco, CA 94101

JAM HANDY
see Prentice-Hall Media

JANUARY PRODUCTIONS
124 Rea Ave.
Hawthorne, NJ 07507

JIMICON RECORDING
Box 536
Portsmouth, RI 02871

JOSHUA TREE PRODUCTIONS
see Bosustow Productions

KCET-TV
475 L'Enfant Plaza S.W.
Washington, DC 20024

KNOWLEDGE AID
see United Learning

LAWREN PRODUCTIONS
Box 1542
Burlingame, CA 94010

LEARNING CORP. OF AMERICA
1350 Ave. of the Americas
New York, NY 10019

LEARNING TREE FILMSTRIPS
Dept. 75
Box 1590
Boulder, CO 80306

LEXINGTON RECORDING COMPANY,
    INC.
29 Marble Ave.
Pleasantville, NY 10570

LIBRARY FILMSTRIP CENTER
3033 Aloma
Wichita, KS 67211

LISTENING LIBRARY, INC.
One Park Ave.
Old Greenwich, CT 06870

LITTLE RED FILMHOUSE
119 S. Kilkea Dr.
Los Angeles, CA 90048

LYCEUM
546 Hofgaarden St.
La Puente, CA 91744

MACMILLAN LIBRARY SERVICES
Div. Macmillan Publishing Co., Inc.
866 Third Ave.
New York, NY 10022

MARSHFILM, INC.
Box 8082
Shawnee Mission, KS 66208

McGRAW-HILL FILMS
Dept. 423
1221 Ave. of the Americas
New York, NY 10020

MEDIA PLUS, INC.
60 Riverside Dr.
New York, NY 10024

MEDIA RESEARCH AND DEVELOP-
MENT
Arizona State University
Tempe, AZ 85281

ARTHUR MERIWETHER, INC.
Educational Resources
Box 457
1529 Brook Ave.
Downers Grove, IL 60515

MILLER-BRODY PRODUCTIONS, INC.
Dept. 78
342 Madison Ave.
New York, NY 10017

ARTHUR MOKIN PRODUCTIONS, INC.
17 W. 60 St.
New York, NY 10023

MOODY INSTITUTE OF SCIENCE
12000 E. Washington Blvd.
Whittier, CA 90606

MOOK & BLANCHARD
546 S. Hofgaarden St.
La Puente, CA 91744

MULTI-MEDIA PRODUCTIONS, INC.
Box 5097
Stanford, CA 94305

BURT MUNK AND CO.
56 E. Walton Place
Chicago, IL 60611

NATIONAL FILM BOARD OF CANADA
680 Fifth Ave.
New York, NY 10019

NATIONAL GEOGRAPHIC
see National Geographic Educational Services

NATIONAL GEOGRAPHIC EDUCATION-
AL SERVICES
Dept. 77
Box 1640
Washington, DC 20013

NEW DAY FILMS
114 Park St.
Brookline, MA 02146

NEWSWEEK EDUCATIONAL DIVISION
444 Madison Ave.
New York, NY 10022

NORCLIFF THAYER, INC.
One Scarsdale Rd.
Tuckahoe, NY 10707

JEFFREY NORTON PUBLISHERS
145 E. 49 St.
New York, NY 10017

NYSTROM
3333 Elston Ave.
Chicago, IL 60618

ODYSSEY PRODUCTIONS, INC.
485 Madison Ave.
New York, NY 10022

PARAMOUNT COMMUNICATIONS
5451 Marathon St.
Hollywood, CA 90038

PARATORE PICTURES
see Random House, Inc.

PARENTS' MAGAZINE FILMS
52 Vanderbilt Ave.
New York, NY 10017

PATHESCOPE EDUCATIONAL MEDIA,
INC.
71 Weyman Ave.
New Rochelle, NY 10802

PATHWAYS OF SOUND, INC.
102 Mt. Auburn St.
Cambridge, MA 02138

LEONARD PECK PRODUCTIONS
Box 3235
Wayne, NY 07470

PERFECTION FORM COMPANY
1000 N. Second Ave.
Logan, IA 51546

PERENNIAL FILMS
Box 855 Rabina
477 Roger Williams
Highland Park, IL 60035

PHILIPS RECORDS, INC.
810 Seventh Ave.
New York, NY 10009

PHOENIX FILMS, INC.
470 Park Ave. S.
New York, NY 10016

PIED PIPER PRODUCTIONS
Box 320
Verdugo City, CA 91406

THE POLISHED APPLE
3742 Seahorn Dr.
Malibu, CA 90265

PRENTICE-HALL MEDIA
150 White Plains Rd.
Tarrytown, NY 10591

PRIMA EDUCATION PRODUCTS, INC.
Irvington-on-Hudson, New York, NY 10533

PROFESSIONAL ARTS
Box 8003
Stanford, CA 94305

PROOF PRESS
Box 1256
Berkeley, CA 94720

PYRAMID FILMS
Box 1048
Santa Monica, CA 90406

Q+ED PRODUCTIONS, INC.
Box 1608
Burbank, CA 91507

RCA EDUCATIONAL DEPT.
Box RCA–1000
Indianapolis, IN 46291

RACISM/SEXISM RESOURCE CENTER
   FOR EDUCATORS
1841 Broadway
New York, NY 10023

RAMSGATE FILMS, INC.
704 Santa Monica Blvd.
Santa Monica, CA 90401

RANDOM HOUSE, INC.
Dept. V–8
400 Hahn Rd.
Westminster, MD 21157

S-L FILM PRODUCTIONS, INC.
Box 41108
Los Angeles, CA 90041

SANDLER INSTITUTIONAL FILMS
1001 N. Poinsettia
Hollywood, CA 90046

MORTON SCHINDEL
see Weston Woods Studios

SCHLOAT
see Prentice-Hall Media

WARREN SCHLOAT PRODUCTIONS
see Prentice-Hall Media

SCHOLASTIC BOOK SERVICES
904 Sylvan Ave.
Englewood Cliffs, NJ 07632

SCHOLASTIC MAGAZINE
50 W. 44 St.
New York, NY 10036

SCIENCE RESEARCH ASSOCIATES, INC.
259 Erie
Chicago, IL 60611

SCIENCE AND MANKIND, INC.
Box 200
Communications Park
White Plains, NY 10602

SCOTT EDUCATION DIVISION
see Prentice-Hall Media

SEE'N EYE PRODUCTIONS, INC.
10084 Westwanda Dr.
Beverly Hills, CA 90210

SHOREWOOD REPRODUCTIONS, INC.
10 E. 53 St.
New York, NY 10022

SIGNET PRODUCTIONS
200 W. 58 St.
New York, NY 10019

SMALL WORLD PRODUCTIONS, INC.
Pomfret Center, CT 06259

SOCIETY FOR VISUAL EDUCATION,
   INC.
1345 Diversey Pkwy.
Chicago, IL 60614

SOUNDWORDS
56–11 217th St.
Bayside, NY 11364

SPECTRA FILMS
see Random House, Inc.

SPOKEN ARTS, INC.
310 North Ave.
New Rochelle, NY 10801

STANFIELD HOUSE
900 Euclid Ave.
Santa Monica, CA 90403

STANTON FILMS, INC.
7943 Santa Monica Blvd.
Los Angeles, CA 90046

SUNBURST COMMUNICATIONS, INC.
Pound Ridge, NY 10576

SUTHERLAND LEARNING ASSOCI-
   ATES, INC.
see Encyclopaedia Britannica Educational
   Corp.

TAPES FOR READERS
5078 Fulton St. NW
Washington, DC 20016

TEACHING RESOURCES FILMS, INC.
Station Plaza, Bedford Hills, NY 10507

TELE-VISUAL PRODUCTIONS, INC.
3377–3379 S.W. Third Ave.
Miami, FL 33143

TEXTURE FILMS, INC.
1600 Broadway
New York, NY 10019

THORNE FILMS, INC.
see Prentice-Hall Media

TIME-LIFE MULTIMEDIA
Rm. 32–48, Time-Life Bldg.
New York, NY 10020

TRANSMEDIA INTERNATIONAL, INC.
1100 17th St. N.W., Suite 1000
Washington, DC 20036

TROLL ASSOCIATES
320 Rte. 17
Mahwah, NJ 07430

UNIT ONE FILM PRODUCTIONS
423 W. 118 St.
New York, NY 10027

UNITED LEARNING
6633 W. Howard St.
Niles, IL 60648

UNIVERSITY OF WISCONSIN R & D CEN-
TER FOR COGNITIVE DEVELOPMENT
see Encyclopaedia Britannica Educational
Corp.

UNIVERSITY FILMS
see McGraw-Hill Films

URBAN MEDIA MATERIALS
212 Minneola Ave.
Roslyn Heights, NY 11577

ADRIAN VANCE PRODUCTIONS, INC.
Box 46456
Hollywood, CA 90046

VIEWLEX EDUCATIONAL MEDIA
Broadway Ave.
Holbrook, NY 11741

VIKING PRESS
625 Madison Ave.
New York, NY 10022

VISUAL EDUCATION CORP.
Box 2321
Princeton, NJ 08540

VISUAL PUBLICATIONS
716 Center St.
Lewiston, NY 14092

WGBH EDUCATIONAL FOUNDATION
Distribution Ctr.
125 Western Ave.
Boston, MA 02134

WNET/13 (GLENN JORDAN) EDUCA-
TIONAL BROADCASTING
356 W. 58 St.
New York, NY 10019

WNVT (NORTHERN VIRGINIA EDUCA-
TIONAL TV)
8325 Little River Tpke.
Annandale, VA 22003

WARDS NATURAL SCIENCE ESTAB-
LISHMENT
Box 1721
Rochester, NY 14603

WARNER EDUCATIONAL PRODUC-
TIONS
Box 8791
Fountain Valley, CA 92708

WESTON WOODS STUDIOS
Weston, CT 06883

H. WILSON CORP.
555 W. Taft Dr.
South Holland, IL 60473

WILSON EDUCATIONAL MEDIA
see H. Wilson Corp.

WINDMILLS LTD. PRODUCTIONS
Box 5300
Santa Monica, CA 90405

WOMBAT PRODUCTIONS, INC.
Box 70
Little Lake, Glendale Rd.
Ossining, NY 10562

XEROX EDUCATIONAL PUBLICATIONS
Box 444
1250 Fairwood Ave.
Columbus, OH 43216